Foundations of MEMS

Chang Liu

*Electrical and Computer Engineering Department
University of Illinois at Urbana–Champaign*

Upper Saddle River, NJ 07458

Library of Congress Cataloging-in-Publication Data on file.

Liu, Chang, Ph.D.
  Foundations of MEMS / Chang Liu.
    p. cm.
  Includes index.
  ISBN 0-13-147286-0
    1. Microeletromechanical systems. I. Title.
621.3--dc22   2005048932

Vice President and Editorial Director, ECS: *Marcia Horton*
Associate Editor: *Alice Dworkin*
Executive Managing Editor: *Vince O'Brien*
Managing Editor: *David A. George*
Production Editor: *Daniel Sandin*
Director of Creative Services: *Paul Belfanti*
Creative Director: *Jayne Conte*
Cover Designer: *Bruce Kenselaar*
Art Editor: *Greg Dulles*
Manufacturing Manager: *Alexis Heydt-Long*
Manufacturing Buyer: *Lisa McDowell*
Senior Marketing Manager: *Holly Stark*

© 2006 Pearson Education, Inc.
Pearson Education, Inc.
Upper Saddle River, New Jersey 07458

All rights reserved. No part of this book may be reproduced, in any form or by any means, without permission in writing from the publisher.

Pearson Prentice Hall™ is a trademark of Pearson Education, Inc.

The author and publisher of this book have used their best efforts in preparing this book. These efforts include the development, research, and testing of the theories and programs to determine their effectiveness. The author and publisher make no warranty of any kind, expressed or implied, with regard to these programs or the documentation contained in this book. The author and publisher shall not be liable in any event for incidental or consequential damages in connection with, or arising out of, the furnishing, performance, or use of these programs.

Printed in the United States of America

10 9 8 7 6 5 4 3 2 1

ISBN 0-13-147286-0

Pearson Education Ltd., *London*
Pearson Education Australia Pty. Ltd., *Sydney*
Pearson Education Singapore, Pte. Ltd.
Pearson Education North Asia Ltd., *Hong Kong*
Pearson Education Canada Inc., *Toronto*
Pearson Educación de Mexico, S.A. de C.V.
Pearson Education—Japan, *Tokyo*
Pearson Education Malaysia, Pte. Ltd.
Pearson Education, Inc., *Upper Saddle River, New Jersey*

To My Family—Lu, Sophia, and My Parents

Contents

PREFACE   xiii

A NOTE TO INSTRUCTORS   xvii

ABOUT THE AUTHOR   xix

NOTATIONAL CONVENTIONS   xxi

Chapter 1 Introduction   1

  1.0 Preview 1
  1.1 The History of MEMS Development 1
  1.2 The Intrinsic Characteristics of MEMS 11
      1.2.1 Miniaturization 12
      1.2.2 Microelectronics Integration 14
      1.2.3 Mass Fabrication with Precision 14
  1.3 Devices: Sensors and Actuators 14
      1.3.1 Energy Domains and Transducers 14
      1.3.2 Sensors 17
      1.3.3 Actuators 19
      Summary 20
      Problems 21
      References 23

Chapter 2 Introduction to Microfabrication   28

  2.0 Preview 28
  2.1 Overview of Microfabrication 28
  2.2 The Microelectronics Fabrication Process 30
  2.3 Silicon-Based MEMS Processes 33
  2.4 New Materials and Fabrication Processes 39
  2.5 Points of Consideration for Processing 40
      Summary 43
      Problems 43
      References 44

Chapter 3 Review of Essential Electrical and Mechanical Concepts   48

  3.0 Preview 48
  3.1 Conductivity of Semiconductors 49
      3.1.1 Semiconductor Materials 49
      3.1.2 Calculation of Charge Carrier Concentration 50
      3.1.3 Conductivity and Resistivity 54

3.2 Crystal Planes and Orientation   58
3.3 Stress and Strain   61
   3.3.1 Internal Force Analysis: Newton's Laws of Motion   61
   3.3.2 Definitions of Stress and Strain   63
   3.3.3 General Scalar Relation Between Tensile Stress and Strain   66
   3.3.4 Mechanical Properties of Silicon and Related Thin Films   68
   3.3.5 General Stress–Strain Relations   70
3.4 Flexural Beam Bending Analysis Under Simple Loading Conditions   73
   3.4.1 Types of Beams   73
   3.4.2 Longitudinal Strain Under Pure Bending   75
   3.4.3 Deflection of Beams   77
   3.4.4 Finding the Spring Constants   78
3.5 Torsional Deflections   83
3.6 Intrinsic Stress   86
3.7 Resonant Frequency and Quality Factor   91
3.8 Active Tuning of the Spring Constant and Resonant Frequency   93
3.9 A List of Suggested Courses and Books   94
Summary   95
Problems   95
References   99

Chapter 4 Electrostatic Sensing and Actuation 103

4.0 Preview   103
4.1 Introduction to Electrostatic Sensors and Actuators   103
4.2 Parallel-Plate Capacitors   105
   4.2.1 Capacitance of Parallel Plates   105
   4.2.2 Equilibrium Position of Electrostatic Actuator Under Bias   108
   4.2.3 Pull-In Effect of Parallel-Plate Actuators   110
4.3 Applications of Parallel-Plate Capacitors   116
   4.3.1 Inertia Sensor   116
   4.3.2 Pressure Sensor   122
   4.3.3 Flow Sensor   127
   4.3.4 Tactile Sensor   130
   4.3.5 Parallel-Plate Actuators   131
4.4 Interdigitated Finger Capacitors   133
4.5 Applications of Comb-Drive Devices   139
   4.5.1 Inertia Sensors   139
   4.5.2 Actuators   143
Summary   145
Problems   145
References   149

Contents   vii

Chapter 5   Thermal Sensing and Actuation   **153**

    5.0   Preview   153
    5.1   Introduction   153
           5.1.1   Thermal Sensors   153
           5.1.2   Thermal Actuators   154
           5.1.3   Fundamentals of Thermal Transfer   154
    5.2   Sensors and Actuators Based on Thermal Expansion   159
           5.2.1   Thermal Bimorph Principle   161
           5.2.2   Thermal Actuators with a Single Material   168
    5.3   Thermal Couples   170
    5.4   Thermal Resistors   173
    5.5   Applications   175
           5.5.1   Inertia Sensors   175
           5.5.2   Flow Sensors   178
           5.5.3   Infrared Sensors   191
           5.5.4   Other Sensors   194
        Summary   199
        Problems   200
        References   204

Chapter 6   Piezoresistive Sensors   **207**

    6.0   Preview   207
    6.1   Origin and Expression of Piezoresistivity   207
    6.2   Piezoresistive Sensor Materials   211
           6.2.1   Metal Strain Gauges   211
           6.2.2   Single-Crystal Silicon   211
           6.2.3   Polycrystalline Silicon   215
    6.3   Stress Analysis of Mechanical Elements   215
           6.3.1   Stress in Flexural Cantilevers   216
           6.3.2   Stress in the Membrane   221
    6.4   Applications of Piezoresistive Sensors   223
           6.4.1   Inertia Sensors   224
           6.4.2   Pressure Sensors   229
           6.4.3   Tactile Sensors   232
           6.4.4   Flow Sensors   235
        Summary   239
        Problems   240
        References   243

Chapter 7 Piezoelectric Sensing and Actuation — 245

- 7.0 Preview 245
- 7.1 Introduction 245
    - 7.1.1 Background 245
    - 7.1.2 Mathematical Description of Piezoelectric Effects 247
    - 7.1.3 Cantilever Piezoelectric Actuator Model 249
- 7.2 Properties of Piezoelectric Materials 252
    - 7.2.1 Quartz 253
    - 7.2.2 PZT 254
    - 7.2.3 PVDF 256
    - 7.2.4 ZnO 256
    - 7.2.5 Other Materials 261
- 7.3 Applications 262
    - 7.3.1 Inertia Sensors 262
    - 7.3.2 Acoustic Sensors 265
    - 7.3.3 Tactile Sensors 268
    - 7.3.4 Flow Sensors 269
    - 7.3.5 Surface Elastic Waves 271
    - Summary 273
    - Problems 273
    - References 276

Chapter 8 Magnetic Actuation — 279

- 8.0 Preview 279
- 8.1 Essential Concepts and Principles 279
    - 8.1.1 Magnetization and Nomenclature 279
    - 8.1.3 Selected Principles of Micromagnetic Actuators 282
- 8.2 Fabrication of Micromagnetic Components 287
    - 8.2.1 Deposition of Magnetic Materials 287
    - 8.2.2 Design and Fabrication of Magnetic Coil 288
- 8.3 Case Studies of MEMS Magnetic Actuators 292
    - Summary 303
    - Problems 304
    - References 305

Chapter 9 Summary of Sensing and Actuation — 307

- 9.0 Preview 307
- 9.1 Comparison of Major Sensing and Actuation Methods 308
- 9.2 Tunneling Sensing 309
- 9.3 Optical Sensing 311
    - 9.3.1 Sensing with Waveguides 311
    - 9.3.2 Sensing with Free-Space Light Beams 312
    - 9.3.3 Position Sensing with Optical Interferometry 313

9.4 Field-Effect Transistors   317
9.5 Radio Frequency Resonance Sensing   320
   Summary   321
   Problems   322
   References   323

Chapter 10 Bulk Micromachining and Silicon Anisotropic Etching   326

10.0 Preview   326
10.1 Introduction   326
10.2 Anisotropic Wet Etching   328
   10.2.1 Introduction   328
   10.2.2 Rules of Anisotropic Etching—Simplest Case   330
   10.2.3 Rules of Anisotropic Etching—Complex Structures   335
   10.2.4 Interaction of Etching Profiles from Isolated Patterns   345
   10.2.5 Summary of Design Methodology   345
   10.2.6 Chemicals for Wet Anisotropic Etching   346
10.3 Dry Etching of Silicon—Plasma Etching   349
10.4 Deep Reactive Ion Etching (DRIE)   352
10.5 Isotropic Wet Etching   353
10.6 Gas-Phase Etchants   353
10.7 Native Oxide   354
10.8 Wafer Bonding   354
10.9 Case Studies   356
   10.9.1 Suspended Beams and Plates   356
   10.9.2 Suspended Membranes   357
   Summary   360
   Problems   361
   References   368

Chapter 11 Surface Micromachining   371

11.0 Preview   371
11.1 Basic Surface Micromachining Processes   371
   11.1.1 Sacrificial Etching Process   371
   11.1.2 Micromotor Fabrication Process—A First Pass   372
   11.1.3 Micromotor Fabrication Process—A Second Pass   374
   11.1.4 Micromotor Fabrication Process—Third Pass   375
11.2 Structural and Sacrificial Materials   376
   11.2.1 Material Selection Criteria   376
   11.2.2 Thin Films by Low-Pressure Chemical Vapor Deposition   378
   11.2.3 Other Surface Micromachining Materials and Processes   380
11.3 Acceleration of Sacrificial Etch   381
11.4 Stiction and AntiStiction Methods   383

11.5 Assembly of 3D MEMS   385
11.6 Foundry Process   389
Summary   390
Problems   391
References   394

Chapter 12 Polymer MEMS   397

12.0 Preview   397
12.1 Introduction   397
12.2 Polymers in MEMS   399
  12.2.1 Polyimide   399
  12.2.2 SU-8   401
  12.2.3 Liquid Crystal Polymer (LCP)   402
  12.2.4 PDMS   403
  12.2.5 PMMA   405
  12.2.6 Parylene   405
  12.2.7 Fluorocarbon   406
  12.2.8 Other Polymers   406
12.3 Representative Applications   407
  12.3.1 Acceleration Sensors   407
  12.3.2 Pressure Sensors   409
  12.3.3 Flow Sensors   413
  12.3.4 Tactile Sensors   415
Summary   417
Problems   418
References   419

Chapter 13 Microfluidics Applications   422

13.0 Preview   422
13.1 Motivation for Microfluidics   422
13.2 Essential Biology Concepts   423
13.3 Basic Fluid Mechanics Concepts   426
  13.3.1 The Reynolds Number and Viscosity   426
  13.3.2 Methods for Fluid Movement in Channels   427
  13.3.3 Pressure Driven Flow   428
  13.3.4 Electrokinetic Flow   430
  13.3.5 Electrophoresis and Dielectrophoresis   431
13.4 Design and Fabrication of Selective Components   434
  13.4.1 Channels   434
  13.4.2 Valves   445
Summary   448
Problems   448
References   450

Chapter 14 Instruments for Scanning Probe Microscopy   455

    14.0 Preview   455
    14.1 Introduction   455
        14.1.1 SPM Technologies   455
        14.1.2 The Versatile SPM Family   457
        14.1.3 Extension of SPM Technologies   459
    14.2 General Fabrication Methods for Tips   460
    14.3 Cantilevers with Integrated Tips   462
        14.3.1 General Design Considerations   462
        14.3.2 General Fabrication Strategies   463
        14.3.3 Alternative Techniques   466
    14.4 SPM Probes with Sensors and Actuators   470
        14.4.1 SPM Probes with Sensors   471
        14.4.2 SPM Probes with Actuators   476
        Summary   481
        Problems   481
        References   482

Chapter 15 Optical MEMS   486

    15.0 Preview   486
    15.1 Passive MEMS Optical Components   487
        15.1.1 Lenses   488
        15.1.2 Mirrors   492
    15.2 Actuators for Active Optical MEMS   495
        15.2.1 Actuators for Small Out-of-Plane Translation   496
        15.2.2 Actuators for Large In-Plane Translation Motion   498
        15.2.3 Actuators for Out-of-Plane Rotation   499
        Summary   502
        Problems   502
        References   505

Chapter 16 MEMS Technology Management   509

    16.0 Preview   509
    16.1 R&D Strategies   509
        Problems   515
        References   516

Appendix A Material Properties   517

Appendix B Frequently Used Formulas for Beams and Membranes   520

Index   522

Preface

Welcome to the world of microelectromechanical systems (MEMS), an emerging research field and industry characterized by the integration of electrical and mechanical engineering, miniaturization, integrative fabrication methods, diverse application reaches, rapid pace of innovation, and vast opportunities for ingenuity. MEMS technology branched off from the integrated circuit industry, from which it inherited semiconductor materials, microfabrication technologies and equipment, and facility infrastructure. The field now sits at a confluent point of many disciplines including electrical engineering, mechanical engineering, material sciences, micro- and nanofabrication, life sciences, chemical engineering, and civil and environmental engineering, to name a few.

A student in the MEMS area is faced with unique challenges.

First, MEMS devices embody concepts from both electrical and mechanical engineering domains. A successful MEMS device cannot be developed without considering both aspects in concert. A reader who has received training in a traditional engineering curriculum must be conversant with the concepts and practices of unfamiliar fields. For example, an electrical engineering student who is trained in the semiconductor device area needs to calculate mechanical bending and stresses. A mechanical engineering student, on the other hand, needs to absorb the basic knowledge about solid-state materials and devices, as well as companion fabrication technologies.

Second, MEMS devices employ microfabrication technology, which is a fast evolving discipline with basic principles unfamiliar to many students and practicing engineers. Even readers with electrical engineering backgrounds are not necessarily familiar with microfabrication technology for integrated circuits. However, a MEMS device cannot be successfully designed and developed without considering *how it will be made later*. The electromechanical design of a sensor or actuator must be made with full cognizance of the opportunities and limitations of microfabrication technology—past, present, and future. At this relatively early stage of MEMS development, design, fabrication, materials, and performances intersect.

Third, the application of MEMS encompasses many fields beyond traditional electrical and mechanical engineering. This presents exciting new opportunities for a student and practitioner of MEMS to become involved in diverse application domains, such as bioengineering, chemistry, nanotechnology, optical engineering, power and energy, and wireless communication, to name a few. The reader must realize that a successful MEMS research project can hardly create the desired impact without developing insight and a grasp of domain-specific knowledge.

Fourth, the performance of a MEMS device is not the only factor that determines its chance of market acceptance. A MEMS device and system must provide combined cost and performance advantages over incumbent and/or competitive technologies if it is to survive and thrive in the real world. No matter how intriguing the miniaturization technology is, a sense of economic and societal reality must never be lost in the excitement of technology creation. The cost (of development and ownership) and the functions of MEMS devices must be carefully considered and optimized in a vast space of possible materials, designs, and fabrication technologies.

Needless to say, a MEMS practitioner must amass a broad knowledge base of materials, designs, manufacturing methods, and industrial trends in order to identify the right problems *and* find the right solutions.

With these challenges in mind, this book is designed to guide the reader in building critical knowledge about the field in a systematic and time-efficient way. The contents and the sequence of the discussions have been fine-tuned as a result of my experience teaching a MEMS course at the University of Illinois at Urbana–Champaign for the past seven years (1997–2003). The students who attended the classes were from both electrical and mechanical engineering departments.

There are four primary objectives of this textbook:

1. Gain critical cross-disciplinary knowledge about designing electromechanical transducers, including sensors and actuators. As a result, the reader should be able to analyze the key performance aspects of simple electromechanical devices and understand the options and challenges associated with a particular design task;
2. Attain a solid background in the area of microfabrication, to the extent that a reader without a prior background in MEMS will be able to critically judge a fabrication process and synthesize a new one for future applications;
3. Become experienced with commonly practiced designs and fabrication processes of MEMS through studies of classical and concurrent cases; and
4. Obtain the analytical and practical know-how to evaluate many intersecting points—design, fabrication, performance, robustness, and cost, among others—involved in successfully developing integrated MEMS devices.

The three main pillars of knowledge for a MEMS engineer are *design, fabrication*, and *materials*. In this book, I will address them in an ascending and widening spiral, with more details and interactions as the book progresses. For example, in Chapters 1 and 2, the reader will be exposed to a general discussion of transduction principles and microfabrication methods. Chapter 3 will discuss the basic electrical and mechanical engineering terms *most commonly* encountered in the everyday practice of MEMS. Chapters 4 through 9 review the various sensing and actuation methods and their uses. More detailed discussions about the device fabrication techniques will be embedded in case studies. In Chapters 10 through 11, a comprehensive treatment of the two most important classes of microfabrication techniques (bulk micromachining and surface micromachining) is presented. Chapter 12 discusses MEMS fabrication techniques related to the polymer materials family.

The final four chapters collectively give the reader an opportunity to integrate various facets of knowledge and to learn about pragmatic methodologies. In Chapters 13 through 15, several major branch areas of MEMS applications are selected for case studies. Various technologies for realizing similar application goals are presented, so that the reader will be able to evaluate the different sets of designs, materials, and technologies. In Chapter 16, I discuss practical issues pertaining to process integration and project management.

I would like to thank my students—past and present—in my *Introduction to MEMS* classes. Their feedback was critical to the layout of this book. I would also like to thank the following research associates and colleagues at Illinois for their encouragement and assistance during the production of this book: David Bullen, James Carroll, Jack Chen, Jonathan Engel, Zhifang Fan, Prof. Yonggang Huang, Prof. David Payne, Kee Ryu, Kashan Shaikh, Edgar Goluch, Loren

Vincent, Xuefeng (Danny) Wang, Alex Zhenjun Zhu, and Prof. Jun Zou. I also would like to thank the following colleagues who provided valuable insights and facts during the writing process: Prof. Roger Howe, Prof. Richard Muller and Prof. Ming Wu (University of California–Berkeley), Prof. Khalil Najafi (University of Michigan), Prof. Ioannis Chasiotis (University of Virginia), Dr. Nancy Winfree (Dominica Inc.), and Prof. George Barbastathis (Massachusetts Institute of Technology).

Finally, I would like to thank the Head and Associate Head of the ECE Department, Prof. Richard Blahut and Prof. Narayan Rao respectively, for their encouragement and support of this project.

CHANG LIU
Urbana, IL

A Note to Instructors

This section is intended to assist instructors who use this book to teach a body of students at the undergraduate or graduate school levels. It summarizes my thoughts on the selection and ordering of materials. I hope it helps instructors fully utilize this book and teach the subject of MEMS effectively.

Materials in this book are presented to facilitate the teaching of MEMS to beginners and to an interdisciplinary body of readers. During the writing process, I tried to maintain a balanced approach.

First and foremost, this book balances the needs of readers and students from a variety of backgrounds. This book is written for an interdisciplinary body of readers and is meant to intellectually satisfy and challenge every student in the classroom, regardless of his or her background. Two extreme feelings of students and readers—*boredom* when a familiar subject is repeated in detail and *frustration* when an unfamiliar subject is not covered sufficiently—should be avoided at all times. To minimize the initial learning curve, only the most vital vocabulary and the most frequently used concepts are introduced.

Secondly, this book presents balanced discussions about design, fabrication, and materials, which are the three pillars of the MEMS knowledge base. Modular case studies were carefully selected to exemplify the intersection of design, materials, and fabrication methods. An instructor may select alternative cases to append to the existing collection.

Third, this book balances practicality and fundamentals. Fundamental concepts are explained and exemplified through text, examples, and homework assignments. Practical and advanced topics related to materials, design, and fabrication are discussed in paragraph-length mini reviews. These are exhaustive, but their length is kept to a minimum to avoid distracting the attention of the reader. I hope this will encourage and facilitate students and instructors who may wish to follow reference leads and explore topics beyond classroom discussions. For the reader's benefit, the references cited in this book are primarily from archival journals and magazines; therefore, they are easily accessible.

This book attempts to provide a logical build-up of knowledge as it progresses from chapter to chapter. A number of important topics, such as mechanical design and fabrication, are discussed in several passes. In terms of design concepts, an instructor can lead students through three steps: (1) learning basic concepts; (2) observing how they are used in real cases; and (3) learning to apply the design methods to homework problems or real applications. In terms of fabrication, three steps can be followed as well: (1) observing how processes work in examples and critically analyzing processes discussed in the case studies; (2) building a detailed knowledge base of processes in a systematic framework; and (3) synthesizing processes in homework problems and for various applications.

Chapters are presented in a modular fashion. Readers and instructors may follow different routes depending on their background and interest. For example, one may choose to review in-depth information about microfabrication (Chapters 10 through 11) before covering transduction principles (Chapters 4 through 9).

A challenge I faced while teaching and when writing this book was how to integrate a rich body of existing work with many points of innovation without making the book cluttered and losing the focus on learning. In other words, the student should feel the excitement of innovation without being diverted from a sense of focus. The contents of this book are carefully organized to achieve this aim. In the first 12 chapters, I review a number of representative applications (cases) with a *consistent* selection throughout the chapters to provide a basis for comparison. When a chapter deals mainly with a transduction principle for sensing, I discuss *inertia* sensors (including acceleration sensors and/or gyros), *pressure* sensors (including acoustic sensors), *flow* sensors, and *tactile* sensors, in that order. These four sensor topics have been carefully chosen out of many possible applications of MEMS. Inertia and pressure sensors are well-established applications of MEMS. Many good research articles are available; many include comprehensive coverage of integrated mechanics and electronics. Flow sensors are unlike inertia and pressure sensors; they generally involve different physical transduction principles, designs, and characterization methods. Tactile sensors must offer a robustness that is better than the three other sensor types; therefore, it will necessitate discussions of unique materials, designs, and fabrication issues. When a chapter deals with a transduction principle that is mainly used for actuation, I will discuss one case of an actuator with small displacement (linear or angular) and another case of an actuator with large displacement, in that order, and if proper examples are available.

I believe the best way to learn a subject is through examples and guided practice. This book offers a large selection of examples and problems for students.

Homework problems cover not only the design and the use of equations. Many aspects of MEMS, including the selection of materials and processes, are beyond the description of a mathematical formula. Many homework problems are designed to challenge a student to think critically about a fabrication process, to review literature, and to explore various aspects of MEMS, either individually or in small cooperative groups.

There are four types of homework exercises—design, review, fabrication, and challenges. A *design* problem helps the student gain familiarity with formulae and concepts for designing and synthesizing MEMS elements. A *review* problem requires the student to search for information outside of the textbook to gain a wider and deeper understanding of a topic. A *fabrication* problem challenges the student to think critically about various aspects of a fabrication process. For example, the student may be required to develop and demonstrate a true understanding of a process by illustrating it in detail or by devising and evaluating alternative approaches. A *challenge* problem stimulates the competitive edge within students. It provides students with opportunities to think at an integrative level by considering many aspects, including physics, design, fabrication, and materials. A challenge problem may be a competitive, research-level question without existing answers, at least at the time of this writing.

Success in science and technology takes more than technical expertise in a narrow area. Teamwork and collaboration are essential for executing a project or building a career. There are a significant number of homework problems throughout this textbook that encourage students to work together in interdisciplinary teams. I believe that teamwork, at this stage, will enhance learning experiences through social and technical interactions with other students and prepare them for success in their future careers.

I hope you will enjoy this book.

About the Author

Chang Liu received M.S. and Ph.D. degrees from the California Institute of Technology in 1991 and 1995, respectively. His dissertation was entitled "The Micromachined Sensors and Actuators for Fluid Mechanics Applications." In January 1996, he joined the Microelectronics Laboratory of the University of Illinois as a postdoctoral researcher. In January 1997, he became an assistant professor with a major appointment in the Electrical and Computer Engineering Department and a joint appointment in the Mechanical and Industrial Engineering Department. In 2003, he was promoted to the rank of Associate Professor with tenure.

Dr. Liu has had 14 years of research experience in the MEMS area and has published 120 technical papers in journals and refereed conference proceedings. He teaches undergraduate and graduate courses covering broad-ranging topics including MEMS, solid-state electronics, electro-mechanics, and heat transfer. He won a campus Incomplete List of Teachers Ranked as Excellent honor in 2001 for developing and teaching the MEMS class, which was a precursor to this book. He received the National Science Foundation's CAREER award in 1998 for his research proposal that focused on developing artificial haircells using MEMS technology. He is currently a Subject Editor of the *IEEE/ASME Journal of MEMS* (sponsored jointly by the ASME) and an Associate Editor of the *IEEE Sensors Journal*. His work has been cited in popular media. Dr. Liu is the co-founder of Integrated Micro Devices (IMD) Corporation (Champaign, IL) and a member of the scientific advisory board of NanoInk Corporation (Chicago, IL). In 2004, he won the University of Illinois College of Engineering Xerox Award for Faculty Research. In the same year, he was elected as a Faculty Associate at the Center for Advanced Studies at the University of Illinois to pursue research in large-format integrated sensors.

Notational Conventions

Author's Note: The design of a MEMS device involves multiple domains of engineering and physics. Symbols and notations have evolved independently in these domains and may overlap with one another. For example, the symbol J corresponds to current density in electrical engineering and torsional moment of inertia in mechanical engineering. The symbol ε often means permittivity to electrical engineers and mechanical strain to mechanical engineers. In this book, *a symbol may represent several different variables*. The exact correlation depends on the specific circumstance of use. I chose against inventing a notation system with no overlap. A lineage of use to different fields is purposefully maintained.

a	acceleration
α	volumetric expansion coefficient
α_r	temperature coefficient of resistance (TCR)
α_s	Seebeck coefficient of a single material
B	magnetic field density
β	linear expansion coefficient
C	concentration
C_{ij}	elements of the stiffness matrix
c_{th}	heat capacity
χ	magnetic susceptibility
D	diffusivity
D	electric displacement
d_{ij}	elements of piezoelectric coefficient matrix
γ	shear strain
E_g	bandgap
E	modulus of elasticity, Young's modulus
E	electric field
ε	permittivity, relative permittivity, dielectric constant
ε	radiative emissivity
F	force
f_r	resonant frequency
G	shear modulus of elasticity
G	gauge factor
H	magnetic field intensity
h	Planck's constant
h	convective heat transfer coefficient
I	current
I	moment of inertia
J	current density
J	torsional moment of inertia
k	force constant
k	Boltzmann constant

κ	thermal conductivity
L	length or characteristic length
M	moment or torque
m	mass
μ	mobility of charge carriers
μ	magnetic permeability
μ	dynamic viscosity
m_n^*	effective mass of electrons
M_p^*	effective mass of holes
n	concentration of electrons
ν	Poisson's ratio
ν	kinematic viscosity
N_d	concentration of donor atoms
N_d^+	concentration of ionized donor atoms
N_a	concentration of acceptor atoms
N_a^-	concentration of ionized acceptor atoms
n_o	concentration of electrons under equilibrium
n_i	concentration of electrons in intrinsic material
p	concentration of holes
p_o	concentration of holes under equilibrium
p_i	concentration of holes in intrinsic material
π_{ij}	component of the piezoresistance tensor
q	electric charge
q''	thermal conduction rate
R	resistance
Re	Reynolds number
r	radius of curvature
R_{th}	thermal resistance
ρ	resistivity
ρ_s	sheet resistivity
ρ_{th}	thermal resistivity
sh	specific heat
S_{ij}	elements of the compliant matrix
s	strain
s_i	elements of the strain tensor matrix
σ	electrical conductivity
σ	normal stress
σ	Stefan–Boltzmann constant
τ	torsion stress, shear stress
τ	fluid shear stress
T	moment or torque
T	temperature
T_i	elements of the stress tensor matrix
U	stored electrical energy
u	distance of undercut
V	voltage
V_p	pull-in voltage

CHAPTER 1

Introduction

1.0 PREVIEW

This chapter will present a broad overview of the microelectromechanical systems (MEMS) field, along with basic vocabularies and concepts necessary for ensuing discussions on topics including design, fabrication, material requirements, and application of MEMS.

In Section 1.1, the reader will have an opportunity to learn about the history of the MEMS field as well as future promises of MEMS. An understanding of the timing and circumstances under which MEMS technology was initiated will help the reader appreciate many characteristics of the technology. Such intrinsic characteristics are summarized in Section 1.2.

A large portion of MEMS applications involve sensors and actuators, collectively known as transducers. (The remaining MEMS applications involve passive microstructures that are not actively addressed or controlled.) In Section 1.3, the reader will be exposed to a broad range of concepts and practices of energy and signal transduction. The section discusses the most important performance metrics for developing sensors and actuators.

Fundamental microfabrication methods for MEMS will be discussed in Chapter 2.

1.1 THE HISTORY OF MEMS DEVELOPMENT

The integrated circuit (IC) technology is the starting point for discussing the history of MEMS. In 1971, the then state-of-the-art Intel 4004 chip consisted of only 2250 transistors. Intel 286 and Pentium III processors, unveiled in 1982 and 1999, had 120,000 and 24 million transistors, respectively. IC technology developed with a level of fierceness rarely matched in other fields. The density of transistor integration has increased twofold every 12–18 months; this growth follows Moore's Law [1], which is named after an observation made by Gordon Moore, one of the co-founders of Intel Corporation. This is a remarkable feat of ingenuity and determination because, at several points in the past several decades, there were deep concerns that the trend predicted—and in some sense, *mandated*—in Moore's Law would not continue, but would run into limits imposed by fundamental physics or engineering capabilities at the time.

The *microfabrication* technology is the engine behind the functional integration and miniaturization of electronics. From the early 1960s to the mid-1980s, the fabrication technology of integrated circuits rapidly matured after decades of research following the invention of the first semiconductor transistor [2].

Many scientific and engineering feats that we take for granted today would not be here without the tremendous pace of progress in the area of *microfabrication* and *miniaturization*. The list includes the exponentially growing use of computers and the Internet, cellular telephony, digital photography (capturing, storing, transferring, and displaying), flat-panel displays, plasma televisions, fuel-efficient automobiles, human genome sequencing (with 3 billion base pairs) [3], rapid DNA sequence identification [4], the discovery of new materials and drugs [5], and digital warfare.

The field of MEMS evolved from the integrated circuit industry. The germination of the MEMS field covers two decades (from the mid-1960s to 1980s), when sparse activities were carried out. For example, anisotropic silicon etching was created to sculpture three-dimensional features into otherwise planar silicon substrates [6]. Several pioneering researchers in academic and industrial laboratories began to use the integrated circuit processing technology to make *micromechanical* devices, including cantilevers, membranes, and nozzles. Crucial elements of microsensors, including piezoresistivity of single-crystalline silicon and polycrystalline silicon, were discovered, studied, and optimized [7–9]. At this stage, the name of the field had yet to be coined. However, both bulk micromachining and surface micromachining technologies were rapidly maturing [10–12].

There are a number of notable early works. In 1967, Harvey Nathanson at Westinghouse introduced a new type of transistor called the resonant gate transistor (RGT) [13]. Unlike conventional transistors, the gate electrode of the RGT was not fixed to the gate oxide but was movable with respect to the substrate. The distance between the gate and the substrate was controlled by electrostatic attractive forces. The RGT was the earliest demonstration of microelectrostatic actuators.

In the 1970s, Kurt Petersen at the IBM research laboratory, along with other colleagues, developed diaphragm-type silicon micromachined pressure sensors. Very thin silicon diaphragms with embedded piezoresistive sensors were made using silicon bulk micromachining. The diaphragm deforms under differential pressures, inducing mechanical stress that was picked up by the piezoresistors. The thin diaphragm allowed greater deformation under a given pressure differential; hence this diaphragm achieved greater sensitivity than conventional membrane-type pressure sensors. The sensors could be micromachined in batches, thereby increasing the uniformity of performance while reducing the costs of production. More details about the design and fabrication of this pressure sensor can be found in Chapter 6. Pressure sensors for applications such as blood pressure monitoring and industrial control provided the earliest commercial success of MEMS technology.

Today, micromachined pressure sensors are built with a variety of structures and fabrication methods. These sensors can be based on capacitive [14], piezoelectric [15], piezoresistive [16], electronics resonance [17], and optical detection [18] techniques. Advanced features for integrated pressure sensors include a built-in vacuum for absolute pressure measurement [14], an integrated telemetry link [19], close-loop control [20], insensitivity to contaminants [21], biocompatibility for integration into micromedical instruments [22], and use of nonsilicon membrane materials (e.g., ceramics, diamonds) for functioning in harsh and high-temperature environments [17, 23, 24].

Now, ink-jet printers offer a low-cost alternative to laser-jet printing and provide high performance and affordable color photographic-quality printing. Canon discovered the ink-jet technology by thermal bubble formation (bubble jet), whereas Hewlett-Packard pioneered the technology of silicon micromachined ink-jet printer nozzles in 1978. Arrays of ink-jet nozzles eject tiny ink droplets ("drop on demand") upon the expansion of liquid volume by thermal bubbles (see Figure 1.1). The collapse of the bubble draws more ink into the ink cavity for the next firing. Color ink-jet printing is achieved by dropping primary subtractive color dyes—cyan, magenta, and yellow (CMY).

Silicon micromachining technology played an enabling role for the ink-jet printing technology [25–27]. Using silicon micromachining, ink-ejection nozzles can be made extremely small and can be packed densely, which is important for realizing a high printing resolution and sharp contrast. Small volume cavities with equally small heaters produce a rapid temperature rise (during ink ejection) and fall, allowing ink-jet printing to reach appreciable speed. In 1995, the number of nozzles per cartridge increased to 300 while the average weight of an ink droplet was only 40 ng. In 2004, ink-jet heads were based on a variety of principles, including thermal, piezoelectric, and electrostatic forces. The volume of each drop is on the order of 10 pl, and inkjet printing can reach a resolution as high as 1000 dpi [28].

Many ink-jet printers on the market today are based on the thermal ink-jet principle and dispense heat-resistant dyes. Alternative ink-jet principles are also possible. Epson-brand ink-jet printers, for example, use piezoelectric ink-jet technology and special ink dyes (since they do not have to be heat resistant). The inks for piezoelectric ink-jet printers dry more quickly to minimize spreading on paper and therefore produce greater resolution.

Today, ink-jet printers compare favorably with laser-jet printing. Ink-jet printers are generally cheaper, although the cost of replacing ink cartridges makes them more expensive to own and use over long periods of time. The technology is being applied beyond text and photo printing. It is now used for direct deposition of organic chemicals [29], elements for organic transistors [30], and biological molecules (such as building blocks of DNA molecules) [31].

In the late 1980s, researchers in the nascent field called **micromachining** mainly focused on the use of *silicon*—either bulk silicon substrate (single-crystalline silicon) or thin film silicon (polycrystalline silicon). These two forms of silicon were readily accessible, as they were used heavily in the integrated circuit industry—bulk silicon is used as the substrate of circuitry, while polycrystalline silicon is used for making transistor gates. Three-dimensional mechanical structures, such as suspended cantilevers or membranes, can be made out of bulk silicon or thin film silicon. In 1984, Petersen published a seminal paper titled "Silicon as a mechanical material" [10]. This paper was widely quoted in the 1990s (and still is) as the field expanded rapidly.

The use of thin film silicon led to surface micromachined mechanisms including springs, gear trains, and cranks, to name a few. In 1989, the first silicon surface micromachined micromotor driven by electrostatic forces was demonstrated by researchers at the University of California at Berkeley [32]. A polysilicon rotor, less than 120 μm in diameter and 1 μm thick, was capable of rotating at a maximum speed of 500 rpm under a three-phase, 350-V driving voltage. This motor, though with limited application at that time, brought the excitement of MEMS to the broader scientific community and the general public. Microrotary motors—which are based on different actuation principles, cover a wider range of scales (even down to nanometers), and have much greater achievable torque and power—have been demonstrated since then [33, 34].

A few years later, the term *microelectromechanical systems (MEMS)* was introduced. It gradually became the internationally accepted name of the field. This name captured the scale

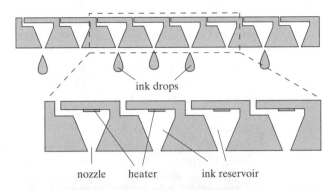

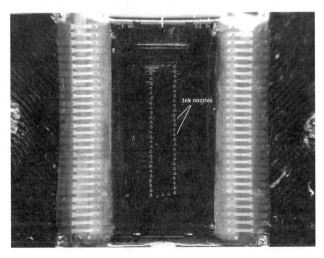

FIGURE 1.1

Micromachined ink-jet printer nozzles. (Top) Schematic view of an ink-jet chip with fluid nozzles. (Middle and bottom) Close-up view of a commercial ink-jet printer head and the silicon chip that has many nozzles. Integrated circuits on chips control nozzle firing.

(micro), practice (electromechanical integration), and aspiration (systems) of the new field. Two subtle facts often elude beginning readers. First, many research results and products of MEMS technology are indeed *components* within a bigger system. Second, the phrase embodies both a unique machining and manufacturing approach (micromachining) *and* a new format of devices and products.

In the 1990s, the field of MEMS entered a period of rapid and dynamic growth worldwide. Government and private funding agencies in many countries funded and supported focused research activities. Early research efforts at several companies started to bear fruit. The most notable examples include the integrated inertia sensors by Analog Devices for automotive air-bag deployment and the Digital Light Processing chip by Texas Instruments for projection display. These two applications are discussed next.

The ADXL series accelerometer made by Analog Devices consists of a suspended mechanical element and signal-processing electronics integrated on the same substrate. The initial development targeted the automotive market [35]. The accelerometer monitors excessive deceleration and initiates air-bag deployment in the event of a life-threatening collision. The mechanical sensing element is a free-moving proof mass suspended by four support springs (Figure 1.2). Movable electrodes in the form of interdigitated fingers are attached to the proof mass. The

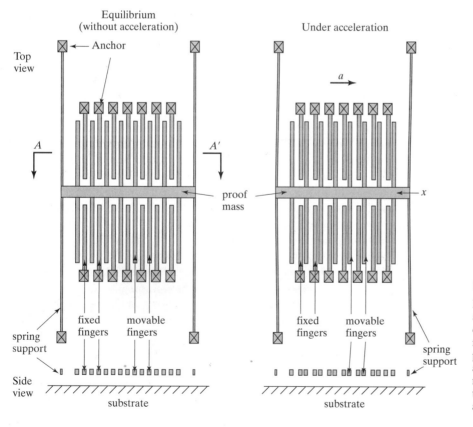

FIGURE 1.2

Mechanical elements of an integrated accelerometer. (Left) The proof mass is at an equilibrium position without acceleration. (Right) The proof mass moves relative to the fixed fingers under an applied acceleration.

fixed and moving electrodes form a bank of parallel-connected capacitors, with the total capacitance depending on the distance between the moving and fixed fingers. If an acceleration (a) is applied to the chip, the proof mass (with mass m) will move under an inertial force ($F = ma$) against the chip frame. This changes the finger distances and therefore the total capacitance. The minute amount of capacitance change is read using on-chip signal-processing electronics. The integration of mechanical elements and electronics is critical for reducing interference noises (stemming from stray electromagnetic radiation) and avoiding parasitic capacitance associated with otherwise long conductor leads.

The MEMS technology offers significant advantages over macroscopic electromechanical sensors, mainly in terms of high sensitivity and low noise. The MEMS approach also decreases the costs of ownership of each sensor, mainly by eliminating manual assembly steps and replacing them with batch fabrication.

Today, one can find a variety of micromachined acceleration sensors on the market based on a number of sensing principles and fabrication technologies. Accelerometers based on capacitive sensing [36, 37], piezoresistivity [38], piezoelectricity [39], optical interferometry [40], and thermal transfer [41, 42] have been demonstrated. Advanced features include integrated three-axis sensing [43], ultra-high sensitivity (nano-g) for monitoring seismic activities [44, 45], increased reliability by eliminating moving mass [42], and integrated hermetic sealing for long-term stability [46].

The technology that produces the accelerometer can be modified to realize rotational acceleration sensors or gyroscopes [47]. Due to their small size, MEMS inertia sensors can be inserted into tight spaces and enable novel applications, including smart writing instruments (e.g., smart pens that detect and transmit handwriting strokes to computers for character recognition), virtual-reality headgears, computer mouses (Gyro mouses), electronic game controllers, shoes that calculate the actual distance they're running, and portable computers that stop the spinning of hard disks if the computer is accidentally dropped.

In the information age, still images and videos are generated, distributed, and displayed in an all-digital manner to maximize quality and lower the distribution cost. Projection display is a powerful tool for digital multimedia presentation, movie theaters, and home entertainment systems. Traditional projection displays are analog in nature, based on liquid crystal display (LCD) technology. The Digital Light Processor (DLP) of Texas Instruments, a revolutionary digital optical projector [48, 49], consists of a light-modulating chip with more than 100,000 individually addressable micromirrors, called digital micromirrors (DMD). Each mirror has an area of approximately $10 \times 10 \ \mu m^2$ and is capable of tilting $+/-7.5°$. The mirror array is illuminated by a light source. Each mirror, when placed at the correct angle, reflects light towards the screen and illuminates one pixel. An array of such mirrors can form an image on a projection screen.

The schematic diagram (top view) of an individual mirror is shown in Figure 1.3. A mirror plate is supported by two torsional support beams and can rotate with respect to the torsion axis. According to the cross-sectional view (along A–A' line), electrodes are located under the mirror to control its position. When one of the electrodes is biased, the mirror will be pulled toward one side by electrostatic attraction force.

Because of the large number and high density of mirrors, they are addressed using a row–column multiplexing scheme. Static random-addressing memory (SRAM) circuits employing 0.8-μm double-level metal CMOS technology for controlling each mirror are embedded on the silicon substrate beneath the layer for mirrors.

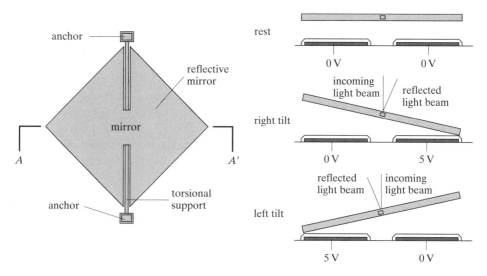

FIGURE 1.3

Diagrams illustrating the structure and operating principles of a single digital micromirror.

The DLP display offers advantages over the incumbent, transmissive LCD projection, including a higher (better) pixel fill factor, greater brightness and black level, greater contrast ratio, more efficient use of light, and stability of contrast and color balance over time. It should be noted that a successful device such as DLP is not an overnight success; it is a result of long-term commitment and development. In fact, the DLP was successfully launched following a string of unsuccessful earlier R&D activities at various companies, which were carried out in a span of 20 years.

Today, digital micromirrors find applications beyond image projection. The technology is being pursued to use for rapid maskless lithography to save the cost of mask making [50], as well as for flexible, *in situ* DNA microarray manufacturing using light-array assisted synthesis [51]. Advanced optical scanning mirrors, such as ones with continuous angular tuning, large displacement range, and more degrees of freedom have been developed for optical communication (Chapter 14).

Besides the acceleration sensor and digital micromirror, many new MEMS device categories were developed in the 1990s, with varying degrees of industrial implementation. A few major spin-off branches of MEMS activities have taken root. A number of these branches along with key technology drivers for them are summarized in Table 1.1.

The optical MEMS area grew rapidly in the late 1990s. Researchers from across the world raced to develop microoptoelectromechanical systems and devices (MOEMS), incorporating components such as binary optical lenses (Figure 1.4a), diffraction gratings (Figure 1.4b), tunable optical mirrors (Figure 1.4c), interferometric filters, and phase modulators for applications including optical display, adaptive optics, tunable filters, gas spectrum analyzers, and network routers [52]. A more detailed review of the optical MEMS field is found in Chapter 15.

In the late 1990s, a large-scale commercialization activity in the optical MEMS area was mounted; it was driven by the anticipated bandwidth bottleneck stemming from the rapid growth of the Internet and personal telecommunication. Free-space optical interconnects between fiber bundles for dynamic routing was the primary focus of many researchers and companies during

TABLE 1.1 Representative Major Branches of MEMS Technology.

Area of research	Perceived drivers of technology
Optical MEMS (See Chapter 15 for details)	Monolithic integration of mechanics, electronics, and optics Unique spatial or wavelength tunability Improved efficiency of optical assembly and alignment accuracy
BioMEMS	Miniaturization (minimal invasion and size matched with biological entities) Rich functional integration within physically small, minimally invasive medical devices
Microfluidics (laboratory-on-a-chip or micro total analysis systems) (See Chapter 13 for details)	Reduced amount of samples and reagents and associated cost Parallel and combinatorial analysis possible Miniaturization, automation, and portability
Radio Frequency (RF) MEMS	Unique performances not found in solid-state RF integrated devices Promises of direct integration of active and passive elements with circuitry
Nano Electromechanical Systems (NEMS)	Unique physical properties due to scaling (e.g., ultra-low mass and ultra-high resonant frequency) Unprecedented sensitivity and selectivity of detection achievable in selected cases

that period of time. Traditionally, optical signals are switched between fiber bundles by following three steps: (1) turning optical signals into electronic signals using optical receiver arrays, (2) using channel shuffling in the electrical domain, and (3) transforming signals in the electronics domain back to the optical domain using laser diode arrays. All-optical interconnect, a new fiber interconnection and switching scheme, offered a promising solution for solving the speed problem associated with the traditional optical-electrical-optical (or OEO) transformation. Using micromachined optical switches, such as the one shown in Figure 1.4c, light beams from one bundle of fibers can be steered directly into a receiving bundle, bypassing the electronic domain and signal transduction links. A great deal of ingenious engineering led to new actuators and fabrication techniques. Many successful products were developed, but they were not manufactured and used on a large scale. This experience provides valuable lessons for the future growth of the MEMS field and high-tech commercialization [53].

BioMEMS encompasses the development and use of MEMS for biological studies, medical diagnoses, and clinical intervention. Because of their miniaturization and rich functional integration, microstructures and microdevices are used in medical applications including retina implants, cochlear implants, embedded physiological sensors, and sensor-enabled smart surgical tools.

As an example, microfabricated neuron probes can facilitate neurobiological studies. Neurons produce spike-like electrical pulses that travel along axons situated in three-dimensional neural tissue. The neuron tissues are extremely complex, with their interconnecting structures and signal-processing algorithms largely unknown to humans. Traditionally, biologists use a single metal wire inserted into the neuron tissue to record neuron spikes. This so-called single-unit recording limits the richness of data and is a major bottleneck for biological discoveries. Arrays of microfabricated neuron recording probes, such as the ones shown in Figure 1.5, promise to replace single-unit neuron recording probes to yield richer neural recording data. Neuron probes with multiple recording sites are now made of silicon or metal. Sophisticated probes provide multiple recording sites, simultaneous chemical elution and collection, integrated pumps and valves for liquid handling, and integrated circuitry for enhancing the signal-to-noise ratio (e.g., [54]).

1.1 The History of MEMS Development

Fresnel lens?

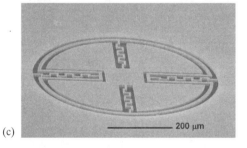

FIGURE 1.4

Micromachined optical components: (a) binary lens, (b) vertical diffraction grating, and (c) two-axis scanning micromirror.

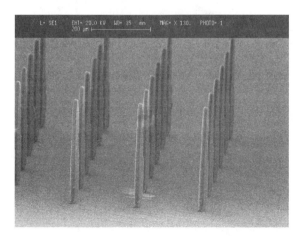

FIGURE 1.5

MEMS neuron probes.

Another excellent example of a BioMEMS application is a drug injection needle array with precisely defined needle dimensions (especially heights). Traditional drug injection induces pain because the needles' tips reach into a layer of skin tissues that are richly populated by nerve bundles. This layer lies 50–70 μm under the skin surface. By using needles that are shorter than this critical length, drug injection can be performed in a subdermal region that is not occupied by neurons; thus, it does not induce pain in patients. Many different types of microneedle arrays have been developed in the past [55, 56]. Researchers have overcome significant challenges involving materials and microfabrication, though a lot of potential solutions are still waiting to be fully tapped.

The MEMS technology can also enable microfluid systems and integrated biological/chemical processors, which can be used for devices such as automated and miniaturized sensors for point-of-care medical diagnosis and distributed environmental monitoring. One example is a microfluid chip that performs cell manipulation, processing, selection, and storage of harvested bovine embryos [57]. Conventionally, these delicate procedures are performed manually, in a very tedious and error-prone manner. This results in a high cost of operations and products, degradation of quality, and waste. A microfluid workstation developed for assisted reproduction performs complex and delicate procedures on individual cells. Cells are moved, held, and manipulated using pneumatic pressure instead of direct probing, in a system diagrammed in Figure 1.6. This increases the speed and efficiency while reducing the chance of cell damage. More information about principles and current research activities in the microfluidics area is provided in Chapter 13.

The MEMS technology also enables innovative components for integrated circuitry, including radio frequency (RF) communication chips. Examples include micromachined relays, tunable capacitors, microintegrated inductors and solenoid coils, resonators and filters, and antennas (Fig. 1.7). A more in-depth discussion of this subject can be found in [58].

The basis of a strong and sustainable industry based on MEMS technology has been formed. In 2003, the yearly sales figures attributed to MEMS technology were $650 M for Hewlett-Packard (ink-jet printing), $353 M for Epson and Lexmark combined (ink-jet printing), $600 M for Texas Instruments (DLP and sensors), $120 M for Analog Devices (sensors), and $108 M for Freescale Semiconductor (pressure and acceleration sensors). Since 2000, many

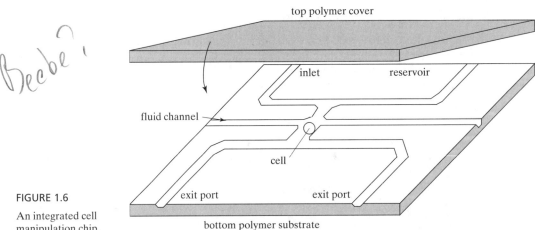

FIGURE 1.6

An integrated cell manipulation chip.

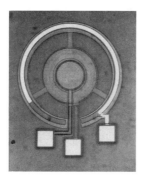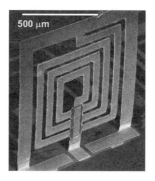

FIGURE 1.7

Micromachined tunable capacitors and inductors.

small companies focusing on MEMS related technologies—including optical MEMS, microfluidics, BioMEMS, and RF MEMS—have been created.

The community of MEMS researchers has been growing rapidly. Each year, several major international conferences are held. The most established conferences include the IEEE International Conference on Solid-State Sensors, Actuators, and Microsystems (the *transducers conference*), the IEEE Annual International Conference on Microelectromechanical Systems (the *MEMS conference*), the Eurosensors conference, the IEEE Workshop on Solid-State Sensors, Actuators, and Systems (held biannually at Hilton Head Island, SC), and the International Conference on Micro Total Analysis (μTAS). Many conferences in specific topic areas, including optical MEMS, actuators, BioMEMS, and MEMS commercialization are held worldwide. Many new journals have been created to address MEMS researchers, including *IEEE/ASME Journal of Microelectromechanical Systems*, the *Sensors and Actuators* journal (by Elsevier BV), the *Journal of Micromechanics and Microengineering*, and *Lab on a Chip* (by the Royal Society of Chemistry). In addition, the following journals frequently publish papers that cover state-of-the-art physics, applications, and related fabrication techniques for micro- and nanoscale devices: *Science*, *Nature*, *Applied Physics Letters*, *Journal of Applied Physics*, *Nano Letters*, *Analytical Chemistry*, and *Langmuir*, among others.

The investigation of coupled electrical and mechanical behavior in a system is not unique to the MEMS field or the microscale. At the macroscopic scale, two subject areas—mechatronics and electromechanics—have been studied for decades. They provide relevant background for MEMS research as well.

In the next 10 to 20 years, the MEMS research field is expected to grow rapidly. Advancements will likely be manifested in several aspects: (1) an increased functional reach of MEMS in interdisciplinary applications; (2) the maturation of design methodology and fabrication technology; (3) the enhancement of mechanical performances such as sensitivity and robustness; (4) the enhancement of electrical functions, low-cost circuit integration, and large-area integration; (5) the seamless integration of heterogeneous materials in a functional system; and (6) a lowered development cost and cost of ownership.

1.2 THE INTRINSIC CHARACTERISTICS OF MEMS

There is no doubt that MEMS will continue to find major new applications in the future. The reason for technology development and commercialization may vary by case. Nevertheless, there are three generic and distinct merits for MEMS devices and microfabrication technologies. These are called the three Ms—*miniaturization*, *microelectronics integration*, and *mass fabrication with precision*.

1.2.1 Miniaturization

The length scale of typical MEMS devices generally ranges from 1 μm to 1 cm. (However, a large array of MEMS devices or an entire system may be larger.) Small dimensions offer many operational advantages, such as soft springs, high resonance frequency, and low thermal mass. For example, the heat transfer to and from a micromachined device is generally fast; the time constant of the ink-jet printer nozzle's droplet ejection is approximately 20 μs. Their small size also allows MEMS devices to be less intrusive in biomedical applications. A case in point is the arrayed neuron probes discussed earlier.

However, not all things work better when miniaturized. Some physical phenomena do not scale favorably when the dimensions are reduced, but certain physical phenomena that work poorly at the macroscale suddenly become very practical and attractive at the microscale. We call these observations *scaling laws*. Scaling laws explain how physics works at different sizes. For example, fleas can jump dozens of times their own height, whereas elephants cannot jump at all. General qualitative observations can be made: smaller things are less affected by gravity, and miniaturization brings about faster speed, higher power density, and efficiency.

A rigorous scaling-law analysis starts with the identification of a characteristic length scale (denoted L) for a device of interest. For example, the length of a cantilever or the diameter of a circular membrane may be denoted as the L of the respective element. The remaining pertinent physical dimensions are assumed to scale linearly with the characteristic length scale with locked ratios.

A performance merit of interest (e.g., stiffness of a cantilever or resonant frequency of a membrane) is expressed as a function of L, with other dimension terms expressed as a fraction or multiple of L. The expression is then simplified to extract the overall effect of L.

Example 1.1 Scaling Law of the Spring Constant

The stiffness of a cantilever is defined by its spring constant. Identify the scaling law governing the stiffness of a cantilever, with length, width, and thickness denoted as l, w, and t, respectively.

Solution. The performance merit of interest is the spring constant. In the small displacement regime, the spring constant is expressed as

$$k = \frac{Ewt^3}{4l^3}, \qquad (1.1)$$

where E is Young's modulus of elasticity, a dimensional-invariant material property. (Chapter 3 will review how this formula is derived.) If we replace the term l with L, w with αL, and t with βL—where both α and β are constants—then Equation (1.1) can be rewritten as

$$k = C\frac{L^4}{4EL^3} \propto L. \qquad (1.2)$$

1.2 The Intrinsic Characteristics of MEMS

The term C is a proportionality constant ($C = \alpha\beta^3$). This scaling law analysis shows that cantilevers with reduced sizes have smaller spring constants.

Example 1.2. Scaling Law of Area-to-Volume Ratio

Derive a scaling law for the ratio of the surface area to the volume of a cube; discuss the consequences for MEMS design.

Solution. A convenient characteristic length of a cube is the length of each edge, designated as L. The volume of the cube is L^3, while the total area is $6L^2$. The ratio of area over the volume is therefore

$$\frac{area}{volume} \propto \frac{L^2}{L^3} = \frac{1}{L}. \tag{1.3}$$

The smaller the L, the greater the ratio of surface area over the volume. This conclusion, also applicable to objects with arbitrary three-dimensional shapes, provides insight on microscale device design. Surface forces, such as van der Waals force, friction, and surface tension force, are very important for the behavior of microscale objects. Volume forces, such as gravitational force, are less dominant.

In many cases, it is important to simultaneously evaluate the scaling laws on several performance aspects; this will help determine the overall merit of scaling based on a combined figure of merit. For example, consider the Analog Devices accelerometer. The following key performance metrics are variable by scale: the spring constant of the support beam (related to sensitivity), the resonant frequency of the support beam (related to bandwidth), and the overall capacitance value (related to sensitivity). Miniaturization generally leads to softer support beams, higher resonant frequency, and higher bandwidth; however, it reduces capacitance value and generally increases circuitry complexity (to accommodate smaller signals).

In recent years, electromechanical devices with characteristic scales between 1 nm and 100 nm were investigated to explore the scaling effect beyond that of traditional MEMS [59, 60]. Such devices and systems are called **nano electromechanical systems** (NEMS). Many NEMS devices are made using nanostructure assembly, such as nanotubes [60, 61] or nanofabricated elements [62]. In certain applications, NEMS offers performance characteristics not available even in MEMS, such as ultra-high mechanical resonance into the GHz range.

NEMS technology is being pursued for chemical sensing due to the small dimensions and unique properties of nanomaterials. The reduced stiffness of cantilevers at the nanoscale, along with unique device features (e.g., chemically functionalized nanoscopic gaps) means such devices can be used to detect interaction with single bacteria, cells, molecules, or even phonons.

NEMS technology for radio frequency devices is actively pursued as a result of favorable scaling laws. A mechanical resonator that is based on the NEMS technology will have lower force constant and greater resonant frequency than a device that is 10 or 100 times larger in each dimension. High-frequency electromechanical resonators and filters have been made using lithography-patterned nanomechanical cantilevers [63, 64]. For example, a NEMS mechanical resonator with a resonate frequency of 1.35 GHz and a quality factor around 20,000–50,000 has been used to validate fundamental quantum mechanical limits imposed by the Heisenberg uncertainty principle [65].

1.2.2 Microelectronics Integration

The ability to seamlessly integrate mechanical sensors and actuators with electronic processors and controllers at the chip level is one of the most unique characteristics of MEMS. This process paradigm is referred to as **monolithic integration**—fabrication of various components on a single substrate in an unbroken, wafer-level process flow. Monolithic processes do not involve hybrid assembly methods, such as robotics pick-and-place, or any manual attachment of individual parts. Accurate dimensions and placement precision are guaranteed by lithography.

Though not all MEMS devices should adhere to the monolithic integration format, it is observed that silicon circuits that are monolithically integrated with mechanical elements have been involved in several successful commercial MEMS applications, such as accelerometers, digital light processors, and ink-jet printer heads.

In the case of the automotive accelerometer, integration allows MEMS sensors to have a significant commercial advantage over pure mechanical products. Systems integration improves signal quality by reducing the length of signal paths and noise.

Circuits integration is the only way that a large and dense array of sensors or actuators can be addressed. In the case of DLP, for example, each mirror is controlled by a CMOS logic circuit that is buried directly underneath. Without circuit integration, it is impossible to address individual mirrors in such a large and dense array.

The performance benefits of monolithic integration should be carefully weighed against other concerns, such as cost, time to market, and flexibility to changes. Chapter 16 presents some words of caution about the practice of monolithic integration.

1.2.3 Mass Fabrication with Precision

MEMS technology can realize two- or three-dimensional features with small dimensions and precision that cannot be reproducible, efficient, or profitable with traditional machining tools. Combined with photolithography, MEMS technology can realize unique three-dimensional features, such as inverted pyramid cavities, high aspect-ratio trenches, through-wafer holes, cantilevers, and membranes. Making these features using traditional machining or manufacturing methods is prohibitively difficult and inefficient. Modern lithography systems and techniques provide finely defined features, as well as uniformity across wafers and batches.

1.3 DEVICES: SENSORS AND ACTUATORS

1.3.1 Energy Domains and Transducers

MEMS technology enables revolutionary sensors and actuators. In general terms, sensors are devices that detect and monitor physical or chemical phenomena, whereas actuators are devices that produce mechanical motion, force, or torque. Sensing can be broadly defined as energy transduction processes that result in perception, whereas actuation can be defined as energy transduction processes that produce actions.

Sensors and actuators are collectively referred to as *transducers*, which transform signals or power from one energy domain to another. There are six major energy domains of interest:

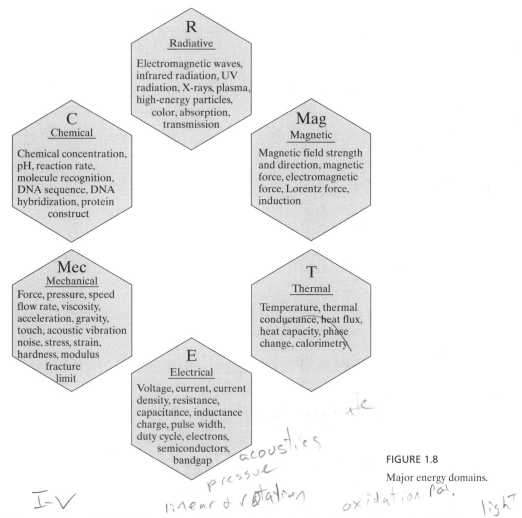

FIGURE 1.8

Major energy domains.

(1) electrical domain (E); (2) mechanical domain (Mec); (3) chemical domain (C); (4) radiative domain (R); (5) magnetic domain (Mag); and (6) thermal domain (T). These energy domains and commonly encountered parameters within them are summarized in Figure 1.8. The total energy within a system can coexist in several domains and can shift among various domains under the right circumstances.

Sensors generally perform two functions. First, they can transform stimulus signals in various energy domains so they are detectable by humans. Second, they can transform stimulus signals into the electrical domain so they can interface with electronic controllers, recorders, or computers. For example, a thermal-couple temperature sensor transforms a thermal signal (temperature) into an electrical signal (voltage) that can be read electronically. Often, more than one sensing principle can be used for a transduction task. Temperature variation can be perceived via resistance changes, volume expansion of fluids, increased radiation power of an object, color change of engineered dyes, shifted resonance frequency of resonant beams, or greater chemical reactivity. More discussion on this topic can be found in Chapter 5.

Energy transduction pathways for particular sensor and actuation tasks do not have to involve only two domains. In fact, the transduction process may incorporate multiple domains. Direct transduction pathways that involve the minimum number of domains do not necessarily translate into simpler devices, lower costs, or better performance.

Energy and signal transduction involves a great deal of research and development, and it is a continuing source of innovation. The desire to discover and implement efficient, sensitive, and low-cost sensing principles transcends the boundary of scientific and technological disciplines. Because many sensing tasks can be achieved in more than one way, either directly (from one energy domain to another) or indirectly (hopping through intermediate energy domains), there is essentially an unlimited number of transduction pathways for achieving one sensor or actuator need. Each transduction pathway entails different sensing materials, fabrication methods, design, sensitivity, responsivity, temperature stability, cross-sensitivity, and cost, etc. A trade-off study must be conducted, taking account of performance, cost, manufacturing ease, robustness, and intellectual property rights.

The development of sensors and actuators is a rich and rewarding research experience. To invent a new sensor principle for a particular application involves selecting or inventing the energy transduction paths, device designs, and fabrication methods that yield simple transduction materials, high performance, and low-cost fabrication. We will discuss a few specific examples of sensors to illustrate the richness of this field and to exemplify the excitement involved with research and development activities. In many cases, new sensing methods result in new device capabilities and industrialization opportunities.

Acceleration sensing (Mec $\rightarrow$ E transduction) Acceleration can be sensed in many different ways. A micromachined proof mass suspended by cantilevers will experience an inertial force under an applied acceleration. The force will cause movement of the suspended proof mass. The movement can be picked up using piezoresistors, which are resistor elements whose resistance changes under applied stress (Mec $\rightarrow$ E). The displacement can also be sensed with a capacitor (Mec $\rightarrow$ E). This is the principle of Analog Device accelerometers. These two methods involve moving mechanical mass. Can we build accelerometers without moving parts? The answer is yes. Inertial force can also move a heated mass, whose ensuing displacement can be picked up by temperature sensors (Mec $\rightarrow$ T $\rightarrow$ E) [42]. Thermal sensing of moving air mass is not as good as capacitive sensing, but the fabrication is readily compatible with integrated circuits. This is the principle of the low-cost acceleration sensor (manufactured by MEMSIC Corporation) designed for low-sensitivity applications (further discussed in Chapter 5). No moving mass is required; thus, concerns of mechanical reliability are eliminated.

Olfactory sensing (C $\rightarrow$ E transduction) Certain molecules are responsible for smell or environmental monitoring. Information about their presence and concentration can be obtained using a number of strategies. A carbon-based material can be designed to absorb certain molecules in the path of surface acoustic wave devices. This alters their electrical resistivity (C $\rightarrow$ E direct transduction), and it can alter mechanical properties such as the frequency of surface acoustic wave transmission (C $\rightarrow$ M $\rightarrow$ E). However, these methods generally involve sophisticated electronics or algorithms. We can build olfactory sensors that are simpler and more intuitive. The binding of chemical molecules can also alter the color of a specially designed chemical compound, which can be detected by low-cost optoelectronics diodes (C $\rightarrow$ R $\rightarrow$ E transduction) [66] or by humans without electronics (C $\rightarrow$ R). Sensors based on this strategy are being made by ChemSensing.

DNA sequence identification (C → E transduction) DNA molecules consist of a chain of base pairs, each with four possible varieties—A, C, G, or T. The sequence of base pairs in a DNA chain determines the code of synthesizing proteins. The ability to decipher base-pair sequences rapidly, accurately, and inexpensively is critical for pharmaceutical and medical applications [67]. There are a wide variety of innovative methods for the detection of the sequence through its telltale binding (hybridization) events. Certain DNA molecules may be chemically modified to incorporate fluorescence (tagged) reporters that light or dim upon binding with another DNA strand. In the most widely practiced cases today, chemical binding events are turned into optical signals before they are transduced to the electrical domain (C → R → E). The fluorescent image is captured using high-power fluorescent microscopes.

However, fluorescent imaging requires sophisticated microscopes and is not suitable for portable field applications. When DNA molecules are attached to gold nanoparticles, they can report the occurrence of hybridization through the aggregation of gold particles, which can result in changes of optical reflectance (C → R → E) [68] or electrical resistivity (C → E) [69]. This detection method provides better sensitivity and selectivity than fluorescence methods; it also eliminates the need for cumbersome fluorescent imaging instruments. It is therefore amenable for miniaturization and remote deployment. This principle is the technological basis of Nanosphere Corporation.

1.3.2 Sensors

Sensors fall into two categories: physical and chemical/biological. Physical sensors measure physical variables such as force, acceleration, pressure, temperature, flow rate, acoustic vibration, and magnetic field strength. Chemical sensors detect chemical and biological variables including chemical concentrations, pH, binding strength of biological molecules, protein–protein interactions, and so forth.

In this textbook, we focus on the discussion of physical sensors. We will explore a number of commonly used principles, including electrostatics, piezoresistivity, piezoelectricity, thermal resistivity, and bimetallic thermal bending. These principles are discussed in greater detail in Chapters 4, 5, 6, 7, and 9.

Many sensing principles might be available for a given application. Sensor developers generally must evaluate a number of transducing pathways and designs according to many performance metrics. The most important sensor characteristics are as follows:

1. *Sensitivity.* The sensitivity is defined as the ratio between the magnitude of the output signal and that of the input stimulus. Note that the sensitivity values may be a function of the input amplitude and frequency, temperature, biasing level, and other variables. When electronic signal amplification is used, it is important to distinguish the sensitivity values before and after amplification.
2. *Linearity.* If the output signal changes proportionally with respect to the input signal, the response is said to be linear. The linear response of a sensor alleviates the complexity of signal processing.
3. *Responsivity.* Responsivity is also known as the accuracy, resolution, or detection limit. This term signifies the smallest signal that a sensor can detect with confidence. It is generally limited by noise associated with the transduction elements and circuits. Noise can be applied to anything that obscures a desired signal. Noise itself can be another signal (interference); most often, however, we use the term to describe the "random" noise of a physical (often thermal)

origin. While interference noise can be corrected or eliminated, such as by careful electrical shielding, random noises are ubiquitous and have much more fundamental origins. In MEMS sensors, the major contribution of noise comes from the following sources:

a. Johnson noise is white noise manifested as open circuit voltage created by a resistor due to random thermal fluctuation of internal electrons and particles. The RMS value of Johnson noise is $V_{noise} = \sqrt{4kTRB}$, where k, T, R, and B are Boltzmann's constant, the absolute temperature, the resistance value, and the bandwidth in hertz, respectively. The amplitude of Johnson noise follows a Gaussian distribution.

b. Shot noise, or the "raindrop-on-a-tin-roof" noise, is another Gaussian and white noise. Its origin is the quantum fluctuation of electric current due to the discrete passage of charges. Shot noise can be estimated as $I_{noise} = \sqrt{2qI_{dc}B}$, where q, I_{dc}, and B are the electron charge, the DC current, and the measurement bandwidth measured in hertz.

c. The 1/f noise, also known as flicker noise or pink noise, is the result of conductance fluctuation when a current passes through a resistor. It derives its name from its characteristic 1/f spectrum dependence. It is possible to optimize sensor design to reduce 1/f noise contributions [70].

d. For many motion-based MEMS sensors (e.g., accelerometers and pressure sensors), the thermal–mechanical noise floor, which is the vibration of microstructures under the impact of Brownian motion mechanical agitation, is another fundamental source of noise [71]. The thermal–mechanical noise is proportional to the square root of Boltzmann's constant, the temperature, the quality factor, the spring constant, the resonant frequency, and the measurement bandwidth [72]. Note that the thermal noise floor can be lowered through ensemble averaging over time or over many identical devices.

4. *Signal-to-noise ratio (SNR).* The SNR is the ratio between the magnitude of signals and noise.

5. *Dynamic range.* The dynamic range is the ratio between the highest and the lowest detectable signal levels. In many applications, a wide dynamic range is desired.

6. *Bandwidth.* Sensors behave differently to constant or time-varying signals. Often, sensors may cease to respond to signals of extremely high frequencies. The effective frequency range is called the bandwidth.

7. *Drift.* Drift may occur because the electrical and mechanical properties of materials vary over time. Sensors with a large drift cannot be used successfully to detect slowly changing signals, such as monitoring stress that builds up in a structure over time.

8. *Sensor reliability.* Sensor performance may change over time, particularly if it is placed under harsh conditions. Sensors developed for military use, for example, need to satisfy the military specification (MIL-SPEC). Reliability and trustworthiness of sensors in a wide temperature range ($-55°C$ to $105°C$) is demanded of such sensors. Many industries have established guidelines and standards involving sensor use.

9. *Cross talk or interference.* A sensor intended for measuring one variable may be sensitive to another physical variable as well. For example, a strain sensor may have finite sensitivity to temperature and humidity. An acceleration sensor with sensitivity in one particular axis may respond to acceleration in other orthogonal axes. Sensor cross talks should be minimized in practical applications.

10. *Development cost and time.* It is always desirable that the sensor development process be inexpensive and fast. Fast time-to-market is important for commercial sensors that are built with custom specifications. Many commercially successful MEMS sensors have been developed over long periods of time and cost millions of dollars. The reduction of cost and development time to the level currently enjoyed by the application-specific integrated circuit (ASIC) industry would be very appealing.

1.3.3 Actuators

Actuators generally transform energy in nonmechanical energy domains into the mechanical domain. For a particular actuation task, there could be several energy transduction mechanisms. For example, we can generate a mechanical movement by using electrostatic forces, magnetic forces, piezoelectricity, or thermal expansion. Several methods commonly used for MEMS are covered in Chapters 4, 5, 7, and 8 and briefly summarized in Table 1.2.

There are many other actuation methods, including pneumatics [73, 74], shape-memory alloys [75–78], thermal expansion [79], phase change [80], electrochemical reactions [81] and energetic combustion [82–84], and friction drag by moving fluids [85]. Microstructure can also be coupled to mesoscopic drivers (e.g., with interlocking mechanisms [86]).

The following are general criteria when considering actuator design and selection:

1. *Torque and force output capacity.* The actuator must provide sufficient force or torque for the task at hand. For example, microoptical mirrors are used to deflect photons. Because photons are lightweight, low levels of force provided by the mirror actuator are sufficient. In some cases, microactuators are used for interacting with the fluid (air or water) to actively control it. Such actuators must provide sufficient force and power to produce appreciable effects.

2. *Range of motion.* The amount of translation or angular movement that the actuator can produce under reasonable conditions and power consumption is an important concern. For example, the DLP micromirrors are required to move within a 15° range. For optical switches used for dynamic network routing, larger angles of displacement (30°–45°) are needed.

TABLE 1.2 Comparison of Actuation Methods.

Mechanism	General description	Comments
Electrostatic actuation	Force generated when an applied electric field acts on induced or permanent charges	Electrodes must be conducting materials
Magnetic actuation	Moment and force due to interaction of magnetic domains with external magnetic field lines	Requires magnetic materials and magnetic sources (solenoid or permanent magnets)
Thermal bimetallic actuation	Differential volume expansion of at least two different materials due to temperature change	Requires materials with different thermal expansion coefficients
Piezoelectric actuation	Change of material dimensions due to an applied electric field	Requires high performance piezoelectric materials

3. *Dynamic response speed and bandwidth.* The actuator must be able to produce a sufficiently fast response. From the viewpoint of actuator control, the intrinsic resonant frequency of an actuator device should be greater than the maximum oscillation frequency.
4. *Ease of fabrication and availability of materials.* To reduce the potential cost of MEMS actuators, there are two important strategies. One is to reduce the cost of materials and processing time. Another is to increase the process yield for a given process in order to produce more functional units in each batch.
5. *Power consumption and energy efficiency.* Many microactuators are envisioned for use in small and mobile systems platforms. The total available power for such systems is generally limited. In this and many other MEMS applications, low-power actuators are preferred to increase the duration of operation.
6. *Linearity of displacement as a function of driving bias.* If the displacement varies with input power or voltage in a linear fashion, the control strategy would be simplified.
7. *Cross-sensitivity and environmental stability:* The actuator must be stable over the long term, against temperature variation, humidity absorption, and mechanical creep. Long-term stability of such actuators is extremely important for ensuring commercial competitiveness and success. A mechanical element may produce displacement, force, or torque in a non-intended axis.
8. *Footprint.* The footprint of an actuator is the total chip area it occupies. In cases of dense actuator arrays, the footprint of each actuator becomes a primary point of consideration.

SUMMARY

This chapter intends to familiarize the reader with the history and current application trends of MEMS technology. We reviewed the advantages associated with the practice of miniaturization and integrated fabrication, and learned that sensors and actuators constitute the majority of MEMS applications. Major points of consideration when developing and selecting sensors and actuators were discussed. These points can be applied beyond MEMS-based sensors and actuators.

The following is a list of major concepts, facts, and skills associated with this chapter.

- The relationship between the microelectronics industry and MEMS.
- Major commercially successful MEMS devices and their competitive advantages over incumbent and competitive technologies.
- Basic principles of commercially successful MEMS devices including accelerometers, digital light processors, and ink-jet printer heads.
- Procedure for performing the scaling law analysis.
- Major energy domains associated with transducer operations.
- Major points of consideration for sensor development.
- Primary sources of sensor noise and their relation to parameters such as temperature and bandwidth of measurement.
- Major points of consideration for actuator development.

PROBLEMS

Problem 1.1: Review

Read the following sections in the paper "Silicon as a mechanical material" by Kurt Petersen: Sections I, II, IV, VI, VIII [10]. The paper can be found in the library or online.

Problem 1.2: Review

Locate the following MEMS-specific journals and conference proceedings in your library or online: (1) *Sensors and Actuators*; (2) *Proceedings of IEEE Annual International Workshop on Microelectromechanical Systems*; (3) *IEEE/ASME Journal of Microelectromechanical Systems*; (4) *Journal of Micromechanics and Microengineering*.

Papers from these journals and proceedings are important to gain further knowledge in the MEMS area beyond the coverage of this textbook. Therefore, it is important to locate these sources.

Identify one particular area of MEMS, and then identify **five** papers from these sources. The papers must come from at least two different sources, and the dates of publication should span at least five years.

Write a two-page, single-spaced summary to compare the contents of these five papers. Summarize the five references in the following format: authors, title, publication source, issue number, page number, and year. Compare the technical elements of these five papers, and explain how these works are related to one another. You may compare the specifications and/or fabrication technology and/or fabrication complexity of these five reported devices. Optionally, you may compare MEMS with other competing technologies in terms of performance, cost, reliability, and customization.

Problem 1.3: Review

Find an online archive of pertinent electronics journals in the MEMS area. Bookmark the following journals: *Science*, *Nature*, *Applied Physics Letters*, *Journal of Applied Physics*, *Proceedings of the National Academy of Science*, *Nano Letters*, *Langmuir*, *Biomedical Microdevices* (Kluwer), and *Lab on a Chip* (Royal Society of Chemistry). Papers related to microfabrication, MEMS, and nanotechnology are frequently published in these journals.

Problem 1.4: Review

Using web search tools, find 10 *university* groups with research programs in the broadly defined MEMS area. At least five of them must be from countries or continents other than your own. Pick four of these research groups and read a recent journal paper (within the last year) from each selected group. Summarize the importance and uniqueness of that work in three to four sentences. You may electronically submit the result to your instructor with links to each of the 10 groups embedded in the file.

Problem 1.5: Review

Find published product specification sheets of three accelerometers; two must be based on MEMS technology and one must be based on non-MEMS technology. Summarize the performance of these three sensor products according to the criteria outlined in Section 1.3.2.

Problem 1.6: Review

Find published product specification sheets for two pressure sensors; one must be based on MEMS technology and another must be based on non-MEMS technology. Summarize the performance of these two products according to the criteria outlined in Section 1.3.2.

Problem 1.7: Review

Find published product specification sheets for a commercial tactile sensor, whether it is MEMS-based or not. Summarize the performance of this product according to the criteria outlined in Section 1.3.2. If certain performance specifications are unpublished, leave them blank or speculate based on your current knowledge.

Problem 1.8: Review

Find published product specification sheets for two flow sensors; one must be based on MEMS technology and another must be based on non-MEMS technology. These two sensors can be based on different principles. Summarize the performance of these two products according to the sensor performance criteria outlined in Section 1.3.2.

Problem 1.9: Review

Find published product specification sheets for a linear solenoid actuator, and summarize its performance specifications according to the criteria outlined in Section 1.3.3.

Problem 1.10: Review

During the life cycle of a biological cell, it undergoes expansion, shear, and tension. Scientists are interested in directly measuring the magnitude of the force inside a cell. Assume the size is less than 2 μm in diameter; find two different methods for measuring forces inside a cell. (*Note*: The measurement device must be located inside the cell and the signals must be able to transmit through the membrane to an outside observer. Use both online resources and scientific literature.)

Problem 1.11: Review

Biology offers many interesting design principles for sensors and actuators. For example, biological haircell receptors are widely found in the animal kingdom. They perform a variety of functions, ranging from hearing and balancing in vertebrate animals to flow sensing in insects and fish. Geckos have feet that firmly attach to walls and yet easily release to allow them to walk on vertical walls and ceilings with ease. Flies have tremendous bidirectional hearing capabilities despite their small size. Discuss and review a biological sensor or actuator. Compare its performance with engineered counterparts (if any) across at least five performance aspects.

Problem 1.12: Design

A resistor is made of a suspended, doped, polycrystalline silicon beam; its resistivity of polysilicon is 5000 Ω. Calculate the resistor's Johnson noise, current shot noise, and mechanical noise when measured in a frequency range of 0 to 100 Hz and 0 to 10 kHz.

Problem 1.13: Design

An old resistor with a nominal value of 10 kΩ lies under room temperature. Estimate the Johnson noise floor at room temperature (27°C) when the measurement bandwidth is 1 kHz. The bias voltage is 5 V.

Problem 1.14: Design

If a nominal voltage of 0.5 volt is applied across the resistor in Problem 1.12, calculate the equivalent noise voltage that corresponds to the shot noise.

Problem 1.15: Design

Plant biologists would like to monitor the growth activities of trees in the rainforest to measure the long-term effects of environment change. One of the parameters of interest is the circumference of tree trunks. Tree trunks grow over the years; however, their size varies (and may decrease) periodically in any given 24-hour cycle. Develop a reliable, low-cost sensor that can measure the circumference of trees in a dense, humid forest. (Keep in mind that frequent services and access to such sensors are limited.) Form groups of three to four students, and identify the most promising method within each group. Present your design, and project the performance and cost of sensors according to the criteria outlined in this chapter.

Problem 1.16: Design

Identify 10 different methods to measure the temperature of an object or liquid. Identify their energy transduction paths. Both engineered and biological sensors can be cited. Grading will be based on the breadth (or variety) of methods you have selected. These methods should have as little overlap with one another as possible.

Problem 1.17: Design

Identify 10 different methods to produce a mechanical force output. Identify their energy transduction paths and succinctly describe them in two to three sentences. Both engineered and biological actuators can be cited. Grading will be based on the breadth of methods you have selected. Try to involve as many energy domains as possible.

Problem 1.18: Design

Derive the scaling law for the static buoyancy force of a solid sphere in a liquid with a density of γ. Assume the sphere is made of a material with a density of γ_s ($\gamma_s < \gamma$).

Problem 1.19: Design

Model the Analog Devices acceleration sensor as a proof mass (m) attached to a cantilever spring. The formula for the spring constant of the cantilever is discussed in this chapter. Derive the scaling laws for static displacement under an acceleration a. Then derive the scaling laws for the resonant frequency of the acceleration sensor. Discuss the advantages and disadvantages for scaling down the size of the sensor.

Problem 1.20: Challenge

Pick five random nouns from a dictionary. Each noun must begin with the same letter that begins your last name. For example, John Doe would choose nouns that begin with the letter "D". Now, add micro- or nano- to the beginning of the noun, and speculate on any potential usefulness of the technology or application. Identify at least one critical research issue related to each case. The grading will be based on the uniqueness and originality of your selections.

Problem 1.21: Challenge

Form a group of three students, and discuss your findings in Problem 1.19. Discuss and select five items from your group, and prepare a group report. For each item, you may discuss the following: (1) relation to an existing body of work; (2) uniqueness and usefulness of this technology; (3) perceived impact; (4) practicality of technology and fabrication.

REFERENCES

1. Schaller, R.R., *Moore's Law: Past, Present, and Future*. Spectrum, IEEE, 1997. p. 52–59.
2. Riordan, M., *The Lost History of the Transistor*. Spectrum, IEEE, 2004. **41**(5): p. 44–49.
3. Rowen, L., G. Mahairas, and L. Hood, *Sequencing the Human Genome*. Science, 1997. **278**(5338): p. 605–607.
4. Pennisi, E., *Biotechnology: The Ultimate Gene Gizmo: Humanity on a Chip*. Science, 2003. **302**(5643): p. 211.
5. Peltonen, L., and V.A. McKusick, *Genomics and Medicine: Dissecting Human Disease in the Postgenomic era*. Science, 2001. **291**(5507): p. 1224–1229.
6. Bean, K.E., *Anisotropic Etching of Silicon*. IEEE Transaction on Electron Devices, 1978. **ED 25**: p. 1185–1193.
7. Smith, C.S., *Piezoresistance Effect in Germanium and Silicon*. Physics Review, 1954. **94**(1): p. 42–49.
8. French, P.J., and A.G.R. Evans, *Polycrystalline Silicon Strain Sensors*. Sensors and Actuators A, 1985. **8**: p. 219–225.
9. Geyling, F.T., and J.J. Forst, *Semiconductor Strain Transducers*. The Bell System Technical Journal, 1960. **39**: p. 705–731.
10. Petersen, K.E., *Silicon as a Mechanical Material*. Proceedings of the IEEE, 1982. **70**(5): p. 420–457.
11. Gabriel, K.J., *Microelectromechanical Systems*. Proceedings of the IEEE, 1998. **86**(8): p. 1534–1535.

12. Angell, J.B., S.C. Terry, and P.W. Barth, *Silicon Micromechanical Devices*. Scientific American Journal, 1983. **248**: p. 44–55.
13. Nathanson, H.C., et al., *The Resonant Gate Transistor*. IEEE Transactions on Electron Devices, 1967. **ED-14**(3): p. 117–133.
14. Chavan, A.V., and K.D. Wise, *Batch-Processed Vacuum-Sealed Capacitive Pressure Sensors*. Microelectromechanical Systems, Journal of, 2001. **10**(4): p. 580–588.
15. Choujaa, A., et al., *AlN/Silicon Lamb-Wave Microsensors for Pressure and Gravimetric Measurements*. Sensors and Actuators A: Physical, 1995. **46**(1–3): p. 179–182.
16. Sugiyama, S., M. Takigawa, and I. Igarashi, *Integrated Piezoresistive Pressure Sensor with Both Voltage and Frequency Output*. Sensors and Actuators A, 1983. **4**: p. 113–120.
17. Fonseca, M.A., et al., *Wireless Micromachined Ceramic Pressure Sensor for High-Temperature Applications*. Microelectromechanical Systems, Journal of, 2002. **11**(4): p. 337–343.
18. Hall, N.A. and F.L. Degertekin, *Integrated Optical Interferometric Detection Method for Micromachined Capacitive Acoustic Transducers*. Applied Physics Letters, 2002. **80**(20): p. 3859–3861.
19. Chatzandroulis, S., D. Tsoukalas, and P.A. Neukomm, *A Miniature Pressure System with a Capacitive Sensor and a Passive Telemetry Link for Use in Implantable Applications*. Microelectromechanical Systems, Journal of, 2000. **9**(1): p. 18–23.
20. Park, J.-S. and Y.B. Gianchandani, *A Servo-Controlled Capacitive Pressure Sensor Using a Capped-Cylinder Structure Microfabricated by a Three-Mask Process*. Microelectromechanical Systems, Journal of, 2003. **12**(2): p. 209–220.
21. Wang, C.C., et al., *Contamination-Insensitive Differential Capacitive Pressure Sensors*. Microelectromechanical Systems, Journal of, 2000. **9**(4): p. 538–543.
22. Gotz, A., et al., *Manufacturing and Packaging of Sensors for Their Integration in a Vertical MCM Microsystem for Biomedical Applications*. Microelectromechanical Systems, Journal of, 2001. **10**(4): p. 569–579.
23. Wur, D.R., et al., *Polycrystalline Diamond Pressure Sensor*. Microelectromechanical Systems, Journal of, 1995. **4**(1): p. 34–41.
24. Zhu, X., et al., *The Fabrication of All-Diamond Packaging Panels with Built-in Interconnects for Wireless Integrated Microsystems*. Microelectromechanical Systems, Journal of, 2004. **13**(3): p. 396–405.
25. Boeller, C.A., et al., *High-Volume Microassembly of Color Thermal Inkjet Printheads and Cartridges*. Hewlett-Packard Journal, 1988. **39**(4): p. 6–15.
26. Petersen, K.E., *Fabrication of an Integrated Planar Silicon Ink-Jet Structure*. IEEE Transaction on Electron Devices, 1979. **ED-26**: p. 1918–1920.
27. Lee, J.-D., et al., *A Thermal Inkjet Printhead with a Monolithically Fabricated Nozzle Plate and Self-Aligned Ink Feed Hole*. Microelectromechanical Systems, Journal of, 1999. **8**(3): p. 229–236.
28. Tseng, F.-G., C.-J. Kim, and C.-M. Ho, *A High-Resolution High-Frequency Monolithic Top-Shooting Microinjector Free of Satellite Drops—Part I: Concept, Design, and Model*. Microelectromechanical Systems, Journal of, 2002. **11**(5): p. 427–436.
29. Carter, J.C., et al., *Recent Developments in Materials and Processes for Ink Jet Printing High Resolution Polymer OLED Displays*. Proceedings of the SPIE—The International Society for Optical Engineering; Organic Light-Emitting Materials and Devices VI, 8–10 July 2002, 2003. **4800**: p. 34–46.
30. Sirringhaus, H., et al., *High-Resolution Inkjet Printing of All-Polymer Transistor Circuits*. Science, 2000. **290**(5499): p. 2123–2126.
31. Hughes, T.R., et al., *Expression Profiling Using Microarrays Fabricated by an Ink-Jet Oligonucleotide Synthesizer*. Nature Biotechnology, 2001. **19**(4): p. 342–347.
32. Fan, L.-S., Y.-C. Tai, and R.S. Muller. *IC-Processed Electrostatic Micro-motors*. In IEEE International Electronic Devices Meeting. 1988.

33. Ahn, C.H., Y.J. Kim, and M.G. Allen, *A Planar Variable Reluctance Magnetic Micromotor with Fully Integrated Stator and Coils*. Microelectromechanical Systems, Journal of, 1993. **2**(4): p. 165–173.
34. Livermore, C., et al., *A High-Power MEMS Electric Induction Motor*. Microelectromechanical Systems, Journal of, 2004. **13**(3): p. 465–471.
35. Eddy, D.S., and D.R. Sparks, *Application of MEMS Technology in Automotive Sensors and Actuators*. Proceedings of the IEEE, 1998. **86**(8): p. 1747–1755.
36. Yazdi, N., K. Najafi, and A.S. Salian, *A High-Sensitivity Silicon Accelerometer with a Folded-Electrode Structure*. Microelectromechanical Systems, Journal of, 2003. **12**(4): p. 479–486.
37. Yazdi, N., and K. Najafi, *An All-Silicon Single-Wafer Micro-g Accelerometer with a Combined Surface and Bulk Micromachining Process*. Microelectromechanical Systems, Journal of, 2000. **9**(4): p. 544–550.
38. Partridge, A., et al., *A High-Performance Planar Piezoresistive Accelerometer*. Microelectromechanical Systems, Journal of, 2000. **9**(1): p. 58–66.
39. DeVoe, D.L., and A.P. Pisano, *Surface Micromachined Piezoelectric Accelerometers (PiXLs)*. Microelectromechanical Systems, Journal of, 2001. **10**(2): p. 180–186.
40. Loh, N.C., M.A. Schmidt, and S.R. Manalis, *Sub-10 cm^3 Interferometric Accelerometer with Nano-g Resolution*. Microelectromechanical Systems, Journal of, 2002. **11**(3): p. 182–187.
41. Dauderstadt, U.A., et al., *Silicon Accelerometer Based on Thermopiles*. Sensors and Actuators A: Physical, 1995. **46**(1–3): p. 201–204.
42. Leung, A.M., Y. Zhao, and T.M. Cunneen, *Accelerometer Uses Convection Heating Changes*. Elecktronik Praxis, 2001. **8**.
43. Butefisch, S., A. Schoft, and S. Buttgenbach, *Three-Axes Monolithic Silicon Low-g Accelerometer*. Microelectromechanical Systems, Journal of, 2000. **9**(4): p. 551–556.
44. Bernstein, J., et al., *Low-Noise MEMS Vibration Sensor for Geophysical Applications*. Microelectromechanical Systems, Journal of, 1999. **8**(4): p. 433–438.
45. Liu, C.-H. and T.W. Kenny, *A High-Precision, Wide-Bandwidth Micromachined Tunneling Accelerometer*. Microelectromechanical Systems, Journal of, 2001. **10**(3): p. 425–433.
46. Ziaie, B., et al., *A Hermetic Glass-Silicon Micropackage with High-Density On-Chip Feedthroughs for Sensors and Actuators*. Microelectromechanical Systems, Journal of, 1996. **5**(3): p. 166–179.
47. Lee, S., et al., *Surface/Bulk Micromachined Single-Crystalline-Silicon Micro-Gyroscope*. Microelectromechanical Systems, Journal of, 2000. **9**(4): p. 557–567.
48. Younse, J.M., *Mirrors on a Chip*. Spectrum, IEEE, 1993. **30**(11): p. 27–31.
49. Van Kessel, P.F., et al., *A MEMS-Based Projection Display*. Proceedings of the IEEE, 1998. **86**(8): p. 1687–1704.
50. Savage, N., *A Revolutionary Chipmaking Technique?* In *IEEE Spectrum*. 2003. p. 18.
51. Singh-Gasson, S., et al., *Maskless Fabrication of Light-Directed Oligonucleotide Microarrays Using a Digital Micromirror Array*. Nature Biotechnology, 1999. **17**(10): p. 974–978.
52. Wu, M.C., and P.R. Patterson, *Free-Space Optical MEMS*. In *MEMS Handbook*, J. Korvink and O. Paul, Ed. New York: William Andrew Publishing, 2004.
53. Senturia, S.D. *Perspective on MEMS, Past and Future: The Tortuous Pathway from Bright Ideas to Real Products*. In The 12th International Conference on Solid-State Sensors, Actuators, and Microsystems. 2003. Boston, MA.
54. Chen, J., et al., *A Multichannel Neural Probe for Selective Chemical Delivery at the Cellular Level*. IEEE/ASME Journal of Microelectromechanical Systems (JMEMS), 1997. **44**(8): p. 760–769.
55. Gardeniers, H.J.G.E., et al., *Silicon Micromachined Hollow Microneedles for Transdermal Liquid Transport*. Microelectromechanical Systems, Journal of, 2003. **12**(6): p. 855–862.

56. Park, J.-H., et al. *Micromachined Biodegradable Microstructures*. In *Micro Electro Mechanical Systems. MEMS-03 Kyoto*. IEEE, The Sixteenth Annual International Conference. 2003.
57. Jo, B.H., et al., *Three-Dimensional Micro-Channel Fabrication in Polydimethylsiloxane (PDMS) Elastomer*. IEEE/ASME Journal of Microelectromechanical Systems (JMEMS), 2000. **9**(1): p. 76–81.
58. Rebeiz, G.M., *RF MEMS: Theory, Design, and Technology*. First ed. 2003: Wiley-Interscience. 483.
59. Wong, E.W., P.E. Sheehan, and C.M. Lieber, *Nanobeam Mechanics: Elasticity, Strength, and Toughness of Nanorods and Nanotubes*. Science, 1997. **277**(5334): p. 1971–1975.
60. Huang, X.M.H., et al., *Nanoelectromechanical Systems: Nanodevice Motion at Microwave Frequencies*. Nature, 2003. **421**: p. 496.
61. Liu, J., et al., *Fullerene Pipes*. Science, 1998. **280**(5367): p. 1253–1256.
62. Carr, D.W., et al., *Measurement of Mechanical Resonance and Losses in Nanometer Scale Silicon Wires*. Applied Physics Letters, 1999. **75**(7): p. 920–922.
63. Cleland, A.N., and M.L. Roukes, *External Control of Dissipation in a Nanometer-Scale Radiofrequency Mechanical Resonator*. Sensors and Actuators A: Physical, 1999. **72**(3): p. 256–261.
64. Yang, Y.T., et al., *Monocrystalline Silicon Carbide Nanoelectromechanical Systems*. Applied Physics Letters, 2001. **78**(2): p. 162–164.
65. LaHaye, M.D., et al., *Approaching the Quantum Limit of a Nanomechanical Resonator*. Science, 2004. **304**(5667): p. 74–77.
66. Rakow, N.A., and K.S. Suslick, *A Colorimetric Sensor Array for Odor Visualization*. Nature, 2000. **406**: p. 710–714.
67. Garner, H.R., R.P. Balog, and K.J. Luebke, *The Evolution of Custom Microarray Manufacture*. IEEE Engineering in Medicine and Biology Magazine, 2002. **21**(4): p. 123–125.
68. Taton, T.A., C.A. Mirkin, and R.L. Letsinger, *Scanometric DNA Array Detection with Nanoparticle Probes*. Science, 2000. **289**(5485): p. 1757–1760.
69. Park, S.-J., T.A. Taton, and C.A. Mirkin, *Array-Based Electrical Detection of DNA with Nanoparticle Probes*. Science, 2002. **295**(5559): p. 1503–1506.
70. Harkey, J.A., and T.W. Kenny, *1/f Noise Considerations for the Design and Process Optimization of Piezoresistive Cantilevers*. Microelectromechanical Systems, Journal of, 2000. **9**(2): p. 226–235.
71. Gabrielson, T.B., *Mechanical-Thermal Noise in Micromachined Acoustic and Vibration Sensors*. Electron Devices, IEEE Transactions on, 1993. **40**(5): p. 903–909.
72. Manalis, S.R., et al., *Interdigital Cantilevers for Atomic Force Microscopy*. Applied Physics Letters, 1996. **69**(25): p. 3944–3946.
73. Van de Pol, F.C.M., et al., *A Thermopneumatic Micropump Based on Micro-Engineering Techniques*. Sensors and Actuators A: Physical, 1990. **21**(1–3): p. 198–202.
74. Rich, C.A., and K.D. Wise, *A High-Flow Thermopneumatic Microvalve with Improved Efficiency and Integrated State Sensing*. Microelectromechanical Systems, Journal of, 2003. **12**(2): p. 201–208.
75. Wolf, R.H., and A.H. Heuer, *TiNi (Shape Memory) Films on Silicon for MEMS Applications*. Microelectromechanical Systems, Journal of, 1995. **4**(4): p. 206–212.
76. Krulcvitch, P., et al., *Thin Film Shape Memory Alloy Microactuators*. Microelectromechanical Systems, Journal of, 1996. **5**(4): p. 270–282.
77. Benard, W.L., et al., *Thin-Film Shape-Memory Alloy Actuated Micropumps*. Microelectromechanical Systems, Journal of, 1998. **7**(2): p. 245–251.
78. Shin, D.D., K.P. Mohanchandra, and G.P. Carman, *High Frequency Actuation of Thin Film NiTi*. Sensors and Actuators A: Physical, 2004. **111**(2–3): p. 166–171.
79. Ohmichi, O., Y. Yamagata, and T. Higuchi, *Micro Impact Drive Mechanisms Using Optically Excited Thermal Expansion*. Microelectromechanical Systems, Journal of, 1997. **6**(3): p. 200–207.

References

80. Bergstrom, P.L., et al., *Thermally Driven Phase-Change Microactuation*. Microelectromechanical Systems, Journal of, 1995. **4**(1): p. 10–17.
81. Neagu, C.R., et al., *An Electrochemical Microactuator: Principle and First Results*. Microelectromechanical Systems, Journal of, 1996. **5**(1): p. 2–9.
82. Rossi, C., D. Esteve, and C. Mingues, *Pyrotechnic Actuator: A New Generation of Si Integrated Actuator*. Sensors and Actuators A: Physical, 1999. **74**(1–3): p. 211–215.
83. DiBiaso, H.H., B.A. English, and M.G. Allen, *Solid-Phase Conductive Fuels for Chemical Microactuators*. Sensors and Actuators A: Physical, 2004. **111**(2–3): p. 260–266.
84. Teasdale, D., et al., *Microrockets for Smart Dust*. Smart Materials and Structures. **10**(6): p. 1145–1155.
85. Konishi, S., and H. Fujita, *A Conveyance System Using Air Flow Based on the Concept of Distributed Micro Motion Systems*. Microelectromechanical Systems, Journal of, 1994. **3**(2): p. 54–58.
86. Chen, Q., et al., *Mesoscale Actuator Device: micro Interlocking Mechanism to Transfer Macro Load*. Sensors and Actuators A: Physical, 1999. **73**(1–2): p. 30–36.

CHAPTER 2

Introduction to Microfabrication

2.0 PREVIEW

A first-time student of MEMS should note two important facts about the microfabrication technology used to prototype and manufacture MEMS devices. First, MEMS fabrication represents a paradigm shift from traditional machining and manufacturing processes. It currently does not involve methods such as milling, lathing, polishing, joining, and welding. A student of MEMS, who is otherwise unfamiliar with the integrated circuit fabrication process, should first be acquainted with a new framework of manufacturing, its unique features, and its limitations. Second, the portfolio of MEMS fabrication techniques (the micromachining "tool box") is rapidly expanding. Through this expansion, the MEMS field is increasing the variety of materials involved, increasing the fabrication efficiency, and reducing the cost of manufacturing.

Section 2.1 presents the general framework of microfabrication using silicon wafers. Section 2.2 discusses a representative process flow for integrated circuits. This discussion is highly relevant because the micromachining process is derived from the IC industry. This is followed by a discussion of bulk and surface micromachining processes based on silicon (Section 2.3). New materials beyond silicon are being actively incorporated into the MEMS research field. Some of these materials and their associated fabrication processes are reviewed in Section 2.4. Section 2.5 discusses major points of consideration when selecting a microfabrication process.

2.1 OVERVIEW OF MICROFABRICATION

MEMS and IC devices are generally made on single-crystal silicon wafers. Figure 2.1 diagrams the overall process; it shows the production of these wafers through the packaging of individual device chips.

However, bulk silicon with crystalline consistency does not exist in nature; it must be prepared through laborious industrial processes. To make bulk crystal silicon, we start with a perfect single-crystal silicon seed. It is dipped into a molten silicon pool and slowly drawn out of the liquid. Silicon crystallizes when drawn into the atmosphere and establishes crystallinity consistent with

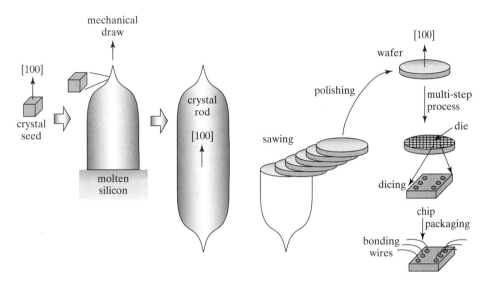

FIGURE 2.1
Wafer process steps.

that of the initial seed. Rods of single crystals with various diameters and longitudinal crystal orientation can be formed this way. The rods are sawed into thin, circular slices and polished to form wafers.

A wafer goes through a multi-step fabrication process in a cleanroom, where dust, particles, and even ions in water are tightly controlled. The cleanliness of air is classified according to the concentration of airborne particles (larger than 0.5 μm). According to the standard method of characterizing the level of a cleanroom, a class 1 cleanroom has fewer than 1 particle and a class 100 cleanroom has fewer than 100 particles per ft^3 of air sampled. As a reference, note that average outdoor air contains more than 400,000 particles per ft^3. Generally speaking, class 1000, 100, 10, 1, and 0.1 cleanrooms can support production down to approximately 4 μm, 1.25 μm, 0.7 μm, 0.3 μm, and <0.1 μm, respectively. State-of-the-art integrated circuits have used linewidths smaller than 0.18 μm since 1999.

Water (for rinsing) and chemical solutions must go through stringent, costly manufacturing and conditioning processes. Ions in water (e.g., sodium ions), even in trace amounts, will migrate into silicon and thin film materials upon direct contact. These ions may become trapped charges in dielectrics and hurt device performance. Deionized water (more broadly speaking, ultrapure water) used in semiconductor manufacturing has resistivity in excess of 18 MΩ, compared with a resistivity of less than 50 kΩ for tap water.

Precision patterns are made using a photolithographic patterning method. Collimated light passes through a mask and an image-reduction lens before hitting the wafer; this process is akin to taking a photograph of an object through a telephoto lens and recording the image on a photosensitive film. In this case, the object being photographed is the mask and the film is the wafer, which is coated with a photosensitive film. Various wavelengths of light can be used. Light with higher energy (and hence a smaller wavelength) is capable of producing smaller line widths. The ultimate resolution is dictated by the diffraction limit.

Using a machine-automated photolithography process called *step-and-repeat*, many identical units can be made on the same wafer with high line-width resolution (0.1 μm or smaller in commercial processes). The machine used for performing the step-and-repeat process is called a stepper. During operation, the stepper prints a reduced image of the mask on a wafer; the wafer is translated by a precise distance, and another exposure is made. Many identical devices are made on one given wafer in a single pass.

This MEMS process differs from conventional manufacturing technology, which generally deals with one part at a time. For example, a 4″-diameter silicon wafer can host more than 50 repeated 1×1 cm^2 units. In this case, at least 50 devices can be made in each process run. An 8″ wafer going through the same process would yield roughly 314 dies of such sizes. To maximize the economy of scale of mass fabrication, devices should be fabricated on large diameter wafers. Nowadays, wafers with a 12″ diameter (300 mm) are used in advanced fabrication facilities (*fabs*) worldwide. In a given process, the percentage of dies with satisfactory performance is called the *process yield*.

These dies have spacings between them so that they can be mechanically cut and separated. Each cut die, called a chip, can then be electrically connected and encapsulated for commercial resale. The process of incorporating a loose die to a housing and a system is called packaging.

In a number of important ways, the MEMS process and the conventional macroscale machining differs drastically:

1. Silicon, a principal substrate material for MEMS and integrated circuits, is mechanically brittle and cannot be shaped by machine cutting tools;
2. MEMS and integrated circuits are made on planar crystalline wafers. The planarity is not just a matter of convenience for automated wafer processing. When lithography patterning is conducted on a planar substrate, a consistent focus distance is ensured, leading to uniformity and resolution. The planarity of a wafer is also important to ensure the entire wafer surface has identical crystal orientation.
3. MEMS chips or parts are generally too small to be confidently grasped, efficiently handled, and dexterously assembled with existing robotics equipment.

This book focuses on the segment of the process flow between a bare wafer and an undiced wafer with fabricated devices.

2.2 THE MICROELECTRONICS FABRICATION PROCESS

A basic understanding of the fabrication technology for integrated circuitry, whose history precedes MEMS, is necessary to understand the micromachining process. A reader who is familiar with the basic microfabrication process of integrated circuits may skip this section.

A fabrication process for integrated circuits generally involves many steps, including material deposition, material removal, and patterning, as illustrated in the following example. The physics and technology of discrete fabrication process steps are not reviewed in this book. Interested readers can refer to Reference [1] for discussion on circuit-specific fabrication technology and Reference [2] for MEMS-specific technologies.

A generic microfabrication process for realizing a field-effect transistor, the building block of modern integrated circuits, is illustrated in Figure 2.2. By repeating the cycle of deposition–lithography–etching, devices with arbitrary complexity can be built. In this particular case, six major cycles transform a bare silicon wafer to one with a metal-oxide-semiconductor (MOS)

2.2 The Microelectronics Fabrication Process

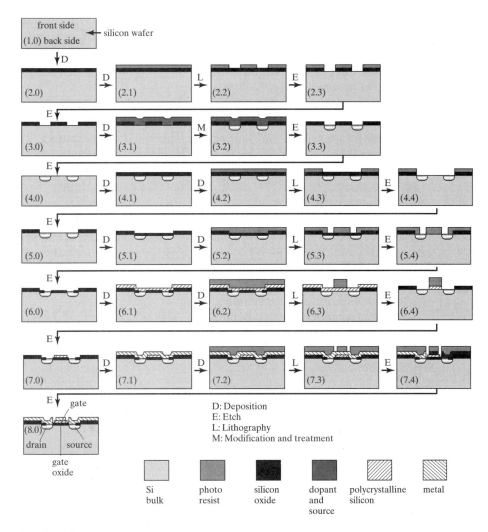

FIGURE 2.2

Fabrication process for an integrated circuit.

field-effect transistor (FET) on the front surface. The figure depicts 30 major steps, showing a cross-sectional view at a single FET level. Steps 1.0, 2.0, 3.0, 4.0, 5.0, 6.0, 7.0, and 8.0, on the leftmost column, are major milestones in the process, whereas process steps $x.y$ ($x = 1 - 7$, $y \neq 0$) are steps that lead to the next major milestone.

The nature of each step is clearly marked. Letters D, L, E, and M denote deposition, lithography (photo exposure and development), etching, and modification/treatment of materials.

A brief description of each step is presented below:

Step 1.0. The figure depicts the cross section of the initial bare silicon wafer. The wafer thickness is not drawn to scale. Material process steps that occur on the back side of a wafer are not drawn in future steps for the sake of simplicity.

Step 2.0. A layer of oxide is deposited. The oxide is patterned using the following subsequent steps (2.1 through 2.3). This oxide layer only serves a transitional purpose. (This point will become obvious later.)
    2.1. A photosensitive resist layer (often called **photoresist** or resist) is deposited on top of the oxide by spin coating.
    2.2. The photosensitive resist is lithographically exposed and developed.
    2.3. The photoresist is used as a mask for etching the oxide.

Step 3.0. The photoresist is removed using organic solvents. The patterned oxide is used as a mask against the impurity doping performed in Steps 3.1 through 3.3.
    3.1. A layer of material containing dopant impurities is deposited.
    3.2. The wafer is thermally treated, causing the dopant to diffuse into silicon in areas not covered by the oxide.
    3.3. The dopant-source layer deposited in Step 3.1 is removed.

Step 4.0. The oxide is removed. Note that many steps and layers of materials are involved to transform a bare wafer (Step 1.0) to a wafer with dopant in selective places (Step 4.0). Additional processes (Steps 4.1 through 4.4) are then performed to produce another layer of patterned oxide.
    4.1. Another layer of silicon oxide is grown.
    4.2. A photosensitive resist is deposited.
    4.3. The resist is lithographically patterned.
    4.4. Using the resist as a mask, the oxide is etched.

Step 5.0. The resist deposited in Step 4.2 is removed. From Step 4.0 to 5.0, the major difference is the oxide cover in undoped regions. An oxide layer is then deposited and patterned (Steps 5.1 through 5.4).
    5.1. A very thin oxide is grown. This so-called gate oxide layer must have very high quality and must be free of contaminants and defects.
    5.2. A resist layer is again deposited.
    5.3. The resist is lithographically patterned.
    5.4. The resist serves as a mask for selectively etching the gate oxide.

Step 6.0. The resist deposited in Step 5.2 is removed. The active regions not covered by oxide will provide electrical contact to metal. A gate electrode, made of polycrystalline silicon, is then deposited and patterned in Steps 6.1 through 6.4.
    6.1. A layer of polycrystalline (doped) silicon is deposited.
    6.2. A layer of photosensitive resist is deposited.
    6.3. The resist is patterned lithographically.
    6.4. The resist serves as a mask for etching the underlying polycrystalline silicon selectively.

Step 7.0. The resist is removed by using organic solvents. The difference between Steps 6.0 to 7.0 is the addition of polycrystalline silicon. Each transistor must be connected with each other and to the outside through low-resistivity metal wires. The metal wires are made in Steps 7.1 through 7.3.
    7.1. A layer of metal is deposited.
    7.2. and 7.3. The metal is coated with resist and lithographically patterned.
    7.4. The resist serves as a mask for etching the metal.

Step 8.0. The resist is removed, realizing a complete field effector transistor.

FIGURE 2.3

Semiconductor processing equipment.

The process used in the industry follows the basic flow diagrammed in Figure 2.2 but involves more detailed steps for quality assurance, functional enhancement, and increasing the yield and repeatability. Many more steps may occur after Step 8.0, as well. A complete process run from start to finish may take three months and 20–40 mask plates.

Several representative machines—a contact aligner for lithography, a metal evaporator for deposition of metal thin film, and a plasma etcher for removal of materials—are shown in Figure 2.3.

2.3 SILICON-BASED MEMS PROCESSES

MEMS devices were first developed on silicon wafers because of the easy availability of mature processing technologies that had been developed within the microelectronics industry, as well as the availability of expertise in process management and quality control.

Silicon actually comes in three general forms: **single-crystal silicon**, **polycrystalline silicon**, and **amorphous silicon**. In a single crystal silicon (SCS) material, the crystal lattice is regularly organized throughout the entire bulk (Figure 2.4). Single-crystal silicon is often encountered in three cases: (1) single-crystal silicon wafers grown from a high-temperature melt/recrystallization process (Figure 2.1); (2) epitaxially grown silicon thin films; (3) single-crystal silicon obtained from recrystallizing polycrystalline or amorphous silicon by global or local heat treatment.

A polycrystalline silicon material (called **polysilicon**, **polySi**, or **poly**) is made of multiple crystalline domains (Figure 2.4). Within each individual domain, the crystal lattice is regularly aligned. However, crystal orientations are different in neighboring domains. Domain walls, also referred to as grain boundaries, play important roles in determining electrical conductivity, mechanical stiffness, and chemical etching characteristics. The polysilicon material can be grown by low-pressure chemical vapor deposition (LPCVD), or by recrystallizing amorphous silicon through global or local heat treatment.

Amorphous silicon, on the other hand, exhibits no crystalline regularity. Amorphous silicon films can be deposited by chemical vapor deposition methods (CVD) at a lower temperature than that required to deposit polysilicon. Due to low temperature, atoms do not have

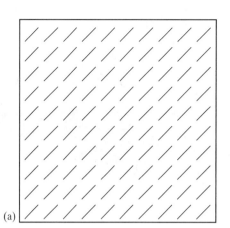 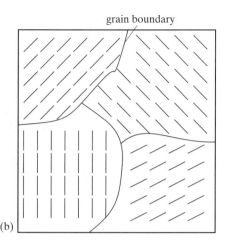

FIGURE 2.4

Crystal structure of single-crystal silicon and polycrystalline silicon.

enough vibrational energy to align themselves after they are incorporated into the solid. Amorphous silicon can be grown using the LPCVD method. (In a typical, horizontal, low-pressure reactor, the transition temperature above which polycrystalline structures form during deposition is 580°C [3].) Amorphous silicon can be formed by plasma-enhanced chemical-vapor deposition (PECVD) method, as well.

The two most fundamental classes of fabrication technologies are **bulk micromachining** [4] and **surface micromachining** [5]. Bulk micromachining processes involve selectively removing the bulk (silicon substrate) material in order to form certain three-dimensional features or mechanical elements, such as beams and membranes. Bulk micromachining may be combined with wafer bonding to create even more complex three-dimensional structures. A review of the bulk micromachining technology can be found in Chapter 10.

An example of a device made using the bulk micromachining method is the earliest micromachined pressure sensor [6]. The fabrication process of this MEMS device is illustrated step by step in three-dimensional drawings (Figure 2.5). It involves two wafers: a bottom wafer is etched to form a cavity, and a top wafer is used to make the membrane. A description for each step in the diagram follows.

Although the process diagrammed in Figure 2.5 may be quite lengthy for beginners, this is still a compressed progress description because many detailed steps are skipped. One of the skipped steps is the routine photolithography sequence, which typically consists of several steps: photoresist coating, cure, exposure, development, subsequent etching of underlying layers using the photoresist as a mask, and removal of the photoresist.

Step A. The process starts with a bare silicon wafer. To create the desired cavity shapes, the wafer must be a certain crystallographic orientation. We will discuss the notation of crystal orientation and the relevance to processing in Chapters 3 and 10. The wafer is cleaned thoroughly to remove any large particles, dirt particles, and invisible organic residues. A combined mechanical wash and oxidizing acid bath may be used, followed by a rinse with ultrapure water.

Step B. The cleaned wafer is placed inside a high-temperature furnace filled with running oxygen gas or water vapor. Oxygen atoms present in the air or dissociated from the water

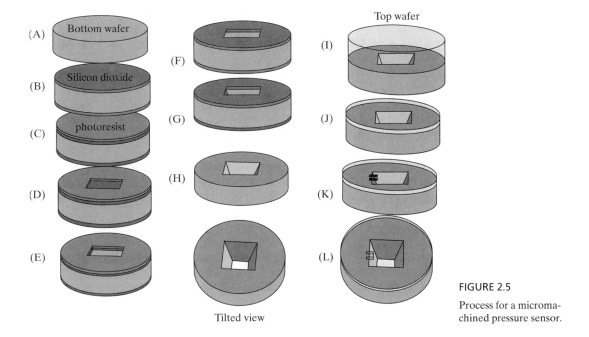

FIGURE 2.5
Process for a micromachined pressure sensor.

molecule will react with silicon to form a protective silicon dioxide thin film. Note the oxide is grown on both sides of the wafer as well as on the edges. (For the sake of completeness, all layers on both sides of the wafer are shown in Figure 2.5.)

Step C. The wafer is removed from the furnace and cooled to room temperature. It will be very clean, because any organic molecules would have decomposed in the high-temperature oxidation step. A layer of thin film photoresist is deposited on the front surface of the wafer. (A chemical called hexamethyldisilazane, or HMDS, is sometimes spin coated or vapor coated to help increase the adhesion between the photoresist and an oxide surface). The photoresist is typically spin coated, by dropping droplets of photosensitive polymers dissolved in solvents on a wafer spinning at high speed. The thickness of the photoresist layer thus formed depends on the spin speed. Alternatively, photoresist thin film can be deposited by vapor coating, mist coating, or electroplating [7].

The wafer is baked in a convection oven to remove some portion of the solvent from the PR layer to establish firmness. This step is generally called "soft bake." Alternatively, the moisture can be driven off with an infrared lamp or vacuum.

Step D. The photoresist is exposed through a mask with a high-energy radiation (such as an ultraviolet ray, electron beam, or X-ray). The light, which has sufficient energy to change chemical and mechanical properties of the resist, is spatially blocked by patterns on the mask. There are two types of photoresists—negative-tone resist (often referred to as negative resist) and positive-tone resist (positive resist). Radiation causes cross-linking polymerization in a negative photoresist. For the positive resist, radiation causes polymer chains to break down.

The entire wafer is then placed inside a developing solution (often called a **developer**) that removes loosely bound photosensitive polymer. In the case of positive photoresist, regions hit by light will be dissolved. In the case of negative photoresist, regions hit by light will stay. The soft-bake process in Step C ensures that the photoresist will not be indiscriminately stripped by the developer.

Step E. The photoresist needs to be baked again, this time at a higher temperature and often for a longer duration than the soft bake. This second baking step, called "hard bake," removes the remaining solvents and makes the photoresist that remains on the wafer stick to the wafer even stronger. The extent of the hard bake will depend on the nature of the subsequent step.

The photoresist mask is used to selectively mask the underlying layer, the silicon oxide, against a hydrofluoric acid etchant bath. An HF etchant attacks oxide within the exposed window, but it has a negligible etch rate on the underlying silicon and the photoresist mask.

Step F. The photoresist is removed using an organic solvent etchant such as acetone (at room temperature or elevated temperatures). The hard-baked photoresist is chemically resistant to the HF etchant but not to acetone. The organic solvent does not etch the oxide and the silicon.

Step G. The silicon wafer is immersed in a wet silicon etchant, which does not attack the silicon oxide. Only the silicon in the open oxide window is etched, resulting in a cavity with sidewalls defined by crystallographic planes. The cavity may reach the other side of the wafer if the open window is large enough for the given wafer thickness.

The wet silicon etching involves an elevated temperature (70–90°C). The etchant would attack hard-baked photoresist, hence it is impossible to use photoresist directly as a mask in this step.

The oxide is removed using HF etchant again.

Step H. The wafer at the end of Step G is tilted to provide a clear view of the through-wafer cavity.

Step I. A second silicon wafer is firmly bound to the front of the bottom wafer processed through Step G. A number of bonding techniques can be used (see Chapter 10). It is important that the environment in which the wafers are processed be very clean, because tiny particles adhering to the bonding surfaces of either wafer will prevent a good bond strength from being reached.

Step J. The bonded top wafer is thinned via mechanical polishing or chemical etching. The remaining thickness of the top wafer determines the thickness of the membrane. Thin membranes should have high sensitivity.

Step K. Strain sensors are then made on the prepared membrane. A thin film layer (e.g., oxide) is deposited and patterned. It serves as a barrier layer to ion implantation. Areas on the silicon wafer that are hit directly by energetic dopant ions will become doped and form a piezoresistor, which changes its resistance upon applied stress (due to membranes bending under the pressure difference). The wafer is tilted and presents another view of the through-wafer cavity (Step L). To keep the description succinct, a few detailed steps are skipped between steps J and K.

Because there is no software to easily draw three-dimensional processing diagrams, researchers often depict a process sequence using the cross-sectional view. Figure 2.6 depicts the same process as Figure 2.5, viewed in a cross section that reveals the membrane and strain

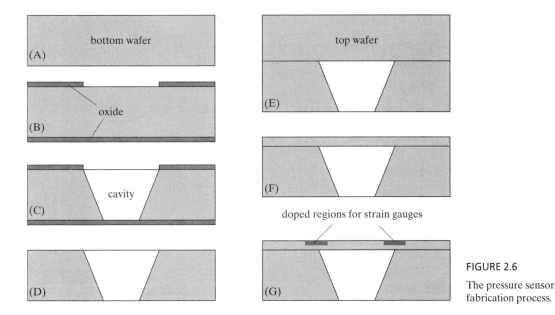

FIGURE 2.6

The pressure sensor fabrication process.

sensors. Select the cross section that reveals major developments associated with key features. Sometimes, multiple cross-sectional views are used to illustrate a complex process.

For processes that involve many steps, it becomes quite tedious to draw the details of routine processes such as photoresist spinning, exposure, development, and removal. An experienced engineer may skip certain process steps to simplify the process drawing at the concept stages. However, this approach should be taken with great caution. Seemingly routine processes may introduce complications in unexpected ways.

A variety of methods can create a membrane pressure sensor with integrated displacement or strain sensors. The membrane can be made of silicon, polysilicon, silicon nitride, silicon dioxide, and polymers (such as Parylene, polyimide, and silicone elastomer). Different materials will yield unique sets of performance characteristics, including sensitivity, allowable membrane size, robustness against excess pressure, optical transparency of the film, dynamic range, and fabrication costs. Various examples of pressure sensors are reviewed in Chapters 4 through 7.

The second class of microfabrication processes for MEMS is surface micromachining. Freestanding mechanical elements can be created by removing an underlying place-holding thin-film layer, instead of the substrate underneath. This spacer layer, called the **sacrificial layer**, constitutes the primary characteristic of a surface micromachining process. This general concept of this process was first envisioned by physicist Richard Feynman [8].

Figure 2.7 illustrates a typical surface micromachining process involving one structural and one sacrificial layer. A sacrificial layer is first deposited and patterned. This is followed by the deposition of a structural layer on top of the sacrificial layer material. Following the fabrication of layered structures, the sacrificial material is selectively removed to free the structure layer on top. For example, cantilevers residing on the surface of a substrate can be made using oxide as a sacrificial layer and polycrystalline thin film as a structural layer. In fact, surface micromachining is so named because micromechanical devices reside within a thin boundary on the front surface of the wafer.

38 Chapter 2 Introduction to Microfabrication

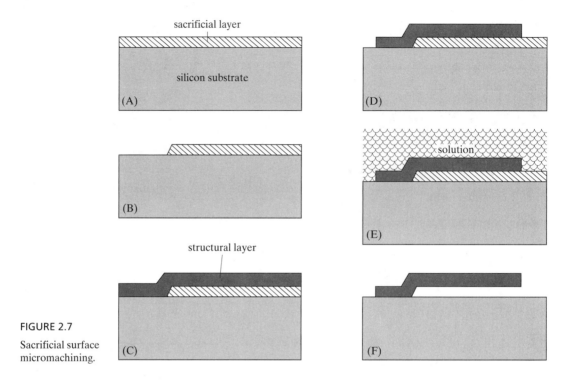

FIGURE 2.7
Sacrificial surface micromachining.

Through the use of microstructures such as hinges, surface micromachined elements can form three-dimensional features, such as the one shown in Figure 2.8. The surface micromachining techniques are discussed in greater detail in Chapter 11.

These two classes of processes—bulk and surface micromachining—are not used to the exclusion of one another, nor are they the only types of processes available for MEMS. Increasingly, bulk and surface micromachining processes are combined to create more complex structures with desired functionalities that cannot be realized using one process. Further, these two major classes can be combined with other classes of processes (such as wafer bonding, laser machining, micromolding, and three-dimensional assembly) to incorporate a variety of bulk and thin-film materials.

For example, three-dimensional metallic or polymer microstructures with thickness on the order of 500 μm to 60,000 μm can be made with a process called LIGA, diagrammed in Figure 2.9 [9]. The word LIGA is an acronym that stands for the main steps of its process: deep X-ray lithography, electroplating (or *galvo* in German), and injection molding (or *abformung*

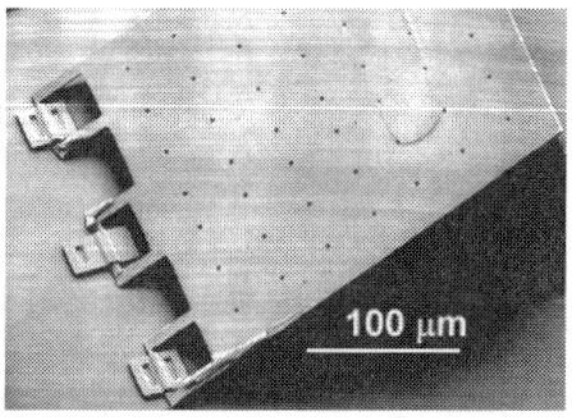

FIGURE 2.8

Surface micromachined hinged structures.

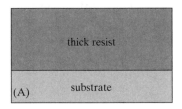
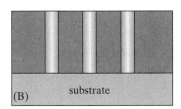
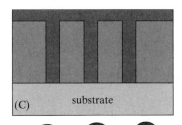
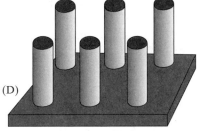

FIGURE 2.9

Demonstration of a LIGA process.

in German). It involves patterning a thick layer of photoresist (e.g., polymethylmethacrylate or PMMA) with high-energy beams produced by a synchrotron radiation (SR) X-ray source (with particle energy on the order of thousands of eV). The photoresist is developed to form structures with deep vertical walls and high aspect ratios. Electroplating is conducted to fill the cavities with metal. The metal piece, harvested by removing the resist, can be used as a precision metal part, or as a mold for the batch fabrication of plastic parts. Today, the original SR-LIGA is being augmented and, in some cases, replaced by the LIGA process based on less energetic ultraviolet (UV) rays. This new process method is known as UV-LIGA or "poor man's" LIGA because it avoids the use of an expensive facility for SR-LIGA.

The monolithic integration of mechanical and circuit elements, if necessary, may follow one of the following major approaches:

1. **Post-processing approach.** Microelectromechanical elements are fabricated on the top surface of a semiconductor wafer with preexisting circuits (for example, see [10]).
2. **Pre-processing approach.** The microelectromechanical elements are fabricated on a wafer first, followed by IC fabrication.
3. **Side-by-side processing approach.** Microelectromechanical elements and integrated electronics are created simultaneously (for examples, see [11–14]).

2.4 NEW MATERIALS AND FABRICATION PROCESSES

In recent years, silicon micromachining techniques are being rapidly augmented with new materials and processes. Silicon, a semiconductor material, is mechanically brittle. It is also expensive or unnecessary for certain applications. New materials, such as polymer and compound semiconductors, can fill the performance gap.

Polymer materials are being incorporated into MEMS because of their unique material properties (e.g., biocompatibility, optical transparency), processing techniques, and low costs (compared with silicon). Polymer materials that have been explored in recent years include silicone elastomers, Parylene, and polyimide, among others. The use of polymer materials is reviewed in Chapter 12.

Many sensors and actuators are needed to operate in harsh conditions, such as direct exposure to environmental elements, high temperatures, wide temperature swings, or high shock. Delicate microstructures made of silicon or inorganic thin-film materials are not suited for such applications. Several inorganic materials are being introduced for MEMS applications in harsh environments. Silicon carbide, in both bulk and thin-film forms, is explored for applications including high-temperature solid-state electronics and transducers [15–17]. Diamond thin films provide the advantage of high electrical conductivity and wear resistance for potential applications including pressure sensors and scanning-electron-microscopy probes [18–20]. Other compound semiconductor materials, including GaAs ([21–24]), are also being investigated.

New material processing techniques are being developed for fabrication on silicon substrates and other materials. New processes for MEMS include laser-assisted etching for material removal and deposition [25], stereolithography for rapid prototyping [26, 27], local electrochemical deposition [28], photoelectroforming [29], high-aspect-ratio deep reactive ion etching [25], micromilling [30, 31], focused ion-beam etching [32], X-ray etching [33] [34], microelectrodischarge [35, 36], ink-jet printing (of metal colloids [37]), microcontact printing [38, 39], *in situ* plasma [40], molding (including injection molding [41]), embossing [42], screen printing [43], electrochemical welding [44], chemical mechanical polishing, and guided and self-directed self-assembly in two or three dimensions [45, 46].

Microfabrication processes can also reach nanoscale resolution to realize nanoelectromechanical systems (NEMS) [27, 47]. Reliable and economical fabrication of electromechanical elements with nanoscopic feature sizes or spacing brings new challenges and methodologies. Traditional lithography does not offer 100-nm resolution readily, at low cost, and with parallelism. A variety of nanostructure patterning techniques, often drastically different from the photolithographic approach, are developed in the physics and chemistry communities for producing nanometer scale patterns. Readers who are interested in exploring these techniques may read about nanoimprint lithography [48], nano whittling [49], and nanosphere lithography [50].

Note that the materials and technologies for microelectronics fabrication have not been standing still either. In fact, traditional photolithography techniques and semiconductor materials associated with integrated circuits are undergoing rapid changes. New processing techniques (such as roll-to-roll printing) are actively pursued for the fast production of large area electronics, photovoltaic generators, and optoelectronic displays [51]. Organic polymer materials replace semiconductor materials for logic [52], storage [53], and optical display [54].

The future of new materials and fabrication methods is bright and exciting. Fabrication and manufacturing technologies such as micromachining, nanofabrication, and microelectronic fabrication have historically been developed in different communities that virtually disregard each other and have independent sets of materials and substrates. As science and technology progresses towards the micro and nanoscopic dimensions, these distinct fabrication methods are being hybridized to create powerful new fabrication methods that will enable new scientific studies and devices.

2.5 POINTS OF CONSIDERATION FOR PROCESSING

A number of general points must be considered when evaluating a fabrication process:

1. The deposition rate and etch rate for material deposition or etching, respectively.

2. The wafer-scale uniformity of deposition rate and etch rate. Since etch or deposition rates are often conducted on a silicon wafer, the uniformity across the wafer is an important measure, especially in industrial processes.
3. The sensitivity to overtime etch. A robust process that is insensitive to overtime etch—which is necessary to account for inevitable process non-uniformity—is more desirable than one that requires precise timing control.
4. The etching selectivity. Etching selectivity is the ratio of etch rate to the intended material and to other materials (such as mask or substrate). The selectivity should be as high as possible at each step. A MEMS student or researcher will develop a working knowledge base of cross-reactivity. Comprehensive reactivity matrices between a large number of materials and etching methods in a consistent laboratory setting have been published [55, 56]. These provide excellent reference for beginning readers. However, material-etch properties depend on specific laboratory conditions and procedures and should be calibrated and documented for each laboratory and facility.
5. The temperature compatibility. The temperature compatibility of all materials on a wafer, whether exposed or hidden, should be considered for each deposition or etching step. Low temperature processes generally allow more choices of materials.
6. The overall processing time and number of steps.

Example 2.1. Chemical and Temperature Compatibility

Discuss the chemical and temperature compatibility related to the process in Figure 2.5.

Solution. A chemical agent or method intended for action on one particular material layer will influence other preexisting materials on the wafer as long as they are exposed to the

TABLE 2.1 Cross-Reactivity Table.

	Etching agents				Deposition methods	
Material	Photoresist developer	Acetone	Hydrofluoric acid (diluted)	Silicon wet etchant	Photoresist spin coating	Thermal oxidation growth
Spin-on photoresist	Yes, low speed erosion	Yes	Compromised integrity	Dissolution	Severe solvent diffusion and thinning	Ashing
Soft-baked photoresist	No	Yes	Compromised integrity	Dissolution	Solvent diffusion and thinning	Ashing
Hard-baked photoresist	No	Yes	No	Dissolution	Moderate thinning	Ashing
Oxide	No	No	Yes	Very slow	No	No
Silicon substrate	No	No	No	Yes	No	Yes, oxidation will occur

chemical. For the successful development of a microfabrication process, the cross reaction between participant materials and the washing and developing agents, etching methods, and deposition steps must be carefully considered to determine the best course of action. Ideally, a fabrication route should be carefully planned such that high selectivity of etching will be achieved at each step.

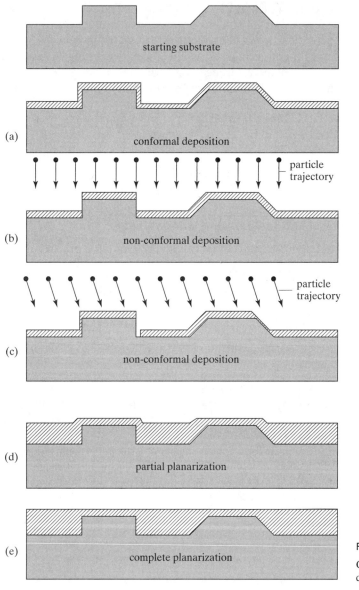

FIGURE 2.10

Generally encountered deposition profiles.

7. The environmental cleanliness requirement. Processes or subset steps that can be performed outside of cleanrooms can result in considerable cost savings.
8. The deposition and etching profiles. Various deposition and etching profiles are available; they are associated with different methods and processing conditions. A number of commonly encountered deposition profiles are summarized in Figure 2.10. The familiarity of these associations is critical for the successful development of MEMS.

SUMMARY

This chapter is a first-pass discussion about microfabrication technology. We introduced the general framework of microfabrication, and we reviewed the process for fabricating representative microelectronic circuit elements and silicon MEMS devices. Future applications of MEMS devices in expanding application areas call for new materials and new processes. We reviewed a number of emerging materials, and we discussed criteria for selecting the process steps in future MEMS development.

The following list summarizes the major concepts, facts, and skills associated with this chapter:

- Reasons for using silicon in a planar wafer form for microelectronics and MEMS.
- The major differences between conventional fabrication and microfabrication, including materials, handling, and tools.
- Major cost elements associated with microfabrication.
- Process steps for making a semiconductor transistor.
- Process steps for making a bulk micromachined pressure sensor.
- Ability to draw a cross-sectional process diagram involving multiple process steps.
- Major points to consider when selecting a process flow involving multiple layers and materials.

PROBLEMS

Problem 2.1: Review
In this practical problem, a student must find the cost of the facilities and equipment related to MEMS industrialization and research. Suppose you are going to build a microfabrication research facility and must buy all the equipment necessary to perform the process depicted in Figure 2.2. Major equipment needed in this process are a photoresist spin coater, a contact aligner or a stepper, a metal evaporator, a thermal oxidation furnace, an LPCVD deposition system for polycrystalline silicon, a light microscope, a plasma etcher, and a surface profile measurement tool (called a surface profilometer). Form a team of three students and research the price of used equipment from a dealer (or dealers). Find the prices of representative systems capable of handling *4"-diameter* wafers. Compile a spreadsheet listing each item, the manufacturer, the model number, and the total cost of all equipment.

Problem 2.2: Review
The equipment listed in the previous problem are related to processing. However, many equipment items are of the behind-the-scene type, for example, those used for sustaining the operation of clean

room environment. Investigate the cost of a new or used system for generating deionized water, which is a critical piece of equipment for maintaining stability of chemistry and products. Consult published price information from used-equipment or new-equipment vendors.

Problem 2.3: Fabrication

Find software that draws a MEMS mask layout and one that draws process diagrams. Install them on a computer for future access. Free software, such as xkic (for circuit layout), is available on the Web. For process flow drawings, try to use dedicated drawing software instead of built-in drawing tools in computer word processors and presentation tools. This will offer better control when creating complex geometries. (*Hint*: An instructor may specify common layout and drawing tools for the entire class.)

Problem 2.4: Fabrication

Draw the fabrication process for the pressure sensor discussed in the text (Figure 2.6). Expand the diagram to include all the distinct steps, including detailed lithography steps. Part of the purpose of this exercise is to identify, acquire, and become familiar with drawing software. Try to represent the geometric profile faithfully by including details such as coverage on sidewalls and slopes.

Problem 2.5: Fabrication

Draw the fabrication process for the surface micromachined cantilever discussed in the text (Figure 2.7). Include all detailed steps, including photoresist spinning, development, and removal. Try to represent the geometric profile faithfully by including details such as coverage on sidewalls and slopes.

Problem 2.6: Fabrication

In the bulk micromachining process for pressure sensors presented in this chapter (Figure 2.5), what is the reason that a photoresist layer must be used to pattern oxide, which then serves as a mask to silicon etching? Is it possible to abbreviate the process by using patterned photoresist as the silicon-etching mask? Why? (Find quantitative evidence and data.)

Problem 2.7: Fabrication

Refer to the process described in Figure 2.5. If silicon nitride is used instead of silicon oxide, what is the proper chemical treatment to reach Steps E and H from D and G, respectively?

Problem 2.8: Fabrication

Draw the cross-sectional process flow of the floating gate transistors discussed in Reference [57]. Include detailed steps, such as photoresist spin coating, development, and stripping. Do not include details of circuit fabrication.

REFERENCES

1. Jaeger, R.C., *Introduction to Microelectronic Fabrication*. 2nd ed. Modular series on solid-state devices, ed. G.W. Neudeck and R.F. Pierret. Vol. V. 2002, Upper Saddle River, NJ: Prentice Hall.
2. Sze, S.M., ed. *Semiconductor Sensors*. 1994, Wiley: New York.
3. Kamins, T., *Polycrystalline Silicon for Integrated Circuits and Displays*. 2nd ed. 1998: Kluwer Academic Publishers.
4. Kovacs, G.T.A., N.I. Maluf, and K.E. Petersen, *Bulk Micromachining of Silicon*. Proceedings of the IEEE, 1998. **86**(8): p. 1536–1551.
5. Bustillo, J.M., R.T. Howe, and R.S. Muller. *Surface Micromachining for Microelectromechanical Systems*. Proceedings of the IEEE, 1998. **86**(8): p. 1552–1574.
6. Petersen, K.E., *Silicon as a Mechanical Material*. Proceedings of the IEEE, 1982. **70**(5): p. 420–457.

7. Pham, N.P., et al., *Photoresist Coating Methods for the Integration of Novel 3-D RF Microstructures*. Microelectromechanical Systems, Journal of, 2004. **13**(3): p. 491–499.
8. Feynman, R., *Infinitesimal Machinery*. Microelectromechanical Systems, Journal of, 1993. **2**(1): p. 4–14.
9. Guckel, H., *High-Aspect-Ratio Micromachining via Deep X-ray Lithography*. Proceedings of the IEEE, 1998. **86**(8): p. 1586–1593.
10. Tea, N.H., et al., *Hybrid Postprocessing Etching for CMOS-Compatible MEMS*. Microelectromechanical Systems, Journal of, 1997. **6**(4): p. 363–372.
11. Yi, Y.-W., et al., *A Microactive Probe Device Compatible with SOI-CMOS Technologies*. Microelectromechanical Systems, Journal of, 1997. **6**(3): p. 242–248.
12. French, P.J., et al., *The Development of a Low-Stress Polysilicon Process Compatible with Standard Device Processing*. Microelectromechanical Systems, Journal of, 1996. **5**(3): p. 187–196.
13. Jiang, H., et al. *A Universal MEMS Fabrication Process for High-Performance On-Chip RF Passive Components and Circuits*. In *Technical Digest: IEEE Sensors and Actuators workshop*. 2000. Hilton Head, SC.
14. Oz, A., and G.K. Fedder. *CMOS-Compatible RF-MEMS Tunable Capacitors*. In *Radio Frequency Integrated Circuits (RFIC) Symposium*. 2003.
15. Mehregany, M., et al., *Silicon Carbide MEMS for Harsh Environments*. Proceedings of the IEEE, 1998. **86**(8): p. 1594–1609.
16. Tanaka, S., et al., *Silicon Carbide Micro-Reaction-Sintering Using Micromachined Silicon Molds*. Microelectromechanical Systems, Journal of, 2001. **10**(1): p. 55–61.
17. Stoldt, C.R., et al., *A Low-Temperature CVD Process for Silicon Carbide MEMS*. Sensors and Actuators A: Physical, 2002. **97–98**: p. 410–415.
18. Wur, D.R., et al., *Polycrystalline Diamond Pressure Sensor*. Microelectromechanical Systems, Journal of, 1995. **4**(1): p. 34–41.
19. Shibata, T., et al., *Micromachining of Diamond Film for MEMS Applications*. Microelectromechanical Systems, Journal of, 2000. **9**(1): p. 47–51.
20. Zhu, X., et al., *The Fabrication of All-Diamond Packaging Panels with Built-in Interconnects for Wireless Integrated Microsystems*. Microelectromechanical Systems, Journal of, 2004. **13**(3): p. 396–405.
21. Zhang, Z.L., and N.C. MacDonald, *Fabrication of Submicron High-Aspect-Ratio GaAs Actuators*. Microelectromechanical Systems, Journal of, 1993. **2**(2): p. 66–73.
22. Adachi, S., and K. Oe, *Chemical Etching Characteristics of (001) GaAs*. Journal of Electrochemical Society, 1983. **123**(5): p. 2427–2435.
23. Chong, N., T.A.S. Srinivas, and H. Ahmed, *Performance of GaAs Microbridge Thermocouple Infrared Detectors*. Microelectromechanical Systems, Journal of, 1997. **6**(2): p. 136–141.
24. Iwata, N., T. Wakayama, and S. Yamada, *Establishment of Basic Process to Fabricate Full GaAs Cantilever for Scanning Probe Microscope Applications*. Sensors and Actuators A: Physical, 2004. **111**(1): p. 26–31.
25. Heschel, M., M. Mullenborn, and S. Bouwstra, *Fabrication and Characterization of Truly 3-D Diffuser/Nozzle Microstructures in Silicon*. Microelectromechanical Systems, Journal of, 1997. **6**(1): p. 41–47.
26. Maruo, S., K. Ikuta, and H. Korogi, *Force-Controllable, Optically Driven Micromachines Fabricated by Single-Step Two-Photon Microstereolithography*. Microelectromechanical Systems, Journal of, 2003. **12**(5): p. 533–539.
27. Teh, W.H., et al., *Cross-Linked PMMA as a Low-Dimensional Dielectric Sacrificial Layer*. Microelectromechanical Systems, Journal of, 2003. **12**(5): p. 641–648.

28. Madden, J.D., and I.W. Hunter, *Three-Dimensional Microfabrication by Localized Electrochemical Deposition.* Microelectromechanical Systems, Journal of, 1996. **5**(1): p. 24–32.
29. Tsao, C.-C., and E. Sachs, *Photo-Electroforming: 3-D Geometry and Materials Flexibility in a MEMS Fabrication Process.* Microelectromechanical Systems, Journal of, 1999. **8**(2): p. 161–171.
30. Friedrich, C.R., and M.J. Vasile, *Development of the Micromilling Process for High-Aspect-Ratio Microstructures.* Microelectromechanical Systems, Journal of, 1996. **5**(1): p. 33–38.
31. Vogler, M.P., R.E. DeVor, and S.G. Kapoor, *Microstructure-Level Force Prediction Model for Micromilling of Multi-Phase Materials.* ASME Journal of Manufacturing Science and Engineering, 2003. **125**(2): p. 202–209.
32. Tseng, A.A., *Recent Developments in Micromilling Using Focused Ion Beam Technology.* Journal of Micromechanics and Microengineering, 2004. **14**(4): p. R15–R34.
33. Feinerman, A.D., et al., *X-ray Lathe: An X-ray Lithographic Exposure Tool for Nonplanar Objects.* Microelectromechanical Systems, Journal of, 1996. **5**(4): p. 250–255.
34. Manohara, M., et al., *Transfer by Direct Photo Etching of Poly(Vinylidene Fluoride) Using X-rays.* Microelectromechanical Systems, Journal of, 1999. **8**(4): p. 417–422.
35. Reynaerts, D., et al., *Integrating Electro-Discharge Machining and Photolithography: Work in Progress.* Journal of Micromechanics and Microengineering, 2000. **10**: p. 189–195.
36. Takahata, K., and Y.B. Gianchandani, *Batch Mode Micro-Electro-Discharge Machining.* Microelectromechanical Systems, Journal of, 2002. **11**(2): p. 102–110.
37. Fuller, S.B., E.J. Wilhelm, and J.M. Jacobson, *Ink-Jet Printed Nanoparticle Microelectromechanical Systems.* Microelectromechanical Systems, Journal of, 2002. **11**(1): p. 54–60.
38. Rogers, J.A., R.J. Jackman, and G.M. Whitesides, *Constructing Single- and Multiple-Helical Microcoils and Characterizing Their Performance as Components of Microinductors and Microelectromagnets.* Microelectromechanical Systems, Journal of, 1997. **6**(3): p. 184–192.
39. Black, A.J., et al., *Microfabrication of Two Layer Structures of Electrically Isolated Wires Using Self-Assembly to Guide the Deposition of Insulating Organic Polymer.* Sensors and Actuators A: Physical, 2000. **86**(1–2): p. 96–102.
40. Wilson, C.G., and Y.B. Gianchandani, *Silicon Micromachining Using In Situ DC Microplasmas.* Microelectromechanical Systems, Journal of, 2001. **10**(1): p. 50–54.
41. Chen, R.-H., and C.-L. Lan, *Fabrication of High-Aspect-Ratio Ceramic Microstructures by Injection Molding with the Altered Lost Mold Technique.* Microelectromechanical Systems, Journal of, 2001. **10**(1): p. 62–68.
42. Shen, X.-J., L.-W. Pan, and L. Lin, *Microplastic Embossing Process: Experimental and Theoretical Characterizations: 1.* Sensors and Actuators A: Physical, 2002. **97–98**: p. 428–433.
43. Walter, V., et al., *A Piezo-Mechanical Characterization of PZT Thick Films Screen-Printed on Alumina Substrate.* Sensors and Actuators A: Physical, 2002. **96**(2–3): p. 157–166.
44. Jackman, R.J., S.T. Brittain, and G.M. Whitesides, *Fabrication of Three-Dimensional Microstructures by Electrochemically Welding Structures Formed by Microcontact Printing on Planar and Curved Substrates.* Microelectromechanical Systems, Journal of, 1998. **7**(2): p. 261–266.
45. Gracias, D.H., et al., *Forming Electrical Networks in Three Dimensions by Self-Assembly.* Science, 2000. **289**(5482): p. 1170–1172.
46. Whitesides, G.M., and B. Grzybowski, *Self-Assembly at All Scales.* Science, 2002. **295**(5564): p. 2418–2421.
47. Despont, M., et al., *VLSI-NEMS Chip for Parallel AFM Data Storage.* Sensors and Actuators A: Physical, 2000. **80**(2): p. 100–107.

48. Chou, S.Y., P.R. Krauss, and P.J. Renstrom, *Imprint of Sub-25 nm Vias and Trenches in Polymers.* Applied Physics Letters, 1995. **67**(21): p. 3114–3116.
49. Zhang, Y., et al., *Electrochemical Whittling of Organic Nanostructures.* Nano Letters, 2002. **2**(12): p. 1389–1392.
50. Haes, A.J., and R.P.V. Duyne, *A Nanoscale Optical Biosensor: Sensitivity and Selectivity of an Approach Based on the Localized Surface Plasmon Resonance Spectroscopy of Triangular Silver Nanoparticles.* Journal of American Chemical Society, 2002. **124**: p. 10596–10604.
51. Shah, A., et al., *Photovoltaic Technology: The Case for Thin-Film Solar Cells.* Science, 1999. **285**(5428): p. 692–698.
52. Sirringhaus, H., et al., *High-Resolution Inkjet Printing of All-Polymer Transistor Circuits.* Science, 2000. **290**(5499): p. 2123–2126.
53. Service, R.F., *Electronics: Organic Device Bids to Make Memory Cheaper.* Science, 2001. **293**(5536): p. 1746a.
54. Kim, C., P.E. Burrows, and S.R. Forrest, *Micropatterning of Organic Electronic Devices by Cold-Welding.* Science, 2000. **288**(5467): p. 831–833.
55. Williams, K.R., and R.S. Muller, *Etch Rates for Micromachining Processing.* Microelectromechanical Systems, Journal of, 1996. **5**(4): p. 256–269.
56. Williams, K.R., K. Gupta, and M. Wasilik, *Etch Rates for Micromachining Processing—Part II.* Microelectromechanical Systems, Journal of, 2003. **12**(6): p. 761–778.
57. Nathanson, H.C., et al., *The Resonant Gate Transistor.* IEEE Transactions on Electron Devices, 1967. **ED-14**(3): p. 117–133.

CHAPTER 3

Review of Essential Electrical and Mechanical Concepts

3.0 PREVIEW

The MEMS research field crosses several disciplines, including electrical engineering, mechanical engineering, material processing, and microfabrication. The successful design of a MEMS device must consider many intersecting points.

For example, when developing a micromachined sensor such as the ADXL accelerometer, a designer must consider both electrical and mechanical aspects. Electrical design aspects include the capacitance value and signal-processing circuits (e.g., analog-to-digital conversion). Mechanical design aspects include the flexibility of the support beams, dynamic characteristics, and intrinsic stresses in the beam.

Often, the need to enhance performance in one particular aspect can only be met at the expense of others. Let's examine another example of the entanglement of various design interests. The impurity level in the semiconductor silicon influences both mechanical and electrical behaviors. Mechanically, the impurity level affects the piezoresistive sensitivity and temperature sensitivity. Electrically, the impurity level sets the electrical conductivity and the noise floor. Reducing the doping concentration, for example, may increase the piezoresistance sensitivity *and* the noise floor at the same time.

This chapter covers the most essential concepts and analytical skills for MEMS. To maintain brevity and balance, only the most important and frequently encountered concepts are reviewed.

Section 3.1 introduces the concept of the semiconductor crystal and doping process, which leads to the discussion of the fundamental procedures for determining the conductivity of semiconductors based on doping concentrations;

Section 3.2 discusses conventions for naming crystal planes and directions;

Section 3.3 covers the basic relation between stress and strain for various materials;

Section 3.4 reviews procedures for calculating the bending of flexural beams under simple loading conditions;

Section 3.5 is dedicated to discussing the deformation of torsional bars under simple loading conditions;

Section 3.6 explains the origin of intrinsic stresses and methods for characterization, control, and compensation;

Section 3.7 discusses the mechanical behavior of microstructures under periodic loading conditions;

Section 3.8 introduces methods for actively tuning the force constant and the resonant frequency of beams.

This chapter touches on materials from several subject areas. For in-depth studies of these and other related topics, the reader is encouraged to refer to a list of suggested reading materials outlined in Section 3.9.

3.1 CONDUCTIVITY OF SEMICONDUCTORS

Silicon, an element commonly found on sandy beaches and in window glass, is the substrate of choice for building integrated circuits. Silicon belongs to a class of materials called semiconductors; they offer remarkable electrical characteristics and controllability for electronic circuits. Semiconductor silicon is the most common material in the MEMS field. Naturally, the electrical and mechanical properties of silicon are of great interest.

Semiconductors are fundamentally different from conductors (e.g., metals) and insulators (e.g., glass or rubber). As its name suggests, a semiconductor is a material whose conductivity lies between that of a perfect insulator and a perfect conductor. However, this only tells half of the story about why semiconductors are used heavily for modern electronics. More important, the conductivity of a semiconductor material can be controlled by a variety of means, such as intentionally introduced impurities, externally applied electric fields, charge injections, ambient light, and temperature variations. These control "levers" lead to many uses of the semiconductor materials. These uses include bipolar junction transistors; field-effect transistors; solar cells; diodes; and sensors for temperature, force, and concentration of chemical species (e.g., chemical field-effect transistors, or ChemFET).

Certainly, the conductivity value of a semiconductor is of tremendous interest. The macroscopic resistance and conductance are related to the microscopic conductivity. One of the most basic exercises in MEMS is to find the conductivity value of a semiconductor silicon piece based on its doping concentrations. This constitutes the major focus of this section.

We assume the reader has a basic knowledge of charge, voltage, electric field, current, and Ohm's law. A review of these topics can be found in the textbooks and courses outlined in Section 3.9.

3.1.1 Semiconductor Materials

The unique electrical properties of semiconductors stem from their atomic structures. In this section, let's look at how their conductivity originates.

Silicon is a Group IV element in the periodic table. Each silicon atom has four electrons in its outer orbit. As a consequence, each silicon atom in the crystal lattice shares four covalent bonds with four neighboring atoms.

Silicon atoms reside in a crystal lattice. The inter-atomic spacing between atoms is determined by the balance of atomic attraction and repulsive forces. The density of silicon atoms in a solid is 5×10^{22} atoms/cm^3 at 300 K.

Electrons that are covalently bonded to the orbits of silicon atoms cannot conduct current and do not contribute to bulk conductivity. The conductivity of a semiconductor material is only related to the concentration of electrons *that can freely move in bulk*. An electron bonded to a silicon atom must be excited with enough energy to escape the outer orbit of the atom for it to participate in bulk current conduction.

The statistically minimal energy needed to excite a covalently bonded electron to become a free charge carrier is called the **bandgap** of the semiconductor material. It corresponds to the energy necessary to break a single covalent bond between two atoms. The bandgap of silicon at room temperature is approximately 1.11 eV, or 1.776×10^{-19} J. For more detailed information, refer to classic textbooks on solid-state electronic devices [1, 2].

Electrons can receive energy and be liberated from their host atoms by a number of means, including lattice vibration (e.g., temperature rise) and absorption of electromagnetic radiation (e.g., light absorption). Temperature plays an important role in determining the concentration of free electrons in a bulk. In fact, semiconductor silicon is insulating at absolute zero temperature (0 K) when no free electrons are available to conduct current. Higher temperatures lead to greater concentrations of free charge carriers and better conductivity.

A metal conductor, in contrast, consists of metal atoms that are linked to one another with metallic bonds, which are generally weaker than covalent bonds. Electrons can readily break free and participate in current conduction. The equivalent bandgap associated with metals is zero. Hence, the conductivity of a metal conductor is always high and not sensitive to changes of temperature and light conditions.

An insulator, on the other hand, involves atomic bonding that is much stronger than that of a semiconductor material, such as an ionic bond. In other words, the bandgap of insulators is much greater than that of semiconductors. Since it is difficult for electrons to break loose from the atom orbits, the concentration of free charge carriers and the conductivity of insulator materials are very low.

Silicon is not the only semiconductor material used for MEMS. Other semiconductor materials, such as germanium, polycrystalline germanium, silicon germanium, gallium arsenide (GaAs), gallium nitride (GaN), and silicon carbide (SiC) have also been used. The bandgaps of these materials are different from that of silicon [3, 4].

Certain organic materials also exhibit semiconductor characteristics. Organic semiconductors are being investigated for flexible circuits and displays. These generally involve different device architecture and fabrication methods than inorganic semiconductors. This topic is beyond the scope of this text; however, interested readers can read the following referenced papers to get acquainted with this emerging research topic [5–8].

3.1.2 Calculation of Charge Carrier Concentration

The electrical conductivity of a semiconductor material is determined by the number of free charge particles in a bulk and their agility under the influence of an electric field. In this section, we will discuss basic formulation for determining the number (or volumetric concentration) of free charge carriers.

There are two types of free charge carriers—electrons and holes. Electrons that are freed by breaking covalent bonds are not the only participants in bulk current conduction. The bond vacancy left by an escaped electron, called a **hole**, can facilitate bulk current movement as a site for successive electron hopping. A hole carries a positive charge. In a given material, the electrons' and holes' abilities to conduct electricity are different, but they are often on the same order of magnitude.

The electron concentration in a semiconductor is denoted as n (units in electrons/cm^3 or cm^{-3}), while the hole concentration is denoted as p (cm^{-3}). The concentrations of electrons and holes under steady-state, thermal equilibrium conditions (i.e., no external current and no ambient light) are commonly referred to as n_0 and p_0, respectively. The subscript 0 indicates values under thermal equilibrium.

The SI unit of the carrier concentration is m^{-3}. Due to historical reasons, the CGS unit system for carrier concentration (cm^{-3}) is prevalently used.

There are two categories of semiconductor bulk—intrinsic and extrinsic. We will discuss the intrinsic semiconductor material first, since it is the simplest form. We will then review extrinsic materials, which are encountered more often in MEMS and microelectronics.

A perfect semiconductor crystal has no impurities or lattice defects in its crystal structure. It is called an **intrinsic semiconductor material**. In this type of material, electrons and holes are created through thermal or optical excitation. When a valence electron receives enough energy, either thermally or optically, it is freed from the silicon atom and leaves a hole behind. This event is called electron-hole pair generation.

The number and concentration of electrons and holes, however, do not increase without a limit over time. Free electrons and holes can recombine and give up energy along the way. This process, called recombination, competes with the generation process. Under steady-state conditions, the generation and recombination rates are identical.

Since the electrons and holes are created in pairs, the electron concentration (n) is equal to the hole concentration (p). For intrinsic materials, their common value is further denoted as n_i, where the subscript i stands for the word *intrinsic*. The relation is summarized as

$$n = p \equiv n_i. \tag{3.1}$$

The magnitude of n_i is a function of the band gap and temperature, according to

$$n_i^2 = 4\left(\frac{4\pi^2 m_n^* m_p^* k^2 T^2}{h^2}\right)^{3/2} e^{-\frac{E_g}{kT}}, \tag{3.2}$$

where m_n^*, m_p^*, k, T, and E_g are the effective mass of electrons, effective mass of holes, Boltzmann's constant, absolute temperature (in degrees Kelvin), and the bandgap. The term h is Planck's constant. For example, the value of n_i for Si at room temperature is only 1.5×10^{10}/cm^3, which is much smaller than the silicon atom density of approximately 5×10^{22}/cm^3.

Most semiconductor pieces are not perfect intrinsic material, however. They often have impurities atoms in them—by accident or by design. The impurity atoms contribute additional electrons or holes, generally in an unbalanced manner. The intentional introduction of impurities, called doping, turns an intrinsic material into an **extrinsic semiconductor material**. Impurities can be introduced in a number of ways, most notably through diffusion and ion implantation. They can also be incorporated into the semiconductor lattice during the growth of the material as well. This process is called *in situ* doping.

The unregulated introduction of metal ions, such as sodium ions from human sweat, can cause a dramatic degradation of transistor performance. Hence, the presence of metal ions is heavily regulated in cleanrooms.

A dopant atom generally displaces a host atom when introduced into the crystal lattice. A dopant atom with more electrons in its outermost shell than a host atom has is able to introduce, or "donate," its extra electrons to the bulk. This step, called ionization, occurs readily under room temperature. This type of dopant is called a donor. For example, a phosphorus (P) atom introduced into silicon bulk is a donor, because phosphorus is a Group V element with five electrons in its outermost shell; however, it needs only four atoms to form covalent bonds with four neighboring silicon atoms. A phosphorus-doped silicon generally has more free electrons than holes ($n_0 > p_0$) at room temperature. When the electron concentration is greater than the hole concentration, the electron is called the **majority carrier** and the semiconductor material is called an **n-type material**. An easy way to memorize this naming convention is to remember that the letter *n* stands for *negative* (as in negative charges), and it corresponds to the symbol for the concentration of electrons; it is the majority carrier in this case.

A dopant atom with fewer electrons in its outermost shell than a host atom has will "accept" electrons from the bulk. This type of dopant is called an acceptor. For example, a boron (B) atom in a silicon bulk is an acceptor, because boron is a Group III element with three electrons in the outer shell. A boron-doped silicon generally has more free holes than electrons ($n_0 < p_0$). When the hole concentration is greater than that of electrons, the hole is the majority carrier and the semiconductor material is called a **p-type**. The letter *p* stands for *positive* (as in positive charges), and it corresponds to the symbol for the concentration of holes; it is the majority carrier in this case.

For a doped semiconductor piece, the concentrations of the carriers (n_0 and p_0) are different from n_i. However, it can be proven that the concentrations of electrons and holes under thermal equilibrium follows a simple relation:

$$n_0 p_0 = n_i^2. \tag{3.3}$$

In MEMS and microelectronics, doped extrinsic semiconductors are used prevalently. One of the most important exercises is to find the free-carrier concentrations from known doping levels. We will review the procedure for a general case below.

For an extrinsic material with a donor concentration of N_d and an acceptor concentration of N_a, the concentrations of electrons and holes can be determined by following the procedure below. (N_d and N_a are generally unequal.) Since the doping process involves injecting neutral atoms into a neutral bulk, the charge neutrality of the bulk is always maintained. The concentration of negative charges in a bulk is made up of electrons and ionized acceptor atoms (N_a^-). The concentration of positive charges in a bulk consists of holes and ionized donor atoms (N_d^+). The condition of charge neutrality gives

$$p_0 + N_d^+ = n_0 + N_a^-. \tag{3.4}$$

3.1 Conductivity of Semiconductors

To calculate the concentration of electrons, we replace p_0 with n_i^2/n_0 and rearrange terms to yield

$$n_0 - \frac{n_i^2}{n_0} = N_d^+ - N_a^- \tag{3.5}$$

or

$$n_0^2 - (N_d^+ - N_a^-)n_0 - n_i^2 = 0. \tag{3.6}$$

An alternative expression in terms of p_o is

$$p_0^2 - (N_a^- - N_d^+)p_0 - n_i^2 = 0. \tag{3.7}$$

Under the common operation temperatures of semiconductors, donors and acceptors are assumed to be fully ionized, i.e., $N_a^- = N_a$ and $N_d^+ = N_d$.

The magnitude of n_0 and p_0 can be found by solving these second-order quadratic equations, once the concentration of dopants are known. Two solutions can be found; the one that makes physical sense should be accepted.

The left-hand side of Equation (3.6) can be simplified if the magnitude of $N_d - N_a$ is much greater than n_i—the term n_i^2 can be ignored. In this case, we can approximate the concentration of electrons by

$$n_0 = N_d^+ - N_a^-. \tag{3.8}$$

Subsequently, the concentration of holes can be found by solving Equation (3.3).

On the other hand, if the acceptor concentration far outweighs that of the donor concentration and $N_a - N_d$ is much greater than n_i, the concentration of holes can be approximated by

$$p_0 = N_a^- - N_d^+ \tag{3.9}$$

using Equation (3.7). The concentration of electrons is found by solving Equation (3.3).

Example 3.1. Calculation of Carrier Concentrations

Consider a piece of silicon under room temperature and thermal equilibrium. The silicon is doped with boron to a doping concentration of 10^{16} atoms/cm^3. Find the electron and hole concentrations.

Solution. We assume that, under room temperature, the boron dopant atoms are all ionized ($N_a = N_a^-$). Since the concentration of ionized acceptor atoms ($N_a^- = 10^{16}$/cm^3) is six orders of magnitude greater than n_i, which is (1.5×10^{10}/cm^3), we can use Equation (3.9), to obtain

$$p_0 = 10^{16}/\text{cm}^3.$$

Therefore, the magnitude of electron concentration is

$$n_0 \approx \frac{n_i^2}{p_0} = 2.25 \times 10^4/\text{cm}^3.$$

Note the concentration of dopants is much smaller than the density of lattice atoms.

3.1.3 Conductivity and Resistivity

The conductivity of a bulk semiconductor is a measure of its ability to conduct electric current. In this section, we will discuss formulas for calculating electrical conductivity when carrier concentrations of both types—electrons and holes—are known. The overall conductivity of a semiconductor is the sum of the conductivities contributed individually by these two types.

Free charge carriers move under the influence of an electric field. This mode of carrier transport is called **drift**. The agility of charge carriers drifting under the influence of a given field affects the conductivity of the bulk. How fast can a free charge carrier move then?

A free charge carrier in a crystal lattice does not reach arbitrarily high speed over time when it is placed in a uniform and constant electric field. Rather, it is frequently slowed or halted when it collides with lattice atoms and other free charge carriers. A charge carrier reaches a statistical average velocity ($\overline{V}$) between collision events. The magnitude of the velocity is the mathematical product of the magnitude of the local electric field (E) and the agility of the carrier, usually represented by a term called the carrier **mobility**, μ. The mobility is defined as

$$\mu = \frac{\overline{V}}{E} \tag{3.10}$$

with the unit being $(m/s)/(V/m) = m^2/V \cdot s$.

The values of the mobility are influenced in a complex manner [1] by the doping concentration, temperature, and crystal-orientation. Certain materials (such as GaAs) offer higher electron and hole mobilities than silicon, and they are used in high-speed electronic circuits.

The statistical average distance a charge carrier travels between two successive collision events is called its **mean free path**, $\overline{d}$. The average time between two successive collision events is the **mean free time**, $\overline{t}$. The average velocity, mean free path, and mean free time are linked by the relation

$$\overline{d} = \overline{V} \cdot \overline{t}. \tag{3.11}$$

Armed with knowledge about carrier concentration and its speed, we can derive the expression for the conductivity of a bulk semiconductor.

We start from the familiar Ohm's law, which states that the bulk resistivity (ρ) associated with a material is the proportionality constant between an applied electric field and the resulting current density J:

$$E = \rho J. \tag{3.12}$$

The current density equals current divided by the cross-sectional area. The conductivity is the reciprocal of the resistivity ($\sigma = 1/\rho$). The relationship between the current density and the applied electric field can be rewritten as

$$J = \sigma E. \tag{3.13}$$

3.1 Conductivity of Semiconductors

The conductivity σ is explicitly defined as

$$\sigma = \frac{J}{E}. \tag{3.14}$$

The relation between J and E can be found by using the model depicted in Figure 3.1. A macroscopic resistor made of a doped semiconductor carries the current I under the applied voltage V. First, a box volume is isolated from the bulk semiconductor resistor. The length of the box is parallel to the direction of the charge movement and that of the external electric field. The length of the box is intentionally chosen to be the mean free path of electrons in the bulk ($\bar{d}$). The current densities of the box and the bulk resistor are identical, since the box is a sampling of a resistor bulk.

The current density associated with the box is the total charge passing through the section in a given period of time, divided by the cross-sectional area (A). The total amount of charge within the volume at any given instant Q equals the product of the carrier concentration, the volume of the imaginary box, and the charge of the unit charged particles q. Since the length of the box is set as the mean free path, these charged particles are bound to pass through the section within the mean free time. The total electric current associated with section A–I_A, is simply the ratio of Q to the mean free time.

The conductivity contributed by electrons is derived by

$$\sigma_n = \frac{J}{E} = \frac{\frac{I_A}{A}}{E} = \frac{\frac{Q}{A\bar{t}}}{E} = \frac{\frac{(nA\bar{d}q)}{A\bar{t}}}{E} = \frac{n\bar{V}q}{E} = n\mu_n q \tag{3.15}$$

where I_A, A, Q, and q are the current, sample cross section, total charge in a volume defined as the product of the A and $\bar{d}$, and charge of a unit carrier ($q = 1.6 \times 10^{-19}$ C), respectively. The term μ_n is the mobility of electrons.

The conductivity associated with holes is

$$\sigma_p = p\mu_p q. \tag{3.16}$$

The term μ_p is the mobility of holes.

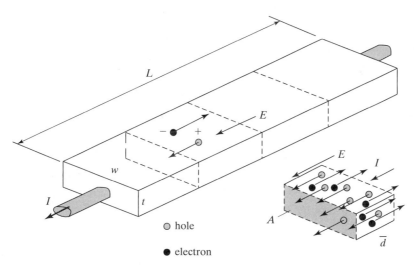

FIGURE 3.1

A semiconductor resistor.

The total conductivity equals the summation of conductivities contributed by electrons and holes. The overall resistivity can be expressed as the functions of the mobility and the carrier concentrations,

$$\rho \equiv \frac{1}{\sigma} = \frac{1}{\sigma_n + \sigma_p} = \frac{1}{q(\mu_n n + \mu_p p)}. \quad (3.17)$$

While the resistivity is a material- and doping-dependent value for semiconductor silicon, the resistance of a resistor element is related to its dimensions. Once the resistivity and dimensions of a resistor are known, the total resistance can be calculated.

The resistance is defined by the ratio of the voltage drop to the current load. The total voltage drop is the product of the electric field and length. Meanwhile, the current is the product of the current density and the cross section ($A = w \times t$). The expression of resistance is

$$R = \frac{V}{I} = \frac{EL}{JA} = \rho \frac{L}{wt} = \frac{\rho}{t} \frac{L}{w} = \rho_s \frac{L}{w}. \quad (3.18)$$

The term ρ_s is the **sheet resistivity**, which equals resistivity divided by the thickness of the resistor or, in the case of a doped semiconductor resistor, the thickness of the doped region. The unit of sheet resistivity is Ω, or $\Omega/\square$. The term $\square$ is called a square. Imagine the pattern of a resistor viewed from the top, formed with square tiles, where the size of the tile equals the width of the resistor, and the number of square tiles needed to lay out the entire resistor pattern equals the ratio between the length and width. Each imaginary tile corresponds to a square. The square notation was invented in the integrated circuit industry as a way of simplifying communication between circuit designers and circuit manufacturers. The sheet resistivity encapsulates the information of doping concentration *and* depth, letting designers focus on geometric layout rather than process details (such as depth).

Example 3.2. Calculation of Conductivity and Resistivity

The intrinsic carrier concentration (n_i) of silicon under room temperature is $1.5 \times 10^{10}/\text{cm}^3$. A silicon piece is doped with phosphorus to a concentration of 10^{18} cm^{-3}. The mobilities of electrons and holes in the silicon are approximately 1350 cm^2/V-s and 480 cm^2/V-s, respectively. Find the resistivity of the doped bulk silicon.

Solution. Following the same procedure as in Example 3.1, we can easily identify the concentration of electrons and holes as

$$n_0 = 10^{18} \text{ cm}^{-3}$$

$$p_0 = \frac{n_i^2}{n_0} = 225 \text{ cm}^{-3}.$$

The resistivity of the doped silicon is calculated by plugging in the following formula:

$$\rho = \frac{1}{\sigma} = \frac{1}{q(\mu_n n_0 + \mu_p p_0)}$$

$$= \frac{1}{1.6 \times 10^{-19} \times (1350 \times 10^{18} + 480 \times 225)}$$

$$= 0.0046 \frac{V \cdot s \cdot cm}{C} = 0.0046 \frac{V \cdot cm}{A} = 0.0046 \, \Omega \cdot cm$$

Example 3.3. Sheet Resistivity

Continue with Example 3.2. If the doped layer is 1 μm thick and has a uniform doping thickness within the layer, find the sheet resistivity of the doped layer. A resistor is defined using the doped layer with the geometries shown in Figure 3.2a. What is the resistance of the resistor? How much heat would be generated by the resistor when a 1 mA current is passed through it? What is the resistance of the resistor shown in Figure 3.2b?

Solution. The sheet resistivity is $\rho_s = \frac{\rho}{t} = \frac{0.0046 \, (\Omega cm)}{10^{-4} \, (cm)} = 46 \, \Omega/\square$.

The resistance can be calculated in two ways. First, resistance is equal to resistivity multiplied by the length and divided by the cross-section area, namely,

$$R = \rho \frac{l}{wt}.$$

The second method is simpler in this case. The resistance is equal to the sheet resistivity multiplied by the number of squares. Ignoring the corners, there are 15 squares within the resistor. The total resistance is

$$R = \rho_s \frac{l}{w} \approx 46 \times 15 = 690 \, \Omega.$$

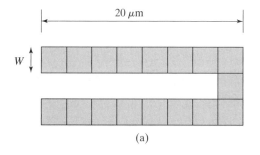
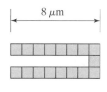

FIGURE 3.2
Layout of two resistors made of doped semiconductor.

When the bias current is 1 mA, the ohmic heating power is

$$P = I^2 R = 0.69 \text{ mW}.$$

Recognizing the fact that the resistor in (b) has the same number of squares as the one in (a), they must have the same resistance value, 690 Ω.

3.2 CRYSTAL PLANES AND ORIENTATION

Silicon atoms in a crystal lattice are regularly arranged in a lattice structure. Material properties (such as Young's modulus of elasticity, mobility, and piezoresistivity) and chemical etch rates of silicon bulk often exhibit orientation dependency. The cross-sectional views of the silicon crystal lattice from several distinct orientations are shown in Figure 3.3. Obviously, the atom packing density varies according to different planes, giving rise to the crystal anisotropy of electrical and mechanical properties and etching characteristics.

A set of common notations, called the Miller Indices, has been developed for identifying and visualizing planes and directions in a crystal lattice.

Let's first discuss the procedure for naming planes. A crystal plane may be defined by considering how the plane intersects the main crystallographic axes of the solid. A rectangular coordinate system is shown in Figure 3.4 with the lattice constant identified as a. A standard procedure used to assign the Miller Indices to a plane in such a coordinate system is discussed next.

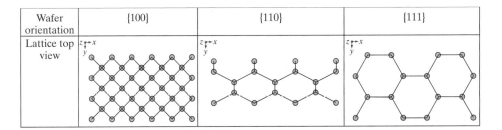

FIGURE 3.3

Lattice cross section of silicon crystal along representative directions.

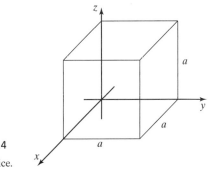

FIGURE 3.4

Cubic lattice.

3.2 Crystal Planes and Orientation

The procedures are most easily illustrated using a rudimentary example. We will first consider the highlighted surface/plane filled with a solid color (Figure 3.5). Two major steps are involved.

Step 1. *Identify the intercepts of the plane with the x-, y-, and z-axes.* In this case, the intercept on the x-axis is at $x = a$, which is the point with coordinates $(a, 0, 0)$. Since the surface is parallel to the y- and z-axes, there are no intercepts on these two axes. We shall consider the intercept to be infinity (∞) for a case when a plane is parallel to an axis. The intercepts on the x-, y-, and z-axes are thus a, ∞, ∞.

Step 2. Take the reciprocals of the three numbers found in Step 1 and reduce them to the smallest set of integers h, k, and l. For the example at hand, the reciprocals of the three intercepts are $1/a$, $1/\infty$ ($= 0$), and $1/\infty$ ($= 0$). We reduce these three values to the smallest set of integers. In this case, the task is achieved by multiplying these numbers with a. The set of integers, enclosed in parentheses according to the form (hkl), constitute the Miller Index for the plane. The Miller Index of the highlighted plane in Figure 3.5 is (100).

From a crystallographic point of view, many planes in a lattice are equivalent. For example, any plane parallel to the shaded plane in Figure 3.5 would also be (100). Three parallel planes in Figure 3.6 are all (100) planes.

Silicon lattice belongs to the cubic lattice family. In a cubic lattice, material properties exhibit rotational symmetry. Hence, (010) and (001) planes in the lattice (Figure 3.6) are equivalent to the (100) plane in terms of material properties. To represent a *family* of such equivalent

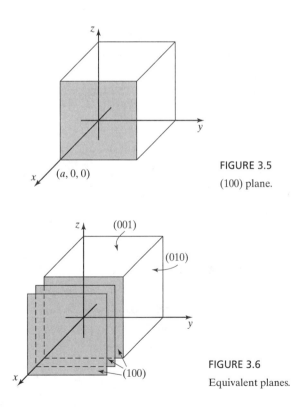

FIGURE 3.5

(100) plane.

FIGURE 3.6

Equivalent planes.

planes, a set of integers is enclosed in braces { } instead of parentheses ( ). For example, crystal planes (100), (010), and (001) are said to belong to the same {100} family (Figure 3.6).

The Miller Index is also used to denote directions in a crystal lattice. The Miller Index of the direction of a vector consists of a set of three integers as well, determined by following the two-step process outlined below:

Step 1. In many cases, a vector of interest does not intercept with the origin of the coordinate system. If this is the case, find the parallel vector that starts at the origin. The Miller Indices of parallel vectors are identical.

Step 2. The three coordinate components of the vector intercepting with the origin are reduced to the smallest set of integers while retaining the relationship among them. For example, the body diagonal of the cube in Figure 3.4 extending from point (0, 0, 0) to point (1a, 1a, 1a) consists of three X, Y, and Z components, all being 1a. Therefore, the Miller Index of the body diagonal consists of three integers (1, 1, and 1) enclosed in a bracket: [111].

In a cubic lattice, a vector with the Miller Index [hkl] is always perpendicular to a plane (hkl).

Vectors that are rotationally symmetric belong to a family of vectors. The notation of a family of vectors consists of three integers enclosed in curly braces, { }.

The most frequently encountered silicon crystal planes in MEMS are illustrated in Figure 3.7. Note the positions of the {100}, {110}, and {111} families of surfaces, along with

TABLE 3.1 Summary of Notation for Planes, Directions, and Their Families.

	Single element notation	Family notation
Directions	[hkl]	<hkl>
Planes	(hkl)	**{hkl}**

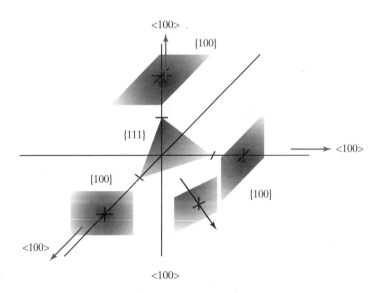

FIGURE 3.7

Identification of important surfaces in silicon.

the corresponding crystal directions, <100>, <110>, and <111>. Any surface in the {111} surface family intercepts the (100) surface at an inclination angle of 54.75°. A (110) plane intercepts a (100) plane at an angle of 45°.

Commercial silicon wafers can be purchased with different front surface "cuts." <100>-oriented wafers are used for metal-oxide-semiconductor (MOS) electronic devices for their low density of interface states. <100> wafers are also prevalently used for bulk silicon micromachining of MEMS devices. <111>-oriented wafers, on the other hand, are commonly used for bipolar junction transistors because of the high mobility of charge carriers in the <111> direction. They have been occasionally used for MEMS applications (see Chapter 10). Surface micromachined microstructures without circuit integration can be performed on wafers of any orientation.

3.3 STRESS AND STRAIN

In this section, we will discuss the basic concepts of stress and strain, as well as their relations. It is assumed that the reader is familiar with the general concepts of force and torque.

3.3.1 Internal Force Analysis: Newton's Laws of Motion

Stress is developed in response to mechanical loading. Therefore, methods for analyzing internal forces inside a micromechanical element based on loading conditions will be discussed first.

Newton's three laws of motion are the foundation for analyzing the static and dynamic behaviors of MEMS devices under loading. Here, these three laws are briefly reviewed.

Newton's Laws	Statement
Newton's First Law of Motion (The Law of Inertia)	Every object in a state of uniform motion tends to remain in the state of motion unless an external force is applied to it.
Newton's Second Law of Motion	The relationship between an object's mass m, its acceleration a, and the applied force F is $F = ma$. Acceleration and force are vectors. The direction of the force vector is the same as the direction of the acceleration vector.
Newton's Third Law of Motion	For every action, there is an equal and opposite reaction.

One of the most frequently encountered consequences of Newton's Laws is that, for any stationary object, the vector sum of forces and moment (torque) on the object *and* on any part of it must be zero.

These laws are used to analyze force distribution inside a material, which gives rise to stress and strain. We will illustrate the procedures for force analysis using the following examples.

Consider a bar firmly embedded in a brick wall with an axial force F applied at the end (Figure 3.8). Since the force is transmitted through the bar to the wall, the wall must produce a reaction according to Newton's Third Law. The wall would act on the left end of the bar with an unknown force. To expose and quantify this force, we would remove the wall and replace it with the actions it imparts on the bar. This **free-body diagram** of the bar clearly reveals that the wall must provide an axial force with equal magnitude but opposite direction to the applied force, so that the total force on the bar is zero to maintain its stationary status (Newton's First Law).

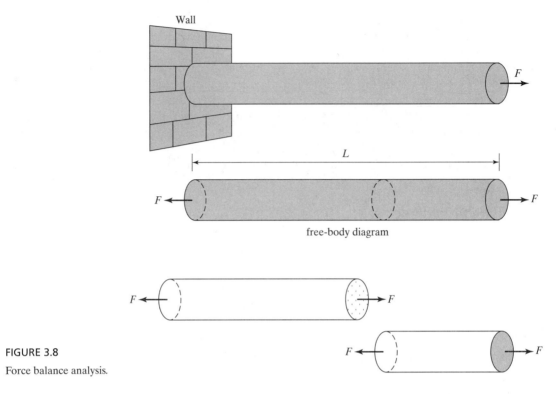

FIGURE 3.8
Force balance analysis.

We can use this technique to expose and quantify hidden forces and stresses at any section. Since the bar is in equilibrium, any part of it must be in equilibrium as well. We can pick an arbitrary section of interest and cut the bar into two halves. (If this section is cut perpendicular to the longitudinal direction of the bar, it is called a cross section.) The convenient way to analyze the force on each of the two pieces is to start from the right-hand piece, since the loading condition on one of its ends is explicitly known.

Since a force is applied at the free end of the bar, an equal but opposite force must develop at the cross section. The two opposite faces at the cut sections must have matched force and moment with opposite signs, as dictated by Newton's Third Law. Hence, a force of magnitude F is believed to act on the right end of the left-hand piece, even if we have no way of measuring it experimentally since the surface is actually hidden.

Now consider the same bar under a force acting in the transverse direction (Figure 3.9). Again, we isolate the bar by imaginarily removing the wall. The sum of forces and moments acting on the isolated bar must be zero. For the net force to be zero, a force of the same magnitude but opposite sign must act on the end of the bar attached to the wall. The pair of forces, however, creates a torque (also referred to as a *couple* or a *moment*, in mechanics) with the magnitude being F times L, the length of the bar. A reactive torque, with the magnitude of F times L but opposite sign, must act on the end of the bar attached to the wall.

To calculate the reactive force and torque at an imaginarily cut section, we can start from the piece to the right, since the loading on one of its ends is known. The face at the cross section would experience an opposite force, as well as a torque to balance the one created by the pair of forces; they are separated by an arm of L'.

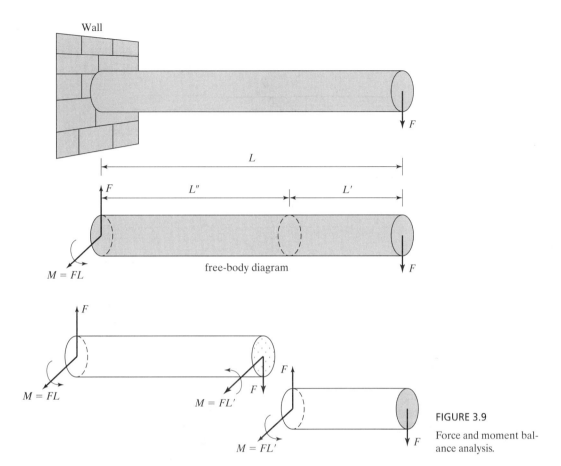

FIGURE 3.9 Force and moment balance analysis.

The imaginarily cut section on the piece to the left would have exactly the opposite force and torque as the opposing surface (according Newton's Third Law). The magnitude of the sum of torques on the left-hand piece is equal to

$$\sum M = FL - FL' = FL'',$$

which equals the force (F) multiplied by the length of the left-hand piece (L''). The net force and torque acting on the left-hand piece are both zero.

3.3.2 Definitions of Stress and Strain

Mechanical stresses fall into two categories—normal stress and shear stress. We will first review the definition of these two stresses using simple cases depicted in Figure 3.10.

For the simplest case of normal stress analysis, consider a rod with a uniform cross-sectional area subjected to axial loading. If we pull on the rod in its longitudinal direction, it will experience tension and the length of the rod will increase (Figure 3.10).

The internal stress in the rod is exposed if we make an imaginary cut through the rod at a section.

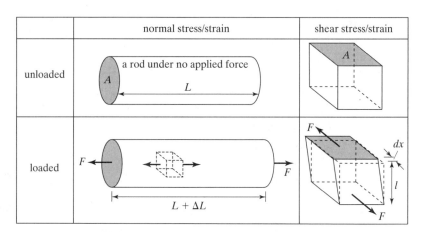

FIGURE 3.10

Normal stress and shear stress.

At any chosen cross section, a continuously distributed force is found acting over the entire area of the section. The intensity of this force is called the stress. If the stress acts in a direction perpendicular to the cross section, it is called **normal stress**. The normal stress, commonly denoted as σ, is defined as the force (F) applied on a given area (A):

$$\sigma = \frac{F}{A}. \tag{3.19}$$

The SI unit of stress is N/m², or Pa.

A normal stress can be tensile (as in the case of pulling along the rod) or compressive (as in the case of pushing along the rod). The polarity of normal stress can also be determined by isolating an infinitesimally small volume inside the bar. If the volume is pulled in one particular axis, the stress is tensile; if the volume is pushed, the stress is compressive.

The unit elongation of the rod represents the strain. In this case, it is called **normal strain** since the direction of the strain is perpendicular to the cross section of the beam. Suppose the steel bar has an original length denoted by L_0. Under a given normal stress, the rod is extended to a length of L. The resultant strain in the bar is defined as

$$s = \frac{L - L_0}{L_0} = \frac{\Delta L}{L_0}. \tag{3.20}$$

The strain is commonly denoted as ε in mechanics. However, it could easily be mistaken with the notation reserved for electrical permittivity (dielectric constant) of a material. In most cases, this is very clear based on the context of the discussion. In others, such as the discussion of piezoelectricity where the constitutive equations involve both strain and permittivity, the strain and permittivity terms must be assigned different notations to avoid confusion. In this textbook, the strain is always denoted as s to prevent any possible confusion.

In reality, the applied longitudinal stress along the x-axis not only produces a longitudinal elongation in the direction of the stress, but a reduction of the cross-sectional area as well (Figure 3.11). This can be explained by the argument that the material must strive to maintain constant atomic spacing and bulk volume. The relative dimensional change in the y

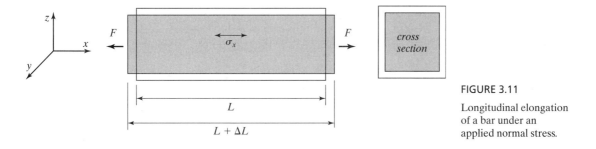

FIGURE 3.11

Longitudinal elongation of a bar under an applied normal stress.

and z directions can be expressed as s_x and s_y. This general material characteristic is captured by a term called **Poisson's ratio**, ν, which is defined as the ratio between transverse and longitudinal elongations, namely,

$$\nu = \left|\frac{s_y}{s_x}\right| = \left|\frac{s_z}{s_x}\right|. \tag{3.21}$$

Stress and strain are closely related. Under small deformation, the stress and the strain terms are proportional to each other according to Hooke's law:

$$\sigma = Es. \tag{3.22}$$

The proportion constant, E, is called the **modulus of elasticity**. The general relation between a stress and strain over a wider range of deformation, however, is much more complicated. More in-depth discussion is found in Section 3.3.3.

The modulus of elasticity, often called **Young's modulus**, is an intrinsic property of a material. It is a constant for a given material, irrespective of the shape and dimensions of the mechanical element. Atoms are held together with atomic forces. If we imagine interatomic force acting as springs to provide restoring force when atoms are pulled apart or pushed together, the modulus of elasticity is the measure of the stiffness of the interatomic spring near the equilibrium point.

Example 3.4. Longitudinal Stress and Strain

A cylindrical silicon rod is pulled on both ends with a force of 10 mN. The rod is 1 mm long and 100 μm in diameter. Find the stress and strain in the longitudinal direction of the rod.

Solution. The stress is calculated by dividing the force by the cross-sectional area,

$$\sigma = \frac{F}{A} = \frac{10 \times 10^{-3}}{\pi \left(\frac{100 \times 10^{-6}}{2}\right)^2} = 3.18 \times 10^5 \text{ N/m}^2.$$

The strain equals the stress divided by the strain,

$$s = \frac{\sigma}{E} = \frac{3.18 \times 10^5}{130 \times 10^9} = 2.4 \times 10^{-6}.$$

Shear stresses can be developed under different force-loading conditions. One of the simplest ways to generate a pure shear loading is illustrated in Figure 3.10, with a pair of forces acting on opposite faces of a cube. In this case, the magnitude of the shear stress is defined as

$$\tau = \frac{F}{A}. \tag{3.23}$$

The unit of τ is N/m². Shear stress has no tendency to elongate or shorten the element in the x, y, and z directions. Instead, the shear stresses produce a change in the *shape* of the element. The original element shown here, which is a rectangular parallelpiped, is deformed into an oblique parallelpiped. Shear strain γ, defined as the extent of rotational displacement, is

$$\gamma = \frac{\Delta X}{L}. \tag{3.24}$$

The shear stress is unitless; in fact, it represents the angular displacement expressed in the unit of radians. The shear stress and strain are also related to each other by a proportional constant, called the shear modulus of elasticity G. The expression of G is simply the ratio of τ and γ:

$$G = \frac{\tau}{\gamma}. \tag{3.25}$$

The unit of G is N/m². The value of G depends on the material, not the shape and dimensions of an object.

For a given material, E, G, and Poisson's ratio are linked through the relationship

$$G = \frac{E}{2(1 + \nu)}. \tag{3.26}$$

3.3.3 General Scalar Relation Between Tensile Stress and Strain

The relationship between tensile stress and strain has been studied both theoretically and experimentally for many materials, especially metals. The relation between normal stress and strain expressed by Equation (3.22) is only applicable over a narrow range of deformation. Their relationship over a wider range of deformation is discussed in this section.

In order to determine the stress–strain relation, a tensile test is commonly used. A rod with precise dimensions, a calibrated crystalline orientation, and a smooth surface finish is subjected to a tension force applied in the longitudinal direction. The amount of relative displacement and the applied stress are plotted on a stress–strain curve until the beam breaks.

A generic stress–strain curve is illustrated in Figure 3.12. At low levels of applied stress and strain, the stress value increases proportionally with respect to the developed strain, with the proportional constant being Young's modulus. This segment of the stress–strain curve is called the elastic deformation regime. If the stress is removed, the material will return to its original shape. This force loading can be repeated many times.

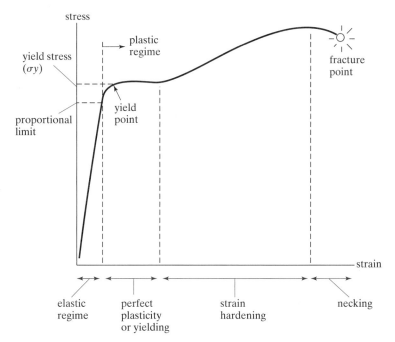

FIGURE 3.12

Broad relation between tensile stress and strain.

As the stress exceeds a certain level, the material enters the plastic deformation regime. In this regime, the amount of stress and strain does not follow a linear relationship anymore. Furthermore, deformation cannot be fully recovered after the external loading is removed.

Bend a metal paper clip wire slightly; it will always return to its original shape. If the wire is bent beyond a certain angle, the clip will never return to its original shape. Plastic deformation is said to have occurred.

Stress–strain curves for materials in compression differ from those in tension [9]. The stress–strain curve has two noticeable points—yield point and fracture point. Before the yield point is reached, the material remains elastic. Between the yield point and the fracture point, the specimen undergoes plastic deformation. At the fracture point, the specimen suffers from irreversible failure. The y-coordinate of the yield point is the **yield strength** of the material. The y-coordinate of the fracture point is the **ultimate strength** (or the fracture strength) of the material.

For many metals, the generic relationship depicted in Figure 3.12 is true. However, not all materials exhibit this generic stress–strain relationship. Some representative curves for different classes of materials are shown in Figure 3.13, including brittle materials (such as silicon) and soft rubber; both are used extensively in MEMS.

There are several common qualitative phrases used to describe materials—strong, ductile, resilient, and tough. These terms can be explained more clearly by relating them to the stress–strain curves.

A material is strong if it has high yield strength or ultimate strength. By this account, silicon is even stronger than stainless steel.

Ductility is an important mechanical property. It is a measure of the degree of plastic deformation that has been sustained at the point of fracture. A material that experiences

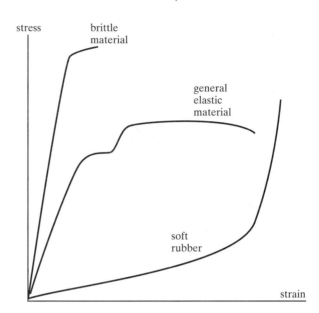

FIGURE 3.13
Stress–strain relations.

very little or no plastic deformation upon fracture is termed brittle. Silicon is a brittle material, which fails in tension with very little elongation after the proportional limit is exceeded. Ductility may be expressed quantitatively as either *percent elongation* or *percent reduction in area*.

Toughness is a mechanical measure of the material's ability to absorb energy up to fracture. For a static situation, toughness may be ascertained from the result of the tensile stress–strain test. It is the area under the stress–strain curve up to the point of fracture. For a material to be tough, it must display both strength and ductility.

Resilience is the capacity of a material to absorb energy when it is deformed elastically, and then to have this energy recovered upon unloading.

3.3.4 Mechanical Properties of Silicon and Related Thin Films

For brittle material like silicon, polysilicon, and silicon nitride, the values of the fracture stress and fracture strain are very important for design. However, the experimental values of fracture stress; Young's modulus; and the fracture strain of single-crystalline silicon, polycrystalline silicon, and silicon are relatively scarce and scattered. They are scarce because the accurate measurement of a miniature sample specimen is more challenging than that of macroscopic samples. They are scattered because the material properties are affected by a variety of subtle factors that are often unreported or not easily traceable (such as exact material growth conditions, surface finish, and thermal treatment history). For macroscopic samples, the variation is less obvious because of the average effect from a large specimen. Unfortunately, many experimental data obtained from macroscopic samples do not apply to microscale samples.

Certain measured material properties, such as fracture strength, quality factor [10], and fatigue lifetime, depend on the size of specimen [11]. The size effects on certain MEMS specimens, such as single-crystal silicon and silicon compound thin films, have been studied extensively

[12–14]. For example, one study found that the fracture strength is size dependent; it is 23–38 times larger than that of a millimeter-scale sample [13]. The fracture behavior of silicon, for example, is governed by the presence of flaws and preexisting cracks. For small single-crystal silicon structures, the devices may exhibit larger elastic strength and strain than predicted for bulk materials due to the lack of flaws in the small volume of silicon involved.

For single-crystal silicon, Young's modulus is a function of the crystal orientation [15]. In the {100} plane, Young's modulus of silicon is the greatest in the [110] direction (168 GPa) [14] and the smallest in the [100] direction (130 GPa). In the {110} plane, Young's modulus of silicon is the greatest in the [111] direction (187 GPa). The shear modulus of silicon is a function of the crystal orientation as well [15]. Poisson's ratio of silicon has a rather broad range, from 0.055 to 0.36, depending on the orientation and measurement configuration [15]. Unfortunately, there is significant data scattering.

For polysilicon thin films, Young's modulus depends on the exact process conditions of the materials, which differ from laboratory to laboratory due to subtle changes of growth conditions. Polysilicon grown by LPCVD exhibits a (100) texture and rather uniform columnar grains [16]. In fact, the mechanical properties of LPCVD polysilicon exhibit angular dependence due to the columnar structure. The value for Young's modulus of polysilicon ranges from 120 GPa to 175 GPa, with the average being approximately 160 GPa. The doping level seems to affect Young's modulus. A comprehensive summary can be found in References [17, 18]. More importantly, it has been observed that the modulus values may differ even if the measurements are conducted on the same material, from the same run, in the same reactor, and in close physical proximity. It appears advisable for any microfabrication facility not to use the properties cited in the literature but to identify the most feasible measurement technique and conduct measurements for every fabrication run [18].

The reported Poisson's ratio ranges from 0.15 to 0.36 [18]. The widely quoted value of Poisson's ratio of polysilicon is 0.22.

The measured fracture strength of polysilicon ranges from 1.0 to 3.0 GPa, obtained on different polysilicon samples and using different techniques (e.g., tension, flexural bending, and nanoindentation) [19]. The fracture strain of polysilicon is temperature related; it increases with rising temperature to a certain point. It is approximately 0.7% (at room temperature) and 1.6% (at 670°C). Young's modulus and fracture strength for polycrystalline materials depends on the crystal microstructure and the size of the devices [20]. Experimental studies have found that the fracture strength of polycrystalline silicon increases as the total surface area of the test section decreases [17].

For thin-film silicon nitride, experimental results tend to deviate from each other and from that of bulk silicon nitride. One report shows that Young's modulus is 254 GPa and the fracture strength is 6.4 GPa, indicating that the fracture strain is 2.5% [21].

The properties of thin films and microstructures are difficult to study directly due to their small dimensions. A number of techniques and structures have been developed to increase the accuracy and efficiency of measurement [11, 20, 22–24]. In some cases, the material properties of microscale structures are most efficiently and accurately studied using microactuators and sensors that are cofabricated with the specimen [11, 25].

Many MEMS applications are potentially subjected to shock loading that may occur during fabrication, deployment, or operation. It is of interest to extract comprehensive guidelines, using both experimental and analytical tools. The behavior of MEMS devices under shock loads has been studied [26]. One strategy for increasing the shock tolerance is to use ductile materials such as polymers. An example of optical scanning with polyimide-based hinges is discussed in Reference [27].

Microstructures may fail under repeated loading even if the magnitude of the stress is below the fracture strength [28]. This phenomenon, called fatigue, can be experienced in everyday life. For example, repeated small angle bending of a paper clip may cause it to develop surface cracks and eventually break over a certain number of cycles. Single-crystal silicon material and micromechanical structures generally exhibit good fatigue life because, as the size of the device becomes smaller, the number of crack-initiating sites on a given structure decreases. Fatigue of single-crystal silicon and other materials have been experimentally studied, often by artificially introducing mechanical defects with controlled dimensions. One study shows that specimen lives ranged from 10^6 to 10^{11} cycles in single-crystal silicon [29]. The fatigue behavior of polysilicon structures has been studied as well [30]. Polysilicon structures can achieve a fatigue life up to 10^{11} cycles. The same study found that the fatigue life of polysilicon structures decreases with increasing stress (uniform or concentrated).

Comprehensive lifetime testing of MEMS devices is rarely done and even more rarely published. Scattered data points do exist. For example, a gold cantilever acting as a MEMS RF switch was able to perform 7 billion cycles of switching [31].

Appendix A contains a table that summarizes representative mechanical and electrical properties of several important MEMS materials.

3.3.5 General Stress–Strain Relations

The formula given in Section 3.3.2 treats stress and strain as scalars. In reality, stress and strain are tensors. Their relationship can be conveniently expressed in matrix form in which stress and strain are expressed as vectors. The general vector representation of stresses and strains will be discussed in this section.

To visualize the vector components of stress and strain, let's isolate a unit cube from inside a material and consider stress components acting on it. The cube is placed in a rectangular coordinate system with axes marked as x, y, and z. For reasons that will become apparent later, the axes x, y, and z are also labeled as axes 1, 2, and 3, respectively.

A cube has six faces. Consequently, there are 12 possible shear force components—two for each face. These are not all independent. For example, each pair of shear stress components acting on parallel faces but along the same axis have equal magnitude and opposite directions for force balance (Newton's First Law). This reduces the number of independent shear stress components to six. These six shear stress components (τ) are graphically illustrated in Figure 3.13. Each component is identified by two subscript letters. The first letter in the subscript indicates the normal direction of the facet on which the stress is applied, while the second letter indicates the direction of the stress component.

Based on torque balance, two shear stress components acting on two facets but pointing towards a common edge have the same magnitude. Specifically, $\tau_{xy} = \tau_{yx}, \tau_{xz} = \tau_{zx}$, and $\tau_{zy} = \tau_{yz}$. In other words, equal shear stresses always exist on mutually perpendicular planes. The independent number of shear stress components is reduced to three.

There are six possible normal stress components—one for each face of a cube. Under equilibrium conditions, the normal stress components acting on opposite facets must have the same magnitude and point to opposite directions. Therefore, there are three independent normal stress components (Figure 3.13). Normal stress components are labeled σ with two subscript letters.

3.3 Stress and Strain

Overall, in a rectangular coordinate system under motion equilibrium, there are three independent normal stresses and three shear ones (Fig. 3.14).

It is quite tedious to write two subscripts for each component. The notations of the six independent components can be further simplified by using the following scheme:

1. Normal stress components σ_{xx}, σ_{yy}, and σ_{zz} are simply noted as T_1, T_2, and T_3, respectively.
2. Shear stress components τ_{yz}, τ_{xz}, and τ_{xy} are simply noted as T_4, T_5, and T_6, respectively.

Correspondingly, there are three independent strains (s_1 through s_3) and three shear strains (s_4 through s_6). The general matrix equation between stress and strain is

$$\begin{bmatrix} T_1 \\ T_2 \\ T_3 \\ T_4 \\ T_5 \\ T_6 \end{bmatrix} = \begin{bmatrix} C_{11} & C_{12} & C_{13} & C_{14} & C_{15} & C_{16} \\ C_{21} & C_{22} & C_{23} & C_{24} & C_{25} & C_{26} \\ C_{31} & C_{32} & C_{33} & C_{34} & C_{35} & C_{36} \\ C_{41} & C_{42} & C_{43} & C_{44} & C_{45} & C_{46} \\ C_{51} & C_{52} & C_{53} & C_{54} & C_{55} & C_{56} \\ C_{61} & C_{62} & C_{63} & C_{64} & C_{65} & C_{66} \end{bmatrix} \begin{bmatrix} s_1 \\ s_2 \\ s_3 \\ s_4 \\ s_5 \\ s_6 \end{bmatrix} \quad (3.27)$$

In short-hand form, the expression is

$$\overline{T} = C\overline{s}. \quad (3.28)$$

The coefficient matrix C is called the stiffness matrix.

The strain matrix is a product of the compliance matrix S and the stress tensor, according to the following matrix expression:

$$\begin{bmatrix} s_1 \\ s_2 \\ s_3 \\ s_4 \\ s_5 \\ s_6 \end{bmatrix} = \begin{bmatrix} S_{11} & S_{12} & S_{13} & S_{14} & S_{15} & S_{16} \\ S_{21} & S_{22} & S_{23} & S_{24} & S_{25} & S_{26} \\ S_{31} & S_{32} & S_{33} & S_{34} & S_{35} & S_{36} \\ S_{41} & S_{42} & S_{43} & S_{44} & S_{45} & S_{46} \\ S_{51} & S_{52} & S_{53} & S_{54} & S_{55} & S_{56} \\ S_{61} & S_{62} & S_{63} & S_{64} & S_{65} & S_{66} \end{bmatrix} \begin{bmatrix} T_1 \\ T_2 \\ T_3 \\ T_4 \\ T_5 \\ T_6 \end{bmatrix} \quad (3.29)$$

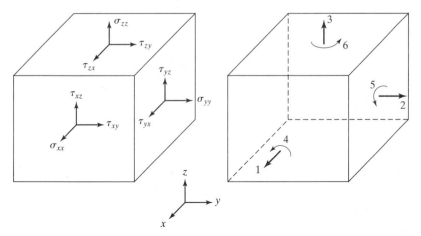

FIGURE 3.14

Principal stress components.

The expression in short-hand form is

$$\bar{s} = S\bar{T}. \tag{3.30}$$

The compliance matrix S is the inverse of the stiffness matrix. In short-hand notation,

$$S = C^{-1}. \tag{3.31}$$

Note the stiffness matrix is denoted by the letter C, whereas the compliance matrix is denoted by the letter S. This is not a mistake. These symbols are commonly used in mechanics.

It is quite tedious to manipulate matrices with 36 components. Fortunately, for many materials of interest to MEMS, the stiffness and the compliance matrices can be simplified. For single-crystal silicon with the coordinate axes along <100> directions, the stiffness matrix is

$$C_{Si,<100>} = \begin{bmatrix} 1.66 & 0.64 & 0.64 & 0 & 0 & 0 \\ 0.64 & 1.66 & 0.64 & 0 & 0 & 0 \\ 0.64 & 0.64 & 1.66 & 0 & 0 & 0 \\ 0 & 0 & 0 & 0.8 & 0 & 0 \\ 0 & 0 & 0 & 0 & 0.8 & 0 \\ 0 & 0 & 0 & 0 & 0 & 0.8 \end{bmatrix} 10^{11} \text{Pa}. \tag{3.32}$$

Example 3.5. Use of the Stiffness Matrix

The stiffness and the compliance matrices incorporate rich information about Young's modulus and Poisson's ratio in three-dimensional space. Find Young's modulus of silicon in the [100] direction based on the stiffness matrix.

Solution. To solve for Young's modulus in the [100] direction, which is the x-direction (axis 1), set all other stress components except in axis 1 to zero, and solve for the strain matrix that would produce only stress in that axis:

$$T_1 = 1.66 \times 10^{11} s_1 + 0.64 \times 10^{11} s_2 + 0.64 \times 10^{11} s_3$$

$$T_2 = 0 = 0.64 \times 10^{11} s_1 + 1.66 \times 10^{11} s_2 + 0.64 \times 10^{11} s_3.$$

$$T_3 = 0 = 0.64 \times 10^{11} s_1 + 0.64 \times 10^{11} s_2 + 1.66 \times 10^{11} s_3$$

Solve these three equations simultaneously to find the relation between T_1 and s_1. We have

$$E_{[100]} = \frac{T_1}{s_1} = 1.66 \times 10^{11} - 2 \frac{0.64 \times 10^{11}}{1.66 \times 10^{11} + 0.64 \times 10^{11}} 0.64 \times 10^{11} = 130 \text{ GPa}.$$

This agrees with the experimental data found in [15].

3.4 FLEXURAL BEAM BENDING ANALYSIS UNDER SIMPLE LOADING CONDITIONS

Flexural beams are commonly encountered in MEMS as spring support elements. Essential skills and common practices for a MEMS researcher include calculating the bending of a beam under simple loading conditions, analyzing induced internal stress, and determining the resonant frequency associated with the element. The reader will be exposed to the following topics in this section:

> Section 3.4.1 discusses types of mechanical beams and boundary conditions associated with supports;
> Section 3.4.2 reviews the distribution of longitudinal stress and strain in a beam under pure bending;
> Section 3.4.3 and Section 3.4.4 explain procedures for calculating the deflection and spring constant of a beam, respectively.

3.4.1 Types of Beams

A beam is a structure member subjected to lateral loads, i.e., forces or moments having their vectors perpendicular to the longitudinal axis. In this textbook, we focus on planar structures—beams that lie in single planes. In addition, all loads act in that same plane and all deflections occur in that plane.

Beams are usually described by the manner in which they are supported. Boundary conditions pertain to the deflections and slopes at the support of a beam. Consider a two-dimensional beam with movement confined to one plane. Each point along the length of the beam can have a maximum of two linear degrees of freedom (DOF) and one rotational degree of freedom. Three possible boundary conditions can exist:

1. The fixed boundary condition restricts both linear DOFs and the rotational DOF. No movement is allowed at the support. At the fixed support, a beam can neither translate nor rotate. Representative examples include the anchored end of a diving board or the ground end of a flagpole.
2. The guided boundary conditions allow two linear DOFs but restrict the rotational DOF.
3. The free boundary conditions provide for both linear DOFS and rotation. At a free end, a point on the beam may translate and rotate. A representative example is the free end of a diving board.

These three distinct types of boundary conditions are graphically represented in Table 3.2.

A flexural beam can be classified according to the combination of the two mechanical boundary conditions associated with it. For example, a beam fixed at one end and free at another is conveniently referred to as a fixed–free beam, commonly called a **cantilever**. In MEMS research, the most frequently encountered types of beams are fixed–free (cantilevers), fixed–fixed (bridges), and fixed–guided beams.

It is important to correctly identify the boundary conditions associated with a beam. The best way to learn and recognize boundary conditions is through examples. Figure 3.15 depicts

74 Chapter 3 Review of Essential Electrical and Mechanical Concepts

TABLE 3.2 Possible Boundary Conditions.

Boundary conditions	Number of linear DOF	Number of angular DOF	Example
Fixed (clamped)	0	0	fixed B.C.
Guided	2	0	guided B.C.
Free	2	1	free B.C.

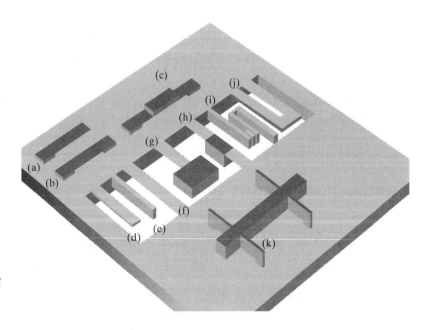

FIGURE 3.15

Various cantilevers of different boundary conditions.

several commonly encountered beam structures on a substrate. The boundary conditions for these beams are summarized as follows:

(a) A fixed–free cantilever parallel to the plane of the substrate, with the free tip capable of moving in a direction normal to the substrate. Lateral, in-plane movement of the free end would encounter more significant resistance.

(b) A fixed–fixed beam (bridge) parallel to the substrate plane.

(c) There are actually two ways to classify this beam. It can be considered a fixed–fixed cantilever parallel to the substrate plane with a thick and stiff part in the middle. Alternatively, this beam can be considered as two fixed–guided beams jointed in parallel to support a rigid part.

(d) A fixed–free cantilever with the free end capable of movement perpendicular to the substrate. Its behavior is similar to that of cantilever (a).

(e) A fixed–free cantilever with the free end capable of movement within the substrate plane. With its thickness greater than its width, the movement of the free end in a direction perpendicular to the substrate plane would encounter greater resistance.

(f) A fixed–fixed beam (bridge).

(g) A fixed–free cantilever carrying a stiff object at the end. The stiff object does not undergo flexural bending due to increased thickness.

(h) This beam is very similar to beam (c) except for the fabrication method.

(i) A fixed–free cantilever with folded length. It consists of several fixed–free beam segments connected in series. The free end of the folded cantilever is capable of movement in a direction parallel to the substrate surface. Movement of the free end perpendicular to the substrate would encounter greater resistance.

(j) Two fixed–free cantilevers connected in parallel. The combined spring is stiffer than any single arm.

(k) Four fixed–guided beams connect to a rigid shuttle, which is allowed to move in the substrate plane but with restricted out-of-plane translational movement.

3.4.2 Longitudinal Strain Under Pure Bending

When a beam is loaded by force or couples, stresses and strains are created throughout the interior of the beam. Loads may be applied at a concentrated location (concentrated load), or distributed over a length or region (distributed load). To determine the magnitude of these stresses and strains, we first must find the internal forces and internal couples that act on cross sections of the beam (see Section 3.3.1).

The loads (either concentrated or distributed) acting on a beam cause the beam to bend (or flex), deforming its axis into a curve. The longitudinal strains in a beam can be found by analyzing the curvature of the beam and the associated deformations. For this purpose, consider a portion of a beam (A–B) in *pure bending* (i.e., the moment is constant throughout the beam) (Figure 3.16). We assume that the beam initially has a straight longitudinal axis (the x axis in the figure). The cross section of the beam is symmetric about the y axis.

It is assumed that cross sections of the beam, such as sections mn and pq, remain plane and normal to the longitudinal axis. Because of the bending deformations, cross sections mn and pq rotate with respect to each other about axes perpendicular to the xy plane. Longitudinal

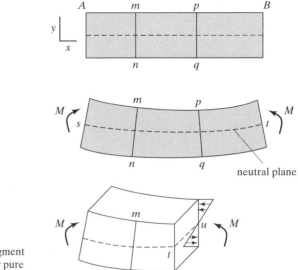

FIGURE 3.16

Bending of a segment of a beam under pure bending.

lines in the convex (lower) part of the beam are elongated, whereas those on the concave (upper) side are shortened. Thus, the lower part of the beam is in tension and the upper part is in compression. Somewhere between the top and bottom of the beam is a surface in which longitudinal lines do not change in length. This surface, indicated by the dashed line *st*, is called the **neutral surface** of the beam. The intersection between the neutral surface with any cross-sectional plane, e.g., line *tu*, is called the **neutral axis** of the cross section. If the cantilever is made of a homogeneous material with a uniform, symmetric cross section, the neutral plane lies in the middle of the cantilever.

For a beam with symmetry and material homogeneity, the distribution of the stress and strain is observed to follow a number of guidelines:

1. The magnitude of stress and strain at any interior point is linearly proportional to the distance between this point and the neutral axis;
2. On a given cross section, the maximum tensile stress and compressive stress occur at the top and bottom surfaces of the cantilever;
3. The maximum tensile stress and the maximum compressive stress have the same magnitude;
4. Under pure bending, the magnitude of the maximum stress is constant through the length of the beam.

The magnitude of stresses at any location in the beam under pure bending mode can be calculated by following the procedure discussed here. At any section, the distributed stress contributes to distributed force, which subsequently gives rise to a reaction moment (with respect to the neutral axis). The magnitude of the normal stress at a distance h to the neutral plane is denoted $\sigma(h)$. The normal forces acting on any given area dA is denoted $dF(h)$. The force contributes to a moment with respect to the neutral axis. The moment equals the force, $dF(h)$, multiplied by the arm between the force and the neutral plane. The area integral of the moment equals the applied bending moment, according to

$$M = \iint_A dF(h)h = \int_w \int_{h=-\frac{t}{2}}^{\frac{t}{2}} (\sigma(h)dA)h. \tag{3.33}$$

Under the assumption that the magnitude of stress is linearly related to h and is the highest at the surface (denoted σ_{max}), one can rewrite the above equation to yield

$$M = \int_w \int_{h=-\frac{t}{2}}^{\frac{t}{2}} \left(\sigma_{max}\frac{h}{\left(\frac{t}{2}\right)}dA\right)h = \frac{\sigma_{max}}{\left(\frac{t}{2}\right)} \int_w \int_{h=-\frac{t}{2}}^{\frac{t}{2}} h^2\, dA = \frac{\sigma_{max}}{\left(\frac{t}{2}\right)} I. \tag{3.34}$$

The term I is called the **moments of inertia** associated with a particular cross section. The maximum longitudinal strain is expressed as a function of the total torque M according to

$$s_{max} = \frac{Mt}{2EI}. \tag{3.35}$$

Practical cases are often more complex. For the simple loading condition depicted in Figure 3.17, the moment along the beam is not constant. Shear stress components are also present. In this case,

$$s_{max} = \frac{Mt}{2EI}.$$

Equation (3.35) can be applied to each individual cross section. For more details, see Section 6.3.1.

3.4.3 Deflection of Beams

This section will cover methods for analyzing the deflection of beams under simple loading conditions.

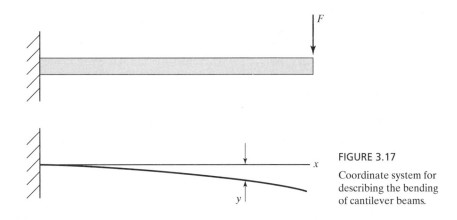

FIGURE 3.17

Coordinate system for describing the bending of cantilever beams.

The general method for calculating the curvature of the beam under small displacement is to solve a second-order differential equation of a beam:

$$EI\frac{d^2y}{dx^2} = M(x), \qquad (3.36)$$

where $M(x)$ represents the bending moment at the cross section at location x, and y represents the displacement at location x. The x-axis runs along the longitudinal direction of the cantilever (Figure 3.17).

The relationship between y and x can be found by solving the second-order differential equation. Solving this equation requires three preparatory steps:

1. Find the moment of inertia with respect to the neutral axis;
2. Find the state of force and torque along the length of a beam;
3. Identify boundary conditions; two boundary conditions are necessary to deterministically find a solution.

The most commonly encountered cross section of a cantilever is a rectangle. Suppose the width and thickness of a rectangle are denoted as w and t, and the moments of inertia with respect to the neutral axis is $I = \frac{wt^3}{12}$ (*provided that the cantilevers bend in the direction of the thickness*). If the cross section of a beam is a circle with radius R, the moment of inertia is $I = \frac{\pi R^4}{4}$.

The torque at any arbitrary location x can be calculated using procedures discussed in Section 3.3.1.

In fact, in many routine MEMS design cases, the interest is to find the maximum displacement associated with a microstructure rather than the deformation profile. The solutions for the maximum angular and transverse displacement for common cases under simple point-loading conditions are summarized in Appendix B.

3.4.4 Finding the Spring Constants

Coiled springs are most commonly found in our daily lives and our macroscale engineering systems. For microscale devices, it is difficult to fabricate and integrate coiled springs. Beams are the most frequently encountered spring element in MEMS. These microbeams serve as mechanical springs for sensing and actuation. The stiffness of these beams is a frequently encountered design concern.

The stiffness is characterized by a term called the **spring constant** (or **force constant**). Here, we will discuss the procedures for calculating the spring constant of various types of beams.

It is necessary to first define the term *spring constant*. We define it here by using familiar coiled springs (Figure 3.18). A coiled spring will extend by an amount x under a pointed loading force F. The displacement and the applied force follow a familiar linear relationship embodied in Hooke's law. The mechanical spring constant is the ratio of the applied force and the resultant displacement

$$k_m = \frac{F}{x}. \qquad (3.37)$$

3.4 Flexural Beam Bending Analysis Under Simple Loading Conditions 79

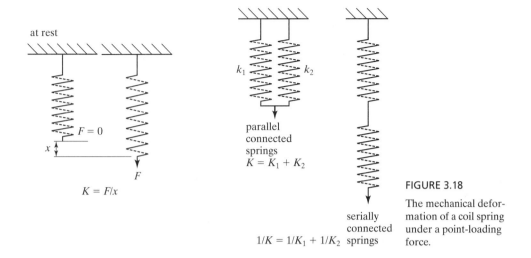

FIGURE 3.18

The mechanical deformation of a coil spring under a point-loading force.

For a cantilever spring, the general expression of a force constant is the force divided by the displacement at a point of interest, commonly the point where the force is applied.

For a cantilever with a point-loading on the free end, the maximum deflection occurs at the free end. For a fixed–fixed bridge with a loading force in the center of the span, the center has the largest deflection.

Let's first focus on analyzing the force constant of a fixed–free beam with a rectangular cross section, one of the most common scenarios encountered in MEMS. The cantilever is diagrammed in Figure 3.19, with the length, width, and thickness of the cantilever beam denoted l, w, and t, respectively. Young's modulus of the beam material along the longitudinal axis is E.

In the case shown in Figure 3.19, the force is applied on the broad surface. The formula for calculating the displacement can be found in Appendix B. The free end of the beam will reach a certain bent angle, θ, with the relationship between θ and F given by

$$\theta = \frac{Fl^2}{2EI}. \tag{3.38}$$

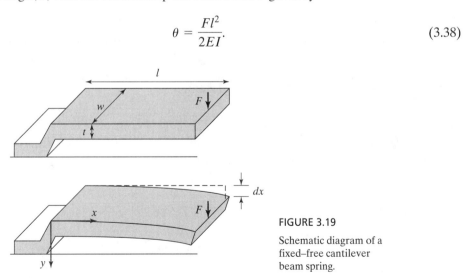

FIGURE 3.19

Schematic diagram of a fixed–free cantilever beam spring.

The resulted vertical displacement equals

$$x = \frac{Fl^3}{3EI}. \tag{3.39}$$

The spring constant of the cantilever is therefore

$$k = \frac{F}{x} = \frac{3EI}{l^3} = \frac{Ewt^3}{4l^3}. \tag{3.40}$$

Apparently, the force constant decreases with increasing length. It is proportional to the width and is strongly influenced by the change in thickness due to the presence of the term t^3.

The stiffness of a cantilever depends on the direction of the bending. If the force is applied longitudinally, the spring constant would be very different. The beam is said to provide compliance in one direction and resistance to movement in another.

Example 3.6. Moments of Inertia of Two Beams

Consider two cantilever beams of the same length and material: one has a cross section of 100 μm by 5 μm, and a second one has a cross section of 50 μm by 8 μm. Which one is more resistant to flexural bending (i.e., stiffer)?

Solution. The moment of inertia of the first beam is

$$I_1 = \frac{wt^3}{12} = \frac{100 \times 10^{-6} \times (5 \times 10^{-6})^3}{12} = 1.04 \times 10^{-21} \text{ m}^4.$$

For the second beam, we have

$$I_1 = \frac{wt^3}{12} = \frac{50 \times 10^{-6} \times (8 \times 10^{-6})^3}{12} = 2.13 \times 10^{-21} \text{ m}^4.$$

Since $I_2 > I_1$, the second beam is stiffer according to Equation (3.40).

Example 3.7 Force Constants of Beams

Find the force constant associated with cases (a) and (b) depicted in Figure 3.20.

Solution. Case (a) is a fixed–free cantilever beam with in-plane transverse loading. The pertinent moment of inertia and spring constant are

$$I = \frac{w^3 t}{12}$$

$$k = \frac{F}{x} = \frac{3EI}{l^3} = \frac{Ew^3 t}{4l^3}.$$

3.4 Flexural Beam Bending Analysis Under Simple Loading Conditions

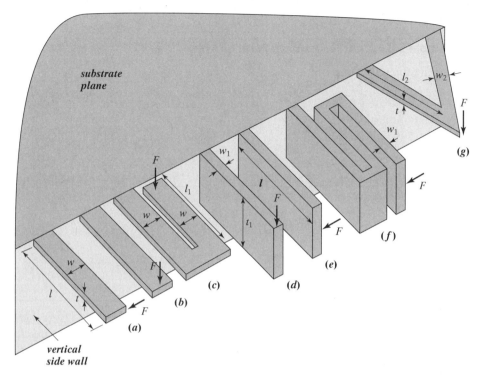

FIGURE 3.20
Calculate spring constant for cases (a) through (g).

Case (b) is a fixed–free cantilever beam with a loading force normal to the substrate. The pertinent moment of inertia and spring constant are

$$I = \frac{wt^3}{12}$$

$$k = \frac{F}{x} = \frac{3EI}{l^3} = \frac{Ewt^3}{4l^3}.$$

In many applications, two or more springs may be connected to form a spring system. They can be connected in two ways: via a parallel or serial connection. If multiple springs are connected in parallel, the total spring constant is the summation of the spring constants of all springs in the system (see Figure 3.18). If multiple coiled springs are connected in series, the inverse of the total spring constant is the summation of the inverse of the spring constants of the constitutive spring elements.

Example 3.8. Cantilevers with Parallel Arms

Find the force constant associated with the case g depicted in Figure 3.20.

Solution. Case (g) consists of two fixed–free cantilevers connected in parallel. The pertinent moment of inertia of each arm is

$$I = \frac{w_2 t^3}{12}.$$

The overall force constant is

$$k = 2\left(\frac{F}{x}\right) = 2\frac{3EI}{l_2^3} = \frac{Ew_2 t^3}{2l_2^3}.$$

Example 3.9. Vertical Translational Plates

Fixed–guided springs are often used to support rigid plates and facilitate their translation. Often, a plate is supported by two or more such beams (Figure 3.21a, b). In these cases, one end of the beam is fixed, with all degrees of freedom limited. Another end of the spring can move in the vertical direction, but no angular displacement is allowed because it is connected to the stiff

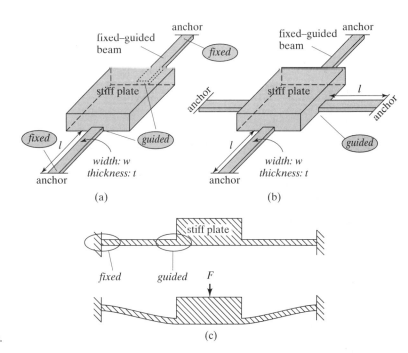

FIGURE 3.21

Commonly encountered plate support configurations.

translational plate, which remains parallel to the substrate under allowable plate movement (Figure 3.21c). Find the expression of the force constant associated with the plate.

Solution. Let's examine the basic formula for the spring constant of a single fixed–guided beam under a transverse force loading F. Once again, the length, width, and thickness of the support beam are l, w, and t, respectively. From Appendix B, the maximum displacement x occurs at the guided end and is related to F by

$$x = \frac{Fl^3}{12EI}. \tag{3.41}$$

The expression for the force constant of each single fixed–guided beam is

$$k = \frac{12EI}{l^3}. \tag{3.42}$$

By reviewing Equation (3.39), it is obvious that for a beam with equal dimensions, the fixed–guided beam would be stiffer than a fixed–free cantilever.

If a force is applied to a plate supported by n cantilevers with equal dimensions and force constants, each spring shares $1/n$th of the total force load. The total force constant experienced by the spring is nk.

The force constant associated with each fixed–guided beam is

$$k = \frac{F}{x} = \frac{12EI}{l^3} = \frac{Ewt^3}{l^3}.$$

For the plate supported by two fixed–guided beams, the equivalent force constant is

$$k = 2\left(\frac{Ewt^3}{l^3}\right).$$

For the plate supported by four fixed–guided beams, the equivalent force constant is

$$k = 4\left(\frac{Ewt^3}{l^3}\right).$$

3.5 TORSIONAL DEFLECTIONS

Whereas beams, especially cantilevers, are frequently used in MEMS for producing linear displacements and small angular rotations, torsional beams are often used to create large angular displacement. A case in mind is the digital micromirror device. The mirror plate is supported by torsional bars to facilitate rotation.

We begin our discussion of torsion by considering a prismatic bar of a circular cross section twisted by torques T acting at the ends. Since every cross section of the bar is identical, and since every cross section is subjected to the same internal torque T, the bar is said to be in pure torsion. It may be proved that cross sections of the bar do not change shape as they rotate about the longitudinal axis. In other words, all cross sections remain plane and circular and all radii remain straight. Furthermore, if the angle of rotation between one end of the bar and the other is small, neither the length of the bar nor its radius will change.

To aid in visualizing the deformation of the bar, imagine that the left-hand end of the bar is fixed in position. Then, under the action of a torque T, the right-hand end will rotate through a small angle ϕ, known as the angle of twist.

Because of this rotation, a straight longitudinal line on the surface of the bar will become a helical curve. The angle of twist changes along the axis of the bar; at intermediate cross sections, it will have a value $\phi(x)$ that is between zero at the left-hand end and ϕ at the right-hand end. The angle $\phi(x)$ will vary linearly between the ends. Point a on the right-hand side will move by a distance d to a new location a'.

The torsion induces shear stress throughout the bar. Stress distribution shows radial symmetry. The magnitude of the shear stress is zero in the center of the cross section and reaches its maximum at the outer surface of the bar. The maximum stress in the bar is denoted τ_{max}. The expression of the maximum shear strain is

$$\gamma_{max} = \frac{d}{L}. \tag{3.43}$$

Further, the magnitude of the stress is linearly proportional to the radial distance to the center. The distribution of shear stress along the radial direction is superimposed on the cross-sectional view of the bar in Figure 3.22.

The relationship between a torque and the maximum shear stress is found by torque balance at any given section

$$T = \int \left(\frac{r}{r_0}\tau_{max}\right) dA \cdot r = \frac{\tau_{max}}{r_0}\int r^2\, dA. \tag{3.44}$$

The surface integral $\int r^2\, dA$ is called the **torsional moment of inertia**, denoted J. For a circular beam $dA = 2\pi r\, dr$, therefore,

$$T = \frac{\tau_{max}}{r_0}\int 2\pi r^3\, dr = \frac{\pi r_0^4}{2}\frac{\tau_{max}}{r_0}. \tag{3.45}$$

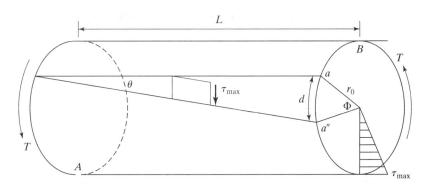

FIGURE 3.22

Torsional bending of a cylinder with circular cross section.

The magnitude of the maximum shear stress is

$$\tau_{max} = \frac{Tr_0}{J}. \quad (3.46)$$

The torsional moment of inertia of a circle with a radius of r_0 is

$$J = \frac{\pi r_0^4}{2}. \quad (3.47)$$

The total angular displacement subtended by the torsional bar segment AB is calculated according to the following:

$$\Phi = \frac{d}{r_0} = \frac{L\theta}{r_0} = \frac{L}{r_0}\frac{\tau_{max}}{G} = \frac{LTr_0}{r_0 GJ} = TL/JG. \quad (3.48)$$

For micromachined devices, torsional bars with rectangular cross sections are often encountered. The moment of inertia for such a torsional bar (with a width and thickness of $2w$ and $2t$, respectively) is

$$J = wt^3\left[\frac{16}{3} - 3.36\frac{t}{w}\left(1 - \frac{t^4}{12w^4}\right)\right] \text{ for } w \geq t. \quad (3.49)$$

The moment of inertia of a beam with a square cross section with the length of each side being $2a$ is

$$J = 2.25a^4. \quad (3.50)$$

Example 3.10. Deformation of Torsional Bars

The suspended beam shown in Fig. 3.23 is under a force $F(F = 10\ \mu N)$. Find the vertical displacement at the end of the beam, assuming the flexural bending of the cantilever beam is negligible. The dimensions of the beam are $L = 40\ \mu m$, $l = 200\ \mu m$, $w = 5\ \mu m$, $t = 2\ \mu m$. Young's modulus of the beam material is $E = 150$ GPa. Poisson's ratio of the beam material is 0.3.

Solution. First find the shear modulus of elasticity and the torsional moment of inertia of the beam. The process is rather straightforward:

$$G = E/2(1 + 0.3) = 57.7 \text{ GPa}$$

$$J = 6.2 \times 10^{-25} \text{ m}^4.$$

Plug the expression for torque, $T = FL$, into the expression for the bending angle to get

$$\phi = \frac{Tl}{2JG} = \frac{FLl}{2JG}.$$

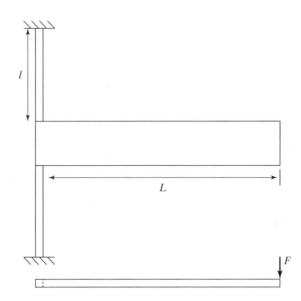

FIGURE 3.23

A torsional supported cantilever.

Plug in the numbers, and we have

$$\Phi = 1.13 \text{ rad} = 65°.$$

The vertical displacement at the free end of the cantilever is equal to the product of the angular displacement and the length of the cantilever. Namely,

$$d_{torsional} = \phi L = \frac{TlL}{2JG} = \frac{FL^2l}{2JG}.$$

3.6 INTRINSIC STRESS

Many thin-film materials experience internal stress even when they are under room temperature and zero external loading conditions. This phenomenon is called the intrinsic stress. Thin-film materials associated with MEMS (such as polysilicon, silicon nitride, and many metal thin films) exhibit intrinsic stresses [32]. The magnitude of the intrinsic stress may be identical or non-uniform throughout the thickness of the thin film. If the stress distribution is non-uniform, a stress gradient is said to be present.

The intrinsic stress is important for MEMS devices because it can cause deformation—damages in excessive cases can affect surface planarity or change the stiffness of a mechanical element. For example, in microoptics applications, flat mirror surfaces are required to achieve desired optical performances. Intrinsic stress may warp the mirror surface and change optical properties. Bent cantilever structures due to the presence of intrinsic stress are shown in Figure 3.24.

Intrinsic stress may affect the mechanical behavior of membranes as well. For a thin-film diaphragm such as the one illustrated in Figure 3.25, the flatness of the membrane is guaranteed when the membrane material is under tensile stress. Excessive tensile stresses in a clamped

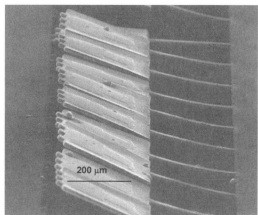

FIGURE 3.24

Microscale devices bent under intrinsic stress.

membrane can cause the membrane to fracture. On the other hand, a film would buckle if a compressive stress were present.

The origin and behavior of intrinsic stress have been widely studied [33, 34]. In many cases related to MEMS structures, the intrinsic stress results from the temperature difference during deposition and use. Thin-film materials are often deposited on a substrate under an elevated

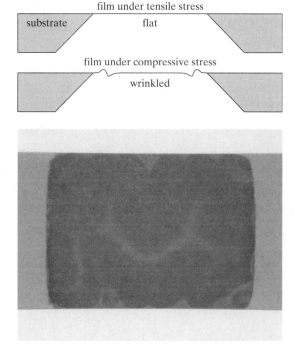

FIGURE 3.25

Cross section of a taut membrane under tensile stress and a warped membrane under compressive stress. An optical micrograph of a warped membrane is shown.

temperature. During the deposition, molecules are incorporated into a thin film with certain equilibrium spacing. However, when the MEMS device is removed from the deposition chamber, the temperature change causes the material to contract at a rate faster or slower than the substrate. As a rule of thumb, the intrinsic stress in a thin film is tensile when a film wants to become smaller than the substrate. Compressive stress results when a film wants to become larger than the substrate.

Intrinsic stress can also result from the microstructure of the deposited film. The incorporation of oxygen atoms into a silicon lattice during a thermal oxidation process, for example, creates compressive intrinsic stress in the oxide film.

Apart from the aforementioned common sources of intrinsic stress, other mechanisms are possible, including phase change of materials and incorporation of impurity atoms (such as dopant atoms).

In certain cases, the bending is desired and intentional. There are a number of application scenarios in which the intrinsic stress can be used to induce out-of-plane deformation to realize unique device architecture.

A common scenario of intrinsic stress-induced beam bending involves a microstructure with two or more structural layers (Figure 3.26). Consider a cantilever consisting of two layers, with the intrinsic stress in Layer 1 being zero. If Layer 2 is under a tensile stress, the beam would bend towards the direction of Layer 2 (Figure 3.26b). On the other hand, if Layer 2 is under a compressive intrinsic stress, the beam would bend towards the direction of Layer 1 (Figure 3.26c).

We will discuss the formulation used to calculate the bending of a two-layer cantilever beam under the difference of intrinsic stress. For simplicity, we assume that the two layers have the same length (l) and width (w). The thickness, intrinsic hydrostatic stress, and Young's modulus of each layer are denoted as t_i, σ_i, and E_i ($i = 1$ or 2). The subscript letters correspond to Layers 1 or 2, with Layer 2 on the bottom.

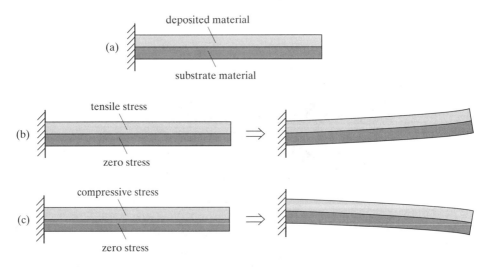

FIGURE 3.26

Beam bending due to intrinsic mechanical stress.

First, the position of the neutral axis is found by

$$\bar{y} = \frac{\frac{1}{2}(E_1 t_1^2 + E_2 t_2^2) + E_2 t_1 t_2}{E_1 t_1 + E_2 t_2}. \quad (3.51)$$

The distance is measured from the bottom of the bottom layer. An effective flexural stiffness is calculated by using

$$I_{eff} E_0 = w\left(E_1 t_1\left(\frac{t_1^2}{12} + \left(\frac{t_1}{2} - \bar{y}\right)^2\right) + E_2 t_2\left(\frac{t_2^2}{12} + \left(\frac{t_2}{2} + t_1 - \bar{y}\right)^2\right)\right). \quad (3.52)$$

The bending moment acting on the cantilever is

$$M = w\left[\left(\frac{t_1^2}{2}\right)\left((\sigma_1(1 - \nu_1) - E_1 \frac{t_1\sigma_1(1 - \nu_1) + t_2\sigma_2(1 - \nu_2)}{E_1 t_1 + E_2 t_2}\right)\right] +$$

$$w\left[\left(\frac{t_2^2 + t_1 t_2}{2}\right)\left((\sigma_2(1 - \nu_2) - E_2 \frac{t_1\sigma_1(1 - \nu_1) + t_2\sigma_2(1 - \nu_2)}{E_1 t_1 + E_2 t_2}\right)\right]. \quad (3.53)$$

The radius of curvature of the beam bending, R, is given by

$$R = \frac{I_{eff} E_0}{M}. \quad (3.54)$$

We would imagine that if a fixed–free cantilever is made of a single homogeneous material, there would be no intrinsic stress nor intrinsic stress-induced bending. This assumption is only true if the intrinsic stress is uniformly distributed throughout the thickness. A gradient of intrinsic stress may cause a cantilever made of a single material to bend, as if the cantilever is made of a large number of thin layers stacked together.

There are three strategies for minimizing undesirable intrinsic bending: (1) use materials that inherently have zero or very low intrinsic stress; (2) for materials whose intrinsic stress depends on material processing parameters, fine tune the stress by calibrating and controlling deposition conditions; (3) use multiple-layered structures to compensate for stress-induced bending.

One of the materials in MEMS that has zero stress is single-crystal bulk silicon. The SCS is a homogeneous material with perfect lattice spacing distribution. Polymer materials are known to have relatively smaller intrinsic stresses than inorganic thin films. Second, the deposition temperature of polymer materials is typically low. Parylene, a polymer material that can be deposited by chemical vapor deposition method under room temperature, has virtually zero stress in it [35].

In many surface micromachining materials, such as polysilicon and silicon nitride, the intrinsic stress is unavoidable (Chapter 10). Although techniques exist to produce stress-free materials, the process is usually delicate, machine specific, and subjected to complex process control. The best practice is to minimize the intrinsic stress-induced bending to an acceptable level. Generally, intrinsic stress in silicon nitride or polycrystalline silicon can be minimized by controlling the pressure, gas mixture, and growth rate [36].

Often, the intrinsic stress of layers cannot be fully cancelled by material processing. It is still possible to produce microstructures free of intrinsic stress-induced bending, by adding stress compensation layers with carefully selected materials, stress, and thickness. For example, in the two-layer beam shown in Figure 3.27, the tensile-stressed upper layer and the compressive-stressed lower layer contribute to a finite amount of beam bending. The overall deformation can be reduced to zero in theory by adding another layer of compressive-stressed material on top.

The intrinsic stress of a material depends strongly on the material deposition conditions. The exact magnitude of intrinsic stress in a given process flow can be experimentally measured by using specially designed testing structures and test protocols [37, 38]. A common technique for measuring the stress is to use a circular plate (such as a wafer) bearing the thin film of interest on one side (Figure 3.28) [39]. The stress in the thin film can be derived from the curvature (bowing) of the plate. A classic formula (called Stoney's formula) has been derived under the conditions that the substrate is thin, elastically isotropic, and flat when bare [39]. For a single-film case, the curvature of the substrate (C) is expressed as

$$C = \frac{1}{R} = \frac{6(1-\nu)\sigma h}{Et^2}, \tag{3.55}$$

where R is the radius of curvature, ν is Poisson's ratio of the substrate, σ is the in-plane stress of the thin film, t is the substrate thickness, and h is the thickness of the thin film. The thin-film

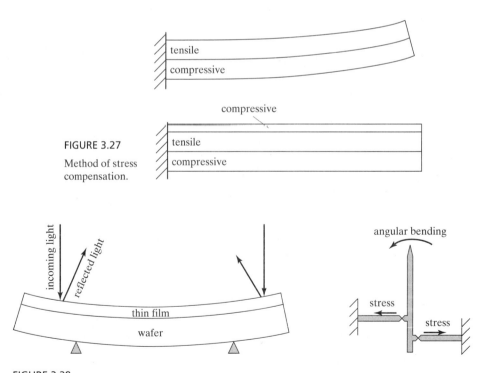

FIGURE 3.27

Method of stress compensation.

FIGURE 3.28

Experimental methods for measuring intrinsic stress.

material should be present on only one side. If the deposition process dictates double-side coating, then the material on one side should be carefully removed.

This method only measures the overall averaged stress level over the entire wafer area. In most cases, the distribution of stress over the wafer is not uniform. The stress within local regions of the wafers is needed for precise control of the process. Surface and bulk micromachined test structures called process monitors are often used. One of the monitoring devices is illustrated in Figure 3.28. It consists of a suspended thin-film dial that is attached to the substrate through two offset pulling beams. The two horizontal pulling beams, when under intrinsic stress, develop a torque that causes the dial to rotate for a certain amount.

3.7 RESONANT FREQUENCY AND QUALITY FACTOR

A typical frequency spectrum of the displacement of a mechanical structure under a periodic loading condition is shown in Figure 3.29. At low frequencies, the displacement remains constant. This represents the low-frequency behavior that applies to the steady loading case. At or near the **resonant frequency** (f_r), the mechanical vibration amplitude sharply increases. The sharpness of the resonant peak is characterized by a term called the **quality factor** (Q). The sharper the resonance peak, the higher the quality factor. The amplitude magnification at the resonant frequency can be beneficial. Operating a microsensor or actuator at the resonant frequency may increase the sensitivity or range of actuation. However, resonance may also lead to the self-destruction of mechanical elements.

At frequencies beyond the resonant frequency, the vibration amplitude decreases (called roll off). A simple physical explanation is that the mechanical structures, with built-in inertia, cannot follow the driving input at frequencies significantly above the f_r.

The quality factor, representing the sharpness of resonance peaks, can be defined in several ways. From the energy point of view, it is the ratio of the total stored energy in a system over the energy lost over each cycle of oscillation. The lower the energy loss per cycle,

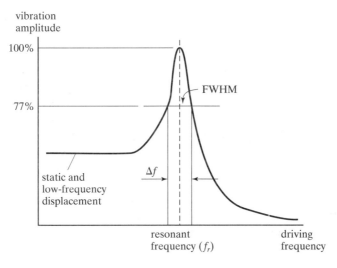

FIGURE 3.29

A typical response spectrum showing the relationship of the amplitude of vibration as a function of the input frequency.

the greater the quality factor. Some common forms of energy loss include the following: (1) mechanical energy dissipation into the substrate through anchors; (2) mechanical damping by surrounding media (e.g., air); and (3) thermal elastic energy dissipation (TED) or internal friction.

Thus, the quality factor of a microdevice can be improved by reducing the operating pressure [10, 40], by altering the operating temperature [10], by improving surface roughness [41], by thermal annealing (up to a 600% improvement demonstrated in [42]), or by modifying the boundary conditions [43, 44]. Q values for micromechanical devices ranging from several hundred [45] to up to 10,000 [44–46] have been demonstrated in micro- or nanomechanical resonators.

Mechanical resonance can be significantly dampened by the presence of air or liquid media surrounding a resonant structure. The resonance curve of an overly dampened element may appear to have no resonance peak at all. If the resonant structure is physically adjacent to a stationary object, significant damping would occur due to the **squeezed film damping** phenomenon. The squeezed film damping can be observed in daily life. If we drop a piece of paper on a flat table, the paper may slide with the trapped air between the paper and the table top, which serves as an air cushion.

Mathematically, the quality factor is related to the full width at half maximum (FWHM), which is the spacing of two frequencies at half power (or 77% amplitude). The ratio of the resonant frequency and the FWHM gives the quality factor:

$$Q = \frac{f_r}{\Delta f}. \tag{3.56}$$

The value of the resonant frequency is of interest in many engineering designs. The simplest case of a mechanical resonant system is a discrete system consisting of a mass m attached to a spring with a force constant k. It is easily proven that the resonant frequency of such a system is given by

$$\frac{1}{2\pi}\sqrt{\frac{k}{m}}.$$

The formula for calculating the resonant frequency of cantilevers under various boundary and loading conditions can be found in Appendix B. For most cases, the first harmonics frequency is the one of most interest.

The resonant frequency of mechanical elements generally increases when scaled down as the value of m decreases rapidly. By down-scaling resonator device dimensions, resonant frequency ranging from several MHz [46], tens of MHz [44], and even to the GHz range [41] has been demonstrated successfully.

Because the resonance frequency is a function of the device dimensions, it is susceptible to changes in temperature. Temperature stabilization is critical for resonators requiring stable frequency. Temperature stability can be achieved by temperature compensation [47].

Additionally, the resonant frequency of microresonators can be precisely tuned by trimming materials (using laser or focused ion-beam etching) or by locally depositing materials [48].

Example 3.11. Resonant Frequency

A mechanical resonator (fixed–fixed, or double-clamped) has been demonstrated using SiC thin-film material. The length (L), width (W), and thickness (T) of the resonator are 1.1 μm, 120 nm, and 75 nm, respectively. Assuming that the resonant frequency found experimentally was 1.014 GHz, and assuming a Young's modulus of 700 GPa, find the density of the SiC material used for the resonator.

Solution. The formula for the resonant frequency according to the table in Appendix B is

$$f_n = \frac{22.4}{2\pi}\sqrt{\frac{EIg}{wL^4}}.$$

The distributed force w is given by

$$w = \frac{\rho(WLT)g}{L} = \rho g WT.$$

By plugging the formula for the resonant frequency into

$$f_n = \frac{22.4}{2\pi}\sqrt{\frac{EIg}{\rho g WTL^4}} = \frac{22.4}{2\pi}\sqrt{\frac{EWT^3 g}{12\rho g WTL^4}} = \frac{22.4}{2\pi}\sqrt{\frac{ET^2}{12\rho L^4}} = 8.6 \times 10^5 \frac{T}{\sqrt{\rho}L^2},$$

we find the value of ρ is

$$\rho = \left(\frac{8.6 \times 10^5}{f_n}\cdot\frac{T}{L^2}\right)^2 = 52.66^2 = 2773 \text{ kg/m}^3.$$

3.8 ACTIVE TUNING OF THE SPRING CONSTANT AND RESONANT FREQUENCY

Mechanical characteristics, such as spring constants and the resonant frequencies of beams and membranes, can be changed by introducing strain. In this section, we focus on methods for tuning characteristics of cantilever beams to exemplify the methodology.

For cantilevers, longitudinal strain can be introduced by axial or transverse-loading forces. We will discuss these two cases separately.

First, an axial tension can alter the spring constant and the resonant frequency. This is analogous to a violinist tuning the sound of a string by adjusting its tension. The longitudinal tuning force can be introduced by lateral electrostatic force or thermal expansion. (By the same token, the shift of resonance frequency can be used to characterize the intrinsic stress of fixed–fixed beams [49].) The resonant frequency of a cantilever under a longitudinal strain ε_s is

$$w = w_0\sqrt{1 + \frac{2L^2}{7h^2}\varepsilon_s}, \tag{3.57}$$

where w_0, h, and L are the untuned resonant frequency, the thickness of the beam, and the length of the beam.

The force constant and resonance frequency of a cantilever can be tuned by transverse forces. To illustrate this case, think about a child standing at the end of a diving board and bouncing lightly. Assume that the board dips below the horizontal level. The child feels the board has a certain mechanical stiffness.

Now, imagine an adult pulling at the end of the diving board. The child will feel that the diving board becomes softer (more displacement under a given weight) when there is an active pulling force. In other words, the diving board appears to be softer to the child. Beam softening results whenever the bias force points in the opposite direction of the cantilever restoring force. The apparent force constant changes from k to k_{eff}. The transverse-loading force may be applied electrostatically [50, 51], thermally [52], or magnetically.

3.9 A LIST OF SUGGESTED COURSES AND BOOKS

For readers who wish to build a more comprehensive foundation of knowledge for pursuing future studies, research, and development in the field of MEMS, we provide a list of 15 closely related courses and representative entry-level textbooks. Information discussed in these courses and books has a strong relevance to MEMS.

Mechanical Engineering

1. Mechanics of materials [9]
2. Theory of elasticity [53]
3. Vibration analysis [54]
4. Thermal transfer [55]
5. Fluid mechanics [56]

Electrical Engineering

1. Fundamentals of electrical and electronic circuits [57]
2. Field and waves [58]
3. General electronic circuit design
4. Physics of semiconductor and solid-state devices [2]
5. Integrated circuit design [59]
6. Electromechanics [60]

Microfabrication

1. Integrated circuit fabrication [61]
2. MEMS fabrication [62, 63]

Material Science

1. Broad and comprehensive discussions of different classes of engineering materials, properties, and processing techniques [64]
2. Material properties of polycrystalline silicon [16]

SUMMARY

This chapter provides a general overview of the major electrical and mechanical aspects that are frequently used in the development of MEMS devices. The goal of this chapter is help readers from different backgrounds reach a common level, and help them become aware of the major issues and major analytical procedures. The reader is encouraged to read textbooks on each individual topic to obtain a more complete understanding.

The following list summarizes the major concepts, facts, and skills associated with this chapter:

- The origin of two types of charge carriers in a semiconductor.
- Procedures for calculating the concentrations of electrons and holes in an extrinsic material when the doping concentrations are known.
- Procedures for finding the conductivity and resistivity of a semiconductor piece.
- The difference and relation between resistivity and sheet resistivity.
- The naming conventions of crystal planes and directions.
- The concepts of normal stress and strain.
- The concepts of shear stress and strain.
- The relationship between normal stress and strain and the various regimes of interest.
- Identification of boundary conditions associated with a beam.
- Procedures for calculating the end deflection of a cantilever under a given force.
- Procedures for calculating the force constant associated with a mechanical beam.
- Procedures for calculating the torsional bending of a torsion bar.
- The origin of intrinsic stress.
- Methods for reducing bending induced by intrinsic stress.
- The formula for calculating the resonant frequency of beams under certain boundary conditions.

PROBLEMS

Problem 3.1: Review
Calculate the volumetric concentration of atoms (atoms/cm^3) and specific mass (density) of single-crystal silicon given the lattice constant of silicon (5.43 Å).

Problem 3.2: Review
Write the compliance matrix for single-crystal silicon.

Problem 3.3: Design
A piece of gold resistor is used as a heater. The thickness of the resistor is 0.1 μm. The width of the resistor is 1 μm. What is the necessary length if a total resistance of 50 ohms is to be achieved? (*Hint*: Find the resistivity of gold from literature or handbooks, and reference the source of information).

Problem 3.4: Design
A phosphorous-doped silicon resistor is 100 μm long, 2 μm wide, and 0.5 μm thick. The doping concentration is 10^{17} atoms/cm^3. The electron mobility (μ_n), which is a function of the doping concentration, is

approximately 1350 cm^2/(V · s). The hole mobility is approximately 480 cm^2/(V · s). What are the concentrations of electrons and holes under thermal equilibrium at room temperature? Find the resistivity of the material and the total resistance.

Problem 3.5: Design
Repeat Problem 3.4 assuming that the resistor is doped with boron. In this case, holes will be the majority carrier.

Problem 3.6: Design
Repeat Problem 3.4 assuming that the doping concentration of phosphorous is 10^{11} atoms/cm^3.

Problem 3.7: Design
A silicon wafer is doped by boron ion implantation and thermal activation. Ion implantation is conducted at a dose of 10^{14} atoms/cm^2. The depth of the doped region (junction depth) is determined by the original depth of high-energy ion penetration and thermal diffusion during the activation step. Assume the junction depth is 1 μm and the concentration is uniform through the depth of the doped layer. What is the doping concentration?

Problem 3.8: Design
A boron-doped silicon resistor has a sheet resistivity of 50 $\Omega/\square$. The length and width of the resistor are 10 μm and 0.5 μm, respectively. Find the total resistance of the resistor. If the thickness of the resistor is known to be 0.3 μm, what is the resistivity and the boron-doping concentration? What are the concentrations of the majority carriers under thermal equilibrium at room temperature?

Problem 3.9: Design
Discuss the detailed steps to name the following shaded planes ({110}, {111}, and {210}).

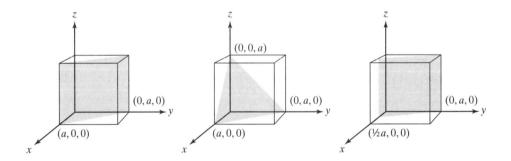

Problem 3.10: Design
Consider two parallel planes that are both parallel to the plane formed by the x and z axes. Prove that their Miller Indices are both (010).

Problem 3.11: Design
Prove that the (111) surface intercepts the (100) surface at an inclination angle of approximately 54.75°.

Problem 3.12: Design
An axial force with a magnitude of U is applied at the end of a bar attached to a wall. Find the reactive force and torque (if any) at the anchored end of the bar and at section A.

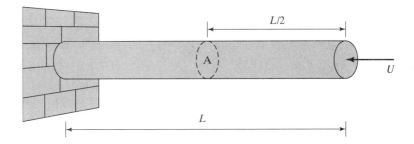

Problem 3.13: Design

Two independent axial forces with magnitudes of U and F are applied at the end of a bar attached to a wall. Find the reactive force and torque (if any) at the anchored end of the bar and at section A. Assume the overall reaction under a combined force is the linear sum of the reactions under two forces acting individually.

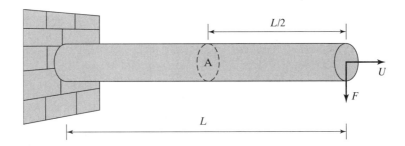

Problem 3.14: Design

An axial torque with a magnitude of M is applied at the end of a bar attached to a wall. Find the reactive force and torque (if any) at the anchored end of the bar and at section A.

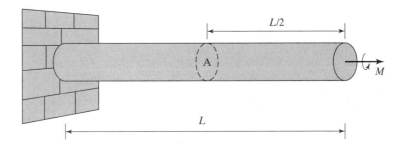

Problem 3.15: Design

A slender silicon beam is under a longitudinal tensile stress. The force is 1 mN, and the cross-sectional area is 20 μm by 1 μm. Young's modulus in the longitudinal direction is 120 GPa. Find the relative elongation of the beam (percentage). What is the force necessary to fracture the beam if the fracture strain of silicon is 0.3%?

Problem 3.16: Design

A silicon cube with a volume of 1 cm³ is placed on a surface. A force of 1 mN is applied vertically on the face. Find the type of stress and the magnitude of stress induced in the direction of the applied force.

Problem 3.17: Design

Assume the width and thickness of a silicon bar are 5 μm and 1 μm, respectively. Find the maximum stress that can be applied to the longitudinal direction of the bar if the maximum fracture strain is 0.2% and Young's modulus is 140 GPa.

Problem 3.18: Design

Find the moment of inertia of the cantilever beams shown below. The material is made of single-crystal silicon. Young's modulus in the longitudinal direction of the cantilever is 140 GPa.

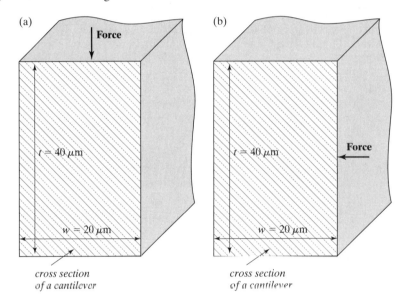

Problem 3.19: Design

Find the force constant of the beam in Problem 3.18 if a force is applied in the longitudinal direction of the cantilever. The beam is 800 μm long in this case.

Problem 3.20: Design

A single-crystal silicon bar is 100 μm long, 5 μm wide, and 1 μm thick at room temperature. The bar is doped to a uniform P concentration of 10^{16} atoms/cm³. Find the moment of inertia and the electrical resistance of the silicon bar. Young's modulus of the silicon bar is 150 GPa and the fracture strain is 0.2%. What is the amount of elongation as a result of a force of 1 mN being applied in the longitudinal direction?

Problem 3.21: Design

The thermal expansion coefficient of single-crystal silicon is approximately 2.6×10^{-6}/°C. A 10°C rise from the room temperature would cause the length and cross section of the beam to change. Calculate the longitudinal strain associated with the lateral elongation, and the equivalent longitudinal stress (considering the changed cross section). What is the percentage change of resistance due to this temperature rise (assuming the resistivity stays the same)? (*Note*: In fact, the resistivity of the silicon will change more because of the change of crystal lattice spacing.)

Problem 3.22: Design

What is the scaling law for the moment of inertia of a flexural beam? What is the implication for MEMS sensors? What is the implication for MEMS actuators?

Problem 3.23: Design

Find the analytical expression of the force constant associated with cases (d), (e), and (f) of Example 3.7.

Problem 3.24: Design

A torsional bar is anchored on two ends with a lever attached in the middle of the bar. A force, $F = 1\ \mu N$, is applied to the end of the lever. Determine the degree of angular bending due to the rotation of the torsional bars. Do not consider the bending of the flexural lever segment. The values of L, w, and t are 1000, 10, and 10 μm, respectively. The beam is made of polycrystalline silicon. (*Hint*: Find Young's modulus and Poisson's ratio from the literature and handbook, and cite the source of information.)

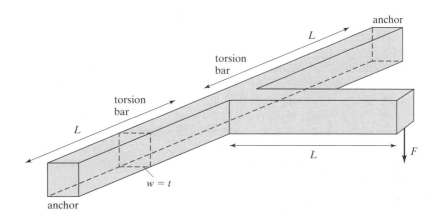

Problem 3.25: Design

Prove that the force constant of a fixed–fixed beam with a central point force loading can be estimated as two fixed–guided beams connected in parallel, each being half as long as the fixed–fixed beam with the same cross-sectional area.

Problem 3.26: Design

To build a 20-μm wide fixed–free cantilever beam with a force constant of 10 N/m and a resonant frequency of 10 kHz out of single-crystal silicon, find the desired length and thickness of the cantilever beam. Young's modulus of silicon is 120 GPa. The density of the silicon material is 2330 kg/m^3. Selected the correct answer from the choices below and explain the reasoning behind your choice:

1. Length = 6.4 mm and thickness = 351 μm
2. Length = 2.9 mm and thickness = 75.7 μm
3. Length = 143 mm and thickness = 3.65 mm
4. None of the above

REFERENCES

1. Streetman, B.G., *Solid State Electronic Devices*. 4th ed. Prentice Hall Series in Solid State Physical Electronics, ed. J. Nick Holonyak. 1995, Englewood Cliffs, NJ: Prentice Hall.

2. Muller, R.S., T.I. Kamins, and M. Chan, *Device Electronics for Integrated Circuits*. 3rd ed. 2003: John Wiley and Sons.
3. Mehregany, M., et al., *Silicon Carbide MEMS for Harsh Environments*. Proceedings of the IEEE, 1998. **86**(8): p. 1594–1609.
4. Stoldt, C.R., et al., *A low-Temperature CVD Process for Silicon Carbide MEMS*. Sensors and Actuators A: Physical, 2002. **97–98**: p. 410–415.
5. Rogers, J.A., *Electronics: Toward Paperlike Displays*. Science, 2001. **291**(5508): p. 1502–1503.
6. Blanchet, G.B., et al., *Large Area, High Resolution, Dry Printing of Conducting Polymers for Organic Electronics*. Applied Physics Letters, 2003. **82**(3): p. 463–465.
7. Sundar, V.C., et al., *Elastomeric Transistor Stamps: Reversible Probing of Charge Transport in Organic Crystals*. Science, 2004. **303**(5664): p. 1644–1646.
8. Mantooth, B.A., and P.S. Weiss, *Fabrication, Assembly, and Characterization of Molecular Electronic Components*. Proceedings of the IEEE, 2003. **91**(11): p. 1785–1802.
9. Gere, J.M., and S.P. Timoshenko, *Mechanics of Materials*. 4th ed. 1997, New York: PWS Publishing Company.
10. Yasumura, K.Y., et al., *Quality Factors in Micron- and Submicron-Thick Cantilevers*. Microelectromechanical Systems, Journal of, 2000. **9**(1): p. 117–125.
11. Van Arsdell, W.W., and S.B. Brown, *Subcritical Crack Growth in Silicon MEMS*. Microelectromechanical Systems, Journal of, 1999. **8**(3): p. 319–327.
12. Wilson, C.J., and P.A. Beck, *Fracture Testing of Bulk Silicon Microcantilever Beams Subjected to a Side Load*. Microelectromechanical Systems, Journal of, 1996. **5**(3): p. 142–150.
13. Namazu, T., Y. Isono, and T. Tanaka, *Evaluation of Size Effect on Mechanical Properties of Single Crystal Silicon by Nanoscale Bending Test Using AFM*. Microelectromechanical Systems, Journal of, 2000. **9**(4): p. 450–459.
14. Yi, T., L. Li, and C.-J. Kim, *Microscale Material Testing of Single Crystalline Silicon: Process Effects on Surface Morphology and Tensile Strength*. Sensors and Actuators A: Physical, 2000. **83**(1–3): p. 172–178.
15. Wortman, J.J., and R.A. Evans, *Young's Modulus, Shear Modulus, and Poisson's Ratio in silicon and Germanium*. Journal of Applied Physics, 1965. **36**(1): p. 153–156.
16. Kamins, T., *Polycrystalline silicon for Integrated Circuits and Displays*. 2nd ed. 1998: Kluwer Academic Publishers.
17. Sharpe, W.N., Jr., et al., *Effect of Specimen Size on Young's modulus and Fracture Strength of Polysilicon*. Microelectromechanical Systems, Journal of, 2001. **10**(3): p. 317–326.
18. Chasiotis, I., and W.G. Knauss, *Experimentation at the micron and Submicron Scale*. In *Interfacial and Nanoscale Fracture*, W. Gerberich and W. Yang, Eds. 2003, Elsevier. p. 41–87.
19. Bagdahn, J., W.N. Sharpe, Jr., and O. Jadaan, *Fracture Strength of Polysilicon at Stress Concentrations*. Microelectromechanical Systems, Journal of, 2003. **12**(3): p. 302–312.
20. Tsuchiya, T., et al., *Specimen Size Effect on Tensile Strength of Surface-Micromachined Polycrystalline Silicon Thin Films*. Microelectromechanical Systems, Journal of, 1998. **7**(1): p. 106–113.
21. Sharpe, W.N., *Tensile Testing at the Micrometer Scale (Opportunities in Experimental Mechanics)*. Experimental Mechanics, 2003. **43**: p. 228–237.
22. Johansson, S., et al., *Fracture Testing of Silicon Microelements In Situ in a Scanning Electron Microscope*. Journal of Applied Physics, 1988. **63**(10): p. 4799–4803.
23. Pan, C.S., and W. Hsu, *A Microstructure for In Situ Determination of Residual Strain*. Microelectromechanical Systems, Journal of, 1999. **8**(2): p. 200–207.
24. Sharpe, W.N., Jr., B. Yuan, and R.L. Edwards, *A New Technique for Measuring the Mechanical Properties of Thin Films*. Microelectromechanical Systems, Journal of, 1997. **6**(3): p. 193–199.

25. Haque, M.A., and M.T.A. Saif, *Microscale Materials Testing Using MEMS Actuators*. Microelectromechanical Systems, Journal of, 2001. **10**(1): p. 146–152.
26. Srikar, V.T., and S.D. Senturia, *The Reliability of Microelectromechanical Systems (MEMS) in Shock Environments*. Microelectromechanical Systems, Journal of, 2002. **11**(3): p. 206–214.
27. Miyajima, H., et al., *A Durable, Shock-Resistant Electromagnetic Optical Scanner with Polyimide-Based Hinges*. Microelectromechanical Systems, Journal of, 2001. **10**(3): p. 418–424.
28. Ritchie, R.O., *Mechanism of Fatigue-Crack Propagation in Ductile and Brittle Solids*. International Journal of Fracture, 1999. **100**: p. 55–83.
29. Muhlstein, C.L., S.B. Brown, and R.O. Ritchie, *High-Cycle Fatigue of Single-Crystal Silicon Thin Films*. Microelectromechanical Systems, Journal of, 2001. **10**(4): p. 593–600.
30. Sharpe, W.N., and J. Bagdahn, *Fatigue Testing of Polysilicon—A Review*. Mechanics and Materials, 2004. **38**: p. 3–11.
31. Chan, R., et al., *Low-Actuation Voltage RF MEMS Shunt Switch with Cold Switching Lifetime of Seven Billion Cycles*. IEEE/ASME Journal of Microelectromechanical Systems (JMEMS), 2003. **12**(5): p. 713–719.
32. Hu, S.M., *Stress-Related Problems in Silicon Technology*. Journal of Applied Physics, 1991. **70**(6): p. R53–R80.
33. Doerner, M., and W. Nix, *Stresses and Deformation Processes in Thin Films on Substrates*. CRC Critical Review Solid States Materials Science, 1988. **14**(3): p. 225–268.
34. Nix, W.D., *Elastic and Plastic Properties of Thin Films on Substrates: Nanoindentation Techniques*. Materials Science and Engineering A, 1997. **234–236**: p. 37–44.
35. Harder, T.A., et al. *Residual Stress in Thin Film Parylene-C*. In *The Sixteens Annual International Conference on Micro Electro Mechanical Systems*. 2002. Las Vegas, NV.
36. Mastrangelo, C.H., Y.-C. Tai, and R.S. Muller, *Thermophysical Properties of Low-Residual Stress, Silicon-Rich, LPCVD Silicon Nitride Films*. Sensors and Actuators A: Physical, 1990. **23**(1–3): p. 856–860.
37. Askraba, S., L.D. Cussen, and J. Szajman, *A Novel Technique for the Measurement of Stress in Thin Metallic Films*. Measurement Science and Technology, 1996. **7**: p. 939–943.
38. Chen, S., et al., *A New In Situ Residual Stress Measurement Method for a MEMS Thin Fixed–Fixed Beam Structure*. Microelectromechanical Systems, Journal of, 2002. **11**(4): p. 309–316.
39. Von Preissig, F.J., *Applicability of the Classical Curvature-Stress Relation for Thin Films on Plate Substrates*. Journal of Applied Physics, 1989. **66**(9): p. 4262–4268.
40. Cheng, Y.-T., et al., *Vacuum Packaging Technology Using Localized Aluminum/Silicon-to-Glass Bonding*. Microelectromechanical Systems, Journal of, 2002. **11**(5): p. 556–565.
41. Huang, X.M.H., et al., *Nanoelectromechanical Systems: Nanodevice Motion at Microwave Frequencies*. Nature, 2003. **421**: p. 496.
42. Wang, K., et al. *Frequency trimming and Q-Factor Enhancement of Micromechanical Resonators via Localized Filament Annealing*. In *Solid State Sensors and Actuators, 1997. Transducers '97 Chicago, International Conference*. 1997.
43. Wang, K., et al. *Q-Enhancement of Microelectromechanical Filters via Low-Velocity Spring Coupling*. In *Ultrasonics Symposium, IEEE Proceedings*. 1997.
44. Wang, K., A.-C. Wong, and C.T.-C. Nguyen, *VHF Free-Free Beam High-Q Micromechanical Resonators*. Microelectromechanical Systems, Journal of, 2000. **9**(3): p. 347–360.
45. Bannon, F.D., J.R. Clark, and C.T.-C. Nguyen, *High-Q HF Microelectromechanical Filters*. Solid-State Circuits, IEEE Journal of, 2000. **35**(4): p. 512–526.
46. Cleland, A.N., and M.L. Roukes, *External Control of Dissipation in a Nanometer-Scale Radiofrequency Mechanical Resonator*. Sensors and Actuators A: Physical, 1999. **72**(3): p. 256–261.

47. Hsu, W.-T., J.R. Clark, and C.T.-C. Nguyen. *Mechanically Temperature-Compensated Flexural-Mode Micromechanical Resonators*. In *Electron Devices Meeting, IEDM Technical Digest. International*. 2000.
48. Joachim, D., and L. Lin, *Characterization of Selective Polysilicon Deposition for MEMS Resonator Tuning*. Journal of Microelectromechanical Systems, 2003. **12**(2): p. 193–200.
49. Chen, S., et al., *A New In Situ Residual Stress Measurement Method for a MEMS Thin Fixed–Fixed Beam Structure*. IEEE/ASME Journal of Microelectromechanical Systems (JMEMS), 2003. **11**(4): p. 309–316.
50. Yao, J.J,. and N.C. MacDonald, *A Micromachined, Single-Crystal Silicon, Tunable Resonator*. Journal of Micromechanics and Microengineering, 1995. **6**(3): p. 257–264.
51. Adams, S.G., et al., *Independent Tuning of Linear and Nonlinear Stiffness Coefficients [Actuators]*. Microelectromechanical Systems, Journal of, 1998. **7**(2): p. 172–180.
52. Syms, R.R.A., *Electrothermal Frequency Tuning of Folded and Coupled Vibrating Micromechanical Resonators*. Microelectromechanical Systems, Journal of, 1998. **7**(2): p. 164–171.
53. Timoshenko, S., and J. Goodier, *Theory of Elasticity*. 3rd ed. 1970, New York: McGraw-Hill.
54. Meirovitch, L., *Elements of Vibration Analysis*. 1986: McGraw-Hill.
55. Incropera, F.P., and D.P. DeWitt, *Fundamentals of Heat and Mass Transfer*. 5th ed. 2002, New York: Wiley.
56. White, F.M., *Fluid Mechanics*. 4th ed. 1999, New York: McGraw-Hill.
57. Irwin, J.D., and D.V. Kerns, *Introduction to Electrical Engineering*. 1995: Prentice Hall.
58. Rao, N.N., *Elements of Engineering Electromagnetics*. 6th ed. Illinois ECE Series. 2004: Prentice Hall.
59. Gray, P.R., et al., *Analysis and Design of Analog Integrated Circuits*. 4th ed. 2001: John Wiley and Sons.
60. Woodson, H.H., and J.R. Melchier, *Electromechanical Dynamics Part I: Discrete Systems*. 1968: John Wiley and Sons.
61. Jaeger, R.C., *Introduction to Microelectronic Fabrication*. 2nd ed. Modular series on solid-state devices, ed. G.W. Neudeck and R.F. Pierret. Vol. V. 2002, Upper Saddle River, NJ: Prentice Hall.
62. Kovacs, G.T.A., *Micromachined Transducers Sourcebook*. 1998, New York: McGraw-Hill.
63. Madou, M.J., *Fundamentals of Microfabrication: The Science of Miniaturization*. 2nd Ed. 2002: CRC Press.
64. Callister, W.D., *Materials Science and Engineering, An Introduction*. 4th ed. 1997, New York: John Wiley and Sons.

CHAPTER 4

Electrostatic Sensing and Actuation

4.0 PREVIEW

Starting with this chapter, we will review several transduction principles commonly used in MEMS sensors and actuators. This chapter focuses on electrostatic sensing and actuation principles and methods. In Section 4.1, we will summarize the relative advantages and drawbacks of electrostatic transduction. Two types of electrostatic transducer configurations, parallel plate and interdigitated comb drives, are reviewed in Sections 4.2 and 4.3, respectively. For each type of transducer configuration, we will discuss specific application cases, including both sensors and actuators.

Other sensing and actuation principles (including thermal, piezoresistive, and piezoelectric energy transduction) will be discussed in Chapters 5 through 7.

4.1 INTRODUCTION TO ELECTROSTATIC SENSORS AND ACTUATORS

A capacitor is broadly defined as two conductors that can hold opposite charges. It can be used as either a sensor or an actuator. If the distance and relative position between two conductors change as a result of applied stimulus, the capacitance value would be changed. This forms the basis of capacitive (or electrostatic) sensing of positions. On the other hand, if a voltage (or electric field) were applied across two conductors, an electrostatic force would develop between these two objects. This is defined as electrostatic actuation.

Electrostatic forces are not often used for driving macroscopic machinery. However, microdevices have large surface-area-to-volume ratios and their masses are generally very small, thus making electrostatic force, which is a *surface force*, an attractive candidate for microactuation.

The electrostatically driven micromotor was one of the earliest MEMS actuators [1]. The motor, schematically diagrammed in Figure 4.1, consists of a rotor that is attached to the

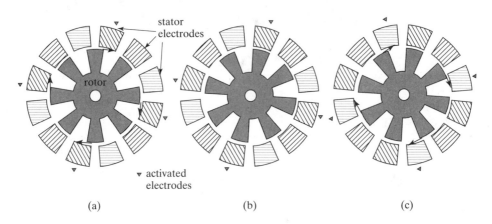

FIGURE 4.1

An electrostatic micromotor.

substrate with a hub, and a set of fixed electrodes, called stators, on the periphery. The stators are grouped together such that each group of four electrodes is electrically biased simultaneously. Three such groups are present for the motor shown in Figure 4.1; they are identified by different fill patterns.

To explain the operation principle of the electrostatic motor, let's begin with the rotor at an arbitrary angular rest position (Figure 4.1a). One group of stator electrodes is first biased. (The electrodes that are biased at a particular stage are identified by arrows placed next to them.) An in-plane electric field develops between any given stator electrode in this group and the closest rotor tooth next to it. This generates an electrostatic attractive force that aligns the tooth with the stator electrode. The torque values are on the order of pico Nm for voltages of the order of 100 V, which is large enough to overcome static friction. A small angular movement of the rotor is made to reach the new configuration diagrammed in Figure 4.1b. The electric bias is shifted to the next group of stator electrodes (Figure 4.1c), resulting in another small angular movement in the same direction. Continuous motion of the rotor can be achieved by activating the stator electrodes by groups in succession. The fabrication process of the motor is discussed in-depth in Chapter 11. Later works in the area of electrostatic actuators have provided new designs for greater power and torque output (e.g., [2]).

In this chapter, we will discuss the principles and governing equations of capacitive sensors and actuators. Major advantages of electrostatic sensing and actuation are summarized in the following paragraphs:

1. **Simplicity.** The sensing and actuation principles are relatively easy to implement; they require only two conducting surfaces. No special functional materials are required. Other sensing methods (such as piezoresistive and piezoelectric sensing) and other actuation methods (such as piezoelectric actuation) require deposition, patterning, and integration of special piezoresistive and piezoelectric materials.
2. **Low power.** Electrostatic actuation relies on differential voltage rather than current. The method is generally considered energy efficient for low-frequency applications. This is especially true under static conditions when no current is involved. At high frequencies, a

time- and frequency-dependent displacement current $i(t)$ will develop in response to a time-varying bias voltage $V(t)$. The magnitude of the current is $i(t) = C\frac{dV(t)}{dt}$. The instantaneous power delivered to a capacitor is $p(t) = i(t) \cdot V(t)$.

3. Fast response. Electrostatic sensing and actuation offers high dynamic-response speed, as the transition speed is governed by the charging and discharging time constants that are typically small for good conductors. For example, the switching time of the mirrors in the digital micromirror display (DMD) array is smaller than 21 μs, which is fast enough to support an 8-bit gray scale display.

Relative disadvantages of electrostatic actuation and sensing should be recognized as well. The high voltage required for static-actuator operation is considered a drawback. The DMD mirrors are switched by voltage on the order of 25 V to achieve $\pm 7.5°$ tilting. A monolithically integrated optical mirror with 9° tilt range requires a 150-V bias voltage [3]. Linear electrostatic actuators requiring hundreds of volts to move a microstructure by several tens of micrometers are rather common. High voltage introduces electronic complexity (for providing high-voltage supply) and material-compatibility issues. Electrodes that are mechanically connected with insulators tend to accumulate charges, especially under DC operation modes. The trapped charges will change the operation characteristics.

There are two major categories of capacitive electrode geometries: parallel-plate capacitors and interdigitated finger (comb-drive) capacitors. The static capacitance values of parallel plates and comb drives are generally very small, being on the order of pF or smaller. For capacitive sensors, the careful design of electrical circuits is needed to register capacitance changes (Δf) on the order of fF or lower [4] in the presence of noise and interference sources.

Capacitive actuators often take advantage of electrostatic *attraction* between two oppositely charged surfaces. Repulsive electrostatic forces, which are used less often, can be developed between two surfaces with same charge polarities [5, 6]. For example, the vertical levitation of interdigitated fingers up to 0.5 μm has been demonstrated under the applied voltage of 20 V for an 18-finger device.

The majority of capacitive devices use electrodes that dissipate charges quickly. Some, however, are made of dielectric materials that develop permanent ordering of molecular dipoles or retain charges. A class of charged material, called electrets, is made of organic polymers such as Teflon or poly(methylmethacrylate) (PMMA) [7]. The electrets can be charged through global [7, 8] or local [9] electron injection to a surface density of 10^{-4} C/m^2 or greater.

4.2 PARALLEL-PLATE CAPACITORS

4.2.1 Capacitance of Parallel Plates

A parallel-plate capacitor is the most fundamental configuration of electrostatic sensors and actuators. Such a capacitor, as its name suggests, consists of two conducting plates with their broad sides parallel to each other. In a broader definition, these two plates are not required to be exactly parallel at all times nor are they required to be planar.

A representative application using the parallel-plate electrostatic actuation principle is the DMD chip made by Texas Instruments [10]. A DMD chip consists of a large array of micromirrors or binary light switches. Each mirror is made of a reflective plate capable of rotating about a torsional support hinge. Two electrodes are embedded below the reflective plate, each covering one half of the plate area on each side of the torsional hinge. Two sets of parallel

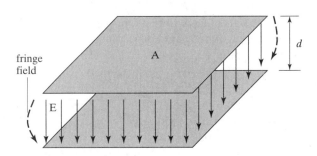

FIGURE 4.2

A parallel-plate capacitor with an overlapping area of A and a spacing of d.

capacitors are thus formed: one between each of the underlying electrodes and one on the mirror surface above. By applying voltages to either bottom electrode, the mirror device tilts under electrostatic attractive force, by $\pm 7.5°$, to deflect incoming light.

Let's examine the simple parallel plate capacitor depicted in Figure 4.2. These two plates have an overlapping area of A and a spacing of d. The dielectric constant, or relative electrical permittivity, of the media between the two plates is denoted ε_r. The permittivity of the media is $\varepsilon = \varepsilon_r \varepsilon_0$, where ε_0 is the permittivity of the vacuum ($\varepsilon_0 = 8.85 \times 10^{-14}$ F/cm).

The value of the capacitance, C, between the two plates is defined as

$$C \equiv \frac{Q}{V}, \tag{4.1}$$

where Q is the amount of stored charge and V is the electrostatic potential. The electric energy stored by a given capacitor U is expressed as

$$U = \frac{1}{2}CV^2 = \frac{1}{2}\frac{Q^2}{C}. \tag{4.2}$$

For a parallel-plate capacitor, electric field lines are parallel to each other and perpendicular to the plate surfaces in the overlapped region. Fringe electric fields reside outside the boundary of electrode plates. The fringe field lines are three-dimensional in nature and should be considered in rigorous design analysis. This book skips the discussion about the fringe electric field.

According to Gauss's law, the magnitude of the primary electric field E is related to Q by

$$E = Q/\varepsilon A. \tag{4.3}$$

The magnitude of the voltage is the electric field times the distance between two plates (d). The capacitance of a parallel-plate capacitor is

$$C \equiv \frac{Q}{V} = \frac{Q}{E \cdot d} = \frac{Q}{\frac{Q}{\varepsilon A}d} = \frac{\varepsilon A}{d}. \tag{4.4}$$

The capacitance value is proportional to A and is inversely proportional to d. It is a function of the electric permittivity ε.

Two parallel plates can move with respect to each other in two ways—normal displacement or parallel-sliding displacement. In this textbook, we are mainly concerned with the cases where two parallel plates move along their normal axis.

The parallel-plate capacitor is a versatile platform for physical, chemical, and biological sensors. By measuring the capacitance value of a parallel capacitor, we can sense the changes of permittivity, A or d. The permittivity can be changed by the temperature and humidity [11] of the capacitor media. The capacitance change via the permittivity route can be used to characterize liquid, air, or even biological particles in the gap. For example, the DNA content of eukaryotic cells can be observed through a linear relation between the DNA content of cells and the change in capacitance that it evokes by the passage of individual cells across a 1-kHz electric field [12]. The overlapped area and distance between plates can be changed by contact force, static pressure [13], dynamic pressure (acoustics) [14], and acceleration [15].

The capacitor can be used as an actuator to generate force or displacement (if at least one of the capacitor plates is suspended or deformable). Let's examine a pair of electrode plates with one plate firmly anchored and another one suspended by a mechanical spring. As a differential voltage is applied between the two parallel plates, an electrostatic attraction force will develop. The magnitude of the forces equals the gradient of the stored electric energy V_s with respect to the dimensional variable of interest. The expression for the magnitude of the force is

$$F = \left| \frac{\partial U}{\partial x} \right| = \frac{1}{2} \left| \frac{\partial C}{\partial x} \right| V^2, \qquad (4.5)$$

where x is the dimensional variable of interest. If the plates move along their normal axis, the gap between electrodes changes. The magnitude of the force can be rewritten as

$$F = \left| \frac{\partial U}{\partial d} \right| = \frac{1}{2} \frac{\varepsilon A}{d^2} V^2 = \frac{1}{2} \frac{CV^2}{d} \qquad (4.6)$$

by replacing the general dimensional coordinate x in Equation (4.5) with d.

Under a constant biasing voltage, the magnitude of the electrostatic force decreases drastically with the increasing gap d. The electrostatic force is considered a short-range force; it is most effective when the gap is on the order of a few micrometers.

The upper limit of the applied voltage for actuation is the breakdown voltage of dielectric media. If the media is air, the breakdown voltage can be predicted following the Paschen curve, which predicts that the breakdown voltage increases for decreasing gap sizes. This is a favorable scaling law for MEMS capacitive devices.

Example 4.1. Calculating Capacitance Value

Consider an air-gap capacitor made with two fixed parallel-planar plates. At rest (zero bias), the distance between two parallel plates is $x_0 = 100 \, \mu m$ and the areas of the plates are $A = 400 \times 400 \, \mu m^2$. The media between the two plates is air. The biasing voltage between these two plates is $V = 5$ Volts. Calculate the numerical value of the capacitance and the magnitude of the attractive force (F). What is the capacitance value if half of the area is filled with water (as the interplate media)?

Solution. To find the capacitance value, we use

$$C = \varepsilon_r \varepsilon_0 \frac{A}{x_0} = 14.17 \times 10^{-15} \, \text{F}.$$

The force can be calculated using

$$F = \frac{\partial V}{\partial d} = \frac{1}{2} \frac{CV^2}{x_0} = 1.8 \times 10^{-9} \, \text{N}.$$

If one half of the plate area is occupied with water and another half by air, each half can be considered separately. The total capacitance consists of two capacitors connected in parallel. The relative dielectric constant of water at room temperature is 76.6. The capacitance associated with the half with water media is

$$C_{water} = \varepsilon_r \varepsilon_0 \frac{A/2}{x_0} = \frac{76.6 \times 8.854 \times 10^{-12} \times 80{,}000 \times 10^{-12}}{100 \times 10^{-6}} = 542.6 \times 10^{-15} \, \text{F}.$$

The capacitance of the half with air media is $C_{air} = 7.08 \times 10^{-15} \, \text{F}$.
The total capacitance is

$$C = C_{air} + C_{water} = 549.6 \times 10^{-15} \, \text{F}.$$

4.2.2 Equilibrium Position of Electrostatic Actuator Under Bias

Many electrostatic sensors and actuators involve at least one deformable plate supported by springs. An important design aspect for such sensors and actuators is to determine the amount of static displacement under a certain biased voltage. In this section, we will discuss the procedures for calculating the equilibrium displacement under static (DC) and quasi-static (low-frequency) biasing conditions.

A parallel-plate capacitor with one movable plate supported by a mechanical spring is diagrammed in Figure 4.3. The top plate is supported by a spring with the force constant being K_m. At rest, the applied voltage, displacement, and the mechanical restoring forces are zero. Gravity does not play an important role in the static analysis of microdevices, because the mass

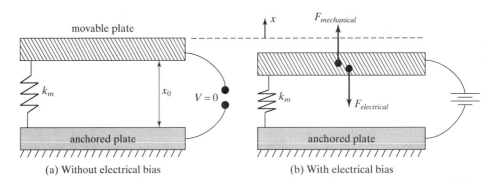

FIGURE 4.3

A coupled electromechanical model.

of plates is generally very small and the gravitational force would not cause appreciable static displacement.

When a voltage is applied, an electrostatic force $F_{electric}$ will be developed. The magnitude of $F_{electric}$ when the movable plate is at its starting position is given by

$$F_{electric} = \frac{1}{2}\frac{\varepsilon A}{d^2}V^2 = \frac{1}{2}\frac{CV^2}{d}. \qquad (4.7)$$

This force will tend to decrease the gap, which gives rise to displacements and the mechanical restoring force. Under static equilibrium, the mechanical restoring force has an equal magnitude but opposite direction as the electrostatic force.

The case of an electrostatic actuator is very intriguing. The magnitude of the electrostatic force itself is a function of the displacement. To make things more complex, the electrostatic force modifies the spring constant as well. From the discussion in Chapter 2, we know that when the electrostatic force acts in the opposite direction as the mechanical restoring force, the spring will be effectively softened. The spatial gradient of the electric force is defined as an *electrical spring constant*,

$$k_e = \left|\frac{\partial F_{electric}}{\partial d}\right| = \left|-\frac{CV^2}{d^2}\right| = \frac{CV^2}{d^2}. \qquad (4.8)$$

As seen, the magnitude of the electrical spring constant changes with position (d) and the biasing voltage (V). On the other hand, the mechanical spring constant (k_m) stays unchanged for small displacements.

The effective spring constant of a structure is the mechanical spring constant minus the electrical spring constant.

Let's derive the equilibrium displacement of a spring-supported electrode plate under a bias voltage V. Suppose the resulting equilibrium displacement is x, with the x-axis pointing in the direction of the increasing gap (Figure 4.3). With the displacement x, the gap between electrodes becomes $d + x$ ($d + x < d$). The electrostatic force at the equilibrium position is

$$F_{electric} = \frac{1}{2}\frac{\varepsilon A}{(x_0 + x)^2}V^2 = \frac{1}{2}\frac{C(x)V^2}{(x_0 + x)}. \qquad (4.9)$$

The magnitude of the mechanical restoring force is

$$F_{mechanical} = -k_m x. \qquad (4.10)$$

Equating the magnitudes of $F_{mechanical}$ and $F_{electrical}$ at x and rearranging terms, we obtain

$$-x = \frac{F_{mechanical}}{K_m} = \frac{F_{electrical}}{K_m} = \frac{C(x)V^2}{2(x_0 + x)K_m}. \qquad (4.11)$$

The equilibrium distance between two capacitor plates can be calculated by solving the preceding quadratic equation with respect to x.

The analysis of the equilibrium position can be visualized graphically as well. We first set up a coordinate system where the horizontal axis represented the space between the two plates and the vertical axis was the *amplitude* of the mechanical or electrostatic forces, irrespective of

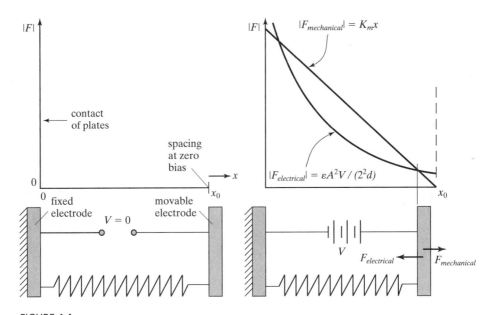

FIGURE 4.4

The magnitude of the electric force and mechanical force as a function of plate spacing.

their directions. A plate located at the origin of the x-axis is rigid, whereas the plate originally placed at a distance x_0 away from the origin is movable.

Two curves, one representing the amplitude of the mechanical restoring force and one representing that of the electrostatic force, are plotted as a function of electrode positions (Fig. 4.4). The mechanical restoring force ($F_{mechanical}$) changes linearly with the position for a fixed-bias voltage V. The electrical force ($F_{electrical}$) increases with x in a *non-linear* fashion, following the expression of Equation (4.9).

In the preceding diagram, there are multiple interception points for the two curves; they correspond to the solutions of Equation (4.11). At each interception point, the magnitude of the electrical force and the mechanical force are identical. The horizontal coordinates of the interception points therefore indicate the equilibrium positions of the movable plate. It is noteworthy that although several interception points are possible, only one of them will be achieved in reality. Specifically, the solution that is closest to the rest position is realized first and is generally the realistic solution.

The graphical method can be used to track the equilibrium position as the bias voltage is changed. The equilibrium positions of the actuator under three representative bias voltages, V_1, V_2, and V_3, can be found graphically (Figure 4.5). As the voltage increases, the family of curves corresponding to the electric forces shifts upwards. The x coordinates of the interception points are moved further away from the rest position.

4.2.3 Pull-In Effect of Parallel-Plate Actuators

At a particular bias voltage, the two curves representing the mechanical restoring force and the electrostatic force intercept at one point tangentially (Figure 4.6). At this interception point, the

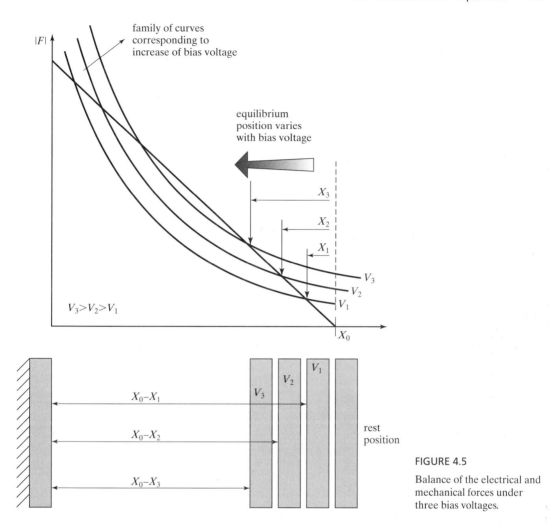

FIGURE 4.5

Balance of the electrical and mechanical forces under three bias voltages.

electrostatic and mechanical restoring forces balance each other. Moreover, the magnitude of the electric force constant (given by the gradient at the intercept point) equals the mechanical force constant. The effective force constant of the spring is zero (i.e., extremely soft). This is a special condition and should be handled carefully. The bias voltage that invokes such a condition is called the **pull-in voltage** or V_P.

If the bias voltage is further increased beyond V_P, the two curves will not have a common interception, thus the equilibrium solution disappears. In reality, the electrostatic force will continue to grow while the mechanical restoring force, increasing only linearly, is unable to catch up with it. The two plates will be pulled against each other rapidly until they make contact, at which point the mechanical contact force will finally balance the electric one. This condition is called **pull in** or snap in.

The voltage and displacement necessary to induce the pull-in condition is therefore quite important for designing electrostatic actuators. Although the pull-in effect can be explained quite easily using graphics, an analytical model is needed to yield the exact values of voltage

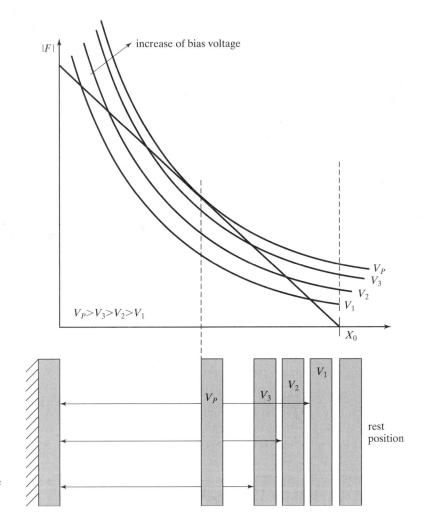

FIGURE 4.6

The electric force and mechanical force balance at the pull-in voltage.

and positions. We will review the pull-in condition for the simple spring-loaded electrode system depicted in Figure 4.3.

At the pull-in voltage, there is one tangential intersection point between $|F_{electrical}|$ and $|F_{mechanical}|$ curves. The magnitudes of the electrical force and the mechanical balance forces are the same. By equating these two forces and rearranging terms, we obtain

$$V^2 = \frac{-2k_m x(x + x_0)^2}{\varepsilon A} = \frac{-2k_m x(x + x_0)}{C}. \tag{4.12}$$

The value of x is negative when the spacing between two electrodes decreases. In addition, the gradients of these two curves at the intersection point are identical, namely,

$$|K_e| = |K_m|. \tag{4.13}$$

The expression for the electric force constant,

$$k_e = \frac{CV^2}{d^2}, \tag{4.14}$$

can be rewritten by plugging in the expression for V^2 (derived in Equation (4.12)). This procedure results in a new expression for the electric force constant,

$$k_e = \frac{CV^2}{(x + x_0)^2} = \frac{-2k_m x}{(x + x_0)}. \tag{4.15}$$

The only solution for x under which Equation (4.13) can be satisfied is

$$x = -\frac{x_0}{3}. \tag{4.16}$$

The preceding equation states that the relative displacement of the parallel plate from its rest position is exactly one third of the original spacing at the critical pull-in voltage. This critical displacement is true irrespective of the actual mechanical force constant and the actual pull-in voltage.

The voltage at pull in can be found easily by plugging the transition distance in to Equation (4.12). This yields

$$V_p^2 = \frac{4x_0^2}{9C} k_m. \tag{4.17}$$

Consequently, the pull-in voltage is found to be

$$V_p = \frac{2x_0}{3} \sqrt{\frac{k_m}{1.5 C_0}}. \tag{4.18}$$

In reality, the pull-in voltage and threshold distance will deviate from the calculation obtained using the idealized case. There are two sources of deviation from the ideal model. First, the fringe capacitance will alter the expression of the electrostatic force. Second, the restoring force provided by the mechanical springs will differ from that predicted by the linear model if the displacement is large.

Comprehensive analysis of the pull-in phenomenon can be carried out for systems with multiple degrees of freedom and hysteresis [16, 17]. More complex cases, such as the pull-in effect of rotational electrostatic actuators based on torsional support [18] and that of electrodes with complicated profiles (e.g., cantilevers and membranes), have been studied in the past [19, 20].

Example 4.2. Calculation of Equilibrium Position

A parallel-plate capacitor is suspended by two fixed–guided cantilever beams, each with its length, width, and thickness denoted by l, w, and t, respectively (Fig. 4.7). The material is polysilicon, with a Young's modulus of 120 GPa ($l = 400$ μm, $w = 10$ μm, and $t = 1$ μm.) The gap x_0 between two plates is 2 μm. The area of the plate is 400 μm $\times$ 400 μm.

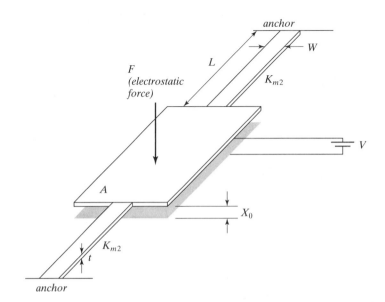

FIGURE 4.7

An electrostatic actuator plate supported by two fixed–guided beams.

Calculate the amount of vertical displacement when a voltage of 0.4 volts is applied. Repeat the calculation of displacement for the voltage of 0.2 volts.

Solution. There are three steps in a standard analysis procedure.

Step 1. Find out the force constant associated with the actuator.

We calculate the force constant of one beam first using a model of a fixed–guided beam. For each beam, the vertical displacement under the applied force F is

$$d = \frac{Fl^3}{12EI},$$

The force constant is therefore

$$K_m = \frac{F}{d} = \frac{12EI}{l^3} = \frac{Ewt^3}{l^3} = \frac{120 \times 10^9 \times 10 \times 10^{-6} \times (1 \times 10^{-6})^3}{(400 \times 10^{-6})^3} = 0.01875 \text{ N/m}.$$

Because the plate is supported by two beams that are connected in parallel, the overall force constant is the sum of two separate force constants, namely,

$$K_m = 0.0375 \text{ N/m}.$$

This is a rather "soft" support.

Step 2. Find out the pull-in voltage.

In order to determine the equilibrium position, we must find the pull-in voltage first. If the applied voltage is greater than the pull-in voltage, the two plates would snap together. We first find the static capacitance value C_0,

$$C_0 = \frac{8.85 \times 10^{-12} (\text{F/m}) \times (400 \times 10^{-6})^2}{2 \times 10^{-6}} = 7.083 \times 10^{-13} \text{ F}.$$

4.2 Parallel-Plate Capacitors

The pull-in voltage is therefore

$$V_p = \frac{2x_0}{3}\sqrt{\frac{k_m}{1.5C_0}} = \frac{2 \times 2 \times 10^{-6}}{3}\sqrt{\frac{0.0375}{1.5 \times 7.083 \times 10^{-13}}} = 0.25 \text{ (volts)}.$$

Thus, when the applied voltage is 0.4 volts, the plate has already reached the pull-in state and the spacing between the two electrodes is zero.

Step 3. Repeat the calculation for the applied voltage of 0.2 volts.

If the applied voltage is 0.2 volts, the pull-in condition is not reached. To find the vertical equilibrium positions, we use

$$V^2 = \frac{-2k_m x(x+x_0)^2}{\varepsilon A} = \frac{-2k_m x(x+x_0)}{C}.$$

We can rewrite this with respect to the value of x:

$$x^3 + 2x_0 x^2 + x_0^2 x + \frac{V^2 \varepsilon A}{2k_m} = 0$$

$$x^3 + 4 \times 10^{-6} x^2 + 4 \times 10^{-12} x + 7.552 \times 10^{-19} = 0.$$

The solution to this equation can be found both numerically and analytically. The analytical solution of

$$x^3 + ax^2 + bx + c = 0$$

can be found by first substituting

$$y = x + a/3.$$

The solutions are

$$y = A + B$$

or

$$x = A + B - \frac{a}{3}.$$

The terms A and B are

$$A = \sqrt[3]{\frac{-q}{2} + \sqrt{Q}}$$

$$B = \sqrt[3]{\frac{-q}{2} - \sqrt{Q}}$$

where

$$q = 2\left(\frac{a}{3}\right)^3 - \frac{ab}{3} + c,$$

and

$$Q = \left(\frac{p}{3}\right)^3 + \left(\frac{q}{2}\right)^2$$

where

$$p = \frac{-a^2}{3} + b.$$

For a third-order polynomial equation, three distinct solutions can be found. For this problem, all the mathematically feasible solutions are listed below:

$x_1 = -2.45 \times 10^{-7}$ μm, $x_2 = -1.2 \times 10^{-6}$ μm, and $x_3 = -2.5 \times 10^{-6}$ μm.

However, the last two numbers are not correct because one of them (x_2) is already into the pull-in range, and x_3 is even beyond the original electrode gap allowance.

Pull in may result in irreversible damage due to short circuits, arcing, and surface bonding. Electric shorting upon pull-in contact may be prevented by depositing dielectric insulators on electrodes.

The pull-in phenomenon prevents the displacement of a parallel capacitor actuator to reach its full-gap allowable range. The displacement is limited to 1/3 of the initial gap size. Full-gap actuation is desired to operate electrostatic actuators with a wider range of motion for many applications. Recent work has shown that full-gap positioning is possible with proper mechanical design and electrical control. A few methods are summarized below:

the use of dynamic control methods, including series capacitor feedback [21];

the use of leveraged actuators [22] or variable height plate design [23];

the use of current drive, instead of voltage drive [24].

4.3 APPLICATIONS OF PARALLEL-PLATE CAPACITORS

Parallel-plate capacitors can be broadly applied in a variety of sensing and actuation applications. We will discuss four representative classes of physical sensors and parallel-plate actuators in this section. Several case studies are presented. A reader who is not familiar with microfabrication technology may refer to Chapters 10 and 11 before reviewing fabrication processes related to these case studies.

4.3.1 Inertia Sensor

According to Newton's Second Law, an acceleration (a) acting on an object with a mass m would produce a reactive inertial force ma relative to its frame. The inertial force may in turn modify capacitive electrode spacing. This constitutes the basic operating principle of electrostatic acceleration sensors. In Case 4.1, the mechanical structure used is a cantilever beam. In Case 4.2, a torsional bar is used instead.

Example 4.3. Capacitance Sensor Response

A parallel capacitor with an area (A) of 100×100 μm^2 is supported by four cantilever beams. The plate is made of polycrystalline silicon that is $t = 2$ μm thick. The distance between the bottom of the plate and the substrate is $d = 1$ μm. Each cantilever beam is 400 μm long (l), 20 μm wide (w), and 0.1 μm thick (t). Find the relative change of capacitance under an acceleration of 1 g.

Solution. The mass of the plate is

$$m = \rho AT = 2330 \times (100 \times 10^{-6})^2 \times 2 \times 10^{-6} = 46.6 \times 10^{-12} \text{ kg}.$$

The magnitude of the force acting on the plate under an acceleration a is

$$F = ma = 46.6 \times 10^{-12} \text{ N}.$$

Suppose Young's modulus of polysilicon is 150 GPa. The force constant associated with all four support beams (fixed–guided) is

$$k = 4 \times \left(\frac{12EI}{l^3}\right) = \frac{48 \times 150 \times 10^9 \times \frac{20 \times 10^{-6} \times (0.1 \times 10^{-6})^3}{12}}{(400 \times 10^{-6})^3} = 0.0001875 \text{ N/m}.$$

The static displacement under the applied acceleration is

$$\delta = \frac{F}{k} = 0.248 \text{ } \mu m.$$

The capacitance value under zero-applied force is

$$C_0 = \varepsilon \frac{A}{d} = \frac{8.85 \times 10^{-12} \times (100 \times 10^{-6})^2}{1 \times 10^{-6}} = 88.5 \times 10^{-15} \text{ F}.$$

The value of the capacitance after the displacement δ is

$$C = \varepsilon \frac{A}{d - \delta} = \frac{8.85 \times 10^{-12} \times (100 \times 10^{-6})^2}{(1 - 0.248) \times 10^{-6}} = 117.7 \times 10^{-15} \text{ F}.$$

The relative change of the capacitance is

$$\frac{C - C_0}{C_0} \times 100\% = 33\%.$$

We will review two examples of micromachined accelerometers; both are realized on silicon wafers with integrated circuits for signal processing. In both cases, the micromachining processes are compatible with the integrated circuits. This compatibility is crucial in order to increase the cost effectiveness of integration and, ultimately, performance of sensors. In the case of accelerometers, the degree to which the electronic and transducer elements can be fabricated during the same or complementary processing steps will determine the practicality and market competitiveness of a particular design.

Case 4.1. Parallel-Plate Capacitive Accelerometer

One of the earliest fully integrated capacitive acceleration sensors is surface micromachined on a wafer with integrated MOS detection circuitry [15]. The sensor consists of a metal-coated oxide cantilever with a 0.35-μg-thick electroplated gold patch at its distal end serving as a proof mass. The length, width, and thickness of the cantilever are 108 μm, 25 μm, and 0.46 μm, respectively. The counter electrode is made of heavily doped p-type silicon. The capacitor gap (C_B) is defined by an epitaxy silicon layer grown on a silicon surface.

A surface micromachining process was developed (Figure 4.8) using thermal oxide as the cantilever structural material and epitaxially grown silicon as the sacrificial layer. The process starts with an n-type, (100) silicon wafer. A heavily boron-doped region (with a concentration of 10^{20}/cm^3) is made using an oxide layer as the doping barrier (step b). Between steps (a) and (b), certain detailed procedures such as oxide growth, deposition and patterning of photoresist, and the subsequent oxide etch and photoresist removal are skipped. An epitaxial silicon layer with a resistivity of 0.5 Ωcm is grown to a thickness of 5 μm over the entire wafer (step c). Another layer of oxide is deposited and patterned (step d), serving as a mask for etching a via hole (step e) and then as a barrier for doping (to form drain, source, and electrical conduction paths on the slopes of the via hole) (step f). The doping is conducted using ion implantation at 100 keV energy and 5×10^{14}/cm^2 dose. During the via hole etch, the heavily doped region will not be attacked because the etchant reduces its etch rate on heavily doped silicon (a topic we will cover in more detail in Chapter 10).

The oxide barrier is then removed (step f). This step is followed by the growth of another layer of thick oxide, which serves as the dielectric insulator, the cantilever, and the etching barrier in regions other than the gate (step g). A layer of metal is deposited and patterned. It provides electrical interconnects to the bottom p^+ electrode, the electrode on top of the oxide cantilever, and the gate of the field-effect transistor (step i). The metal layer consists of 20-nm-thick Cr followed by 40-nm-thick gold, with the Cr used to satisfy the critical requirement of enhancing the adhesion between the gold and the substrate. Finally, a wet silicon etch is performed to undercut epitaxial silicon beneath the oxide cantilever (step j).

The released cantilever is naturally bent due to intrinsic stress present in the metal and oxide thin films. The upward bending is approximately 1.5° at the end of the cantilever. Under the influence of applied acceleration, the beam will further deform from the stationary profile. For these reasons, the cantilever surface is not perfectly parallel to the substrate. The capacitance value between a curved cantilever and the doped counter electrode can be estimated in a piece-wise fashion, by summing incremental capacitances contributed by longitudinal segments of the cantilever. By ignoring fringe capacitance, the total capacitance is estimated as

$$C_B = \int_0^L C_x \, dx = \int_0^L \frac{\varepsilon_0 b}{d + \delta x} dx, \qquad (4.19)$$

4.3 Applications of Parallel-Plate Capacitors

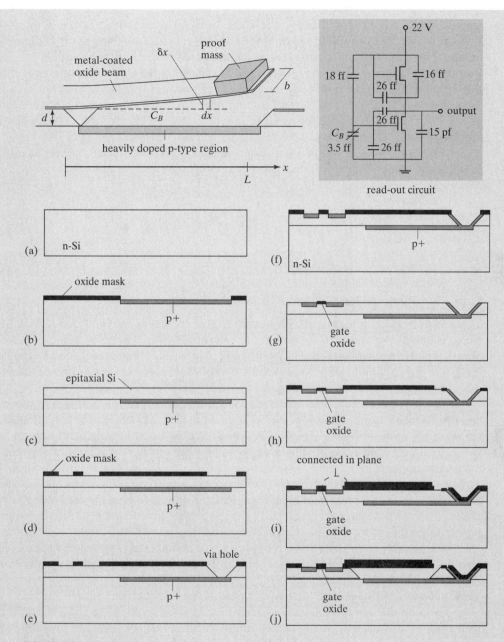

FIGURE 4.8
Capacitive accelerometer.

where L and b are the length and width of the cantilever, respectively, and ε_0 is the permittivity of the air medium. Other terms in the equation are indicated in Figure 4.8. The calculation of the beam-bending curvature is complicated by the variable stiffness associated with different regions of the cantilever—the area overlapped with the proof mass is much thicker and considered a rigid body.

The capacitance change is read using a relatively simple impedance converter. The sensor is capable of 2.2 mV/g of acceleration sensitivity, corresponding to a beam displacement of 68 nm/g. The mechanical resonant frequency of the cantilever is 22 kHz.

This sensor was developed early, and it used silicon and silicon-related thin films heavily. In Chapter 11, we will discuss several alternatives to the materials and processes discussed in this case.

Case 4.2. Torsional Parallel-Plate Capacitive Accelerometer

The sensor previously discussed showcased a process where micromechanical and electronic elements are fabricated in an interwoven process flow. However, this practice is not always possible nor advantageous. The monolithic integration raises issues such as material and process compatibility. For example, the composite beam consisting of oxide and metal exhibit undesirable intrinsic bending.

Case 4.2 illustrates a different strategy for integrating electronics with mechanical elements [25]. The authors developed surface micromachined sensing elements that can be added to silicon wafers after the standard processing of electronic elements is completed.

Exposing a silicon wafer with circuitry to high temperature for a prolonged duration may cause dopants in active regions of the circuits (e.g., source and drain) to out-diffuse. This will irreversibly change electrical characteristics or introduce device failure in extreme cases. The top-level surface micromachined structures need to be deposited and processed under relatively low temperatures. The oxide structural layer used in Case 4.1 requires a high temperature to be deposited and is therefore not suited here.

The new device consists of a flat nickel-top plate supported by torsional bars. Counter electrodes are located on the substrate surface (Figure 4.9). Since the plate weight is asymmetrically distributed with respect to the rotational axis, acceleration along the normal axis to the substrate will cause the top plate to rock in one direction or another.

The total mass of the top plate is determined by the geometric design. A nickel plate with a size of 1×0.6 mm^2 in area and 5 μm in thickness is used. Devices covering various acceleration ranges can be made. In fact, many elements with varying characteristics can be made on the same die to increasing the overall dynamic range of the sensor. For a 25-g device, the torsion bars are 8 μm wide, 100 μm long, and 5 μm thick. The mass of the suspended plate is 6.9×10^{-7} g. The capacitance value is approximately 150 pF at rest.

The process starts with a silicon wafer that has already gone through the complete cycle of IC fabrication. Conducting electrode patches on the IC wafer serve as the bottom

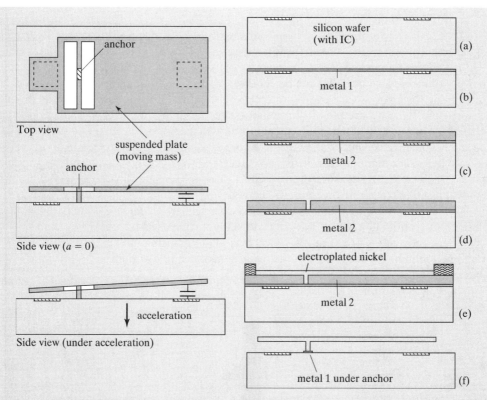

FIGURE 4.9

Schematic diagram of a surface micromachined parallel-plate capacitor serving as an accelerometer.

electrodes (step a). First, a conductive layer, metal 1, is deposited over the substrate surface (step b). This serves as a seed layer for subsequent electroplating. A second layer of conductive metal (metal 2) is deposited and patterned, forming bottom electrode patterns (step c). The combined thickness of these two metal layers is 5 μm. Next, a photoresist layer is deposited and patterned, opening windows to reach the metal 1 layer (step d). Electroplating of nickel takes place in the open window to define the movable plate (step e). The thickness of the electroplated nickel determines the thickness of the movable plate and that of the torsional bar. The photoresist is removed, followed by the etching of the sacrificial metal. The bottom conductive layer (seed layer) is etched as well. It is important to make sure that the metal 1 layer underneath the anchor is not removed (step f). All steps, including deposition and etching, take place under room temperature.

The sensor's response is calibrated within a broad temperature range (−55°C to 125°C) as dictated by military and automotive industry specifications. The shift is within 200 ppm per °C within this temperature range. Here, capacitive sensing is advantageous over other modes of sensing (e.g., piezoresistive) as the temperature sensitivity is relatively low.

4.3.2 Pressure Sensor

Pressure sensors are widely used in automotive systems, industrial process control, medical diagnostics and monitoring, and environmental monitoring. The membrane thickness is a primary factor in determining the pressure sensor sensitivity. MEMS technology allows membranes to be very thin compared to what can be achieved with conventional machining. The uniformity of membrane thickness is very good using microfabrication. Pressure sensors based on piezoresistive sensing elements are the most popular (see Chapter 5); however, ones based on capacitive sensing are also used rather widely. Membrane-based pressure sensors are ideally suited for parallel capacitive sensing. Capacitive pressure sensors offer the advantage of greater pressure sensitivity, lower temperature sensitivity, and reduced power consumption compared with piezoresistive pressure sensors. Piezoresistive pressure sensors, discussed in Chapter 5, are self-consistent and do not rely on counter electrodes or matching surfaces.

We will review two examples of pressure sensors to illustrate unique and innovative designs and fabrication steps. Case 4.3 is dedicated to a capacitive pressure sensor with a sealed chamber to provide a pressure reference. It is made by using bulk micromachining and wafer-bonding techniques. Case 4.4 is an acoustic sensor made by combining bulk micromachining and surface micromachining steps.

Case 4.3. Membrane Parallel-Plate Pressure Sensor

A membrane pressure sensor can detect pressure differential across the membrane. Two pressure ports are typically required. To simplify the pressure sensor design and use, absolute pressure sensors are often desirable. In such sensors, the reference pressure at one side of the membrane is integrated. One popular choice is to provide a zero-pressure reference (vacuum) by hermetic sealing.

A pressure sensor with a batch-processed, hermetically sealed vacuum chamber is illustrated below [26]. The use of a vacuum avoids the expansion of trapped air and increases the bandwidth by eliminating air damping inside the cavity. According to the authors, the device must retain resolution over a temperature range from −25°C to 85°C. However, the need for vacuum packaging and integration with integrated circuits presents a critical challenge.

The sensor cross section is shown in the last step of Figure 4.10. A membrane made of doped silicon serves as the pressure-sensing element and one electrode. The counter electrode consists of patterned metal thin film on the bottom substrate (made of glass in this case). A fabrication process was developed where the micromachined silicon membrane was transferred to a glass substrate. The process begins with a (100) silicon wafer (step a). An oxide mask is deposited and patterned (step b), serving as a chemical barrier during a wet anisotropic etching of silicon (using KOH solutions) (step c). A 9-μm-deep recessed region is created with the slopes being {111} surfaces (step d). Another layer of oxide is grown, this time as a conformal coating. The oxide is then photolithographically patterned (step f).

4.3 Applications of Parallel-Plate Capacitors 123

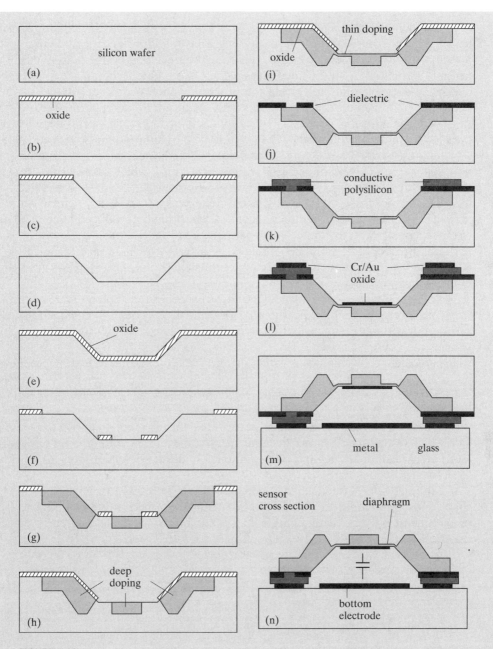

FIGURE 4.10

Fabrication process of pressure sensor with sealed cavity.

The depositing of photoresist film over a wafer with recessed cavities poses a challenge. The uniformity of the spin-coated resist layer will be negatively impacted by the presence of surface topology. Further, photo exposure action on the bottom of the cavity, which is 9 μm away from the ideal focus plane of the photolithography exposure tool, reduces line-width resolution. During the process, caution should be exercised whenever photolithography is performed on wafers with significant topographic features.

A boron diffusion step at 1175°C is conducted (step g); this forms doped regions as thick as 15 μm. This is followed by the stripping of the oxide and the growth and patterning of yet another oxide layer (step h). The second oxide layer is used in a subsequent boron diffusion step to define a doped region (depth = 3 μm), which becomes the thickness of the membrane diaphragm (step i). A layer of silicon oxide is deposited and patterned to form a dielectric insulation (step j). The researchers patterned via holes in the oxide, which allow a subsequently deposited polysilicon to contact the boron-doped region and provide electrical contact with the membrane later (step k). A short diffusion session (at 950°C) is performed to dope the polysilicon; this is followed by a chemical mechanical polishing (CMP) step to increase the top surface smoothness. The polishing step enhances the yield of the sealing step later.

A layer of metal (consisting of Cr and Au) is deposited and patterned, with the gold facing the front side of the wafer (step l). An oxide layer is deposited and patterned to reside on the bottom of the cavity to provide electrical isolation in case the top membrane touches the bottom electrode. The researchers flip bonded the wafer onto a glass wafer, which is coated with a composite Ti-Pt-Au layer. Wafer-level anodic bonding to the glass is performed in a vacuum (1 $\times$ 10^{-6} torr) at 400°C for 30 minutes (step m). The back side of the silicon wafer is etched in an anisotropic silicon etchant to dissolve the silicon rather than the heavily doped, raised membrane (step n).

Due to the large gap, the sensor has a wide dynamic range (500–800 torr) along with a very high resolution (25 mtorr, the equivalent of the altitude difference of one foot at sea level) after readout and digital compensation. The device yielded a pressure sensitivity of 25 fF/torr (or 3000 ppm/torr).

Alternatively, hermetically sealed cavities can be formed by chemical vapor deposition under a vacuum to encapsulate strategically placed etch holes [27]. Capacitive pressure sensors with or without hermetic sealing can also be made using surface micromachining processes [28].

Case 4.4. Membrane Capacitive Condenser Microphone

Next, we will discuss a condenser microphone, which is a pressure sensor for measuring acoustic pressure fronts created when sound waves travel through air or liquid. Sound waves are oscillating pressure waves. The classic definition of the strength of sound is sound pressure level, SPL (in decibels or dB). The expression for SPL is

$$\text{SPL} = 20 \log\left[\frac{p_1}{p_0}\right], \quad (4.20)$$

where p_1 is the sound pressure and p_0 is a reference pressure. There is no universally accepted standard reference pressure; however, the pressure of 0.0002 microbar (or 2×10^{-5} N/m^2) is commonly used in air acoustics and has been used in underwater noise acoustics.

A condenser microphone consists of a parallel-plate capacitor, plus one solid plate (called a diaphragm) that moves under incoming acoustic waves while another one is being perforated (called a backplate). The perforation reduces the amount of plate deformation. A capacitor formed between these two electrodes would therefore change its value in response to incoming sound waves. The monolithic integration of the capacitor with integrated circuitry is key to realizing high resolution and miniaturization.

The condenser microphone discussed in this paper [29] does not involve wafer bonding as in Case 4.3. The schematic diagram of the microphone is shown in the final step of Figure 4.11. Mechanically, the device consists of a perforated plate made of polyimide thin film, and a solid plate made of the same material. Metal conducting thin films are integrated with both plates. The capacitor is electrically connected to the on-chip integrated circuitry.

The fabrication process of this device combines surface and bulk micromachining steps. It starts with a silicon wafer that contains fully processed integrated circuitry elements. The active device regions are made in a p-type epitaxial layer. Both n- and p-channel field-effect transistors are made on the same substrate; two of each kind are shown on two ends of the cross-sectional view of the wafer (step a). The wafer is covered with a passivation oxide dielectric.

A composite metal thin film (consisting of chromium, platinum, and chromium) is deposited and photolithographically patterned. A layer of photo-patternable polyimide is deposited and patterned, overlapping with the metal thin film below. The researchers used a Cr layer to increase adhesion between the platinum layer to the surrounding structural layers (oxide below and polyimide above) (step b). A layer of aluminum is deposited on top of the polyimide; its thickness defines the gap of the future parallel-plate capacitor (step c). On top of the aluminum, a composite metal thin film (Cr/Pt/Cr) is deposited and patterned to form a conducting plate with perforation holes (step d).

Another layer of polyimide is deposited and patterned; it has proper registration to the perforated electrode plate below (step e). Next, a layer of chromium is deposited on the backside of the wafer and patterned. The chromium layer provides sufficient selectivity during the subsequent deep reactive ion etching, which etches through the backside of the wafer, exposing the backside of the first Cr/Pt/Cr composite layer (step f). The aluminum sacrificial layer is then etched away, resulting in a finished device (step g).

In the finished device, the diaphragm exhibits a tensile intrinsic stress of 20 MPa, which is ideal for keeping the diaphragm flat. The thickness of the diaphragm is 1.1 μm, whereas the thickness of the backplate is 15 μm. The gap of the capacitor is 3.6 μm. The size of the membrane is 2.2 mm $\times$ 2.2 mm, and the size of acoustic holes and the spacing between them is 30×30 μm^2 and 80 μm, respectively.

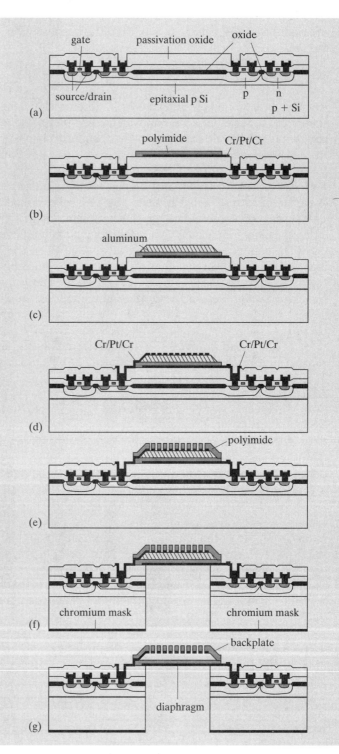

FIGURE 4.11
Fabrication process of condenser microphone.

Circuit features include a Dickson-type DC–DC voltage converter and a MOS buffer amplifier. The voltage converter is a charge pump in which charges are built up at the output by a two-phase oscillator. The voltage converter yielded an output voltage of 14.3 V for a supply voltage of 1.9 V. The integrated microphone has a sensitivity of 29 mV/Pa for a supply voltage of 1.9 V and a bandwidth of 27 kHz.

Alternatively, capacitive sensors for pressure and acoustics signals have also been made using surface micromachining processes in the past (e.g., Reference [30]).

4.3.3 Flow Sensor

Flow sensors broadly encompass devices for measuring point-fluid speed, volume-flow rate, shear stress at the wall, and pressure. Microintegrated flow sensors have been an area of active research since the early days of the MEMS field. Micromachined flow sensors offer the following advantages: (1) small size and less disturbance to the flow field of interest; (2) high sensitivity stemming from compliant mechanical elements or circuit integration; and (3) potential for achieving a large array of sensors with uniform performance. We will focus on discussing a fluid shear stress sensor in Case 4.5.

Case 4.5. Capacitive Boundary-Layer Shear Stress Sensor

Fluid flowing past a solid surface introduces a boundary layer, inside which the flow velocity is reduced. Inside the boundary layer, the velocity varies with the distance to the wall surface (y). The shear stress is defined as the velocity gradient at the boundary multiplied by the viscosity of the fluid:

$$\tau_w = \mu \frac{du}{dy}. \tag{4.21}$$

The term μ is the dynamic viscosity with the unit being kg/(m·s).

Shear stress sensors reveal critical fluid-flow conditions at the bottom of the boundary flow; these conditions are difficult to measure conventionally. The area integral of shear stress produces drag force. The shear stress information can be used for active control of turbulent flow field, for actively monitoring fluid drag, and for achieving drag reduction.

Techniques for measuring fluid shear stress fall into two categories: direct measurement and indirect measurement. Two popular techniques are the hot-wire/hot-film anemometer (indirect measurement) and the floating-element technique (direct measurement).

A floating-element shear stress sensor was the first MEMS shear stress sensor developed [31]. It determines the magnitude of local shear stress directly by measuring the drag force it experiences. As shown in Figure 4.12, a suspended floating element is flush mounted on the surface of a wall. The displacement of the floating element due to the shear

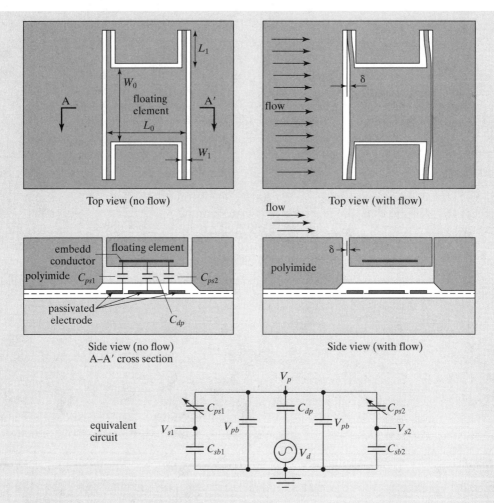

FIGURE 4.12

A floating-element shear stress sensor.

force (drag force) acting on the plate is transduced into plate displacement, which can be measured by a variety of techniques, including electrostatics (discussed here), piezoresistivity (Chapter 6), piezoelectricity (Chapter 7), and optical sensing.

The capacitive floating element consists of a plate (with an area of W_0 times L_0) suspended by four fixed–guided cantilevers, each with a length, width, and thickness of L_1, w_1, and t_1, respectively. The plate is assumed to be a rigid body. A distributed drag force is applied on the element as well as on the cantilevers.

Under a given flow shear stress τ_w, a distributed force P acting on the floating plate and a distributed force q (N/m) on the four fixed–guided beams are given as

$$P = \frac{1}{2}\tau_w W_0 L_0 \tag{4.22}$$

and

$$q = \tau_w W_1. \tag{4.23}$$

To estimate the total displacement, we assume a force of $P/4$ applied as a point load at the guided end of one fixed–guided beam, along with a distributed load. The total displacement, under the combined forces, is a linear summation of displacement under each applied force.

The electronic detection uses a differential capacitance readout scheme. Three passivated electrodes are located on the surface of the wafer underneath the element and a thin conductor is embedded in the polyimide. The coupled capacitances between the drive electrode in the center and the two symmetrically placed sense electrodes are modified by the motion of the floating element. This change in capacitance is transduced by connecting the sense electrodes to a pair of matched depletion-mode MOSFET's on-chips. (Fringe capacitances are ignored in the analysis.) The drive voltage V_d is coupled to the plate (V_p) through C_{dp}. The sense capacitance C_{ps1} and C_{ps2} vary linearly with the deflection of the plate δ according to

$$C_{ps1} = C_{ps0}\left[1 - \frac{\delta}{fW_s}\right] \tag{4.24}$$

and

$$C_{ps2} = C_{ps0}\left[1 + \frac{\delta}{fW_s}\right]. \tag{4.25}$$

The expression of the output voltage is given by

$$V_{s1} - V_{s2} = \left[\frac{C_{ps1}}{C_{ps1} + C_{sb}} - \frac{C_{ps2}}{C_{ps2} + C_{sb}}\right]V_p. \tag{4.26}$$

The formula can be expanded and rearranged to reveal a linear relation between the output voltage and the displacement,

$$V_{s1} - V_{s2} = -\left[\frac{C_{ps0}}{C_{sb}}\right]\left[\frac{C_{dp}}{C_{dp} + C_t}\right]\left[\frac{2V_d}{fW_s}\right]\delta. \tag{4.27}$$

The fabrication process for the device begins with a silicon wafer with MOS circuits already defined. The entire wafer is passivated with a 750-nm-thick atmospheric chemical vapor deposition of silicon dioxide and a 1-μm polyimide layer (Dupont 2545). The polyimide is added between the passivated electrodes and the sacrificial layer to eliminate stress cracking in the silicon-dioxide layer. A 3-μm-thick aluminum layer is evaporated as the sacrificial layer. It is patterned photolithographically. A 1-μm-thick polyimide layer is coated again and cured. This is followed by the evaporation of a 30-nm thick chromium layer, which serves as the floating electrode. A 30-μm-thick polyimide layer is applied in seven coating steps. A layer of aluminum is deposited on the top layer and serves as a mask for etching the polyimide to define the plate and the cantilever. The undercut is 6 μm corresponding to a 30 μm deep etching. The aluminum is removed using a mixture of phosphoric acid, acetic acid, nitric acid, and water. It takes approximately 2 hours to completely release a 500 μm × 500 μm floating element.

Each support beam is 1 mm long, 10 μm wide, and 30 μm thick. Young's modulus of the polyimide is 4 GPa. The cantilevers are loaded in axial tensile stress due to the residual stress in the polyimide plate. Detailed fluid mechanics characterization is performed to demonstrate a 52-μV/Pa sensitivity, which 40 dB above the expected noise floor.

4.3.4 Tactile Sensor

Another application of the parallel capacitive sensors involves the tactile sensors [32], which are critical components for robotics applications. In order to accurately measure tactile information, the sensors must have a high density of integration and a high sensitivity, preferably in multiple axes. MEMS technology provides the basis for miniaturization and functional integration. In Case 4.6, a tactile sensor capable of measuring force input in multiple axes is discussed. This device uses multiple electrodes and capacitors.

Case 4.6. Multi-Axis Capacitive Tactile Sensor

In this case, a compact sensor capable of measuring normal contact and shear contact in two axes was made [32]. A parallel-plate capacitor is formed by bonding a silicon wafer with a glass one (Fig. 4.13). One piece consists of a cone-shaped silicon mesa suspended by a circular silicon membrane with a thickness of t and a radius of a. The glass piece consists of a recessed region in which electrodes are patterned. Four electrodes, each with an area of L^2, are arranged in a quad configuration. Four capacitors are formed, between the four electrodes and the suspended plate. These are denoted C_1 through C_4.

If a normal force is applied perpendicular to the substrate, the distance between the movable mass and the bottom electrodes is reduced uniformly for all four capacitors. The capacitance change is related to the displacement by

$$\Delta C = \frac{\varepsilon_r \varepsilon_0 L^2}{d^2} \Delta d. \tag{4.28}$$

However, if a shear force is applied to induce rotational movement of the silicon mass, the changes of capacitance for the four capacitors will be different. Two capacitances increase while the other two decrease, with almost the same degree of change. Under a tilting angle θ, the total capacitance of the tilted-plate capacitor is estimated as

$$C_\theta = \frac{\varepsilon_r \varepsilon_0 L^2}{d + 0.5\,\theta L}. \tag{4.29}$$

The processing of the silicon part began with a standard p-type <100> silicon wafer, which is polished on both sides (Figure 4.6a). All standard IC processes were performed on the front side. First, a buried n-type layer (3.5 μm deep) was formed by doping. Then, a 6-μm-thick n-type epitaxial silicon layer was grown (Figure 4.14b). The buried n-type layer

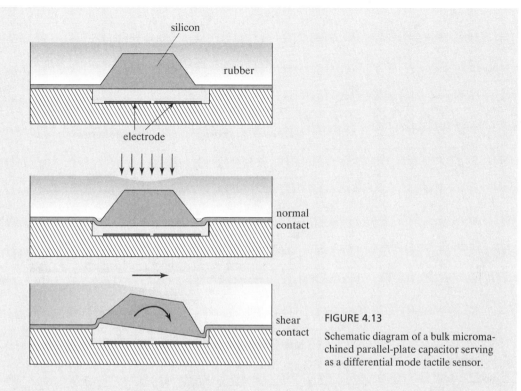

FIGURE 4.13

Schematic diagram of a bulk micromachined parallel-plate capacitor serving as a differential mode tactile sensor.

and the epitaxial layer constitute the thickness of a flexible membrane. A deep p-type diffusion doping is performed to electrically isolate each capacitor electrode (not shown). A composite layer of silicon oxide followed by silicon nitride is grown on both sides. The silicon nitride and oxide on the backside is patterned to serve as an etch mask for wet anisotropic etching (Figure 4.14d). After the etching, a contact pad on top of a membrane is formed by anisotropic silicon wet etch (Figure 4.14e). The silicon nitride and oxide layers are then removed using wet chemical etchants. The silicon wafer is then bonded to a glass wafer, which consists of a recessed region (3 μm deep) with patterned electrodes on the bottom. Anodic bonding is achieved at 400°C with a voltage bias of 1000–1200 V.

The sensor output with respect to applied calibration forces has been characterized. In the range of 0 to 1 gram, normal forces cause a capacitance change of 0.13 pF, whereas shear forces causes a differential capacitance of 0.32 pF. The capacitance change is linearly proportional to the calibration force within this range.

4.3.5 Parallel-Plate Actuators

Parallel-plate capacitors can be used for microactuation. Most common applications involve linear displacement vertical to the plane of electrodes or rotational displacement. When designing an electrostatic microactuator, it is important to keep in mind the trade-off between the

132 Chapter 4 Electrostatic Sensing and Actuation

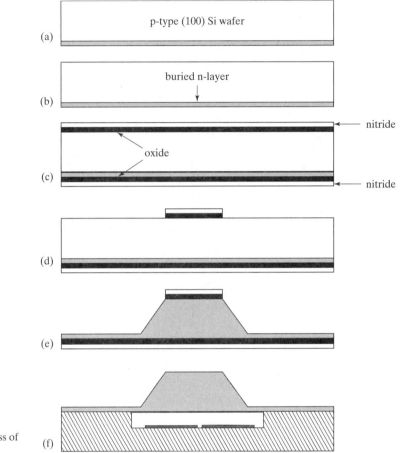

FIGURE 4.14

Fabrication process of tactile sensor.

range of movement and the available force. The amount of displacement that can be achieved with a parallel-plate capacitive actuator is limited by the initial gap spacing. Increasing the initial spacing allows for longer range movement but limits the magnitude of forces. Examples of such actuators used for optical applications are discussed in Chapter 15.

Parallel-plate actuators have been used to achieve a long range in plane movement by employing a device called the scratch-drive actuator (SDA) [33–35]. The principle of the SDA is shown in Figure 4.15. Each SDA consists of a parallel plate with a bushing along one edge. Under no applied voltage bias, the parallel plate is parallel to the substrate. When a bias voltage is applied, one edge of the parallel plate will contact the substrate first (Figure 4.15b). As the bias voltage gradually increases, the contact area between the top plate and the bottom substrate increases. This is often referred to as a "zipping" motion (Figure 4.15c). As the zipping motion progresses towards the edge with the bushing, the bushing is forced to rotate and "skids" when the lateral force caused by the zipping motion exceeds the friction force. Upon the removal of the bias voltage, the parallel-plate capacitor returns to the horizontal plane, but the plate travels by a small in-plane increment that is anchored by the landed bushing. Rapid succession of

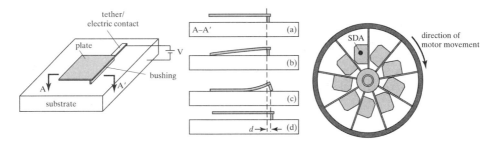

FIGURE 4.15

The operation principle of scratch drive actuators.

periodic actuation causes the scratch drive to achieve high-speed linear displacement. Velocities of SDA drives can reach 80 μm/s at a 1000-kHz activation frequency, with the speed linearly proportional to the frequency at a lower range. The linear output force of SDA is known to increase significantly, from 10 μN to 60 μN, when the voltage peak increases from 68 to 112 V. The maximum output force of SDA may reach 100 μN [34]. Using a series of SDA actuators harnessed to translational stages or rotary devices, researchers have been able to achieve continuous rotating motors [36] (Figure 4.15).

In the third stage of the SDA actuation, large angular displacement is manifested at the end with bushing. Electrostatic drives with zipping motion are capable of generating large angular displacement in this fashion. This capability is useful for generating large, off-substrate angular displacement [37]. (See Chapter 15 for additional discussions.)

4.4 INTERDIGITATED FINGER CAPACITORS

While parallel-plate capacitors generate sensing and actuation across planar electrodes facing each other, a different class of capacitors take advantage of capacitance generated from sidewalls of electrodes. Such capacitors provide alternative fabrication and operation modes compared with parallel-plate capacitors. They involve **interdigitated fingers** (IDT) to increase the edge coupling length (Figure 4.16) [38]. Two sets of electrodes are placed in the same plane parallel to the substrate. Generally, one set of fingerlike electrodes is fixed on-chip while a second set is suspended and free to move in one or more axes. Since the interdigitated fingers are shaped like the teeth on a comb, such configuration is commonly referred to as the **comb-drive** device.

In a generic configuration (Figure 4.16), two sets of fingers are in the same plane and their comb fingers are engaged with an overlapping distance of l_0. The length of the fingers is denoted as L_c. For simplicity, we assume the fingers in both sets have an identical thickness (t) and width (w_c). The distance between a fixed-comb finger and an immediately neighboring movable finger is d. The thickness of fingers corresponds to that of the conductive thin film.

The capacitance between a pair of electrode fingers is contributed by vertical surfaces of the fingers in the overlapped region, as well as by fringe capacitance fields. Capacitances derived from multiple finger pairs are connected in parallel. Hence, the total capacitance is a summation of capacitance contributed by neighboring fingers.

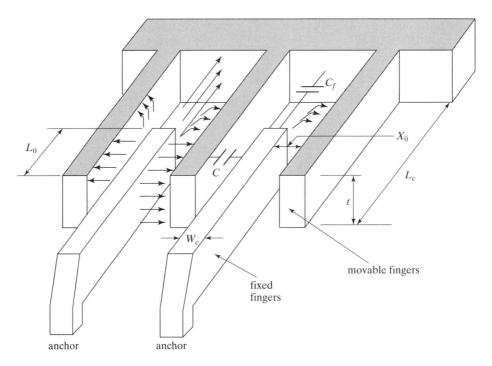

FIGURE 4.16

Perspective view of comb-drive sensors and actuators.

Opposite walls of comb fingers in the overlapped region form a parallel-plate capacitor and contribute a capacitance C (Figure 4.16). The magnitude of C between two immediate neighboring fingers is

$$C = \varepsilon_r \varepsilon_0 \frac{l_0 t}{d}. \tag{4.30}$$

The fringe capacitance, C_f, is difficult to estimate. There is no established compact analytical formula for estimating the fringe capacitance. The most accurate way to estimate the fringe capacitance is by using the finite element method (FEM) [39] (Figure 4.17). However, FEM analysis is time consuming and does not directly provide trends of influence by design variables. A simplistic way to estimate the fringe capacitance is to assume it is a fixed fraction of the capacitance developed in the overlapped regions [40]; however, this is not entirely accurate and the proportional constant depends on the dimensions of electrodes.

In this textbook, the fringe capacitance is lumped into a term C_f. When calculating the force sensitivity (for sensors) and actuation force (for actuators), the contribution of the fringe capacitance is ignored except in specially noted cases.

Two types of comb-drive devices are commonly encountered, depending on the relative movement of the two sets of comb fingers allowed by their mechanical suspensions. The first is a transverse comb-drive device, as shown in Figure 4.18. The set of free fingers moves in a direction

4.4 Interdigitated Finger Capacitors 135

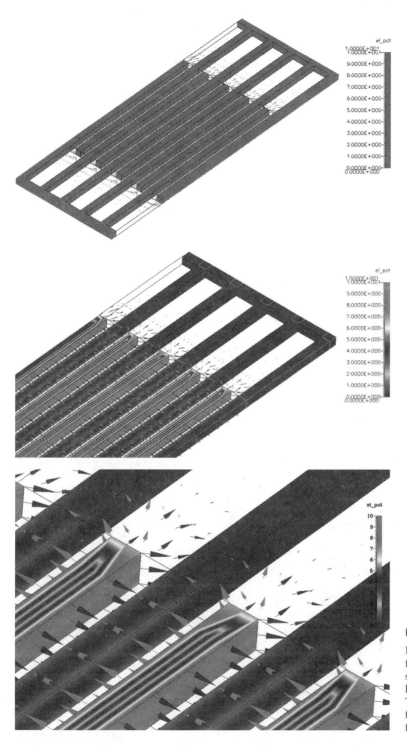

FIGURE 4.17

Distribution of electric-field lines between two sets of fingers, with three levels of magnification. The simulation is performed using software from CFDRC.

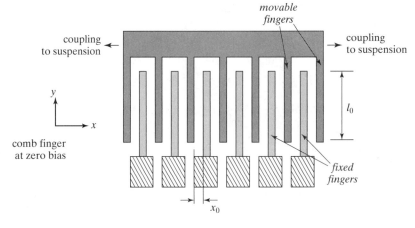
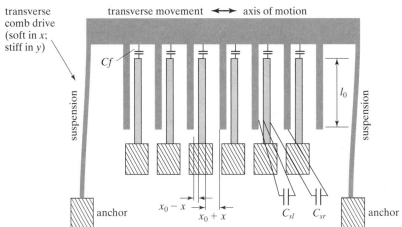
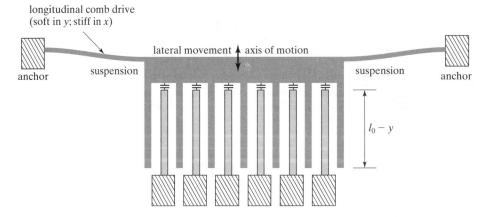

FIGURE 4.18

Schematics of (a) a comb drive, (b) a transverse comb drive, and (c) a longitudinal comb drive.

perpendicular to the longitudinal axis of comb fingers. Accelerometers for automotive air-bag deployment made by Analog Devices use transverse comb-drive configurations (see Chapter 1).

Let's focus on a single fixed finger and its two neighboring moving fingers. According to the earlier analysis, there are two major capacitances associated with each finger pair, one to the left-hand side of the finger, called C_{sl}, and one to the right-hand side, called C_{sr}. At rest, the values of these two capacitances are

$$C_{sl} = C_{sr} = \frac{\varepsilon_0 l_0 t}{x_0}. \tag{4.31}$$

When the free finger moves by a distance of x, the capacitance values of these two capacitors become

$$C_{sl} = \frac{\varepsilon_0 l_0 t}{x_0 - x} \tag{4.32}$$

and

$$C_{sr} = \frac{\varepsilon_0 l_0 t}{x_0 + x}. \tag{4.33}$$

The total value of capacitance is

$$C_{tot} = C_{sl} + C_{sr} + C_f. \tag{4.34}$$

When the transverse comb is used as a sensor, the displacement sensitivity (S_x) can be obtained by taking the derivative of C_{tot} with respect to x, namely,

$$S_x = \frac{\partial C_{tot}}{\partial x}. \tag{4.35}$$

If the transverse comb drive is used as an actuator, the magnitude of the force can be calculated by taking a derivative of the total stored energy with respect to x,

$$F_x = \left|\frac{\partial U}{\partial x}\right| = \left|\frac{\partial}{\partial x}\left(\frac{1}{2}C_{tot}V^2\right)\right|. \tag{4.36}$$

A second type of comb-drive device is called the longitudinal comb drive (Figure 4.18c). The direction of relative movement is along the longitudinal axis of the fingers, which is allowed by the suspension. With a lateral movement y, the capacitances associated with a single finger changes to

$$C_{sl} = C_{sr} = \frac{\varepsilon_0 (l_0 - y) t}{x_0}.$$

The displacement sensitivity can be obtained by taking the derivative of C_{tot} with respect to y:

$$S_y = \frac{\partial C_{tot}}{\partial y}. \tag{4.37}$$

Limited modeling results that incorporate the fringe electric field of comb drives are available. Earlier research has revealed more accurate models of the displacement sensitivity of longitudinal comb drives by considering fringe capacitance. An approximate model for the longitudinal comb drive, obtained by fitting data of finite element simulation, is [39]

$$\frac{dC}{dy} = C_1 + C_2 \tanh(C_3 + C_4 y) + C_5(L_f - y)^{-2}.$$

If the longitudinal comb drive is used as an actuator, the force can be calculated by taking a derivative of the total stored energy with respect to y:

$$F_y = \frac{\partial E}{\partial y} = \frac{\partial}{\partial y}\left(\frac{1}{2} C_{tot} V^2\right). \tag{4.38}$$

Early research has revealed the compact expressions for the force term when the fringe electric field is considered. According to Reference [41], the total force associated with each finger pair, considering the fringe capacitance, is

$$F = \frac{V}{2}\left[2\varepsilon_0 \frac{V}{\pi}\left\{\ln\left(\left(\left(\frac{w_c}{x_0} + 1\right)^2 - 1\right)\left(1 + \frac{2x_0}{w_c}\right)^{1+\frac{w_c}{x_0}}\right) + \frac{2\pi t}{x_0}\right\} - 2(x_0 + w_c)\frac{\varepsilon_0 V^2}{2\pi l_0}\right]. \tag{4.39}$$

However, this formula has not been matched with experimental data to prove its validity.

The sensing and actuation characteristics are strongly influenced by the mechanical supports connected to the set of fingers. Mechanical design issues have been studied. For example, the design of flexural suspensions can be optimized to generate large deformations while minimizing stress concentration [42].

Coplanar transverse and longitudinal comb drives are prevalent in MEMS. However, there are many different configurations and geometries of comb-drive capacitors that deviate from these two major configurations. Some comb drives are designed to produce vertical displacement, even when they initially reside in the same plane. This is achieved by taking advantage of the fringe-capacitance fields [6]. This levitation force is rather small, but it can be useful for applications such as optical-phase tuning. Reflective surfaces only need to be moved by up to one wavelength.

Comb fingers that do not reside in a common plane initially can be used to create out-of-plane forces or moments. The out-of-plane arrangement of comb fingers can be achieved through novel fabrication sequences or by using intrinsic stress-induced bending to lift one set of in-plane fingers angularly. Several applications are reviewed in 15.2.3 [43, 44]. For example, an optical modulator capable of 1.8 μm displacement has been made with 15 V driving voltage [45].

Nearly all existing comb fingers have large length-to-width aspect ratios and appear rectangular when viewed from the top or side. Alternative profiles are possible. Comb-drive fingers with curved profiles [46] can provide important benefits such as maximum force, increased linearity, or tailored force-displacement profiles [47].

4.5 APPLICATIONS OF COMB-DRIVE DEVICES

Comb-drive devices can be used in a variety of sensing and actuation applications. Comb-drive acceleration sensors and linear actuators are discussed below.

4.5.1 Inertia Sensors

Inertia sensors based on comb drives can be realized in a variety of ways. After all, one of the most classical MEMS sensors, the Analog Devices ADXL accelerometer, is based on coplanar transverse comb-drive fingers. The original ADXL accelerometer measures acceleration in only one in-plane axis. An acceleration sensor with a different sensitive axis (one pointing perpendicular to the substrate) is reviewed in Case 4.7. It utilizes the movement of fingers in the vertical direction of the substrate.

Case 4.7. Comb-Drive Accelerometer

A vertical comb acceleration sensor has an inertia mass connected to anchor frames by two torsional bars [48]. The transduction principle is similar to that of Case 4.2, except for one major difference: electrode configurations. In Case 4.2, a parallel capacitor is formed between the moving mass and the substrate. In this case, two sets of comb fingers that move out of plane with respect to each other are used (Figure 4.19). A change in capacitance is caused by the variation in the interelectrode overlap area.

For small angle displacement, the change in capacitance under an angular displacement θ is given by

$$\Delta C \cong \frac{n\varepsilon_r l_f}{d}(2l_m + l_f)\theta, \quad (4.40)$$

where l_m, l_f, d, and n are the length of the inertia mass, the length of the sensing finger, the gap distance, and the number of sense fingers. The rotational angle is related to the torque M by the expression

$$\theta = \frac{Ml_m l_b}{4\alpha G t w_b^3} a \quad (4.41)$$

where a and α are the acceleration and a correctional factor (0.281) accounting for the rectangular cross section of the torsion beam. The length, width, and thickness of the torsional beams are l_b, w_b, and t, respectively.

The fabrication process for the device is comparably simple when there is no requirement for IC integration. A process for prototyping is illustrated in Figure 4.20. The silicon wafer is heavily doped by boron to a depth of 12 μm (Figure 4.20a), which corresponds to

140 Chapter 4 Electrostatic Sensing and Actuation

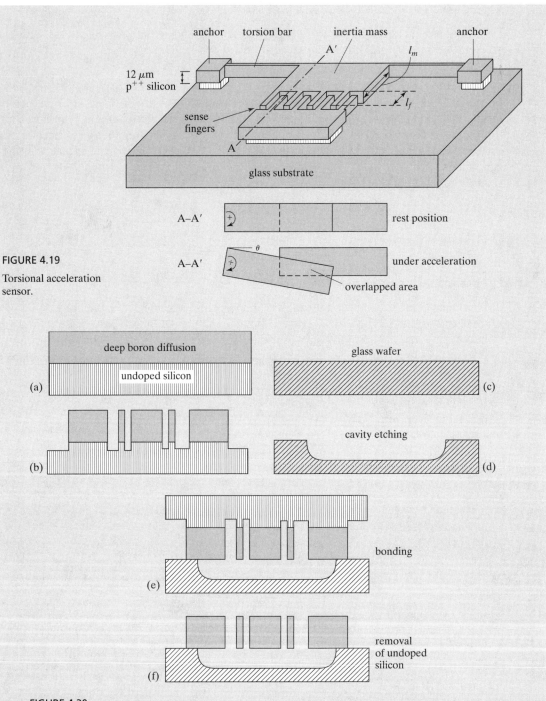

FIGURE 4.19
Torsional acceleration sensor.

FIGURE 4.20
Fabrication process of torsional acceleration sensor.

the thickness of the top electrode plates. Electroplated nickel is used (not shown) to mask against a deep reactive ion etching through the deeply doped region to define the fingers and mass (Figure 4.20b). A piece of glass is etched to make a cavity in a selective region (Figure 4.20d). The silicon wafer and the glass are bonded (Figure 4.20e). The undoped silicon layer is dissolved in silicon wet etchants (Figure 4.20f). The etching solution does not attack the deeply doped silicon and the glass significantly. A recessed cavity below the bank of the moving comb fingers allows a larger range of finger displacement and reduced aerodynamic damping.

Compared with parallel-plate capacitors, the capacitance between two neighboring sets of fingers is relatively small. However, we can achieve a large capacitance and force by increasing the number of comb pairs. The sensor is tested using a switched capacitance integrated circuit with a gain of 15 mV/fF. The sensitivity of the device is 300 mV/g.

Example 4.4. Sensitivity of the Accelerometer

A surface micromachined accelerometer is shown in Figure 4.21. The proof mass is supported by two cantilevers with length L, width w, and thickness t. The comb fingers have an overlapped length of l_0, thickness of t, and spacing of d. What is the sensitive axis of the sensor? Derive an expression for the acceleration sensitivity (the change of capacitance as a function of the applied acceleration).

Solution. The proof mass would experience strong resistance to movement in the y direction because it would be in the longitudinal direction of the support beams.

The support cantilever is wider than its thickness. Therefore, the resistance to movement in the x axis is greater than the resistance in the z axis. The sensor is most sensitive to acceleration along the z axis. The deformed shape of the sensor is shown in Figure 4.22.

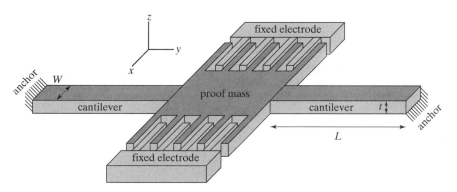

FIGURE 4.21

An out-of-plane accelerometer with a sensitive axis normal to the substrate plane.

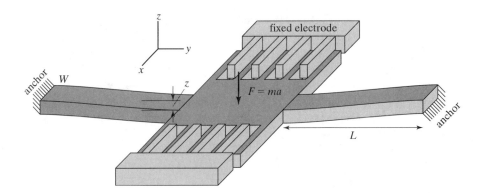

FIGURE 4.22

Profile sensor when deformed.

The mass of the proof mass is made up of two parts: the proof mass shuttle and comb fingers. The force constant associated with the mass is twice that of each individual fixed–guided cantilever. The overall force constant is

$$k = 2 \times \frac{12EI}{L^3}.$$

The total capacitance at rest is contributed by eight fixed electrodes and therefore 16 vertical wall capacitors. The value of the total capacitance is

$$C(t) = 16\left(\frac{\varepsilon_0 l_0 t_0}{d}\right).$$

The displacement in the z axis, z, causes the effective thickness (t) to change. Upon a displacement z, the capacitance becomes

$$C(t) = 16\left(\frac{\varepsilon_0 l_0 (t_0 - z)}{d}\right),$$

where the vertical displacement z is a function of the applied acceleration a,

$$z = \frac{ma}{\frac{24EI}{L^3}} = \frac{maL^3}{24EI}.$$

The relative change of capacitance with respect to a is

$$\frac{\partial C}{\partial a} = \frac{\partial}{\partial a}\left(16\left(\frac{\varepsilon_0 l_0 \left(t_0 - \frac{maL^3}{24EI}\right)}{d}\right)\right) = \frac{2\varepsilon_0 l_0 mL^3}{3dEI}.$$

4.5.2 Actuators

Comb-drive actuators are frequently used for generating in-plane or out-of-plane displacement. The amplitude of displacement under DC or quasi-static biasing is always rather limited. Here, we review two cases of longitudinal comb drives for achieving large-distance movement. In Case 4.8, large linear displacement is achieved by using relatively large comb-drive fingers. In Case 4.9, however, large rotational (or linear displacement) stems from mechanical gear engagement.

Case 4.8. Comb-Drive Actuator for Optical Switching

Large deflection electrostatic actuators have been used for realizing a 1×4 optical switch for fiber communication systems [49]. The switch consists of four linear stages, each moved with a lateral comb-drive device (Fig. 4.23). The actuator is capable of moving a mirror in and out of the designated optical path. Depending on which mirror is moved into the optical path, the light from an optical fiber can be reflected into one of four receiving channels. Using a 660-nm light with a typical fiber and lens, the beam remains collimated over a distance of 4 mm, and it has 98% of its optical power within a 140-μm diameter. Therefore, each actuator must travel a total distance of at least 140 μm, and the actuators for all four channels must fit within a 4-mm path.

An area-efficient comb-drive actuator with deflection ranges and force is therefore needed. A unique comb drive with linearly graded comb teeth is implemented. The engagement length varies across the comb finger set. The deflection measurements were repeatable to within ± 1 μm using this method. For equal-length combs, the displacement is 100 μm at a voltage of less than 150 V. For an actuator with comparable supports but linearly engaging combs, the displacement is greater than the equal-length comb case.

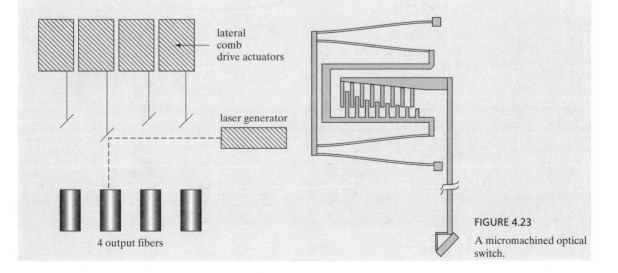

FIGURE 4.23
A micromachined optical switch.

Case 4.9. Comb-Drive Actuator with Large Displacement

The displacement achievable in a comb drive is rather limited, whether it is based on longitudinal or transverse combs. However, a larger displacement of mechanical elements can be realized using a number of designs. For example, comb drives with limited travel range can be used to achieve a large angular or linear displacement if the motion of the comb fingers is coupled to an object through stick-and-release mechanisms or gears. A number of strategies, including ones that do not use the electrostatic actuation principle, are reviewed in Section 15.2.2.

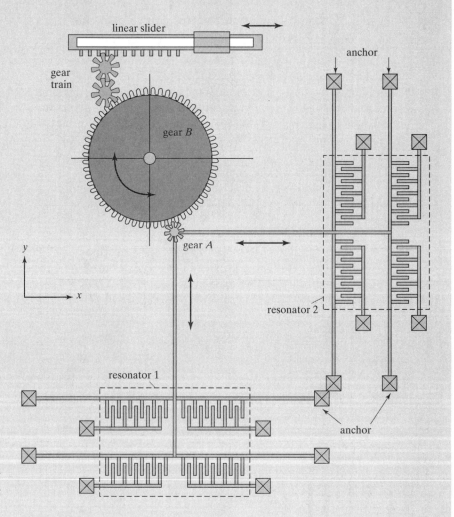

FIGURE 4.24

A micromachined gear movement mechanism.

We will examine a gear-train mechanism made by researchers at the Sandia National Laboratories (Figure 4.24). The gear train receives driving power from two sets of lateral comb drivers, labeled resonator 1 and resonator 2. Resonator 1 and resonator 2 both consist of lateral comb fingers connected in parallel to increased force output. Resonator 1 and 2 drive the gear A in y and x directions, respectively. The movement in the x and y axes is phase locked, so that gear A follows an elliptical path. Gear A therefore engages gear B intermittently, causing gear B to continuously rotate in one direction. The motion of gear B is further translated through the gear train to achieve linear movement of a linear slider.

SUMMARY

This chapter is dedicated to electrostatic sensing and actuation. The following list summarizes the major concepts, facts, and skills associated with this chapter:

- Two types of capacitive electrode configuration.
- Formula for estimating the capacitance and force between two parallel capacitor plates.
- Procedures for calculating the equilibrium position between two parallel capacitor plates when one of them is suspended by a mechanical spring support.
- The definition of the pull-in effect.
- The implication of the pull-in effect on the operational characteristics of parallel-plate capacitors.
- Procedure for estimating the pull-in voltage and pull-in distance.
- Different configurations of interdigitated finger capacitors and their relative pros and cons in terms of maximum displacement, linear/angular displacement, and force output.

PROBLEMS

Problem 4.1: Design
A parallel-plate capacitor has an area of 100×100 μm^2. Calculate the capacitance values for two distances between the electrode plates: 1 and 0.5 μm. The medium is air.

Problem 4.2: Design
For a cantilever beam with length l, width w, and thickness t, which one of the following is NOT true if the beam dimensions (length, width, and thickness) are each reduced 1000 times?

1. Reduced force constant and more flexible beam
2. Increased resonant frequency
3. Reduced fracture toughness
4. Increased surface area/volume ratio

Problem 4.3: Design

Two parallel capacitor plates have an area of 1 mm² and a spacing of 1 μm (air gap). One is fixed, and another one is suspended by a mechanical spring with a force constant of 1 N/m. What is the capacitance value (C_0) when no voltage is applied between the two plates?

1. $C_0 = 8.85 \times 10^{-12}$ f
2. $C_0 = 8.85 \times 10^{-9}$ f
3. $C_0 = 8.85 \times 10^{-6}$ f
4. $C_0 = 0$ f

Problem 4.4: Design

A parallel-plate capacitor with four silicon support beams is shown below. The top plate has an area of 1×1 mm². The four support beams are each 500 μm long, 5 μm wide, and 0.3 μm in thickness. What is the force constant K_m experienced by the parallel-plate capacitor? (*Note*: The diagram is not drawn to scale.) Young's modulus of silicon is 120 GPa.

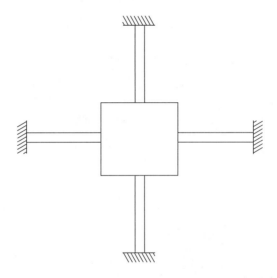

1. 0.1 N/m
2. 0.4 N/m
3. 0.00013 N/m
4. 0.00052 N/m
5. 0.00003 N/m

Problem 4.5: Design

If the support beams in Problem 4.6 become 0.25 mm long (i.e., half as long as before), which of the following statements would be correct?

1. The pull-in voltage would increase by four times.
2. The plate would pull in when the displacement of the beam exceeds $x_0/6$.

3. The electric force constant (k_e) at the pull-in point will be increased by eight times.
4. The resonant frequency of the device will decrease.

Problem 4.6: Design

According to the conditions given in Problem 4.6, under a bias voltage of 0.3 volt, what is the distance between the two plates? The original spacing between the two plates is 1 μm.

Problem 4.7: Design

Derive the analytical expression of capacitive sensitivity with respect to changes in x for the transverse comb-drive device. Discuss at least two strategies for increasing the sensitivity.

Problem 4.8: Design

Derive an analytical expression of capacitive sensitivity with respect to changes in y for the longitudinal comb-drive device. Discuss at least two strategies for increasing the sensitivity.

Problem 4.9: Design

For the silicon-oxide cantilever described in Case 4.1, calculate the resonant frequency of the cantilever if the metal thin film on the silicon oxide is ignored. Compare the results to the published resonant frequency.

Problem 4.10: Review

For the sensor discussed in Case 4.1, determine the sensitivity with respect to minimal tip displacement at the free end of the cantilever.

Problem 4.11: Design

Consider the silicon oxide/metal composite cantilever described in Case 4.1 at rest (zero bias voltage). The bending due to intrinsic stress results in an elevation difference of $\delta(x)$ measured at two ends of the beam. The bent beam assumes the shape of a curve with a radius of curvature of R, with the angle at the end of the cantilever being 1.5°. Derive an expression for the total capacitance between the beam and the substrate.

Problem 4.12: Fabrication

For the fabrication process of the silicon-oxide cantilever described in Case 4.1, find the etch rate of the silicon anisotropic etchant used in step (j) on single-crystal silicon and all other materials exposed to it. Identify all materials that are exposed and vulnerable to the silicon etchant. Use the table in Reference [50].

Problem 4.13: Design

Calculate the area of the bottom electrode in Case 4.2, assuming the fringe capacitance can be ignored.

Problem 4.14: Design

Using the information given in Case 4.2, find the maximum vertical displacement of the torsion plate under an acceleration of 1 g. (*Hint*: Use the torsional bar displacement analysis.)

Problem 4.15: Design

Explain the origin of sensitivity to temperature for the acceleration sensor discussed in Case 4.1.

Problem 4.16: Review

The pressure sensor in Case 4.3 has a membrane with a diameter of 1 mm. Use information given in the process flow to calculate the precise spacing between the two electrodes; be accurate to within 5%. Find the nominal value of the capacitance.

Problem 4.17: Review

Referring to Problem 4.19, build a spreadsheet to calculate the relative cost of materials and processing. Consider human labor hours for material preparation, processing, lithography, and raw materials (such as silicon wafers, etc.). Form a group of three to four students and finish this assignment as a group.

Problem 4.18: Design

In the device discussed in Case 4.4, what is the nominal capacitance value of the capacitor? What is the implication for the performance and fabrication if the area of the capacitor is increased?

Problem 4.19: Design

For the shear stress sensor device in Case 4.5, derive an analytical expression of the displacement d under a given shear stress; ignore the distributed drag force on the cantilevers. In other words, consider only the drag force on the floating plate. Based on the expression, identify three key strategies for increasing the sensitivity. Discuss the impact on the fabrication associated with each strategy.

Problem 4.20: Design

Derive Equations (4.28) and (4.29) of the tactile sensor operation discussed in Case 4.6.

Problem 4.21: Fabrication

For the fabrication process of the silicon-oxide cantilever described in Case 4.1, which steps need to be changed if we want to obtain a silicon-nitride cantilever via low-pressure chemical vapor deposition? What would be the companion etching method in those steps affected?

Problem 4.22: Fabrication

Form a group of three or four students; discuss an alternative method of forming an integrated vacuum (below 100 mTorr) using a microfabrication process. Discuss how this particular method may be used for making a pressure sensor with a vacuum reference similar to the one in Case 4.3.

Problem 4.23: Fabrication

In step (n) of the process for realizing the device discussed in Case 4.3, discuss the etch selectivity of the etchant used to dissolve the silicon wafer on all materials that are exposed to it. Discuss potential precautions for the process and at least one way to improve the process robustness.

Problem 4.24: Fabrication

In step (g) of the fabrication process of the microphone of Case 4.4, discuss the etch-rate selectivity between the etchant and all other materials that are directly exposed to it.

Problem 4.25: Challenge

Discuss how we can design a practical electrostatic actuator to reach the second, infrequently used equilibrium position given graphically in Figure 4.4 and analytically in Equation (4.11).

Problem 4.26: Challenge

Design an XY platform with an area of 200 μm $\times$ 200 μm; it must be capable of an independent displacement of 5 μm in both axes. The voltage required should be minimized. Form a team of three to four students, and come up with an analytical design, a device layout, and a fabrication process for such a device.

Problem 4.27: Challenge

Repeat Problem 4.26. This time, try to limit the driving voltage to below 36 V.

Problem 4.28: Challenge

Repeat Problem 4.26. This time, integrate nine sharp tips (for scanning tunneling microscopy or nanolithography) on the stage.

Problem 4.29: Challenge

Repeat Problem 4.26. This time, assume that six conducting wires must be connected to devices located on the platform for sensing and actuation. For example, the conducting wires may be used to address nanolithography tips on the platform. The conducting wires will complicate the mechanical beam design if the wires run along the mechanical suspension.

Problem 4.30: Challenge

The translation stages of XY axes are widely used for alignment and scanning. Let's examine issues related to building an on-chip XY translation stage for in-plane scanning. It is desired to move a 2 mm × 2 mm shutter plate using in-plane XY directions. An electrostatic actuator is used to move a silicon shuttle plate by 10 μm under a bias voltage of lower than or equal to 200 V. Assume the comb drives and suspensions are made in 500-μm-thick silicon wafers using deep-reactive ion-etching processing, and the minimal linewidth and spacing is 5 mm. Determine whether a practical design can allow sufficient force, force constant, and resonance frequency (100 Hz).

REFERENCES

1. Fan, L.-S., Y.-C. Tai, and R.S. Muller, *IC-Processed Electrostatic Micromotors*. Presented at IEEE International Electronic Devices Meeting, 1988.
2. Livermore, C., et al., *A High-Power MEMS Electric Induction Motor*, Microelectromechanical Systems, Journal of, 2004. **13**: p. 465–471.
3. Greywall, D.S., et al., *Crystalline Silicon Tilting Mirrors for Optical Cross-Connect Switches*, Microelectromechanical Systems, Journal of, 2003. **12**: p. 708–712.
4. Kung, J.T., and H.-S. Lee, *An Integrated Air-Gap-Capacitor Pressure Sensor and Digital Readout with Sub-100 Attofarad Resolution*, Microelectromechanical Systems, Journal of, 1992. **1**: p. 121–129.
5. Lee, K.B., and Y.-H. Cho, *Laterally Driven Electrostatic Repulsive-Force Microactuators Using Asymmetric Field Distribution*. Microelectromechanical Systems, Journal of, 2001. **10**: p. 128–136.
6. Tang, W.C., M.G. Lim, and R.T. Howe, *Electrostatic Comb Drive Levitation and Control Method*, Microelectromechanical Systems, Journal of, 1992. **1**: p. 170–178.
7. Zou, Q., et al., *A Novel Integrated Silicon Capacitive Microphone-Floating Electrode "Electret" Microphone (FEEM)*, Microelectromechanical Systems, Journal of, 1998. **7**: p. 224–234.
8. Hsieh, W.H., T.-Y. Hsu, and Y.-C. Tai, *A Micromachined Thin-Film Teflon Electret Microphone*. Presented at Solid State Sensors and Actuators, Transducers '97 Chicago. International Conference, 1997.
9. Jacobs, H.O., and G.M. Whitesides, *Submicrometer Patterning of Charge in Thin-Film Electrets*, Science, 2001. **291**: p. 1763–1766.

10. Van Kessel, P.F., et al., *A MEMS-Based Projection Display*, Proceedings of the IEEE, 1998. **86**: p. 1687–1704.
11. Dokmeci, M., and K. Najafi, *A High-Sensitivity Polyimide Capacitive Relative Humidity Sensor for Monitoring Anodically Bonded Hermetic Micropackages*, Microelectromechanical Systems, Journal of, 2001. **10**: p. 197–204.
12. Sohn, L.L., et al., *Capacitance Cytometry: Measuring Biological Cells One-by-One*, Proc. National Academy of Sciences, 2000. **97**: p. 10687–10690.
13. Mastrangelo, C.H., X. Zhang, and W.C. Tang, *Surface-Micromachined Capacitive Differential Pressure Sensor with Lithographically Defined Silicon Diaphragm*, Microelectromechanical Systems, Journal of, 1996. **5**: p. 98–105.
14. Hall, N.A., and F.L. Degertekin, *Integrated Optical Interferometric Detection Method for Micromachined Capacitive Acoustic Transducers*, Applied Physics Letters, 2002. **80**: p. 3859–3861.
15. Petersen, K.E., A.S. Hartel, and N.F. Raley, *Micromechanical Accelerometer Integrated with MOS Detection Circuitry*, IEEE Transactions on Electron Devices, 1982. **ED-29**: p. 23–27.
16. Nemirovsky, Y., and O. Bochobza-Degani, *A Methodology and Model for the Pull-In Parameters of Electrostatic Actuators*, Microelectromechanical Systems, Journal of, 2001. **10**: p. 601–615.
17. Rocha, L.A., E. Cretu, and R.F. Wolffenbuttel, *Analysis and Analytical Modeling of Static Pull-In with Application to MEMS-Based Voltage Reference and Process Monitoring*, Microelectromechanical Systems, Journal of, 2004. **13**: p. 342–354.
18. Degani, O., et al., *Pull-In Study of an Electrostatic Torsion Microactuator*, Microelectromechanical Systems, Journal of, 1998. **7**: p. 373–379.
19. Loke, Y., G.H. McKinnon, and M.J. Brett, *Fabrication and Characterization of Silicon Micromachined Threshold Accelerometers*, Sensors and Actuators A: Physical, 1991. **29**: p. 235–240.
20. Osterberg, P., et al., *Self-Consistent Simulation and Modelling of Electrostatically Deformed Diaphragms*. Presented at Micro Electro Mechanical Systems, MEMS '94 Proceedings, IEEE Workshop, 1994.
21. Chan, E.K., and R.W. Dutton, *Electrostatic Micromechanical Actuator with Extended Range of Travel*, Microelectromechanical Systems, Journal of, 2000. **9**: p. 321–328.
22. Hung, E.S., and S.D. Senturia, *Extending the Travel Range of Analog-Tuned Electrostatic Actuators*, Microelectromechanical Systems, Journal of, 1999. **8**: p. 497–505.
23. Zou, J., C. Liu, and J. Schutt-aine, *Development of a Wide-Tuning-Range Two-Parallel-Plate Tunable Capacitor for Integrated Wireless Communication System*, International Journal of RF and Microwave CAE, 2001. **11**: p. 322–329.
24. Nadal-Guardia, R., et al., *Current Drive Methods to Extend the Range of Travel of Electrostatic Microactuators Beyond the Voltage Pull-In Point*, Microelectromechanical Systems, Journal of, 2002. **11**: p. 255–263.
25. Cole, J.C., *A New Sense Element Technology for Accelerometer Subsystems*. Presented at Digest of Technical Papers, 1991 International Conference on Solid-State Sensors and Actuators, 1991.
26. Chavan, A.V., and K.D. Wise, *Batch-Processed Vacuum-Sealed Capacitive Pressure Sensors*, Microelectromechanical Systems, Journal of, 2001. **10**: p. 580–588.
27. Liu, C., et al., *A Micromachined Flow Shear Stress Sensor Based on Thermal Transfer Principles*, IEEE/ASME Journal of Microelectromechanical Systems (JMEMS), 1999. **8**: p. 90–99.
28. Jin, X., et al., *Fabrication and Characterization of Surface Micromachined Capacitive Ultrasonic Immersion Transducers*, Microelectromechanical Systems, Journal of, 1999. **8**: p. 100–114.

29. Pederson, M., W. Olthuis, and P. Bergveld, *High-Performance Condenser Microphone with Fully Integrated CMOS Amplifier and DC–DC Voltage Converter*, Microelectromechanical Systems, Journal of, 1998. **7**: p. 387–394.

30. Ladabaum, I., et al., *Surface Micromachined Capacitive Ultrasonic Transducers*, Ultrasonics, Ferroelectrics and Frequency Control, IEEE Transactions on, 1998. **45**: p. 678–690.

31. Schmidt, M.A., et al., *Design and Calibration of a Microfabricated Floating-Element Shear-Stress Sensor*, Electron Devices, IEEE Transactions on, 1988. **35**: p. 750–757.

32. Chu, Z., P.M. Sarro, and S. Middelhoek, *Silicon Three-Axial Tactile Sensor*, Sensors and Actuators A: Physical, 1996. **54**: p. 505–510.

33. Akiyama, T., and K. Shono, *Controlled Stepwise Motion in Polysilicon Microstructures*, Microelectromechanical Systems, Journal of, 1993. **2**: p. 106–110.

34. Akiyama, T., D. Collard, and H. Fujita, *Scratch Drive Actuator with Mechanical Links for Self-Assembly of Three-Dimensional MEMS*, Microelectromechanical Systems, Journal of, 1997. **6**: p. 10–17.

35. Akiyama, T., D. Collard, and H. Fujita, *Correction to Scratch Drive Actuator with Mechanical Links for Self-Assembly of Three-Dimensional MEMS*, Microelectromechanical Systems, Journal of, 1997. **6**: p. 179.

36. Donald, B.R., et al., *Power Delivery and Locomotion of Untethered Microactuators*, Microelectromechanical Systems, Journal of, 2003. **12**: p. 947–959.

37. Chiou, J.-C., and Y.-C. Lin, *A Multiple Electrostatic Electrodes Torsion Micromirror Device with Linear Stepping Angle Effect*, Microelectromechanical Systems, Journal of, 2003. **12**: p. 913–920.

38. Tang, W.C., T.C.H. Nguyen, and R.T. Howe, *Laterally Driven Polysilicon Resonant Microstructures*, Sensors and Actuators A, 1989. **20**: p. 25–32.

39. Judy, M.W., *Micromechanisms Using Sidewall Beams*. In Ph.D. Thesis, Electrical Engineering and Computer Science Department. Berkeley, CA: University of California at Berkeley, 1994, p. 273.

40. Fedder, G.K., *Simulation of Microelectromechanical Systems*. In Ph.D. Thesis, Electrical Engineering and Computer Science Department. Berkeley, CA: University of California at Berkeley, 1994, p. 298.

41. Johnson, W.A., and L.K. Warne, *Electrophysics of Micromechanical Comb Actuators*, Microelectromechanical Systems, Journal of, 1995. **4**: p. 49–59.

42. Pisano, A.P., and Y.-H. Cho, *Mechanical Design Issues in Laterally-Driven Microstructures*, Sensors and Actuators A: Physical, 1990. **23**: p. 1060–1064.

43. Conant, R., *Thermal and Electrostatic Microactuators*. In Electrical Engineering and Computer Sciences. Berkeley: University of California at Berkeley, 2002.

44. Krishnamoorthy, U., D. Lee, and O. Solgaard, *Self-Aligned Vertical Electrostatic Combdrives for Micromirror Actuation*, Microelectromechanical Systems, Journal of, 2003. **12**: p. 458–464.

45. Lee, A.P., et al., *Vertical-Actuated Electrostatic Comb Drive with In Situ Capacitive Position Correction for Application in Phase Shifting Diffraction Interferometry*, Microelectromechanical Systems, Journal of, 2003. **12**: p. 960–971.

46. Jensen, B.D., et al., *Shaped Comb Fingers for Tailored Electromechanical Restoring Force*, Microelectromechanical Systems, Journal of, 2003. **12**: p. 373–383.

47. Ye, W., S. Mukherjee, and N.C. MacDonald, *Optimal Shape Design of an Electrostatic Comb Drive in Microelectromechanical Systems*, Microelectromechanical Systems, Journal of, 1998. **7**: p. 16–26.

48. Selvakumar, A., and K. Najafi, *Vertical Comb Array Microactuators*, Microelectromechanical Systems, Journal of, 2003. **12**: p. 440–449.
49. Grade, J.D., H. Jerman, and T.W. Kenny, *Design of Large Deflection Electrostatic Actuators*, Microelectromechanical Systems, Journal of, 2003. **12**: p. 335–343.
50. Williams, K.R., K. Gupta, and M. Wasilik, *Etch Rates for Micromachining Processing—Part II*, Microelectromechanical Systems, Journal of, 2003. **12**: p. 761–778.

CHAPTER 5

Thermal Sensing and Actuation

5.0 PREVIEW

This chapter investigates design, materials, and fabrication issues pertaining to the development of micromachined thermal sensors and actuators. Thermal sensors have two connotations: (1) sensors *for* measuring thermal properties such as temperature and heat, and (2) sensors *based on* thermal transfer principles. General thermal transfer principles are discussed in Section 5.1. Section 5.2 is devoted to sensors and actuators based on thermal expansion of materials. More specifically, we will focus on sensors and actuators based on the thermal bimorph configuration, and actuators based on thermal expansion of structures made of a single material. In Sections 5.3 and 5.4, two mechanisms for temperature sensing are discussed: thermal couples and resistive temperature sensors (or thermal resistors). Finally, we review a number of thermal sensor applications in Section 5.5.

5.1 INTRODUCTION

5.1.1 Thermal Sensors

The measurement of temperature and heat is widely practiced and can be achieved using many different principles. This chapter focuses on discussing just a handful of principles that are widely used in and unique to the MEMS field. These include (1) thermal bimorph sensors (Section 5.2); (2) thermal couples (Section 5.3); and (3) thermal resistive sensors (Section 5.4).

Temperature sensing is not only useful for analysis of thermal behavior. The thermal transfer process—through conduction, convection, or radiation—depends on certain physical variables, such as the spacing between objects, travel velocity of media, and the properties of materials and media. Many measurement tasks can be achieved, including the sensing of distance [1], acceleration [2], flow speed [3], and materials characteristics [4].

5.1.2 Thermal Actuators

Actuation of microscale devices and structures can be achieved by injecting or removing heat. Changes in temperature profile in turn result in mechanical displacement or force output, through thermal expansion [5], contraction, or phase change. The temperature of a microstructure can be raised by absorption of electromagnetic waves (including light), ohmic heating (joule heating), conduction, and convection heating. Cooling can be achieved via conduction dissipation, convection dissipation, radiation dissipation, and active thermoelectric cooling.

Thermal actuation is used in commercial MEMS products. Many ink-jet printers today eject ink droplets using thermal expansion of liquid inks. A schematic diagram of a single thermal ink-jet printer nozzle is shown in Figure 5.1. It consists of a micromachined fluid chamber with an opening. A microheater is embedded on the substrate within the cavity. A pulse of electrical current raises the temperature of the heater and produces a vapor bubble that squeezes the liquid ink and ejects a droplet. Upon cooling, the vapor bubble collapses and the cavity is refilled with ink for next firing.

Due to the use of the microfabrication method, the volume of ink and the thermal mass associated with the ink are small. The heating and cooling of the fluid can be performed at high speed. For many commercial ink cartridges, it takes 1 μs or less for the vapor bubble to initiate, and 15 μs for the ink to be ejected. The refill of the cavity takes approximately 24 μs.

The successful design of a reliable ink-jet nozzle involves electrical, mechanical, and thermal aspects. For example, the peak pressure within the cavity can reach 14 ATM with the peak temperature at the surface around 330°C. Both the positive and negative (cavitation) pressures are so intense that they can cause cracking of material layers over time. This warrants special design considerations. In order to eliminate cracking stemming from repeated thermal expansion and contraction, the heater is engineered from a proprietary metal oxide material that has a thermal expansion coefficient of nearly zero.

5.1.3 Fundamentals of Thermal Transfer

Temperature is manifested at the microscale by the vigorousness of atomic vibration. Heat transfer occurs whenever a temperature gradient is present in a material. Successful design of thermal actuators and sensors requires familiarity with heat transfer processes.

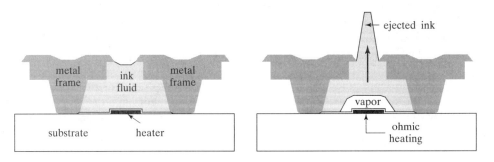

FIGURE 5.1
The principle of ink ejection.

There are four possible mechanisms for heat to move from one point to another:

1. **Thermal conduction** is the transfer of heat through a solid media in the presence of a temperature gradient.
2. **Natural** (passive) **thermal convection** is the transfer of heat from a surface into a stationary body of fluid. A temperature gradient in a fluid induces local flow movement through buoyancy. The movement of fluid mass facilitates heat transfer.
3. **Forced thermal convection** is the transfer of heat to a body of moving fluid. The bulk fluid movement provides enhanced heat transfer compared with that of natural convection.
4. **Radiation** represents the loss or gain of heat through the electromagnetic radiation propagating in vacuum or air.

The governing equations relating the heat transfer rate and the temperature gradient for the four heat transfer mechanisms are

$$\text{conduction: } q_{cond}'' = -k \frac{dT}{dx} \tag{5.1}$$

$$\text{natural and forced convection: } q_{conv}'' = h(T_s - T_\infty) \tag{5.2}$$

$$\text{radiation: } E = \varepsilon \sigma T_R^4. \tag{5.3}$$

In these equations, q_{cond}'' is conduction heat flux (W/m^2) along a given axis (designated x), κ the thermal conductivity, q_{conv}'' the convection heat flux (W/m^2), h the convective heat transfer coefficient (W/m^{2}K), T_s and T_∞ the temperatures of the surface and fluid, respectively, E the emissive power or the rate at which energy is released per unit area by radiation (W/m^2), T_R the absolute temperature (K) of a surface, σ the Stefan–Boltzmann constant ($\sigma = 5.67 \times 10^{-8}$ W/(m^{2}K^4)), and ε the radiative emissivity (with values in the range $0 \leq \varepsilon \leq 1$).

The convective heat transfer coefficient is influenced by surface geometries, fluid velocity, viscosity, and thermal diffusivity. Typical values of the convective heat transfer coefficients are summarized in Table 5.1.

These four heat transfer principles can be found everywhere in our daily lives. Let's examine heat flow pathways associated with the case of boiling a pot of water on an electric stove. Heat is generated at the heating coil by passing current and is eventually lost to the ambient background, which is assumed to be at a constant room temperature. Major heat transfer pathways and the direction of heat flux are identified by arrows (Figure 5.2).

TABLE 5.1 Typical Values of Convection Transfer Coefficients.

		h(W/m$^2 \cdot k$)
Natural convection	Gases	2–25
	Liquids	50–1000
Forced convection	Gases	25–250
	Liquids	100–20,000
Convection with phase change (boiling or condensation)		2500–100,000

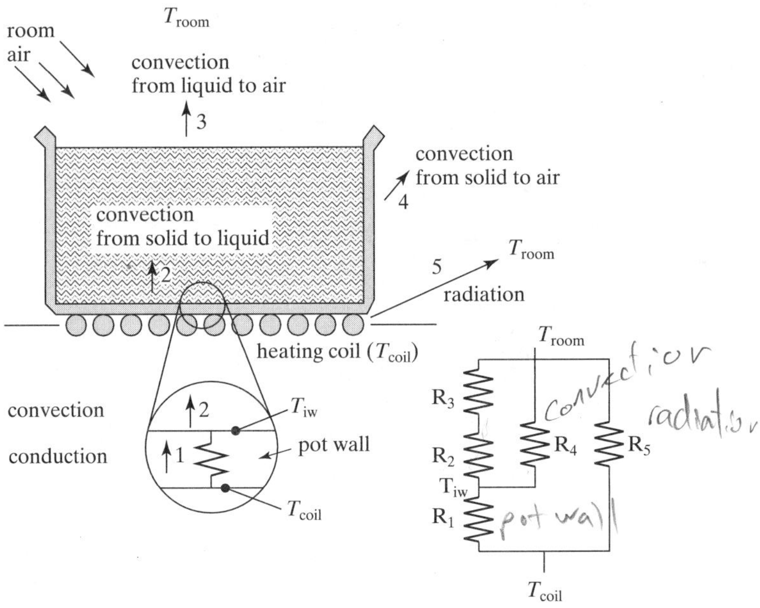

156 Chapter 5 Thermal Sensing and Actuation

FIGURE 5.2

Heat transfer processes associated with pot heating.

The heat produced by the heating coil first travels through the walls of the pot to reach the body of water inside. Heat transfer occurs within the thickness of the wall via *thermal conduction*. Once the heat reaches the interior wall of the pot, heat transfer to the liquid within begins. The liquid mass closest to the wall warms up and begins to rise, setting up a *natural convection*, which brings heat away from the pot wall to the interior of the body of water. If the liquid in the pot were stirred, the heat transfer from the inside wall of the pot to the liquid could become stronger as natural convection is replaced by *forced convection*.

The body of water is exposed to air at the top. If the air outside is still, heat will transfer from the water to the air through natural convection. On the other hand, if the air were moving (e.g., stirred by a fan), heat would travel from the hot liquid to the air by forced convection.

Meanwhile, a person standing nearby feels the heat wave coming from the heating coil. Heat is said to move through the air by radiation. Assuming the heating coil is at a significantly high temperature, the radiation heat transfer can be quite strong. Certainly, the radiative dissipation reduces the energy efficiency of heating the water.

Even for such a simple example, the heat transfer pathways are quite complex. The thermal transfer pathway illustrated in Figure 5.2 is simplified and only serves the purpose of illustrating the difference between various heat transfer principles. In fact, there are many secondary heat transfer pathways pertaining to this case. For example, heat is lost to the air from the outer pot wall through radiation and convection. Temperatures of objects are considered spatially discrete here; in fact, the temperature profile varies continuously.

A heat flow will result between two points of different temperatures. The ability of a media or an object to transfer heat between two points is quantified by its **thermal resistance**. The greater the thermal resistance between two points, the better the thermal isolation and the smaller the heat transfer rate under a given temperature difference (thermal driving force).

5.1 Introduction

The concept of thermal resistance can be better understood and appreciated by examining the analogy between heat transfer and electrical current flow. Temperature difference is the driving force for heat flow and transfer. The difference of temperature between two points sets up a heat flow, much like an electrical voltage gives rise to organized movement of charges (electric current). The electrical resistance is defined as the ratio of voltage and current. Likewise, the thermal resistance can be defined as the ratio of temperature difference and the heat flow.

Figure 5.2 shows an equivalent thermal resistance network associated with the water-heating example. Heat is transferred between the coil (T_{coil}) and the ambient (T_{room}). The subscripts for each thermal resistance component are related to the heat transfer pathway diagrammed.

We will focus on discussing thermal resistance associated with the thermal conduction process. The value of the thermal resistance of a microstructure is influenced by its dimensions, as well as by its thermal properties of the media.

The analytical expression of conductive thermal resistance under one-dimensional heat transfer is the simplest. The conductive thermal resistance of longitudinal objects with the uniform cross section (such as beams) can be easily calculated.

For a longitudinal rod with a length of l, a constant cross section of A, and a thermal resistivity of ρ_{th} ($=1/\kappa$), the heat flow through it with the difference of terminal temperatures being ΔT is given by

$$q_{cond} = q_{cond}'' \cdot A = -kA \frac{\Delta T}{l}. \tag{5.4}$$

Using the thermal-electrical analogy, the thermal resistance is

$$R_{th} = \left| \frac{\Delta T}{q_{cond}} \right| = \frac{1}{\kappa} \frac{l}{A} = \rho_{th} \frac{l}{A}. \tag{5.5}$$

Recall that for an electrical resistor with a length of l, cross section of A, and an electrical resistivity of ρ, the total resistance is

$$R = \rho \frac{l}{A}. \tag{5.6}$$

It is obvious that Equation (5.5) and Equation (5.6) have very similar forms.

For two-dimensional thermal conductors (e.g., a membrane heated in the center) and three-dimensional ones (e.g., a volume heated from the center), the effective thermal resistance is much more difficult to estimate. In these cases, computer numerical simulation or direct experimental methods may be needed. We will discuss the methodology of experimentally measuring the thermal resistance associated with a micromachined resistive temperature sensor in Section 5.4.

Information about the thermal resistance associated with the convection and radiation processes can be found elsewhere [6]. The heat transfer coefficient of convective heat transfer is dependent on the geometries and the flow rate regimes. The thermal resistance associated with radiation heat loss is actually a function of the temperatures.

Example 5.1. Thermal Resistance of a Suspended Bridge

A resistive ohmic heater is located in the middle of a suspended fixed–fixed cantilever. The beam is made of silicon nitride and the metal leads are made of aluminum. Find the numerical value of the thermal resistance experienced by the heater. Assume the width of the beam is 10 μm, $L = 100$ μm, and the thickness of both aluminum (t_m) and silicon nitride (t_b) is 0.2 μm. If the input power to the heater is 1 mW, what is the steady temperature assuming the silicon bulk stays at 27°C?

Solution. The heater is connected to the substrate frame by four parallel connected thermal resistors, two of them contributed by the metal layer and two by the silicon beam. The substrate frame has much greater thermal mass than the bridge—its temperature is assumed to be a constant. The substrate frame is called a heat sink. Heat flow occurs between the heater and the substrate.

The thermal conductivity of aluminum and silicon nitride are 240 and 5 W/mK, respectively. The thermal resistance associated with each metal thermal resistor is

$$R_m = \rho_{th,m} \frac{L}{W_m t_m} = \frac{1}{240} \times \frac{200 \times 10^{-6}}{0.2 \times 10 \times 10^{-12}} = 4.17 \times 10^5 \text{ K/W}$$

The thermal resistance associated with each silicon nitride beam is

$$R_{SiN} = \rho_{th,SiN} \frac{L}{W_{SiN} t_{SiN}} = \frac{1}{5} \times \frac{200 \times 10^{-6}}{0.2 \times 10 \times 10^{-12}} = 2 \times 10^7 \text{ K/W}.$$

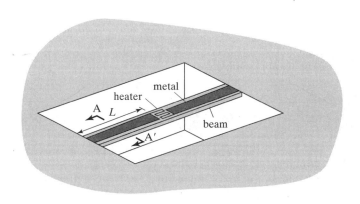

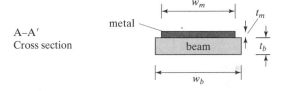

A–A′
Cross section

These four resistors are connected in parallel. The total thermal resistance R is found by

$$\frac{1}{R} = \frac{2}{R_m} + \frac{2}{R_b}.$$

The total resistance is

$$R = 2.042 \times 10^5 \text{ K/W}.$$

If we ignore radiation heat loss and conduction/convection from the suspension to ambient air, the heat generated (1 mW) is entirely conducted through the bridge. Therefore, the total heat flux from the four parallel resistors is 0.1 mW. The temperature difference between the heater and the frame is

$$\Delta T = R \times 0.0001 \text{ W} = 20.4 \text{ K}.$$

In many devices, the speed at which the temperature of a microstructure can rise or fall is critical. The heating and cooling speed of an ink-jet printer determines the maximum printing speed, for example. The thermal resistance influences the dynamic response speed of a thermal sensor or actuator.

The relationship between stored thermal energy (Q) and temperature change is

$$Q = sh \cdot m \cdot \Delta T = C_{th} \cdot \Delta T, \tag{5.7}$$

where sh (J/kgK) is the **specific heat**, which is the amount of heat per unit mass required to raise the temperature of an object by one degree Celsius or Kelvin. The term C_{th}(J/K) is called the **heat capacity**, which is the equivalent of electrical capacitance in a thermal–electrical analogy. The general expression of the time constant associated with the heating or cooling of a microstructure is

$$\tau = R_{th}C_{th} = R_{th} \cdot sh \cdot m. \tag{5.8}$$

The small mass of microstructures favors the reduction of time constant.

5.2 SENSORS AND ACTUATORS BASED ON THERMAL EXPANSION

Thermal expansion is an omnipresent behavior of materials. The dimensions and volume of structures made of semiconductors, metals, and dielectric materials would increase upon temperature rise. The **volumetric thermal expansion coefficient** (TCE), commonly denoted as α, is the ratio between the relative change of volume to the degree of temperature variation,

$$\alpha = \frac{\frac{\Delta V}{V}}{\Delta T}. \tag{5.9}$$

The **linear expansion coefficient** is the change of only one dimension of an object due to temperature variation,

$$\beta = \frac{\frac{\Delta L}{L}}{\Delta T}. \tag{5.10}$$

The volumetric and linear expansion coefficients are related by

$$\alpha = 3\beta. \tag{5.11}$$

The linear thermal expansion coefficients of representative organic and inorganic materials are summarized in Table 5.2. Apparently, the value of the thermal expansion coefficient for most solid materials is small. The extent of expansion is rather limited for practical values of temperature rise. For instance, upon a temperature rise of 100°C, a 1-mm-long silicon cantilever will only elongate by a distance of $2.6 \times 10^{-6} \times 100 \times 10^{-3} = 2.6 \times 10^{-7}$ m $= 0.26$ μm.

It should be noted that for thin-film materials such as silicon nitride and polycrystalline silicon films, thermal properties such as the thermal expansion coefficient depend on the exact composition of the material, determined by specific process settings, equipment settings, and thermal treatment history. Data on thermal properties collected from literature vary quite widely. Furthermore, many data points in Table 5.2 are collected from bulk materials and should only serve as a reference when dealing with microscopic samples. Representative thermal properties of commonly used metals are summarized in Table 5.3.

Actuators based on the thermal expansion of liquid or air have been achieved in the past [5, 7]. The thermal expansion coefficients of many liquids are greater than those of solids. The value of α for water above 40°C is approximately 400 ppm/K. Some special engineered fluids have an even greater α. For example, the volumetric thermal expansion coefficients of 3M performance fluids are in the range of 0.16%/K [5].

The thermal expansion of air has been used to move liquid drops in microfluid channels. For example, trapped air volume on the order of 100 nL heated by tens of degrees Celsius can generate air pressure on the order of 7.5 kPa [7].

TABLE 5.2 Selected Thermal Properties of Common Materials.

Material	Thermal conductivity (W/cmK)	Linear temp. coefficient of expansion
Aluminum	2.37	25
Aluminum oxide	0.36	8.7
Aluminum oxide	0.46	–
Carbon	0.016	–
Carbon	23	–
Cr	0.94	6
Cu	4.01	16.5
GaAs	0.56	5.4
Ge	0.6	6.1
Au	3.18	14.2
Si	1.49	2.6
SiO$_2$ (thermal)	0.0138	0.35
SiN (silicon)	0.16	1.6
Polyimide (Dupont PI 2611)	–	3
Polysilicon	0.34	2.33
Ni	0.91	13
Ti	0.219	8.6

5.2 Sensors and Actuators Based on Thermal Expansion

TABLE 5.3 Thermal and Electrical Properties of Metal Thin Films.

Material	Resistivity ($\mu\Omega cm$)	Thermal conductivity (W/cmK)	TCR (ppm/°C)	Coefficient of thermal expansion (ppm/K)
Aluminum (Al)	2.83	2.37	3600	25
Chromium (Cr)	12.9	0.94	3000	6.00
Copper (Cu)	1.72	4.01	3900	16.5
Gold (Au)	2.40	3.18	8300	14.2
Nickel (Ni)	6.84	0.91	6900	13
Platinum (Pt)	10.9		3927	8.8

The thermal expansion of gases due to temperature change can be derived from the ideal gas law. For an ideal gas, the relation between the volume and temperature is given by

$$PV = nRT = NkT, \qquad (5.12)$$

where P is the absolute pressure, V the volume, T the absolute temperature, n the number of moles, N the number of molecules, R the universal gas constant (R = 8.3145 J/molK), and k the Boltzmann constant (1.38066×10^{-23} J/K). Recall the relationship between k and R is given by

$$k = R/N_A, \qquad (5.13)$$

where N_A is Avagadro's number ($N_A = 6.0221 \times 10^{23}$).

The thermal conductivity, resistivity, TCR, and coefficient of thermal expansion of commonly used metal materials are summarized in Table 5.3.

5.2.1 Thermal Bimorph Principle

α – volume expansion

The thermal bimetallic effect is a very commonly used method for sensing and actuation. This mechanism allows the temperature variation in microstructures to be exhibited as the transverse displacement of mechanical beams.

The thermal bimorph consists of two materials joined along their longitudinal axis serving as a single mechanical element. (Often, a thermal bimetallic actuator may consist of more than two layers of materials. This book focuses on the analysis of two-layered architectures only.) Figure 5.3 shows a composite beam with two layers, made of materials 1 and 2, having the same length (L) but different coefficients of thermal expansion (CTE) ($\alpha_1 > \alpha_2$). The subscript refers to the material layer. Likewise, Young's modulus, width, and thickness of the two layers are denoted E_i, w_i, and t_i (i = 1 or 2). With a uniform temperature rise of ΔT, the length of two sections changes unequally. Because the two-layered materials are tightly joined at the interface, the beam must curve toward the layer made of the material with a lower CTE value. A transverse beam bending is therefore produced.

Next, we analyze the formula for calculating the displacement of a bimetallic beam. Under a uniform temperature change of ΔT, the beam curves and assumes the shape of a section of an

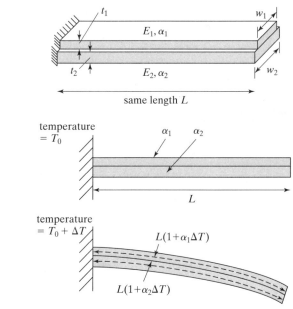

FIGURE 5.3

Thermal bimetallic bending ($\alpha_1 > \alpha_2$).

arc with the length of the arc being L. The radius of curvature of the arc r can be calculated using this formula

$$\frac{1}{r} = \frac{6w_1w_2E_1E_2t_1t_2(t_1+t_2)(\alpha_1-\alpha_2)\Delta T}{(w_1E_1t_1^2)^2 + (w_2E_2t_2^2)^2 + 2w_1w_2E_1E_2t_1t_2(2t_1^2+3t_1t_2+2t_2^2)}. \tag{5.14}$$

The arc is a section of a circle with the radius of curvature being denoted r, spanning an arc angle θ.

The value of θ is determined by

$$\theta = \frac{l}{r}. \tag{5.15}$$

Once the radius of curvature is found, the vertical displacement of the free end of the beam can be determined by trigonometry according to Figure 5.4.

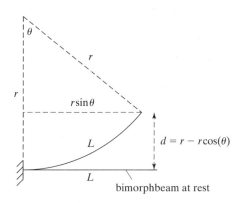

FIGURE 5.4

Geometry of a bent beam.

The vertical displacement at the free end of the cantilever is therefore

$$d = r - r \cos \theta. \tag{5.16}$$

If the overall bending angle is small, the magnitude of the vertical displacement can be estimated by replacing $\cos \theta$ with the first two terms in its Taylor series expansion:

$$d = r - r\left(1 - \frac{1}{2}\theta^2 + O(\theta^4)\right) \approx \frac{1}{2}r\theta^2. \tag{5.17}$$

Example 5.2. Displacement of a Bimorph Actuator

A bimorph cantilever beam is made of two layers of different lengths. The layer on top is made of aluminum (material 2), whereas the layer on the bottom is made of silicon nitride (material 1). The width of both layers is 20 μm. The length of the segment between points A and B is 100 μm, so is the length of the segment from point B to C. Young's modulus of aluminum and silicon nitride are E_2 = 70 GPa and E_1 = 250 GPa, respectively. The thickness of aluminum and silicon nitride sections is t_2 = 0.5 μm and t_1 = 1 μm, respectively. The thermal expansion coefficients of aluminum and silicon nitride are α_2 = 25 ppm/°C and α_1 = 3 ppm/°C, respectively. At room temperature, the cantilever is straight.

Find the radius of the curvature (r) of the cantilever beam when the beam is uniformly heated to 20°C above the room temperature. Determine the amount of vertical displacement at the free end of the beam under this condition.

Solution. The composite beam consists of two segments. The segment spanning points A and B undergoes curvature bending due to the thermal bimorph effect. The segment spanning points B

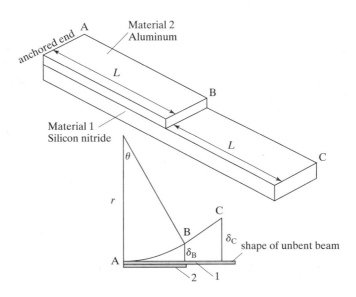

and C consists of only a single layer of material. It will not undergo curvature bending, but it will follow the angular bending that takes place at point B.

For segment A–B, we can write the expression for the radius of curvature as

$$\frac{1}{r} = \frac{6w_1w_2E_1E_2t_1t_2(t_1 + t_2)(\alpha_1 - \alpha_2)\Delta T}{(w_1E_1t_1^2)^2 + (w_2E_2t_2^2)^2 + 2w_1w_2E_1E_2t_1t_2(2t_1^2 + 3t_1t_2 + 2t_2^2)}$$

$$= \frac{6(400 \times 10^{-12})(70 \times 250 \times 10^{18})0.5 \times 10^{-12}(1.5 \times 10^{-6})(22 \times 10^{-6})20}{2.5 \times 10^{-11} + 1.225 \times 10^{-13} + 7(4 \times 10^{-12})}$$

$$= \frac{1.386 \times 10^{-8}}{5.3 \times 10^{-11}} = 260.9 \text{m}^{-1}.$$

The radius of curvature in the segment A–B is

$$r = 0.00383 \text{ m}.$$

The angle spanned by the curved section is

$$\theta = \frac{L}{r} = \frac{0.0001}{0.00383} = 0.026 \text{ rad} = 1.49°.$$

The vertical displacement at point B is

$$\delta_B = r - r\cos(\theta) = 1.3 \times 10^{-6} \text{ m}.$$

The vertical displacement at point C is

$$\delta_C = \delta_B + L\sin(\theta) = 3.9 \times 10^{-6} \text{ m}.$$

The thermal bimorph displacement can be used for both sensing and actuation purposes. It is in fact the mechanism for many household electromechanical thermostats. A conventional thermostat consists of a spiral bimorph metal coil. The end of the coiled beam is attached to an electrical relay, in the form of a sealed glass vial containing a drop of mercury. When the environmental temperature changes, the end of the coil tilts and triggers movement of the mercury-drop relay to regulate current flow in heating/cooling circuitry.

Several examples of micromachined bimorph temperature sensors are discussed in Section 5.5. In the remainder of this section, we will focus on thermal bimorph actuators.

As a component of an actuator, a thermal bimaterial beam can produce angular and linear displacements. It is interesting to draw a comparison between the two actuation methods we have studied so far—electrostatic actuation and thermal actuation. Both electrostatic and thermal bimetallic actuation methods are commonly used in MEMS. The relative merits and disadvantages of both methods are summarized in Table 5.4.

Two examples of thermal bimetallic actuators are reviewed below; both are used for transporting micro objects. The actuator in Case 5.1 does not incorporate integrated circuits for control purposes, whereas the one in Case 5.2 is integrated with on-chip electronics.

5.2 Sensors and Actuators Based on Thermal Expansion

TABLE 5.4 Characteristics of Electrostatic and Thermal Bimorph Actuation.

	Electrostatic actuation	Thermal bimorph actuation
Advantages	1. Low power operation at low frequencies 2. High response speed	1. Relatively large range of movement can be achieved 2. Small actuator footprint for comparable displacement
Disadvantages	1. Relatively small range of motion 2. Requires large area and footprint in order to generate large force and displacement	1. Moderate- to high-power operation as current is used to generate ohmic heating 2. Lower response speed as the time constant is governed by thermal heating and dissipation

Case 5.1. Bimorph Artificial Cilia Actuator

In microelectronics manufacturing, the handling and assembly of small objects such as dies cut from a wafer is generally labor intensive and inefficient. A technology is needed for small chips to be transported and oriented efficiently in the production line. An array of thermal actuators, mimicking biological cilia, have been developed to carry and transport a small object laterally in a plane [8]. According to Figure 5.5, an object is supported by the array of out-of-plane cilia. These actuators are divided into two groups that are activated out of phase with respect to each other. Power is applied to these two groups of actuators at a certain clock frequency.

At the start of a cycle, both groups of actuators are elevated to hoist a small object (Figure 5.5a). One group is actuated and lowered first (Figure 5.5b). The other group is lowered

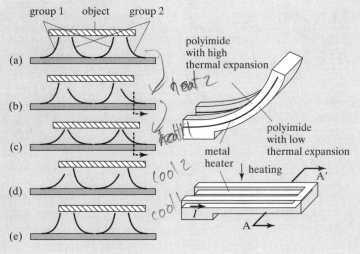

FIGURE 5.5 Artificial cilia array for object transport.

later, causing the object it carries to move forward by a small amount (Figure 5.5c). The group that was lowered first is returned to the elevated position, moving the object by a small distance again (Figure 5.5d). The second group of actuators is then elevated, returning the system to its starting configuration. Actuators belonging to the first group are released to complete a cycle (Figure 5.5e). Through each cycle, the object is moved by a small incremental distance. Long-distance movement of the object results through repeated cycles.

Each cilium is a curved bimorph cantilever, 500 μm long, 100 μm wide, and 6 μm thick (Figure 5.5). The top view of the cantilever reveals a folded wire loop. The resistance of wire associated with each wire loop is 30–50 Ω. The cross section consists of three major layers—a polyimide layer with a high thermal expansion coefficient, a gold heater, and another bottom polyimide layer with a lower thermal expansion coefficient. The polyimide layers as well as metal layers exhibit tensile intrinsic stress to raise the cilia tip by 250 μm at rest. When electrical current passes through the resistor, the gold and polyimide layers are heated. Because the top polyimide layer has a greater thermal expansion coefficient, the cantilever bends downward. Large displacement can be achieved. For example, a vertical displacement of 250 μm and a companion horizontal displacement of 80 μm result when a 22.5-mA drive current is used, corresponding to a power dissipation of 33 mW for each actuator.

The fabrication process can be carried out on bare silicon or glass wafer. A four-mask fabrication process (Figure 5.6) begins with deposition of a 1.6-μm-thick aluminum thin film, serving as a sacrificial layer (step a). A layer of polyimide is spin coated and patterned to form a 2.2-μm-thick film (step b). Metal thin film patterns (200-nm-thick gold and 100-nm-thick nickel) are formed by evaporation and patterned by wet etching (step c). Adhesion between the gold and polyimide is enhanced significantly by the use of the nickel adhesion layer. The authors coated another layer of polyimide again to a thickness of 3.6 μm (step d).

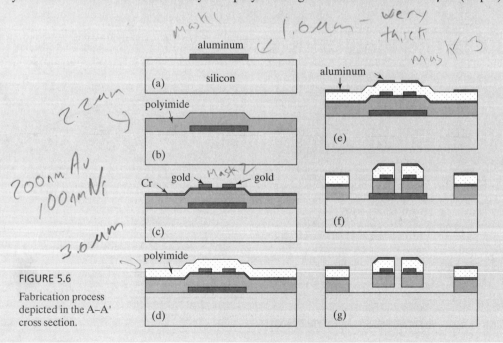

FIGURE 5.6

Fabrication process depicted in the A–A' cross section.

This polyimide layer needs to be patterned. One option is to use photo-definable polyimide. The authors did not use this approach. One possible reason is that photo-patternable polyimide with suitable TCE value cannot be found.

Instead, a polyimide layer is deposited and later patterned by etching. It was determined that plasma etching is the best course for minimizing undercut. In order to pattern this layer of polyimide, it needs to be masked by a protective layer. Both metal thin films and photoresist may serve as the protective layer. Photoresist, however, does not provide sufficient etch selectivity. As a consequence, the top polyimide is coated with a layer of thin-film metal. The thin-film metal is patterned photolithographically and etched (step e). It then serves as a mask for a subsequent etching of the polyimide layer using oxygen plasma to define the polyimide beams (step f). The sacrificial layer is removed in selective regions to release the polyimide beams (step g).

Demonstration of the cilia array carrying a 2.4-mg silicon wafer traveling at 27–500 μm/s with 4-mW input for each actuator has been successfully made.

Case 5.2. Bimorph Actuators for Object Transport

The cilia actuators remain in down positions for a significant fraction of the cycle. The cilia transport chip discussed in Case 5.1 requires constant power input to keep a cilia in its down position. This translates into rather significant power consumption.

Researchers later developed an improved version of the thermal actuator for transporting objects in a two-dimensional plane [9] (Figure 5.7). Each flap-shaped cilia actuator

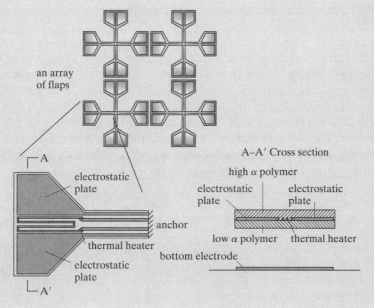

FIGURE 5.7
Polyimide thermal actuator.

consists of a metal resistive heater as well as a thin-film electrode plate. The two materials of the different thermal expansion coefficients are both polymer materials. One polymer is PIQ-L200 with a TCE value of 2.0 ppm/°C. Another one is PIQ-3200 with a TCE value of 54 ppm/°C. The metal film is sandwiched between two polyimide layers, serving as a heater and one electrode of a parallel capacitor. A counter electrode is located on the substrate surface.

Each flap is bent out of plane under zero applied power because of intrinsic stress in the materials. Upon applied ohmic heating, a flap bends towards the substrate surface. Instead of using a constant current to keep the cilia down, electrostatic actuation can be used to hold the flap position. When the flap is parallel to the substrate plane, the magnitude of the electrostatic force is the greatest due to close proximity of electrodes. The electrostatic force is energy efficient for static position holding, consuming virtually no power.

Later, the group changed the geometries of the thermal manipulator and integrated them on a substrate with CMOS controller electronics [10]. The design of each actuator is considerably simpler than the previous generation—the electrodes for electrostatic holding is removed, perhaps because it is expensive to integrate parallel capacitors with a large footprint onto CMOS chips. The heater is made of Ti–W heater resistors with 1 kΩ resistance each. A DC-power input of 35–38 mW produces a vertical displacement of 95 μm accompanied by a 17-μm lateral displacement. An array was programmed to perform simple linear and diagonal translations and squeeze-, centering-, and rotating-field manipulation of silicon pieces of various shapes.

5.2.2 Thermal Actuators with a Single Material

Thermal bimorph bending is convenient for creating out-of-plane linear or angular displacement. It can produce in-plane displacement provided that the layered thermal bimorph materials are stacked on vertical surfaces. However, the fabrication process for such stacked structures is rather difficult.

Thermal actuators based on a single material for generating in-plane motion has been demonstrated by using bent-beam electrothermal actuators [11]. The schematic diagram of a representative bent-beam actuator is shown in Figure 5.8. A bent beam is made of silicon doped to a certain concentration; it serves as the ohmic heater as well. Current passing through the bent beam causes two branches to expand, resulting in tip motion in the transverse direction. The beam is capable of producing peak output force on the order of 1–10 nN with a driving voltage of less than 12 V. Static displacement on the order of 10 μm at 79-mW input power has been demonstrated with beams that are 410 μm long, 6 μm wide, and 3 μm thick.

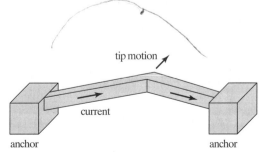

FIGURE 5.8

A bent-beam electrothermal actuator.

Alternative configurations of thermal actuators based on a single material are discussed in Case 5.3.

The output force can be further increased by connecting many actuators in parallel [12].

Case 5.3. Lateral Thermal Actuators

Some lateral driven thermal actuators are based on the asymmetrical thermal expansion of a microstructure with two arms made of the same conductive material. These two arms have different heating power and thermal expansion, resulting in a difference of longitudinal expansion when a current passes through them. Two strategies have been demonstrated (Fig. 5.9):

1. An actuator may consist of two beams with the same cross-sectional area but different lengths [13]. The longer arm is associated with greater electric resistance and thermal resistance. By passing an identical current through the loop, the longer arm will produce a higher temperature and greater expansion, causing the tip to bend towards the direction of the shorter arm. An actuator was tested, with the long beam being 500 μm long and the short one 300 μm long. The width and thickness of the beams are 2.8 and 2 μm, respectively. Lateral displacement as large as 9 μm has been demonstrated under a bias voltage of 14 V. The resistivity, Young's modulus, Poisson's ratio, and density of the arms are 5×10^{-6} Ωcm, 150 GPa, 0.066, and 232 kg/m^3, respectively.

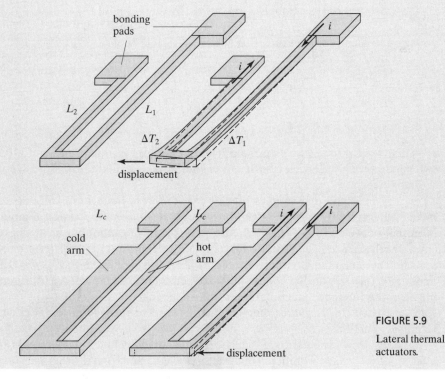

FIGURE 5.9

Lateral thermal actuators.

2. An actuator may contain two beams with same length but different cross-sectional areas [14–17]. The arm with a smaller cross section is called the hot arm. The one with a larger section is called a cold arm. Under the same current, the hot arm reaches a high temperature and therefore greater linear expansion. An actuator with a hot arm being 2.5 µm wide and 240 µm long and a cold arm being 16 µm wide and 200 µm long was able to generate a 16-µm displacement at 3 V and 3.5 mA for a 2-µm-thick polysilicon layer. Modeling has been performed [17]. The thermal actuator can be connected together to generate a large force [15] for various applications including optical aligners.

5.3 THERMAL COUPLES

In 1823, the German physicist Seebeck discovered that a voltage was developed in a conductor loop containing two dissimilar materials when the temperature at the joint between two materials was increased above room temperature (Figure 5.10a and b). The voltage is capable of maintaining a current if the loop is closed. (In fact, the voltage was noticed only when the current was able to deflect a magnetic compass needle nearby.)

Two wires made of dissimilar materials joined at one point constitute a **thermal couple**. A thermal couple is most commonly used to measure the temperature difference between the joined sensing junction and a reference one. It is also used for generating electricity from a temperature gradient between their two junctions.

Seebeck tested a number of materials, including naturally found semiconductors ZnSb and PbS. When the temperature difference ΔT is applied, it will be accompanied by an electric voltage ΔV. He found that the open circuit voltage is linearly proportional to the difference of temperature. The ratio between the developed open circuit voltage and the temperature difference is the Seebeck coefficient:

$$\alpha_s = \frac{\Delta V}{\Delta T}. \tag{5.18}$$

The term α_s is the **Seebeck coefficient** of a thermal couple, unique to the combination of two materials. The Seebeck coefficient is alternatively called the thermoelectric power, or just thermopower.

Why must thermal couples involve two different materials? Although a single piece of metal is theoretically capable of exhibiting a voltage difference when a temperature difference is present between its two ends (junctions), this voltage cannot be easily measured or used, because another piece of conducting material must be present to interrogate or communicate this voltage. The conducting wire inevitably exhibits the Seebeck effect as well. If the conducting material is the same as the conducting material under test, no voltage difference would be detected at the reference junction because the Seebeck voltage of the two pieces cancel. As a result, two pieces of *dissimilar* metals must be used to provide an open circuit voltage at the reference junction.

Seebeck coefficients are in fact associated with individual metal elements. If the Seebeck coefficient of the two constitutional materials (labeled *a* and *b*) in the thermal couple in Figure 5.10

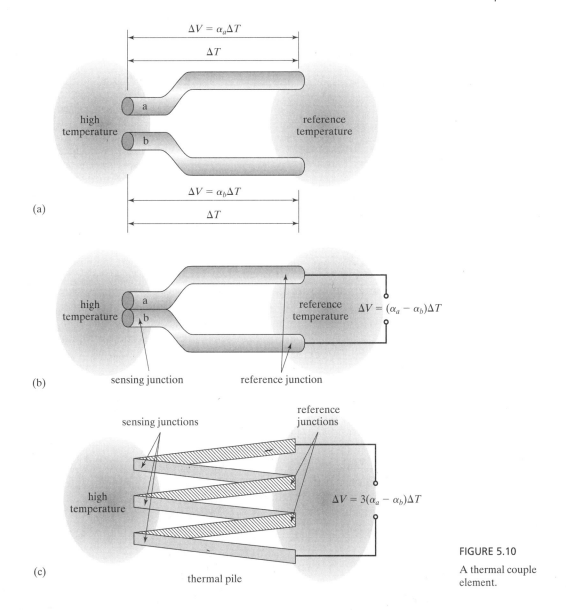

FIGURE 5.10

A thermal couple element.

are denoted α_a and α_b, respectively, the Seebeck coefficient of the thermal couple is defined as following:

$$\alpha_{ab} = \alpha_a - \alpha_b. \tag{5.19}$$

Although thermal couples can be made of a seemingly unlimited large number of material combinations, most widely used thermal couples are limited and well characterized. The difference between α_a and α_b should be as great as possible. The material make-up and properties of some commonly encountered industrial thermal couples are presented in Table 5.5.

TABLE 5.5 Seebeck Coefficient of Common Industrial Thermal Couple Materials.

Type	Metal 1	Metal 2	Temperature range (°C)	Sensitivity (μV/°C); comment
E	Chromel	Constantan	−270–900	68; high sensitivity, non-magnetic
J	Iron	Constantan	−210–1200	55; low temperature range
K	Chromel	Alumel	−270–1250	41; low cost, general purpose
T	Copper	Constantan	−270–400	55; low temperature range
R	Platinum	Rhodium 13% Rhodium	−50–1450	10; low sensitivity, high cost, suitable for high temperature measurement
S	Platinum	Rhodium 6% Rhodium	−50–1450	10; same as R type

Thermal couples offer several distinct advantages over thermal resistors or other temperature-sensing methods. The thermal couple provides an output without offset and offset drift. It does not suffer from interference from any physical or chemical signals except for light. The thermal couple does not require any electrical biasing and is *self-powered*.

Semiconductor materials often show a thermoelectric effect that is larger than that observed in metals [18]. For non-degenerate silicon, the Seebeck coefficient is derived from three main effects: (1) with increasing temperature, a doped silicon becomes more intrinsic; (2) with increasing temperature, the charge carriers acquire a greater average velocity, leading to a charge buildup on the cold side of the semiconductor; (3) the temperature difference in a piece of silicon causes a net flow of phonons from hot to cold ends. A transfer of momentum from acoustic phonons to the charge carriers can occur under certain conditions.

Both doped single-crystal silicon [19] and polycrystalline silicon [20, 21] are promising candidates for building thermal couples. The values of Seebeck coefficients have been characterized under various doping concentration and dopant types.

The Seebeck coefficient of silicon is a function of the doping level and conductivity. The value for single-crystal silicon, at the level of -1000 μV/k, greatly exceeds that previously tabulated for various metals. For polycrystalline silicon, the Seebeck coefficient is on the order of 100 μV/K when the bulk resistivity is 10 $\mu\Omega$m. For practical design purposes, it is convenient to approximate the Seebeck coefficient as a function of electrical resistivity; thus,

$$\alpha_s = \frac{26k}{q} \ln(\rho/\rho_0),$$

where ρ_0 is a reference resistivity (5×10^{-4} Ωcm), k the Boltzmann constant, and q the elementary electric charge.

The output voltage from thermal couples will be increased when multiple thermal couples are connected in an end-to-end fashion with the hot and cold junctions aligned. This configuration is called a **thermal pile** (Figure 5.10c). It is analogous to many batteries lined up together to provide a higher ouput voltage. The output voltage equals that of a single thermal couple multipled by the number of thermal couples in the system.

Micromachined thermal couples have been made for temperature sensing based on a variety of materials [18]. For example, surface micromachined scanning thermal couple probes have been made using the following sets of thermal couple materials: Ni and W (with a Seebeck coefficient of 22.5 µV/K per junction), and chromel and alumel (37.5 µV/K per junction) [22]. Thermal couple probes have also been made by exploiting the junction between Au–Ni (14 µV/K per junction) and Au–platinum (5 µV/K per junction). In many cases, the thermal electric power is lower than the corresponding bulk value [23].

It is noteworthy that the Seebeck effect belongs to a broad family of thermalelectric effects. A decade after the Seebeck effect was discovered, a companion phenomenon was discovered by a French scientist, Peltier. He found that electrons moving through a solid can carry heat from one side of the material to the other side. The true nature of the Peltier effect was explained later by Lenz. Upon the flow of electric current, heat is absorbed or generated at the junction of two conductors. Lenz also demonstrated freezing a drop of water at a bismuth–antimony junction and melting the ice by reversing the current.

In a Peltier thermalelectric device, it is the flow of charge carriers that pumps heat from one side of the material to another. The ratio of heat flow to electric current for a particular material is known as the Peltier coefficient, Π. In fact, the Peltier and Seebeck coefficients are related, according to

$$\Pi = \alpha_s T, \tag{5.20}$$

where T is the absolute temperature.

5.4 THERMAL RESISTORS

The resistance value of a resistor is a function of the resistivity ρ and its dimensions, including length l and the cross-sectional area A. It is given as

$$R = \rho \frac{l}{A}. \tag{5.21}$$

Both the resistivity and the dimensions are functions of temperature. As a result, the resistance value is sensitive to temperature. A **thermal resistor** is an electrical resistor with appreciable temperature sensitivity.

The resistance of a thermal resistor, R, is related to the ambient temperature in the relationship shown below:

$$R_T = R_0(1 + \alpha_R(T - T_0)), \tag{5.22}$$

where R_T and R_0 are the resistance at temperature T and T_0, respectively. The term α_R is called the **temperature coefficient of resistance** (TCR). This equation is valid for moderate temperature excursion. If the temperature difference is greater, non-linear terms may be needed to yield an accurate expression of R_T.

Thermal resistors can be made of metal or semiconductors. In both cases, the dimensions of a resistor change with temperature. The electrical resistivity of both metals and semiconductors varies with temperature, but the principles of such change are quite different for metals and semiconductors. For metal resistors, temperature rises introduce enhanced lattice vibration, which tends to impede the movement of charge carriers. In the case of semiconductors, temperature affects the lattice spacing and, in turn, the effective mass and mobility of charge carriers. For semiconductor thermal resistors, the resistivity-temperature relation is influenced by the doping concentration and dopant types.

Metals are used as thermal resistors because of their simplicity of processing. Platinum is widely used as a thermal resistive material because the resistivity is very linear with respect to temperature. It has a reasonably high TCR (39.2×10^{-4}/K) but a low resistivity; this tends to limit its use in microdevices since a large aspect ratio (length over width) design is required in order to generate an appreciable base resistance value, R_0.

The term **thermistor** is generally used to refer to semiconducting thermoresistors. Semiconductor thermistors have the advantage of large resistivity and ease of miniaturization. However, their TCR is typically lower than that of metals. The doped polysilicon is often used as a temperature-sensitive resistive element. Its TCR value depends on the concentration of doping [24]. The TCR for p-type doped polysilicon ranges from 0.1%/°C–0.4%/°C with concentrations ranging from 10^{18} to 10^{20} cm^{-3} [25].

Example 5.3. Thermistor Resistance

A thermal resistor is made of doped p-type single-crystal silicon, with the nominal resistance (R_0) of 2 kΩ. Assume the TCR of the material is 100 ppm/°C. Predict the resistance of the device at a temperature 50°C above the ambient.

Solution. Given the temperature coefficient of resistance, the resistance value at 50°C above the room temperature is

$$R = R_0(1 + \alpha_R \Delta T) = 2000 \times (1 + 100 \times 10^{-6} \times 50) = 2010 \, \Omega.$$

The temperature coefficient of resistance can be measured simply by heating up a resistor using a temperature-controlled stage and monitoring the resistance value. The temperature of the stage should be increased slowly in small increments, allowing sufficient time between increases to ensure thermal equilibrium. The bias voltage and current must be kept low in this experiment; this minimizes the contribution of the electric heating power.

The current and voltage used to interrogate the resistance value of a thermoresistor may introduce heat to it. This phenomenon is called **self-heating**. The heating power of a resistor under current I is

$$P = I^2 R. \tag{5.23}$$

The self-heating of a resistor by the interrogation currents may change its temperature and the resistance value.

The temperature and hence resistance value of a resistor under the self-heating condition is dependent on the rate of heat input and dissipation. The ohmic heating energy may be dissipated through conduction and convection. If the convective heat loss is dominant, a self-heated resistor can be used to measure the convective heat loss rate and flow rates. This forms the basis of hot-wire anemometry, further discussed in Case 5.6 [3]. Under zero flow movement, the temperature of the heated resistive wire reaches a steady-state value. When the resistive wire is subjected to a moving fluid, heat is convectively moved from the element,

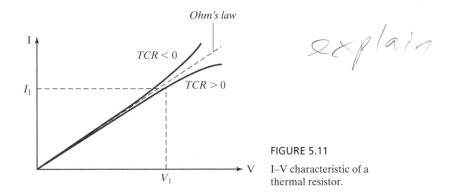

FIGURE 5.11
I–V characteristic of a thermal resistor.

reducing its temperature. The changes of the temperature of the sensing element can be inferred by measuring the real-time resistance value. Similarly, the self-heating effect can be used for measuring flow shear stress in the boundary layer (Case 5.7) [26, 27].

The value of the thermal resistance associated with a thermal resistor can be easily obtained from the current-voltage characteristics (I–V). According to Ohm's law, the current in a resistor will increase linearly with respect to the input voltage. A linear current-voltage plot is often encountered, with the slope of the I–V characteristic curve representing the inverse of the resistance of the resistor.

If self-heating occurs, the resistance of the resistor is changed. Hence, the slope of the I–V curve will change. If the TCR value is positive, the resistance will increase at an elevated power input level. A representative I–V curve of a thermal resistor with TCR > 0 is shown in Figure 5.11. The slope of the I–V curve decreases with the self-heating at high power input levels. On the other hand, if the TCR value of a resistor is negative, the slope of the I–V curve increases at high power input levels.

The degree of bending of the I–V curve of a resistor correlates to its thermal isolation. For two identical thermal resistors, the one with better thermal isolation will reach a greater temperature and a higher degree of resistance change under a given input (self-heating) power. On the other hand, the device with poor thermal isolation will have a lower equilibrium temperature. Its I–V curve will experience less bending. (The fact that I–V curve bending is never casually noticed in discrete bulk resistors is partially attributed to the fact that these resistors are generally not thermally insulated at all).

5.5 APPLICATIONS

The thermal transfer principle can be used to measure a variety of physical variables, including acceleration, position, displacement, and flow rate. We discuss a few representative examples of sensors in this section.

5.5.1 Inertia Sensors

Almost all existing accelerometers use piezoresistive or capacitive sensing principles to detect the movement of proof masses. Piezoresistive accelerometers show appreciable temperature dependence, whereas capacitive accelerometers have problems with electromagnetic interference.

Thermal accelerometers, on the other hand, turn displacement into changes of temperature or heat flow. Here, we discuss two examples of thermal accelerometers; one is based on moving mass (Case 5.4) and one has no moving mass (Case 5.5).

Case 5.4. Accelerometer Based on Thermal Transfer Principle

The principle of a thermal accelerometer with a silicon moving mass is shown in Figure 5.12 [28]. It consists of a heat source (realized by a heating resistor) with the temperature T and a heat sink (the package) with a temperature set at T_0. Two temperature sensors, based on the thermal pile principle, are symmetrically located with respect to the heater. The heater and the temperature sensors are located on a thin membrane, which restricts lateral heat flow to the package frame (heat sink). The thin membrane increases thermal resistance and allows the heater to reach an appreciable temperature without wasting heating power.

The lateral heat flow is designated Q_1 in the figure. Heat generated by the heat also travels through the air gap to the suspended proof mass above it (Q_3). In addition, heat also moves by conduction in air to the thermal pile sensors (Q_2). When the applied acceleration is zero, the temperature of the thermal piles reaches a certain steady-state temperature upon a given heat dissipation.

An external acceleration causes the relative distance between the proof mass and the heater to change. If the distance is reduced, as shown in Figure 5.12, the heat flux elements Q_2 and Q_3 are enhanced relative to Q_1, changing the steady-state temperature distribution and the temperature readings of the thermal piles.

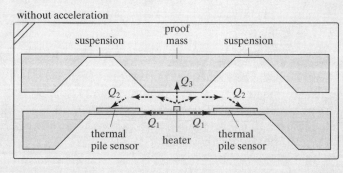

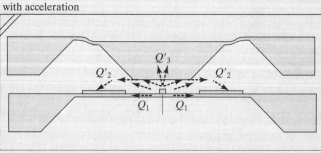

FIGURE 5.12
Principle of a thermal accelerometer with moving mass.

In this paper, an expression of the temperature at a given location on the membrane as a function of the distance between the proof mass and the membrane is derived. However, the accuracy was not validated. The temperature sensitivity of the device is relatively low. The output of the thermal pile in the absence of the proof mass changed only by 1% in the temperature range of 20–100°C, according to the authors. There is certainly room for improvement with better material selections and designs. The sensor exhibited a sensitivity of 9–25 mV/g, a sensing range of 0.4–0.8 g (which is relatively narrow), and a frequency response of up to 300 Hz.

Case 5.5. Thermal Accelerometer with No Moving Mass

Nearly all existing MEMS accelerometers incorporate moving inertia mass. These freestanding structures involve unique fabrication processes that cannot easily be run in integrated circuit processing plants, thus preventing MEMS accelerometers from leveraging the enormous and economical infrastructure of the IC industry. Freestanding structures also increase the degree of difficulty for handling and packaging.

It will be tremendously important if accelerometers can be made without free moving parts, thus increasing their compatibility with the integrated circuit foundry and improving the robustness of such sensors.

A simple acceleration sensor without moving mass has been made (Fig. 5.13). It consists of an ohmic heater and at least two temperature sensors symmetrically placed with respect to the heater [29]. The chip is placed in a hermetically sealed package with air inside. The heater heats up a pocket of air. Under rest conditions, the spatial profile of the hot air pocket is symmetric such that the two temperature sensors produce identical temperature readings. If an acceleration is applied on the ceramic package, the chip would move slightly in the direction of the applied acceleration. The air mass would lag behind due to inertia, causing an asymmetric distribution of the temperature profile in the air. The readings of the two temperature sensors would become different, with the difference corresponding to the magnitude of the applied acceleration.

This principle has been implemented using microfabrication technology [2] and with full CMOS signal processing integration [30]. The heater and sensors are fabricated on a semiconductor substrate with preexisting signal conditioning and processing electronics. The fabrication process of heaters and temperature sensors are highly compatible with integrated circuits. No moving parts are necessary.

The sensitivity of the first microfabricated device of this kind [2] was characterized by using earth's gravity as a reference. By changing the direction of the device's sensitivity axis relative to gravity, the output of the device changes linearly. The equivalent noise floor was 0.5 mg with a 20-mW biasing power in atmospheric pressure. The sensitivity of this accelerometer is linearly proportional to the Grashof number Gr given by

$$Gr = \frac{a\rho^2 \chi^3 \beta(\Delta T)}{\mu^2}, \qquad (5.24)$$

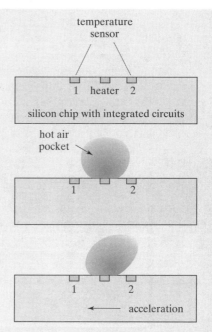

FIGURE 5.13

Principle of a thermal-transfer-based accelerometer.

where a is the acceleration, ρ the density of the gas, χ the linear dimension, β the coefficient of expansion, ΔT the temperature of the heater, and μ the viscosity.

Later studies of the sensitivity showed dependence on the pressure of the gas medium [31]. At a low pressure range, a square dependence was found. For higher pressures, different optimum sensitivities were obtained according to the distance between the heater and detector. The sensitivity can increase a thousand fold if the device can be packaged with a high pressure.

Sensors based on similar designs have been made by other groups for the detection of acceleration or tilting [30], with a sensitivity of 115 μV/g (for thermal pile configuration) and 25 μV/g (for thermistor configuration) measured with operation frequency up to several hundred Hz. The Johnson noise (thermal noise) floor was estimated according to

$$V_n = \sqrt{4kTRB}, \tag{5.25}$$

where k, T, R, and B are the Boltzmann's constant, the temperature, the resistance value, and the bandwidth of interest. For the thermopile device, the resistance was 64 kΩ, giving a noise voltage of 32.6 nV/$\sqrt{\text{Hz}}$. For the thermistor device, the resistance was 4 kΩ, with noise flow of 8.14 nV/$\sqrt{\text{Hz}}$.

5.5.2 Flow Sensors

The transfer of mass by fluid flow and the transfer of heat are intricately related. Flow sensors based on the thermal transfer principles are very popular. We will discuss hot-wire anemometers, which measure flow rate by the convective heat transfer effect they create (Case 5.6). On the other hand, a hot-wire element lying on the surface of a substrate can measure the flow shear stress. A surface micromachined shear stress sensor based on the thermal transfer principle is discussed in Case 5.7. The discussion of this device also leads to the description of an experimental technique to measure thermal resistance associated with micromachined structures.

Case 5.6. Hot-Wire Anemometer

Hot-wire anemometry (HWA) is a well-studied technique for measuring the velocity of fluid flow. It utilizes a thermal element that serves as both a resistive heater and a temperature sensor. The temperature and resistance of the thermal resistor, biased in the self-heating regime, changes with the speed of flow movement.

The HWA is noted for its low cost, fast response (in the kHz range), small size, and low noise. Conventional HWA sensors are assembled individually by mounting a thin wire made of platinum or tungsten onto support prongs. The wires may be thinned (e.g., by etching in acidic solutions) until the desired dimensions are reached (typically a few mm long and a few μm in diameter). This active portion of the sensor is then mounted on a long probe with an electrical connection for ease of handling. The schematic diagram of a typical finished device is shown in Figure 5.14. Conventional HWAs suffer from two major shortcomings, however. First, the fabrication and assembly process is delicate and does not guarantee uniformity of performance. Secondly, it is prohibitively difficult to form large arrays of HWA for measuring flow field distribution.

The research discussed in [32] was motivated by the need to reduce the cost of HWA, to fabricate and manufacture hot wire anemometers efficiently, and to produce arrayed HWAs on potentially flexible substrates.

Two strategic approaches are taken to achieve these goals. First, HWA sensors are made by using surface micromachining in conjunction with three-dimensional assembly

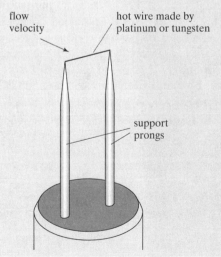

FIGURE 5.14

Schematic diagram of a conventional hot-wire anemometer.

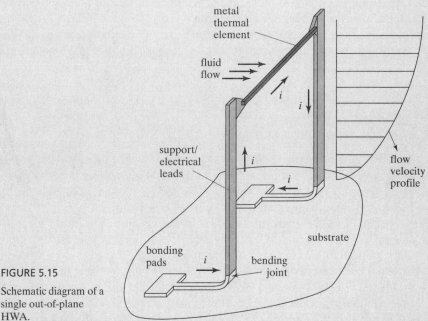

FIGURE 5.15

Schematic diagram of a single out-of-plane HWA.

methods. This circumvents the need to use the bulk micromachining method, which requires a relatively long etching time and complex processing. Bulk etching using anisotropic wet etchants frequently poses concerns of material compatibility as all materials on a given substrate are required to sustain wet etching for long periods (several hours to etch through typical silicon wafers). This method reported in [32] enables more efficient assembly and allows the formation of large arrays of HWAs. Secondly, hot wires were made with thin-film metal instead of polycrystalline silicon. This could reduce costs as the use of silicon bulk (as a substrate) or thin film (as hot wire) is not required. By eliminating silicon doping and bulk etching steps, the fabrication process can be realized in a more efficient manner.

The schematic diagram of the new out-of-plane anemometer is shown in Figure 5.15. A thermal element is elevated from the substrate to a predetermined height that corresponds to the length of the support prongs. By elevating the thermal element away from the bottom of the velocity boundary layer, the thermal element experiences greater fluid flow velocity and exhibits greater sensitivity. The thermal element is electrically connected to the substrate through the support prongs as well.

The hot wire is made of temperature-sensitive metal thin films. The polyimide support is used because it provides the hot wire with needed structural rigidity without increasing the cross-sectional area and thermal conductivity. The thermal conductivity of polyimide is low, almost two orders of magnitude lower than that of a metal, e.g., nickel (see Table 5.2).

In this work, the thickness of the polyimide thin film is roughly 2.7 μm. If the thickness is much lower than this value, the mechanical rigidity will likely be degraded. On the other hand, if the thickness is much greater, there is concern that the polyimide support will decrease the frequency response of the HWA due to added thermal mass.

The fabrication process utilizes an efficient 3D assembly method called the Plastic Deformation Magnetic Assembly (PDMA), which is discussed in detail in Chapter 11 and in Reference [33]. A brief discussion of this method is provided in the following paragraphs. The PDMA process utilizes surface micromachined structures that are anchored to substrates with cantilever beams made of ductile metal materials (e.g., gold and aluminum). The microstructure is attached to pieces of electroplated ferromagnetic material (e.g., Permalloy). By applying an external magnetic field, the ferromagnetic material is magnetized and interacts with the field to bend the microstructure out of plane. If the amount of bending is significant, the cantilever support hinges will be plastically deformed, resulting in permanently bent microstructures even after the magnetic field is removed. The process is very efficient and can be realized in parallel on the wafer scale.

The overall fabrication process is shown in Figure 5.16. The starting wafer is silicon. However, the process can be performed on glass or polymer substrates as the overall temperature of the process is intentionally kept low. First, a chrome/copper/titanium metal stack is evaporated and patterned as the sacrificial layer (Figure 5.16a). A 10-nm-thick chrome film serves as the adhesion layer. A 250-Å-thick titanium thin film reduces the in-process oxidation of the 2500-Å-thick copper film. A 2.7-μm-thick photo-definable polyimide (HD-4000) is

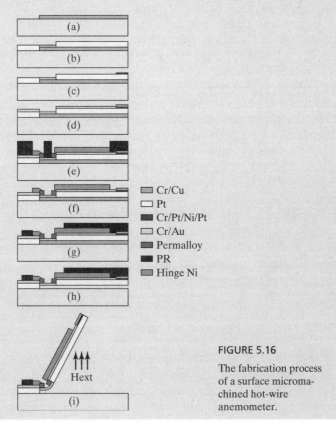

FIGURE 5.16

The fabrication process of a surface micromachined hot-wire anemometer.

spun-on, patterned via lithography, and cured at 350°C (Figure 5.16b) for two hours. This polyimide layer forms the support prong and part of the hot wire. A Cr/Pt/Ni/Pt film is then evaporated and patterned to form the thermal element (Figure 5.16c). The thickness of the Cr thin film, an adhesion layer, is 200 Å. An 800-Å-thick Ni resistor is sandwiched between two 200-Å-thick Pt films, which are used to reduce the possible oxidation of Ni while in operation because Pt is relatively inert at high temperatures. The authors then evaporate and pattern a 5000-Å-thick Cr/Au film (Figure 5.16d) to serve as a mechanical bending element as well as electrical leads of the hot-wire filament. They electroplate a 4-μm-thick Permalloy thin film on portions of the cantilever support prongs (Figure 5.16e).

Sacrificial layer release is performed by using a solution containing acetic acid and hydrogen peroxide to selectively remove the copper thin film. PDMA assembly is carried out to lift the entire sensor out of plane (Figure 5.16i) by placing a permanent magnet (field strength 800 Gauss) at the bottom of the substrate. To finish the process, the device chip is then rinsed in deionized water and dried.

The adhesion between the Au layer and the polyimide layer is very important for device integrity. Without adequate adhesion, the polyimide and the gold film would separate during the PDMA assembly. One of the ways to improve adhesion was to use Cr as an adhesion layer and treat the polyimide layer by using O_2 reactive ion etching (RIE) before the metal deposition. Cr seems to be the adhesion layer of choice, and the RIE treatment creates a hydrophilic structure on the polyimide surface that enhances adhesion. Ti was initially used as the adhesion material for both the Au and the hot-wire element due to its good stability in chemical etchants and higher electrical resistivity. However, it failed to provide sufficient adhesion.

For certain applications, it may be advantageous to be able to strengthen the bent hinge such that the HWA can operate at high flow speed. The mechanical rigidity can be reinforced by electroplating metals (Figure 5.16g).

Thermal and electrical properties of materials used in this device are summarized in Table 5.6. All values are cited from Reference [34] except for the thermal conductivity of polyimide, which is cited from the product guidelines of HD Microsystems. The steady-state response of the sensor to air velocity has been experimentally obtained up to 20 m/s under both constant current and constant temperature modes. Frequency response up to 10 kHz has been demonstrated with the small thermal mass.

TABLE 5.6 Table of Material Property.

	TCR (ppm)	Resistivity (Ω-cm) $\times 10^{-6}$	Thermal conductivity, k (W/cm°C)
Tungsten	4500	4.2	1.73
Platinum	3927	10.6	0.716
Nickel	6900	6.84	0.91
Polyimide (PI 2611)	—	—	0.001 ~ 0.00357

Case 5.7. Thermal Transfer Shear Stress Sensor

Flow shear stress in the boundary layer can be measured by direct and indirect methods. A shear stress sensor based on the direct measurement principle has been discussed in the previous chapter. In future chapters, a few direct-measurement shear stress sensors based on other position-sensing methods will be reviewed.

Indirect measurement of flow shear stress can be achieved using thermal transfer principles. This case study will examine the design, fabrication, and characterization of a surface micromachined, indirect shear stress sensor [35]. The sensor consists of a resistive element located on a substrate residing at the bottom of a velocity boundary layer. Heat generated by the resistive element is lost to the fluid (by convection) and to the substrate (by conduction). The rate of heat loss from a heated resistive element to the moving fluid is dependent on the velocity profile in the boundary layer.

The heated element creates a thermal gradient in the fluid as well. A thermal boundary layer is characterized as a region where temperature gradients are present in the fluid. Within the thermal boundary layer, fluid temperature decreases with increasing distance away from the heated element until the temperature reaches that of the mean stream flow. Typical profiles of thermal and velocity boundary layers are illustrated in Figure 5.17.

The steady-state temperature of the resistor at constant input power corresponds to the rate of heat loss to the fluid. The analytical relationship between the shear stress (τ) and the temperature of the heated resistor can be found under a few assumptions. These assumptions include: (1) the thermal boundary layer of the thermal element lies within the velocity boundary layer; (2) the thermal transfer in the direction of the span can be ignored; (3) effects of natural convection are much smaller than forced convection.

The temperature of the resistive element is inferred from its instantaneous resistance. The resistance R at temperature T is expressed as

$$R = R_0 \left(1 + \alpha(T - T_0)\right), \tag{5.26}$$

where R_0 is the resistance at the room temperature T_0 and α is the TCR of the resistor.

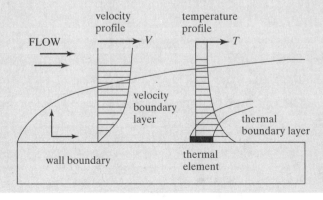

FIGURE 5.17
Velocity and thermal boundary layers.

A fraction of the ohmic heating input power is transferred to the flow while the rest is lost through the substrate. The power balance is expressed as

$$i^2 R = \Delta T (A(\rho \tau)^{1/3} + B), \qquad (5.27)$$

where the term A equals $0.807 A_e C_p^{1/3} k_T^{1/3} / L^{1/3} \mu^{1/3}$, and B is pertinent to the conduction heat loss to the substrate. Here, A_e is the effective area of the thermal element, C_p the heat capacity of the fluid, k_T the thermal conductivity of the fluid, L the stream-wise length of the resistor, and μ the viscosity. At a given shear stress and power input, the greater the term B, the smaller the temperature difference (ΔT) will result. Increased heat loss to the substrate therefore contributes to a lower sensitivity and detection limit. The heat loss to the substrate should be minimized in the design.

Figure 5.18 shows the schematic top and side views of the shear-stress sensor. The heating and heat-sensing element is made of phosphorous-doped polysilicon. The resistor is 2-μm wide and 0.45-μm thick; its length ranges from 20 to 200 μm. The resistors are uniformly doped to a low sheet-resistance value of 50 $\Omega/\square$ with typical resistances between 1.25–5 kΩ at room temperature for the range of resistor lengths indicated above. Each resistor is located at the center of a cavity diaphragm, which is typically 200×200 μm^2 in area and 1.5 μm in thickness. Two metallization wires, each 10 μm wide, connect the polysilicon resistor to the external electronics.

The novel aspect of the sensor is that the diaphragm lies on top of a vacuum cavity, which minimizes the heat conduction from the diaphragm to the substrate through the gap. The diaphragm is separated from the bottom of the cavity by approximately 2 μm, with the pressure inside the cavity being lower than 300 mtorr. This design feature offers effective thermal isolation between the heated element and the substrate.

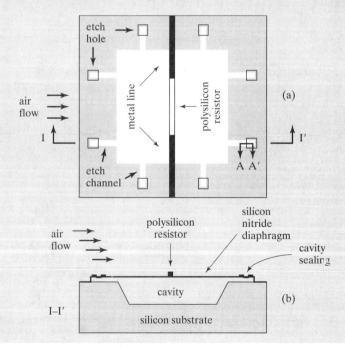

FIGURE 5.18
Schematic diagram of the thermal shear stress sensor.

The heat isolation can be further improved by increasing the depth of the cavity, thus enhancing the thermal resistance from the diaphragm to the substrate. However, this would make the fabrication process more complex.

The micromachining fabrication process is illustrated in Figure 5.19. First, a 0.4-μm silicon nitride layer is deposited by low-pressure chemical vapor deposition (LPCVD) and photolithographically patterned to define the location and shape of the cavities. Each future cavity is defined as a 200×200 μm^2 square window, in which the silicon nitride material is removed using plasma etching (with SF_6 gas), exposing the underlying silicon substrate (step 1 in Figure 5.19). The silicon is etched by plasma etching or isotropic wet etching to reach a certain depth. The ideal depth of each cavity, measured from the silicon nitride surface to the bottom of the silicon surface, is carefully controlled to be 0.73 μm.

The etched trench is filled back with thermal oxidation growth (step 2 in Figure 5.19). During a thermal oxidation process, oxygen atoms react with silicon in the exposed windows and convert silicon into silicon dioxide. The silicon/silicon dioxide interface moves further into the substrate. At the end of this oxidation process, 44% of the total oxide thickness will be contributed by oxidation below the original silicon surface.

Thermal oxidation is a self-limiting process. As the oxide thickness increases, oxygen atoms find it much more difficult to penetrate the existing oxide to reach the buried oxide–silicon interface. Under practical processing time for the industry, the thickness of silicon oxide is on the order of 1.3 μm.

Thus, 1.3-μm-thick silicon dioxide is grown using thermal oxidation at 1050°C in four hours. About 0.73 μm, or 56% of the oxide thickness, occurs above the original air–silicon interface. This is the reason why the cavity depth is designed to be 0.73 μm.

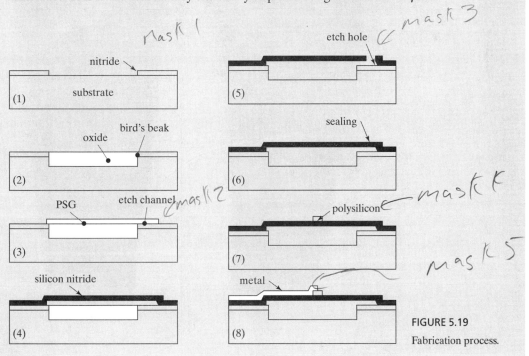

FIGURE 5.19
Fabrication process.

A 500-nm LPCVD sacrificial phosphosilicate glass (PSG) is then blanket deposited. The wafer is annealed at 950°C for an hour. The PSG layer is then patterned using photolithography to define the sacrificial layer and the etching channels overlying the etch cavity (step 3 in Figure 5.19). Unmasked PSG is etched away with buffered hydrofluoric acid within 20 seconds.

Following the removal of the photoresist material, a 1.2 μm-thick low-stress silicon nitride is then deposited as the diaphragm material (step 4 in Figure 5.19). The silicon nitride material is selectively removed with SF6 plasma to expose the underlying sacrificial PSG. Both sacrificial PSG and the thermal oxide are completely etched away using (49%) hydrofluoric acid in 20 minutes. HF solution also etches silicon nitride, but at a very slow rate of approximately 40 Å/min.

After etching, the wafer is thoroughly rinsed in deionized (DI) water for 1 hour to purge HF from within the empty cavity through out-diffusion. The water within cavities is then removed by spin drying the wafer at 7 krpm rotation speed; this is followed by convection-oven baking at 120°C for an hour to evaporate moisture inside the cavities.

A second LPCVD silicon nitride layer (400 nm thick) is deposited at approximately 300 mTorr (0.04 Pa) and 850°C to seal the cavities under vacuum. Because there are still water molecules inside the cavities after the baking, the deposition chamber is purged in nitrogen ambient at 600°C for 30 minutes before deposition starts. This step completely removes residue moisture before the high-vacuum nitride deposition begins.

The deposition of silicon nitride on the front is more pronounced than inside etch channels. The deposition profile at the entrance of etch holes is shown in Figure 5.20. The two deposition fronts eventually meet to permanently seal the cavity under low vacuum. The sealing performance has been experimentally investigated to reflect the influence of sealing materials (including silicon nitride, polysilicon, and oxide by LPCVD) and the geometries of etch holes [36].

To form the resistor, a 450-nm LPCVD polysilicon layer is deposited at 620°C. The polysilicon film deposited at this temperature is completely crystallized with crystal grain sizes on the order of 600 Å. Polysilicon doping is done by ion implantation with phosphorus using a total dose of 1×10^{16} cm^2 at 40 keV of energy. The wafer is then annealed at 1000°C for 1 hour to activate the dopant and to reduce intrinsic stress in the as-deposited polysilicon material. The measured sheet resistivity of the polysilicon is 50 $\Omega/\square$. The polysilicon is then patterned and plasma is etched to form individual resistors. Following the

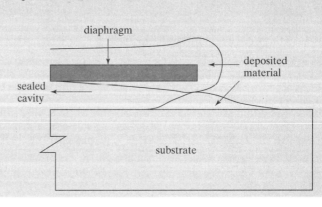

FIGURE 5.20

A close-up cross-sectional view at the etch hole opening (A–A' cross section depicted in Figure 5.18 of a sealed etch channel opening).

removal of the photoresist, another 100-nm layer of LPCVD silicon nitride is deposited to passivate the polysilicon resistor. This film prevents resistance from long-term drifting due to spontaneous oxidation of the polysilicon resistor in air.

Contact holes are patterned and etched in plasma to allow for access to the polysilicon resistor through the last silicon nitride layer. Finally, aluminum for wire leads is deposited and patterned.

Micrographs of the fabricated devices are shown in Figure 5.21. The area of the cavity is $200 \times 200~\mu m^2$; the resistor is 40 μm long and 2 μm wide. Since the cavity is held under vacuum, the diaphragm is bent down by the external atmospheric pressure so that optical interference patterns (Newton Rings) can be seen under the microscope. Figure 5.22 is a scanning electron micrograph of the polysilicon resistor.

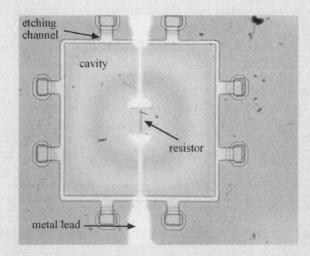

FIGURE 5.21

Optical micrograph of a shear stress sensor with a sealed cavity.

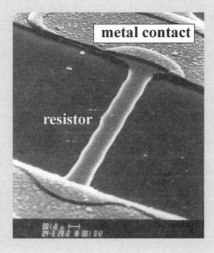

FIGURE 5.22

A scanning electron micrograph of a polysilicon resistor.

188 Chapter 5 Thermal Sensing and Actuation

Comprehensive steady-state and dynamic measurement of the sensor has been performed. Responses to fluid flow have been measured and matched with theoretical models. The frequency response of the sensors under constant current biasing is 9 kHz.

Shear stress sensors with similar architecture but different materials (e.g., parylene as a membrane and metal as thermal resistors) have been made [37].

We mentioned earlier that the I–V characteristics can be used to experimentally measure the thermal resistance associated with a thermal resistor. In this section, we will exemplify this procedure by using the shear stress sensor and variants.

The I–V characteristics of three resistive elements, including the one on the shear stress sensor, are measured experimentally. The structures of the three elements are summarized below and shown in Figure 5.23. The first one, called a Type-1 device, is exactly the shear stress sensor discussed earlier. It consists of a thin diaphragm that is separated from the bulk substrate by a distance of 2 μm. The cavity is hermetically sealed under low pressure. The second device, Type-2, is identical to the first one, except that its cavity has a small hole in it. The cavity is therefore placed under atmospheric pressure. The third heater is placed directly on top of a room temperature bulk substrate.

A schematic diagram of a thermal resistance network is shown in Figure 5.24. The thermal energy generated by ohmic heating is released to the air above and the substrate underneath. A portion of the heat is released into the surrounding air directly. For the three cases, the amount of heat released to the air is identical. The second path of heat transfer is through the substrate. For the Type-1 device, the heat must be conducted through the thin diaphragm first. As the cavity is well sealed under a low temperature with few air molecules for conducting heat, the heat transfer through the cavity to the substrate is minimal. For the Type-2 device, the heat can be conducted through the air molecules in the cavity and into the substrate. The thermal resistance associated with the Type-2 device is expected to be

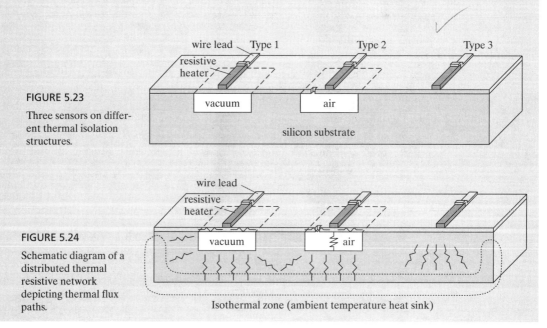

FIGURE 5.23

Three sensors on different thermal isolation structures.

FIGURE 5.24

Schematic diagram of a distributed thermal resistive network depicting thermal flux paths.

smaller than that of the Type-1 device. In the third case, the heat is transferred directly into the substrate, which is considered a heat sink. The thermal resistance is conjectured to be the smallest among all three cases.

The I–V characteristics of a thermal resistive element can be used to determine the thermal resistance associated with it. Let us now review actual experimental results. The representative I–V characteristics of three resistors are shown in Figure 5.25. For the Type-1 and Type-2 devices, the curvature of I–V curves at high input power (the product of current and voltage) is obvious. The deviation from the straight line is caused by resistance changes, which are induced by ohmic heating under the measurement bias. The amount of curvature for the first case is greater than that for the second case. The Type-3 resistor shows an almost linear relationship through the measurement range, indicating that the self-heating effect is not significant.

One can find the relationship between the resistance and the applied ohmic heating power from the I–V characteristics. The following procedure will transform the I–V characteristics' relation to a relation between temperature and power input.

Each data point along the I–V curve corresponds to a current I_d and a voltage V_d. For each measurement point, the ratio of the horizontal coordinate (voltage) and the vertical coordinate (current) represents the instantaneous resistance, i.e.,

$$R_d = \frac{V_d}{I_d}. \tag{5.28}$$

The product of these two coordinates, on the other hand, produces the figure for the input power, according to

$$P_d = V_d I_d. \tag{5.29}$$

One can therefore translate the I–V characteristics' curve, point by point, into a curve representing the relation between resistance and input (R vs. P plot) (Figure 5.26).

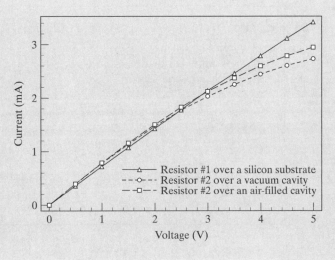

FIGURE 5.25
I–V characteristics of the three cases.

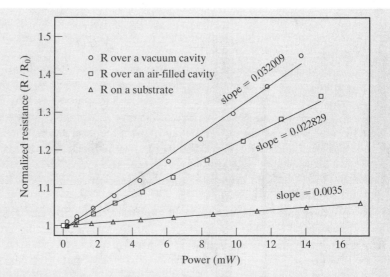

FIGURE 5.26
Resistance vs. power.

The resistance increases with the temperature and the input power; thus, linear R–P curves are expected. The greater the self-heating effect, the greater is the slope of the curve. It is shown that the Type-1 device, with vacuum cavity thermal isolation, exhibits the greatest rise of resistance, indicating the strongest thermal isolation.

If the TCR value of a thermal resistor is known, the temperature of a device can be deduced from its resistance measurement. The temperature difference between the resistor and the ambient can be inferred from the resistance reading, according to

$$\Delta T_d = \frac{\dfrac{R_d - R_0}{R_0} - 1}{\alpha_r}. \tag{5.30}$$

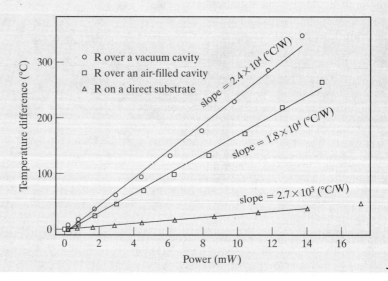

FIGURE 5.27
Temperature vs. input power.

Therefore, the R vs. P curves can be further translated to plots showing the relationships between the temperature and the input power (T vs. P plot, Figure 5.27). The slope of a T vs. P line conveniently corresponds to the thermal resistance of each case.

The thermal resistance associated with a Type-2 device is 6.6 times greater than that of a Type-3 device, indicating the usefulness of the membrane. The thermal resistance associated with a Type-1 device is 8.9 times greater than that of a Type-3 device, confirming the benefits of vacuum sealing.

5.5.3 Infrared Sensors

Thermal sensors are uniquely useful for the detection of radiation, especially in the infrared (IR) spectrum. In this section, we discuss micromachined infrared detectors, a key technology in many military and civilian applications, including night vision, environmental monitoring, biomedical diagnostics, and non-destructive testing. Detectors of IR radiation can be broadly classified into two categories: photonic and thermal. Photonic sensors have been made with materials that have energy bandgaps sufficiently small (e.g., 0.1 eV) to absorb IR radiation with a 8–14 μm wavelength. (The photon energies corresponding to wavelengths of 8 and 14 μm are 0.15 eV and 0.089 eV, respectively). However, the small bandgap makes such devices susceptible to thermal noise. Such sensors are generally cooled under cryogenic conditions at the expense of equipment weight and complexity. Many military and civilian applications of infrared sensors would be hampered by the need to carry cryogenic cooling liquids.

Another broad category of IR sensors that addresses this deficiency is based on photothermal heating—they generally convert infrared radiation to heat by using a heat absorber. The absorbed heat increases the temperature of the absorber and its carrier. The temperature rise is in turn sensed in a number of ways, including using thermoresistors with high TCR values and proper thermal isolation. This constitutes a large class of infrared sensors called **bolometers** [38, 39]. The temperature rise can also be detected using thermal couples [40] and the mechanical displacement of thermal bimetallic beams [39].

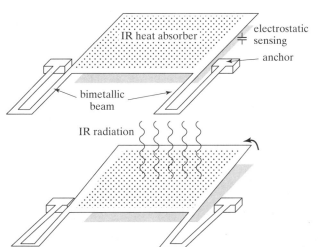

FIGURE 5.28

Infrared sensor based on the capacitive sensing of the bending of thermal bimetallic beams under temperature changes due to absorbed IR power.

Sensors based on thermal bimetallic beams can be further classified based on methods for detecting the bending, including piezoresistive [41], capacitive (Figure 5.28), and optical deflection techniques [42]. Such sensing principles offer a noise-equivalent temperature difference (NETD) on the same order of magnitude as cryogenically cooled infrared detectors (3 mK) [43, 44]. The bimetallic beams associated with the IR absorber must have large thermal resistance, i.e., a large length and small cross section. Micromachining technology is ideal for realizing this class of sensors. In order to reach the theoretical limit of detection, the thermal bimetallic beam must be optimized with respect to noise, thermal isolation, and response speed [45].

The bending induced by thermal absorption can be reported in many ways. We discuss an infrared sensor using the thermal bimetallic beam and optical readout in Case 5.8.

Case 5.8. Bimetallic Structure for Infrared Sensing

A microoptomechanical infrared receiver with a unique optical readout contains a focal plane array (FPA). Each pixel in the array consists of a bi-material cantilever beam and a large plate, according to Figure 5.28 [44]. Absorption of the incident IR radiation by each cantilever beam raises its temperature, resulting in angular bimetallic deflection. An optical system is used to simultaneously measure the deflections of all the cantilever beams of the FPA and collectively project a visible image formed by reflecting off deflected plates. Using such an array, an IR scene is directly converted into a projected image in the visible spectrum. The direct-view optical readout eliminates the need for electronic and wire leads.

Each pixel consists of a plate made of two layers: silicon nitride and gold (Figure 5.29). The silicon nitride material has an absorption peak in the 8–14 μm range. It has a much lower thermal conductivity and thermal expansion coefficient than Au. On the other hand, gold is used as a reflector of visible light. The overall length of the cantilever is 200 μm.

The design of the support beams must balance several primary performance aspects:

1. The thermal isolation between the absorber and the substrate should be as large as possible;
2. The bending of the beam due to intrinsic stresses in the films should be as small as possible to make the cantilever as flat as possible at zero influx;
3. The sensitivity of the displacement vs. the temperature change should be as large as possible.

Let us examine the thermal resistance design aspect more closely. In designing each pixel, the high thermal resistance of each pixel should be targeted to maximize the temperature rise of each beam for a given IR flux (q[W/m^2]) and area A. Assuming the IR energy is 100% absorbed, the temperature rise is

$$\Delta T = (qA)R_T, \qquad (5.31)$$

where R_T is the effective thermal resistance associated with each absorber plate. The effective thermal resistance from the IR absorber to the ambient consists of three parts

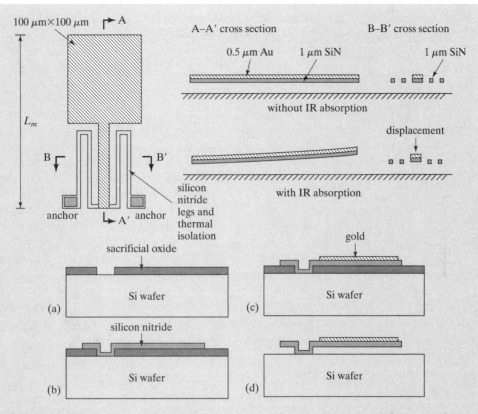

FIGURE 5.29
Infrared thermal sensor.

connected in parallel: (1) thermal resistance associated with the cantilever ($R_{T,l}$); (2) the conductance via the surrounding air ($R_{T,g}$); (3) thermal resistance associated with radiation ($R_{T,r}$).

The cantilever is made of silicon nitride and does not contain any metal. Therefore, the expression for $R_{T,l}$ is

$$R_{T,l} = \rho_{\text{SiN}} \frac{l}{wt},$$

where l, w, and t are the length, width, and thickness of the cantilever, and the ρ_{SiN} is the thermal resistivity.

The heat loss associated with gas conduction can be minimized by operating the pixel in a vacuum.

The thermal resistivity associated with radiation must consider the fact that radiation energy loss occurs on both front and back surfaces. Since radiation energy is the focus of sensing, the thermal resistance due to radiation must be considered. The effective thermal resistance associated with this membrane is

$$R_{T,R} = \frac{\Delta T}{EA} = \frac{\Delta T}{\sigma(\varepsilon_{top} + \varepsilon_{bottom})A\Delta T^4} = \frac{1}{\sigma(\varepsilon_{top} + \varepsilon_{bottom})A\Delta T^3}.$$

A simplified fabrication process is shown in Figure 5.29. The fabrication begins with a boron-doped silicon wafer with resistivity of 10 to 20 Ωcm. The first step is the deposition of a 5-μm-thick phosphorous silicate glass film as the sacrificial layer. The sacrificial layer is patterned and etched (Figure 5.29). A low-stress silicon nitride layer is deposited (with 1 μm thickness) and patterned, followed by the thermal evaporation of gold thin film (0.5 μm thick). A layer of chromium with a 10 nm thickness is deposited between the gold and the silicon nitride to enhance adhesion. The PSG layer is removed to release the beams.

It is found that gold and chromium thin films exhibit intrinsic tensile stress and pull the cantilever out of plane by about 10 μm. This is undesirable because it shifts the optical FPA array out of focus. There are many ways to minimize or eliminate intrinsic bending. One idea discussed in the paper is to vary the deposition recipe of the silicon nitride so it actually consists of two distinct layers with different morphology and intrinsic stress [44]. The bottom layer has strong tensile stress to compensate for the stress of the gold/chromium layer. Although this method works in theory, it is difficult in practice as the film stress and thickness of gold and chromium changes from run to run.

5.5.4 Other Sensors

Here, we discuss two specific applications to illustrate the power and versatility of thermal-transfer based sensing. The first one focuses on microcantilevers with integrated thermal control and temperature sensing elements for data storage and retrieval (Case 5.9). The second one is an end-point detector for the CVD deposition of Parylene thin film (Case 5.10).

Case 5.9. Cantilevers for Data Storage and Retrieval

Arrays of atomic force microscope cantilevers have been used for combined thermomechanical writing and thermal-based reading [46–48]. Digital information can be stored as surface topological features. For example, depressed surfaces represent a digital state while surfaces of normal elevation represent another state. Writing with sharp scanning probe microscope tips provides for a potentially high density of data storage (with a dot spacing of 250 nm) and a high writing speed (heat pulse frequency greater than 100 kHz and tip traveling speed greater than 2.4 mm/s). The bit writing has been performed in thin polymethyl methacrylate (PMMA) films, with the cantilever reaching 350°C for a bit writing (Figure 5.30). The writing demonstrated in Reference [47] is done using a 1-μm-thick, 70-μm-long, two-legged silicon cantilever. The heat can be provided by laser

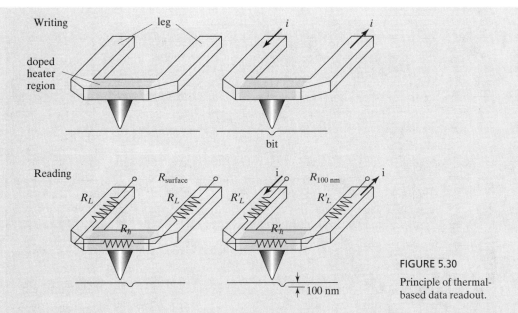

FIGURE 5.30
Principle of thermal-based data readout.

or integrated ohmic heaters. The resistive heater region at the tip is formed by heavy ion implantation of the cantilever legs with the tip region being lightly doped.

The heater cantilever used for writing was given the additional function of a thermal readback sensor by exploiting the temperature-dependent resistance. The principle is discussed in the following paragraphs. Imagine that a steady heater power is sent to the SPM probe as it is scanned across a surface at constant elevation. Heat from the SPM probe is conducted through the tip, the tip-substrate air media, and the substrate. The distance between the end of a tip to the substrate (considered a heat sink) changes as the tip is moved across the substrate surface. When a tip is directly hovering above a region with flat topology, the distance between the end of a tip and the substrate is rather small. However, when the tip is directly located over a pit, the distance between the tip and the substrate is increased. The increased distance translates into greater thermal resistance, and hence a greater temperature of the tip and cantilever.

The electrical resistance, mechanical stiffness and resonant frequency, and thermal transfer characteristics of the device are determined by the length and cross section of the two legs. Long legs increase the parasitic heating resistance R_L (undesirable), reduce the mechanical spring stiffness (desirable), lower the resonant frequency (undesirable), and increase the thermal resistance and thermal response time (undesirable). An optimal design must balance the consideration of intersecting points based on sufficiently accurate models.

The cantilever optimized for data writing is not necessarily optimized for thermal data reading. This work aims to optimize the cantilevers concurrently for data writing and reading. The cantilever sensitivity for data reading of a 100-nm-deep bit is characterized. The thermal reading sensitivity is as high as 4×10^{-4}/nm in terms of vertical displacement, with the resistance per cantilever being 5.1 kΩ.

Case 5.10. Process Monitor for Parylene Deposition

The chemical vapor deposition process of Parylene (poly-para-xylylene) is well established. Parylene is a dielectric polymer material that may be grown by chemical vapor deposition methods at room temperature. It is an exceptionally conformal coating material with a very low pinhole density. Parylene has been used to fabricate microelectromechanical devices such as microfluidic circuits, microinjectors, and valves/pumps, to name a few. Since the film is used to function as a mechanical structure, the thickness becomes an important parameter that determines the performance specifications of sensors and actuators.

It is difficult to predict and control the deposition thickness. The method for controlling the thickness is to preload the deposition system with a controlled amount of dimer material. An *in situ* end-point detector for accurately monitoring the deposition thickness of Parylene is needed.

One candidate for the film thickness sensor is an acoustic resonator whose resonance frequency changes with mass loading by deposited materials. This is commonly used for monitoring the thickness of deposited metal thin films. However, the method works less ideally for Parylene, with a potentially large deposition thickness and heavy mass loading. The sensor also requires somewhat sophisticated electronics for driving the resonator and interpreting the frequency shift.

The schematic diagram of the Parylene sensor is shown in Figure 5.31a and b [49]. The sensor consists of a heating element and a temperature sensor. The heater and the temperature sensor are located at distal ends of two diving-board-type cantilever beams. The distance between the distal ends of the two cantilever beams, denoted d, is well defined in the mask layout. Using microlithography, the size of the gap can be accurately defined.

Parylene is deposited in a low-pressure environment, with the typical deposition pressure ranging from 20 to 40 mtorr. When a sensor with an open gap is placed in a vacuum, the thermal conduction through the gap is negligible. As Parylene is deposited in a conformal fashion, the distance between the two distal ends of cantilevers is gradually reduced (Figure 5.31c). When the Parylene thickness reaches $d/2$, the two Parylene fronts will meet, thereby filling the gap and completing a thermal conduction path (Figure 5.31d). As the gap is filled with Parylene, a thermally conducting medium, heat can be transferred by both the first and the second transfer modes. Heat generated by the heater now has a "thermal shortcut" to reach the temperature sensor. This change of the thermal transfer characteristic is used to infer the process end point. A single sensor with a gap d can indicate when the Parylene thickness reaches $d/2$.

Microfabrication technology is essential for the successful implementation of such an end-point detector. Optical lithography and micromachining allow the cantilever beams to be narrow and thin, thereby reducing the heat transfer from the heater to the substrate. This increases the time constant associated with the second heat-transfer mode, allowing the presence of the first transfer modes to be detected easily. It also reduces unnecessary heat loss and power consumption. Furthermore, optical lithography is critically important to precisely define the gap distance d.

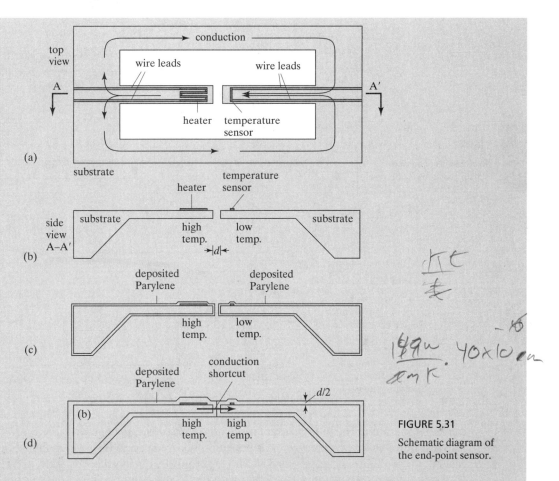

FIGURE 5.31

Schematic diagram of the end-point sensor.

The configuration of a typical sensor is shown in Figure 5.32. The thickness of the cantilever beams is 40 μm. The value of the effective thermal resistances associated with the temperature sensor and the heater are 1.3×10^7 W/K, assuming that the thermal conductivity of silicon is 149 W/m·K. Generally speaking, it is advantageous to use thin and narrow beams to increase the thermal resistances. However, metal leads exhibit intrinsic stress, which may bend the supporting cantilever beams. The bending could become significant compared with the gap spacing (d) if the silicon beam is overly thin. This would alter the effective distances between the heater and sensor. On the other hand, if the beams are overly thick, the heat transfer associated with the heater and temperature sensor will be reduced. The device would require more power to operate.

To avoid distortion of gap spacing due to intrinsic stress, single-crystal silicon was used for the cantilever beams. Single-crystal silicon material has very little intrinsic stress. The thickness of the beams is carefully controlled in process.

Patterned metal resistors located on silicon beams perform heating and sending. The temperature coefficient of resistance (TCR) is found to be 0.14%/°C. If power is applied to the heater for an extended period of time, the substrate of the sensor will be heated to a

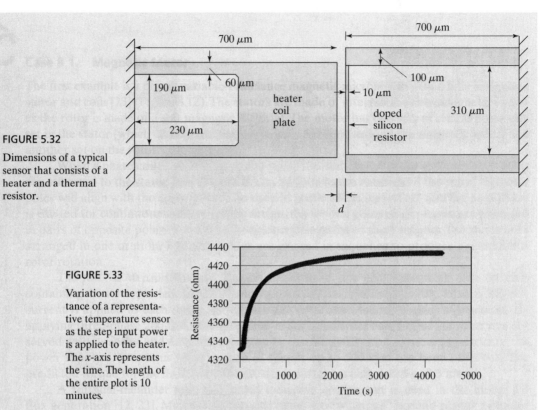

FIGURE 5.32

Dimensions of a typical sensor that consists of a heater and a thermal resistor.

FIGURE 5.33

Variation of the resistance of a representative temperature sensor as the step input power is applied to the heater. The x-axis represents the time. The length of the entire plot is 10 minutes.

higher temperature. The time constant associated with substrate heating is first characterized. The resistance value of a thermistor as power pulses are applied is shown in Figure 5.33. The magnitude of the input voltage is 5 V. The input power during the ON state is 0.1 W. It is shown that the time constants associated with the heating and cooling processes are 336.8 s and 145 s, respectively.

Square-wave pulses with a constant magnitude (5 V) and pulse width (5 s) are applied periodically throughout a run. According to [49], the average interval between pulses is 5 minutes. During a typical run, multiple output peaks are registered. The general appearance of the peaks before and after the gap is closed is different. A typical waveform selected out of the peaks before the gap closure is shown in Figure 5.34. As the power is suddenly increased, the resistance (and therefore temperature) of the sensor increases exponentially with a measured time constant on the order of hundreds of seconds. After the power is cut off, the resistance value gradually returns to the original level.

A representative plot of the last five peaks is shown in Figure 5.35. It is obvious that the resistance of the temperature sensor changes rapidly upon application of the power. This rapid change is caused solely by the fact that the heat pulse travels directly across the gap, now bridged by Parylene, to the temperature sensor. The rate of resistance change then slows, indicating that the substrate heating effect has taken over. After the power is turned

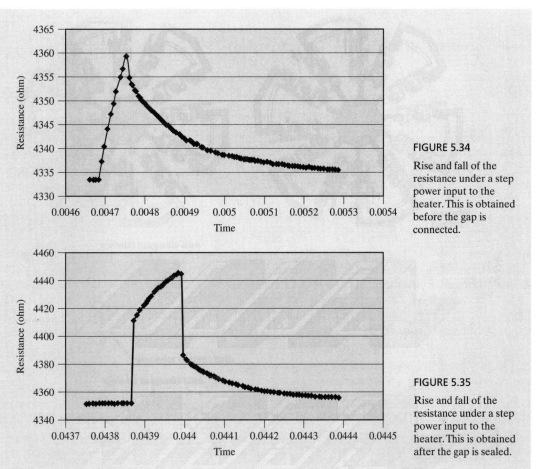

FIGURE 5.34
Rise and fall of the resistance under a step power input to the heater. This is obtained before the gap is connected.

FIGURE 5.35
Rise and fall of the resistance under a step power input to the heater. This is obtained after the gap is sealed.

off, it is again seen that the resistance decreases rapidly before the substrate heating effect catches up.

SUMMARY

At the end of this chapter, the reader should understand the following concepts and facts, and be able to perform the following analyses:

- Physical properties and first-order mathematical description of major temperature sensing methods: thermal beam bending, thermal couples, and thermal resistors.
- Three major forms of heat transfer and their governing equations.
- Definition of thermal resistance in the case of conduction, convection, and radiation.
- Calculation of conductive thermal resistance associated with a microstructure.
- Calculation of thermal bimetallic bending based on simple geometries.
- Procedures for experimentally estimating the temperature coefficient of the resistance of a thermal resistor.

- Method for directly extracting the value of thermal resistance associated with an ohmic-heated thermal resistive element.
- Principle of thermal couple measurement.

PROBLEMS

For homework exercises in this chapter, use the parameters in this table unless instructed specifically otherwise.

Category	Parameter	Value
Young's modulus	Silicon	120 Gpa
	Silicon nitride	385 Gpa
	Gold	57 GPa
Fracture strain	Silicon	0.9%
	Silicon nitride	2.0%
Thermal expansion coefficient	Gold (Au)	14.2 ppm/°C
	Aluminum (Al)	25 ppm/°C
	Nickel	13 ppm/°C
	Silicon and polysilicon	2.33 ppm/°C
Density (kg/m^3)	Silicon	2330
	Silicon nitride	3100
ε_0		8.854×10^{-12} F/m

Problem 5.1: Review
Derive the analytical expressions for thermal resistance associated with convection and radiation cases.

Problem 5.2: Design
A bimetallic cantilever beam is made of two components with the same length. The material on top (denoted Material 2) is gold, whereas the material on the bottom (Material 1) is SCS, single-crystal silicon. The width of both segments is 10 μm. The length of both segments is 1 mm. Young's modulus of gold and silicon are $E_2 = 57$ GPa and $E_1 = 150$ GPa, respectively. The thicknesses of the gold and silicon sections are $t_2 = 0.5$ μm and $t_1 = 1.5$ μm, respectively. The thermal expansion coefficients of gold and silicon are $\alpha_2 = 14$ ppm/°C and $\alpha_1 = 2.33$ ppm/°C, respectively. Find the radius of curvature (r) of the cantilever beam when the beam is uniformly heated to 20°C above room temperature. Determine the amount of vertical displacement at the free end of the beam under this condition.

Problem 5.3: Design
Two pieces of metal wire are connected in a thermal-couple configuration as follows. The temperature at point 1 is higher than the temperatures at points 2 and 3. There are three possible metals: (1) chromel, with Seebeck coefficient = 30 μV/K; (2) alumel, with Seebeck coefficient = -11 μV/K; and (3) iron, with Seebeck coefficient = 10 μV/K. Which of the following statements would be correct? Which one gives the best temperature sensitivity? Show your analysis steps.

1. Metal 1 = chromel, Metal 2 = chromel, produces maximum possible thermal-couple sensitivity of 60 μV/K
2. Metal 1 = alumel, Metal 2 = chromel, produces maximum possible thermal-couple sensitivity of 30 μV/K

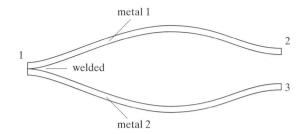

3. Metal 1 = iron, Metal 2 = alumel, produces maximum possible thermal-couple sensitivity of 21 μV/K
4. Metal 1 = chromel, Metal 2 = alumel, produces maximum possible thermal-couple sensitivity of 41 μV/K

Problem 5.4: Design

For a 1-mm-long gold–silicon composite beam (fixed–free), the vertical displacement at the end of the beam is 20 μm when the temperature of the beam is heated by 10 degrees above room temperature. Estimate the angular and vertical displacement if the temperature is raised to 30°. The formula for calculating the bending radius of curvature is

$$k = \frac{1}{r} = \frac{6w_1 w_2 E_1 E_2 t_1 t_2 (t_1 + t_2)(\alpha_1 - \alpha_2)\Delta T}{(w_1 E_1 t_1^2)^2 + (w_2 E_2 t_2^2)^2 + 2w_1 w_2 E_1 E_2 t_1 t_2 (2t_1^2 + 3t_1 t_2 + 2t_2^2)}.$$

Show the analysis steps in the answer. (*Hint*: Use the Taylor series expansion to approximate the function $\cos(\theta)$ when θ is small.)

1. vertical displacement = 60 μm, angular displacement = 6.8°
2. vertical displacement = 49 μm, angular displacement = 2.3°
3. vertical displacement = 49 μm, angular displacement = 15°
4. vertical displacement = 60 μm, angular displacement = 15°

Problem 5.5: Design

Consider three polysilicon thermal resistors (next page) with the TCR values of (1) $\alpha = 1000$ ppm/°C, (2) $\alpha = 2000$ ppm/°C, and (3) $\alpha = -1000$ ppm/°C. An I–V curve measurement is conducted on these three resistors. Which of the following I–V and R–P curves (A through F) are likely to be true? Explain your analysis.

1. A
2. B
3. C
4. D
5. E
6. F

Problem 5.6: Design

Derive the analytical expression of the output force generated by a bimetallic cantilever beam under a given temperature change ΔT.

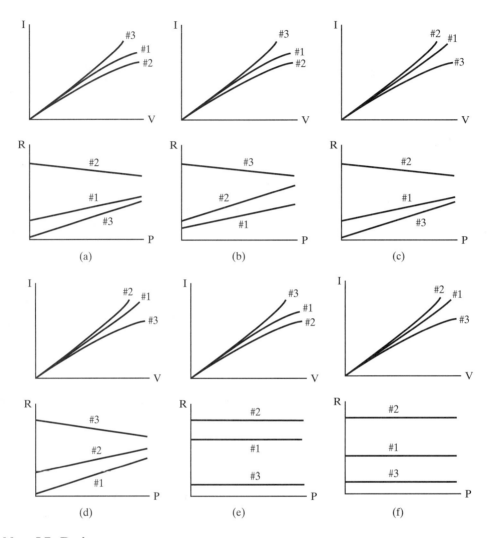

Problem 5.7: Design

A 100-μm-long longitudinal aluminum rod is subjected to a uniform change of temperature. The temperature is raised 20°C above the ambient. Calculate the amount of lateral elongation. What if a temperature gradient is present, being 20°C above the ambient at one end and at the ambient temperature at another?

Problem 5.8: Design

Develop an analytical formula to estimate the vertical displacement of the cantilever discussed in Reference [9] of Case 5.1. Although the displacement is large and the linear model no longer applies, try to use the linear model nonetheless and obtain an estimate, and compare it with experimental results.

Problem 5.9: Fabrication

For Case 5.1, draw mask patterns for each step that requires a mask. The drawing of elements such as resistors, cantilevers, and joints does not have to be precise but should be kept roughly to scale according to information provided in the paper.

Problem 5.10: Fabrication

For Case 5.1, draw the cross-sectional view of the fabrication process showing the flow at the anchor region [9].

Problem 5.11: Design

For Case 5.2, estimate the electrostatic force and voltage necessary to hold the bimetallic flap in the down position, based on geometries given in Reference [9].

Problem 5.12: Fabrication

For Case 5.2, draw the detailed fabrication process according to Figure 5.5 of Reference [10].

Problem 5.13: Fabrication

For Case 5.2, draw a detailed fabrication process with a cross-sectional view of an individual flap. Find the etch rate selectivity between an etchant (or development agent) in each step and other materials present.

Problem 5.14: Design

For Case 5.3, polysilicon is used as the structural layer. Can you develop a fabrication process for realizing a lateral actuator using the Polyimide PIQ-3200 [the one used in Case 5.2] as the structural layer? What are the related design considerations? Find ways to integrate heating elements without causing out-of-plane intrinsic bending.

Problem 5.15: Fabrication

For Case 5.3, find the total power applied to the arm when the applied voltage is 14 V. Related design parameters are outlined in Reference [13].

Problem 5.16: Challenge

For Case 5.3, derive an analytical formula to calculate the horizontal bending due to the expansion of thermal stress for the case where the two arms have different lengths but identical cross sections [13]. Compare your analytical formula with the experimental results.

Problem 5.17: Design

For Case 5.6, determine the thermal resistance associated with the elevated heating elements, if the element is 1 mm long. The multi-layer nature of the support beam should be taken into consideration, as well as the fact that the heating element is hoisted by two beams. Limit your calculation to four materials: polyimide, electroplated Permalloy, Pt, and Nickel.

Problem 5.18: Fabrication

For Case 5.5, develop a fabrication processes by which the heater and the sensors can be fully integrated with a chip consisting of integrated circuits. The heater and the sensors are to be suspended with high resistivity bridges to reduce the thermal consumption. Draw a compressed fabrication process showing critical steps.

Problem 5.19: Challenge

Design a pressure sensor based on a thermal sensing principle. Develop a fabrication process for such a sensor.

Problem 5.20: Fabrication

For Case 5.7, determine the reason why polysilicon is used as the thermal resistor. If the resistor was made of metal, what changes would be necessary in terms of design and fabrication?

Problem 5.21: Fabrication

For each step in the process depicted in Figure 5.29, find the correct etching agent and identify and etch rate selectivity on all materials exposed to it.

Problem 5.22: Design

For Case 5.8, if the amount of intrinsic bending at the folded beams attached to the mirror is 10 μm, find the magnitude of the intrinsic stress in the metal layer, assuming the stress is contributed by the chromium layer alone (i.e., ignoring the contribution of the gold layer). The intrinsic stress is silicon nitride is assumed to be zero.

The Young's modulus, thickness, Poisson's ratio and width of the chromium layer are 279 GPa, 10 nm, 0.21, and 3 μm.

The Young's modulus, thickness, Poisson's ratio, and width of the silicon nitride layer are 385 GPa, 1 μm, 0.29, and 3 μm.

Problem 5.23: Design

Design an accelerometer using the thermal sensing principle discussed in Case 5.9 as a fundamental sensing principle. Devise the analytical formula for its sensitivity, assuming the resistance sensitivity is given in Case 5.9. Draw a complete, realistic, and robust fabrication process.

Problem 5.24: Design

Refer to the design of the Parylene end-point sensors discussed in Case 5.10. Calculate the thermal resistance associated with the two beams on which heater and temperature sensors are located.

Problem 5.25: Challenge

Design an arrayed tactile sensor using the thermal sensing principle discussed in Case 5.9. Develop a realistic individual address method. Develop a fabrication process. The tactile sensor must involve elastomer material (e.g., polydimethylsiloxane or Parylene) as the final cover. Discuss the power consumption of a 10 × 10 tactile pixel array (excluding electronics).

Problem 5.26: Challenge

Develop an alternative design for an *in situ* Parylene thickness monitor. Compare the relative advantages and disadvantages of the sensor compared with Reference [49].

Problem 5.27: Challenge

Review technologies for low-cost, robust fingerprint sensors. Discuss a fingerprint sensor based on thermal sensing technology, and discuss relative merits and disadvantages. Develop a fabrication process for the device. Discuss a series of characterization tasks to perform on the device.

REFERENCES

1. Vettiger, P., et al., *Ultrahigh Density, High-Data-Rate NEMS-Based AFM Storage System*. Microelectronic Engineering, 1999. **46**(1–4): p. 101–104.
2. Leung, A.M., Y. Zhao, and T.M. Cunneen, *Accelerometer Uses Convection Heating Changes*. Elecktronik Praxis, 2001. **8**.
3. Chen, J., et al., *Two Dimensional Micromachined Flow Sensor Array for Fluid Mechanics Studies*. ASCE Journal of Aerospace Engineering, 2003. **16**(2): p. 85–97.
4. Engel, J.M., J. Chen, and C. Liu. *Development of a Multi-Modal, Flexible Tactile Sensing Skin Using Polymer Micromachining*. In *The 12th International Conference on Solid-State Sensors, Actuators and Microsystems*. 2003. Boston, MA.
5. Yang, X., C. Grosjean, and Y.-C. Tai, *Design, Fabrication, and Testing of Micromachined Silicone Rubber Membrane Valves*. Microelectromechanical Systems, Journal of, 1999. **8**(4): p. 393–402.
6. Incropera, F.P., and D.P. DeWitt, *Fundamentals of Heat and Mass Transfer*. 5th ed. 2002, New York: Wiley.

7. Handique, K., et al., *On-Chip Thermopneumatic Pressure for Discrete Drop Pumping*. Analytical Chemistry, 2001. **73**(8): p. 1831–1838.

8. Ataka, M., et al., *Fabrication and Operation of Polyimide Bimorph Actuators for a Ciliary Motion System*. Microelectromechanical Systems, Journal of, 1993. **2**(4): p. 146–150.

9. Suh, J.W., et al., *Organic Thermal and Electrostatic Ciliary Microactuator Array for Object Manipulation*. Sensors and Actuators A: Physical, 1997. **58**(1): p. 51–60.

10. Suh, J.W., et al., *CMOS Integrated Ciliary Actuator Array as a General-Purpose Micromanipulation Tool for Small Objects*. Microelectromechanical Systems, Journal of, 1999. **8**(4): p. 483–496.

11. Que, L., J.-S. Park, and Y.B. Gianchandani, *Bent-Beam Electrothermal Actuators—Part I: Single Beam and Cascaded Devices*. Microelectromechanical Systems, Journal of, 2001. **10**(2): p. 247–254.

12. Lott, C.D., et al., *Modeling the Thermal Behavior of a Surface-Micromachined Linear-Displacement Thermomechanical Microactuator*. Sensors and Actuators A: Physical, 2002. **101**(1–2): p. 239–250.

13. Pan, C.S., and W. Hsu, *A Microstructure for in situ Determination of Residual Strain*. Microelectromechanical Systems, Journal of, 1999. **8**(2): p. 200–207.

14. Guckel, H., et al. *Thermomagnetic Metal Flexure Actuators*. In *Solid-State Sensor and Actuator Workshop,* 1992. 5th Technical Digest, IEEE. 1992.

15. Comtois, J.H., and V.M. Bright, *Applications for Surface-Micromachined Polysilicon Thermal Actuators and Arrays*. Sensors and Actuators A: Physical, 1997. **58**(1): p. 19–25.

16. DeVoe, D.L. *Thermal Issues in MEMS and Microscale Systems*. Components and Packaging Technologies, IEEE Transactions on [see also Components, Packaging and Manufacturing Technology, Part A: Packaging Technologies, IEEE Transactions on], 2002. **25**(4): p. 576–583.

17. Huang, Q.-A., and N.K.S. Lee, *Analysis and Design of Polysilicon Thermal Flexure Actuator*. Journal of Micromechanics and Microengineering, 1999. **9**: p. 64–70.

18. Van Herwaarden, A.W., et al., *Integrated Thermopile Sensors*. Sensors and Actuators A (Physical), 1990. **A22**(1–3): p. 621–630.

19. Geballe, T.H., and G.W. Hull, *Seebeck Effect in Silicon*. Physical Review, 1955. **98**(4): p. 940–947.

20. Von Arx, M., O. Paul, and H. Baltes, *Test Structures to Measure the Seebeck Coefficient of CMOS IC Polysilicon*. Semiconductor Manufacturing, IEEE Transactions on, 1997. **10**(2): p. 201–208.

21. Sarro, P.M., A.W. van Herwaarden, and W. van der Vlist. *A Silicon–Silicon Nitride Membrane Fabrication Process for Smart Thermal Sensors*. Sensors and Actuators A: Physical, 1994. **42**(1–3): p. 666–671.

22. Li, M.-H., J.J. Wu, and Y.B. Gianchandani. *Surface Micromachined Polyimide Scanning Thermocouple Probes*. Microelectromechanical Systems, Journal of, 2001. **10**(1): p. 3–9.

23. Luo, K., et al., *Sensor Nanofabrication, Performance, and Conduction Mechanisms in Scanning Thermal Microscopy*. Journal of Vacuum Science & Technology B: Microelectronics and Nanometer Structures, 1997. **15**(2): p. 349–360.

24. Schafer, H., V. Graeger, and R. Kobs, *Temperature Independent Pressure Sensors Using Polycrystalline Silicon Strain Gauges*. Sensors and Actuators, 1989. **17**: p. 521–527.

25. Kanda, Y., *Piezoresistance Effect of Silicon*. Sensors and Actuators A: Physical, 1991. **28**(2): p. 83–91.

26. Huang, J.B., et al., *Improved Micro Thermal Shear-Stress Sensor*. IEEE Transactions on Instrumentation and Measurement, 1996. **45**(2): p. 570–574.

27. Jiang, F., et al. *A Flexible MEMS Technology and Its First Application to Shear Stress Sensor Skin*. In *IEEE International Conference on MEMS*. 1997.

28. Dauderstadt, U.A., et al., *Silicon Accelerometer Based on Thermopiles*. Sensors and Actuators A: Physical, 1995. **46**(1–3): p. 201–204.

29. Dao, R., et al., *Convective Accelerometer and Inclinometer.* In *United States Patents.* 1995, REMEC Inc.: USA.
30. Milanovic, V., et al., *Micromachined Convective Accelerometers in Standard Integrated Circuits Technology.* Applied Physics Letters, 2000. **76**(4): p. 508–510.
31. Mailly, F., et al., *Effect of Gas Pressure on the Sensitivity of a Micromachined Thermal Accelerometer.* Sensors and Actuators A: Physical, 2003. **109**(1–2): p. 88–94.
32. Chen, J., and C. Liu, *Development and Characterization of Surface Micromachined, Out-of-Plane Hot-Wire Anemometer.* IEEE/ASME Journal of Microelectromechanical Systems (JMEMS), 2003. **12**(6): p. 979–988.
33. Zou, J., et al., *Plastic Deformation Magnetic Assembly (PDMA) of Out-of-Plane Microstructures: Technology and Application.* IEEE/ASME Journal of Microelectromechanical Systems (JMEMS), 2001. **10**(2): p. 302–309.
34. Kovacs, G.T.A., *Micromachined Transducers Sourcebook.* 1998, New York: McGraw-Hill.
35. Liu, C., et al., *A Micromachined Flow Shear Stress Sensor Based on Thermal Transfer Principles.* IEEE/ASME Journal of Microelectromechanical Systems (JMEMS), 1999. **8**(1): p. 90–99.
36. Liu, C., and Y.-C. Tai, *Sealing of Micromachined Cavities using Chemical Vapor Deposition Methods: Characterization and Optimization.* Microelectromechanical Systems, Journal of, 1999. **8**(2): p. 135–145.
37. Fan, Z., et al., *Parylene Surface-Micromachined Membranes for Sensor Applications.* Microelectromechanical Systems, Journal of, 2004. **13**(3): p. 484–490.
38. Richards, P.L., *Bolometers for Infrared and Millimeter Waves.* Journal of Applied Physics, 1994. **76**(1): p. 1–24.
39. Eriksson, P., J.Y. Andersson, and G. Stemme, *Thermal Characterization of Surface-Micromachined Silicon Nitride Membranes for Thermal Infrared Detectors.* Microelectromechanical Systems, Journal of, 1997. **6**(1): p. 55–61.
40. Chong, N., T.A.S. Srinivas, and H. Ahmed, *Performance of GaAs Microbridge Thermocouple Infrared Detectors.* Microelectromechanical Systems, Journal of, 1997. **6**(2): p. 136–141.
41. Datskos, P.G., et al., *Remote Infrared Radiation Detection Using Piezoresistive Microcantilevers.* Applied Physics Letters, 1996. **69**(20): p. 2986–2988.
42. Varesi, J., et al., *Photothermal Measurements at Picowatt Resolution Using Uncooled Micro-optomechanical Sensors.* Applied Physics Letters, 1997. **71**(3): p. 306–308.
43. Perazzo, T., et al., *Infrared Vision Using Uncooled Micro-optomechanical Camera.* Applied Physics Letters, 1999. **74**(23): p. 3567–3569.
44. Zhao, Y., et al., *Optomechanical Uncooled Infrared Imaging System: Design, Microfabrication, and Performance.* Microelectromechanical Systems, Journal of, 2002. **11**(2): p. 136–146.
45. Lai, J., et al., *Optimization and Performance of High-Resolution Micro-optomechanical Thermal Sensors.* Sensors and Actuators A: Physical, 1997. **58**(2): p. 113–119.
46. Mamin, H.J. and D. Rugar, *Thermomechanical Writing with an Atomic Force Microscope Tip.* Applied Physics Letters, 1992. **61**(8): p. 1003–1005.
47. Binnig, G., et al., *Ultrahigh-Density Atomic Force Microscopy Data Storage with Erase Capability.* Applied Physics Letters, 1999. **74**(9): p. 1329–1331.
48. King, W.P., et al., *Design of Atomic Force Microscope Cantilevers for Combined Thermomechanical Writing and Thermal Reading in Array Operation.* Microelectromechanical Systems, Journal of, 2002. **11**(6): p. 765–774.
49. Sutomo, W., et al., *Development of an End-Point Detector for Parylene Deposition Process.* IEEE/ASME Journal of Microelectromechanical Systems (JMEMS), 2003. **12**(1): p. 64–70.

CHAPTER 6

Piezoresistive Sensors

6.0 PREVIEW

Piezoresistivity is a common sensing principle for micromachined sensors. Doped silicon, in particular, exhibits remarkable piezoresistive response characteristics among all known piezoresistive materials [1, 2]. In this chapter, we will first review the origin of and general expression for piezoresistivity (Section 6.1). A number of representative piezoresistive materials, including single-crystal silicon and polycrystalline silicon, are reviewed in Section 6.2. Piezoresistive elements respond to internal strains in mechanical elements. In Section 6.3, we will discuss methods and formula for estimating the magnitude of internal strain in beams and membranes under simple loading conditions. The design, fabrication process, and performance of a few representative piezoresistive sensors are discussed in detail in Section 6.4.

6.1 ORIGIN AND EXPRESSION OF PIEZORESISTIVITY

First discovered by Lord Kelvin in 1856, the piezoresistive effect is a widely used sensor principle. Simply put, an electrical resistor may change its resistance when it experiences a strain and deformation. This effect provides an easy and direct energy/signal transduction mechanism between the mechanical and the electrical domains. Today, it is used in the MEMS field for a wide variety of sensing applications, including accelerometers, pressure sensors [3], gyro rotation rate sensors [4], tactile sensors [5], flow sensors, sensors for monitoring structural integrity of mechanical elements [6], and chemical/biological sensors.

The resistance value of a resistor with the length l and the cross-sectional area A is given by

$$R = \rho \frac{l}{A}. \tag{6.1}$$

The resistance value is determined by both the bulk resistivity (ρ) and the dimensions. Consequently, there are two important ways by which the resistance value can change with applied strain. First, the dimensions, including the length and cross section, will change with strain.

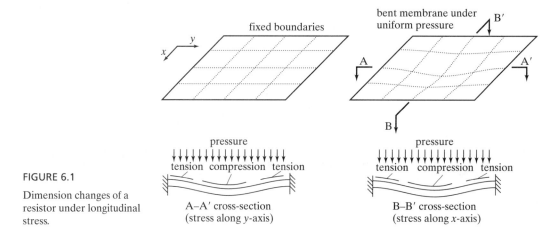

FIGURE 6.1

Dimension changes of a resistor under longitudinal stress.

This is easy to understand, though the relative change in dimensions is generally small. Note that transverse strains may be developed in response to longitudinal loading. For example, if the length of a resistor is increased, the cross section will likely decrease under finite Poisson's ratios (Figure 6.1). Secondly, the resistivity of certain materials may change as a function of strain. The magnitude of resistance change stemming from this principle is much greater than what is achievable from the first one.

By strict definition, piezoresistors refer to resistors whose *resistivity* changes with applied strain. Metal resistors change their resistance in response to strain mainly due to the shape deformation mechanism. Such resistors are technically called strain gauges. The resistivity of semiconductor silicon changes as a function of strain. Silicon is therefore a true piezoresistor. In this chapter, both semiconductor piezoresistors and metal strain gauges are discussed.

The fact that the resistivity of semiconductor silicon may change under applied strain is fascinating. The reason for the strain dependence of resistivity is explained as follows. Recall from Chapter 3 that the resistivity of a semiconductor material depends on the mobility of charge carriers. The formula for the mobility is

$$\mu = \frac{q\bar{t}}{m^*}, \tag{6.2}$$

where q is the charge per unit charge carrier, $\bar{t}$ the mean free time between carrier collision events, and m^* the effective mass of a carrier in the crystal lattice. Both the mean free time and the effective mass are related to the average atomic spacing in a semiconductor lattice, which is subject to changes under applied physical strain and deformation. The quantum-physical explanation of the piezoresistive effect is further explained in Reference [7].

Let us now focus on the macroscopic description of the behavior of a piezoresistor under a normal strain. The change in resistance is linearly related to the applied strain, according to

$$\frac{\Delta R}{R} = G \cdot \frac{\Delta L}{L}. \tag{6.3}$$

The proportional constant G in the above equation is called the **gauge factor** of a piezoresistor. We can rearrange the terms in this equation to arrive at an explicit expression for G,

$$G = \frac{\frac{\Delta R}{R}}{\frac{\Delta l}{l}} = \frac{\Delta R}{\varepsilon R}. \tag{6.4}$$

The resistance of a resistor is customarily measured along its longitudinal axis. Externally applied strain, however, may contain three primary vector components—one along the longitudinal axis of a resistor and two arranged 90° to the longitudinal axis and each other. A piezoresistive element behaves differently towards longitudinal and transverse strain components.

The change of measured resistance under the longitudinal stress component is called longitudinal piezoresistivity. The relative change of measured resistance to the longitudinal strain is called the **longitudinal gauge factor**. On the other hand, the change of resistance under transverse strain components is called transverse piezoresistivity. The relative change of measured resistance to the transverse strain is called the **transverse gauge factor**.

For any given piezoresistive material, the longitudinal and transverse gauge factors are different. For polycrystalline silicon, the transverse gauge factor is generally smaller than the longitudinal one. The absolute magnitude and difference depend on the dopant type and doping concentrations [8].

It is important to realize that longitudinal and transverse strains are often present at the same time though one of them may play a clearly dominating role. The total resistance change is the summation of changes under longitudinal and transverse stress components, namely

$$\frac{\Delta R}{R} = \left(\frac{\Delta R}{R}\right)_{longitudinal} + \left(\frac{\Delta R}{R}\right)_{transverse} = G_{longitudinal} \cdot s_{longitudinal} + G_{transverse} \cdot s_{transverse}. \tag{6.5}$$

Three cases of piezoresistive force sensors are schematically illustrated below to exemplify longitudinal and transverse piezoresistor configurations. Strain gauges, represented by the resistor symbol, are bonded to the outer surfaces of rods which are subjected to external loading forces. Different resistor orientations and external force loading directions are presented (Figure 6.2). In the case illustrated in Figure 6.2a, the longitudinal piezoresistance dominates. In the cases illustrated in Figure 6.2b and c, transverse piezoresistances dominate.

Resistance changes are often read using the Wheatstone bridge circuit configuration. A basic Wheatstone bridge consists of four resistors connected in a loop. An input voltage is applied across two junctions that are separated by two resistors. Voltage drop across the other two junctions forms the output. One or more resistors in the loop may be sensing resistors, whose resistances change with the intended variables. In the bridge shown in Figure 6.3a, one resistor (R_1) is variable by strain. The other resistors—R_2, R_3, and R_4—are made insensitive to strains by being located in regions where mechanical strain is zero, such as on rigid substrates.

The output voltage is related to the input voltage according to the following relationship,

$$V_{out} = \left(\frac{R_2}{R_1 + R_2} - \frac{R_4}{R_3 + R_4}\right) V_{in}. \tag{6.6}$$

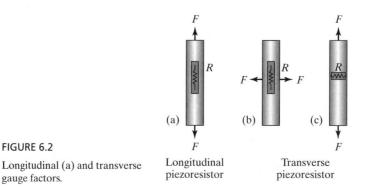

FIGURE 6.2

Longitudinal (a) and transverse gauge factors.

(a) Longitudinal piezoresistor

(c) Transverse piezoresistor

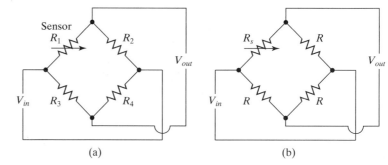

FIGURE 6.3

Wheatstone bridge circuits.

In many practical applications, all four resistors share an identical nominal resistance value. A representative case is shown in Figure 6.3b. In this case, the resistance of the variable resistor (sensor) is represented as

$$R_s = R + \Delta R, \tag{6.7}$$

whereas the nominal resistance values of other three resistors are denoted R. The output voltage is linearly proportional the input voltage according to

$$V_{out} = \left(\frac{-\Delta R}{2R + \Delta R} \right) V_{in}. \tag{6.8}$$

Most piezoresistors are temperature sensitive. For the purpose of eliminating the effect of changing environmental temperature on the output, the Wheatstone bridge is particularly effective. Variation of environmental temperature would cause changes to all resistances in the bridge with the same percentage. Hence, the temperature variation would cause the numerator and the denominator of the right-hand terms of Equation (6.8) to be scaled by an identical factor. The temperature effect is therefore cancelled out.

6.2 PIEZORESISTIVE SENSOR MATERIALS

6.2.1 Metal Strain Gauges

Metal strain gauges are commercially available, often in the form of metal-clad plastic patches that can be glued to surfaces of mechanical members of interest. Resistors are etched into the metal cladding layer. Typical strain gauge patterns are shown in Figure 6.4. A zigzagged conductor path is commonly used to effectively increase the length of the resistor and the amount of total resistance under a given area.

Some of the criteria that are applied when selecting a metal strain gauge include accuracy; long-term stability; cyclic endurance; range of operational temperature; ease of installation; tolerable amounts of elongation; and stability in a harsh environment. To satisfy these requirements, commercial metal strain gauges are often not made of pure metal thin films but of tailored metal alloys.

For micromachined sensors, the sizes of devices are very small. It is impractical to bond or attach discrete strain gauges to devices. Instead, strain gauges are fabricated on mechanical beams and membranes using monolithic integration processes.

Metal resistors are generally deposited (by evaporation or sputtering) and patterned. Elemental metal thin films can be used as strain gauges in MEMS, with their gauge factors ranging from 0.8 to 3.0. The sources and procedures for depositing elemental metal films are readily available.

Strain gauges made of thin-film metals do not compare favorably with semiconductor strain gauges in terms of piezoresistive gauge factors. However, metal thin films provide sufficient performance for many applications. Using metal instead of a semiconductor eliminates the need for doping and lengthy process steps. Also, metal resistors can be deposited and processed under temperatures much lower than what would be needed for doping semiconductors. Metal can generally sustain a much greater elongation before fracture. As such, metal resistors can be placed on polymer materials for polymer MEMS devices (e.g., tactile sensors [9]) and provide improved mechanical robustness compared with silicon counterparts.

6.2.2 Single-Crystal Silicon

Semiconductor strain gauges are made by selectively doping silicon [1, 2, 10]. A schematic diagram of a representative process used for doping selective regions of silicon with dopant atoms

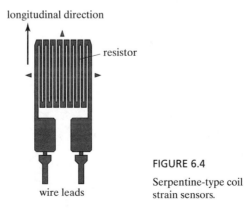

FIGURE 6.4

Serpentine-type coil strain sensors.

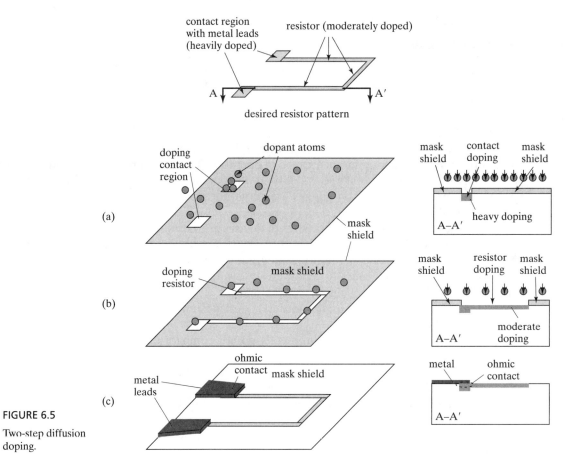

FIGURE 6.5

Two-step diffusion doping.

is shown in Figure 6.5. The desired shape of the resistor is shown in the top-most figure. The resistor feature should be moderately doped (with a concentration ranging from 10^{15} to 10^{18} cm^{-3}). The two ends of the resistor should have a higher doping concentration, on the order of 10^{19} to 10^{20} cm^{-3} in order to form ohmic contacts with metal leads.

The diffusion of dopants in silicon follows Fick's law, which states that the concentration C is a function of the time (t) and spatial location (x),

$$\frac{\partial C(x,t)}{\partial t} = -\frac{\partial J(x,t)}{\partial x} = \frac{\partial}{\partial x}\left(D\frac{\partial C(x,t)}{\partial x}\right), \tag{6.9}$$

where $J(x,t)$ is the flux of dopant per unit area at location x and time t. The term D is the diffusivity of dopants, which is dependent on the temperature according to

$$D = D_0 \exp\left(-\frac{E}{kT}\right). \tag{6.10}$$

The term D_0 is a reference factor (with units being cm²/s), E the activation energy (in eV), T the temperature (in degrees Kelvin), and k the Boltzmann constant (in eV/K). For boron and phosphorus, the values of D_0 are 0.76 and 3.85 cm²/s in single-crystal silicon, respectively. The corresponding values of E are 3.46 and 3.66 eV, respectively.

The solution of Equation (6.9) is given by

$$C(x,t) = C_s erfc\left(\frac{x}{2\sqrt{Dt}}\right). \tag{6.11}$$

The reduction of the substrate temperature during the diffusion process can drastically decrease the diffusivity and the spatial extent of dopant spreading.

The piezoresistive coefficients of a doped single-crystal silicon piezoresistor are influenced by its relative orientation to crystallographic directions. If we consider a rectangular coordinate system having arbitrary orientation with respect to the crystallographic axes of a homogeneous semiconductor, the electric field components E_i and current density components i_i are related by a symmetric resistivity matrix as in

$$\begin{pmatrix} E_x \\ E_y \\ E_z \end{pmatrix} = \begin{bmatrix} \rho_1 & \rho_6 & \rho_5 \\ \rho_6 & \rho_2 & \rho_4 \\ \rho_5 & \rho_4 & \rho_3 \end{bmatrix} \begin{pmatrix} i_x \\ i_y \\ i_z \end{pmatrix}. \tag{6.12}$$

This is the familiar Ohm's law expressed in a matrix form. From routine applications of Ohm's law, we are used to thinking of the relation between potential differences and current as involving only a scalar constant of proportionality. This is obvious for a one-dimensional conductor such as a link in a circuit. However, for electrical conduction in a three-dimensional single-crystalline medium, the current density and the potential gradient will not in general have the same direction as the latter.

What is the relation between changes of resistivity (ρ_i, $i = 1 - 6$) and the applied stress and strain? Recall from Chapter 3 that there are six independent stress components in a 3-D space: three normal stresses (σ_{xx}, σ_{yy}, and σ_{zz}) and three shear stresses (τ_{xy}, τ_{yz}, and τ_{zx}). Their notations have been further unified and simplified according to the scheme $\sigma_{xx} \rightarrow T_1$, $\sigma_{yy} \rightarrow T_2$, $\sigma_{zz} \rightarrow T_3$, $\tau_{yz} \rightarrow T_4$, $\tau_{xz} - T_5$, and $\tau_{xy} \rightarrow T_6$. Changes to the six independent components of the resistivity matrix, ρ_1 through ρ_6, are related to the six stress components.

In the case of silicon, if the x-, y-, and z-axes are aligned to the <100> crystal axes of silicon, the relation between resistivity and stress expressed in a matrix equation format is

$$\begin{pmatrix} \Delta\rho_1/\rho_0 \\ \Delta\rho_2/\rho_0 \\ \Delta\rho_3/\rho_0 \\ \Delta\rho_4/\rho_0 \\ \Delta\rho_5/\rho_0 \\ \Delta\rho_6/\rho_0 \end{pmatrix} = [\pi][T] = \begin{bmatrix} \pi_{11} & \pi_{12} & \pi_{12} & 0 & 0 & 0 \\ \pi_{12} & \pi_{11} & \pi_{12} & 0 & 0 & 0 \\ \pi_{12} & \pi_{12} & \pi_{11} & 0 & 0 & 0 \\ 0 & 0 & 0 & \pi_{44} & 0 & 0 \\ 0 & 0 & 0 & 0 & \pi_{44} & 0 \\ 0 & 0 & 0 & 0 & 0 & \pi_{44} \end{bmatrix} \begin{pmatrix} T_1 \\ T_2 \\ T_3 \\ T_4 \\ T_5 \\ T_6 \end{pmatrix}, \tag{6.13}$$

where ρ_0 is the isotropic resistivity of the unstressed crystal and the terms π_{ij} the component of the piezoresistance tensor. There are three independent piezoresistive coefficient matrices: π_{11}, π_{12}, and π_{44}.

The piezoresistive coefficients of single-crystal silicon are not constants but are influenced by the doping concentration [11, 12], type of dopant [11, 12], and temperature of the substrate [2, 12]. Different elements of the π matrix (π_{11}, π_{12}, and π_{44}) are affected differently by temperature and doping concentrations. For both p- and n-type silicon, the value of the piezoresistive coefficient decreases with increasing temperature and doping concentrations. The values of π_{11}, π_{12}, and π_{44} for single-crystalline silicon under certain doping concentration and dopant types have been experimentally characterized. Several typical values for selected doping concentrations are listed in Table 6.1.

However, all 36 of the coefficients in the coefficient matrix $[\pi]$ may be nonzero when referring to a Cartesian system of arbitrary orientation relative to the crystallographic axes [13]. In the case of silicon, the components of the π matrix change if the x-, y-, and z-axes are not aligned to <100> directions.

Table 6.2 summarizes the effective longitudinal and transverse piezoresistive coefficient for most commonly occurring cases, when the piezoresistor points in <100>, <110>, or <111> directions [2, 14].

The effective piezoresistive gauge factors attributed to each case in Table 6.2 are determined by multiplying the piezoresistive coefficient with Young's modulus in the direction of the applied strain. Recall that Young's modulus of silicon is a function of the crystal directions as well, as outlined in Chapter 2.

TABLE 6.1 Table of Piezoresistivity Components for Single-Crystal Silicon under Certain Doping Values.

Piezoresistance coefficient (10^{-11}Pa^{-1})	n-type (resistivity = 11.7 Ωcm)	p-type (resistivity = 7.8 Ωcm)
π_{11}	−102.2	6.6
π_{12}	53.4	−1.1
π_{44}	−13.6	138.1

TABLE 6.2 Formula for Transverse and Longitudinal Gauge Factors for Various Commonly Encountered Resistor Configurations.

Direction of strain	Direction of current	Configuration	Piezoresistive coefficient
<100>	<100>	Longitudinal	π_{11}
<100>	<010>	Transverse	π_{12}
<110>	<110>	Longitudinal	$(\pi_{11} + \pi_{12} + \pi_{44})/2$
<110>	<1$\bar{1}$0>	Transverse	$(\pi_{11} + \pi_{12} - \pi_{44})/2$
<111>	<111>	Longitudinal	$(\pi_{11} + 2\pi_{12} + 2\pi_{44})/2$

6.3 Stress Analysis of Mechanical Elements

For high-precision sensor applications, it is important to keep in mind that piezoresistive sensitivity is not exactly constant. A second-order correction term can be added to Equation (6.3) if necessary [2, 11]. However, the treatment of this topic is beyond the scope of this text.

Example 6.1. Longitudinal and Transverse Piezoresistivity

A longitudinal piezoresistor is embedded on the top surface of a silicon cantilever near the anchored base. The cantilever points in the <110> direction. The piezoresistor is p-type doped with resistivity of 7.8 Ωcm. Find the longitudinal gauge factor of the piezoresistor.

Solution. The longitudinal piezoresistive coefficient is

$$(\pi_{11} + \pi_{12} + \pi_{44})/2 = \frac{(6.6 - 1.1 + 138.1) \times 10^{-11}}{2} = 71.8 \times 10^{-11} \text{Pa}^{-1}.$$

Young's modulus of single-crystal silicon is 168 GPa in the <110> orientation. The effective gauge factor is

$$G = 71.8 \times 10^{-11}(1/\text{Pa}) \times 168 \times 10^9(\text{Pa}) = 120.6.$$

Appropriate doping concentrations must be carefully selected when designing silicon piezoresistors. A successful design must balance the needs to have an appreciable resistance value, to maximize the gauge factor, and to minimize temperature effects. The doping concentration affects these three performance concerns. The strain gauge factor is a function of the doping concentration. The temperature coefficient of resistance (TCR) of a piezoresistor ideally should be as small as possible to minimize effects of temperature variation. For piezoresistors made of doped silicon, the TCR is a function of the doping concentration (see Section 5.4).

6.2.3 Polycrystalline Silicon

For MEMS piezoresistors, polysilicon offers a number of advantages over single-crystalline silicon, including the ability to be deposited on a wide range of substrates [8]. The polycrystalline silicon also exhibits piezoresistivity, but the gauge factor is much smaller than that of single crystalline. The gauge factor is not dependent on the orientation of the resistor within the substrate plane. However, it is influenced by growth and annealing conditions. The gauge factors for n- and p-type polycrystalline silicon substrates are strongly influenced by the doping concentration (Figure 6.6). The changes in the piezoresistive gauge factor with respect to the doping concentration do not follow monotonic curves. Rather, the gauge factors reach peak magnitude at a particular doping concentrations—around 10^{19} cm^{-3} for p-type polysilicon and 3×10^{19} cm^{-3} for n-type.

6.3 STRESS ANALYSIS OF MECHANICAL ELEMENTS

It is important to be able to analyze the stress and strain distribution in a mechanical element under a given applied force or torque. The loading condition can be simple (e.g., a point force acting at one location of the structure) or complex (e.g., simultaneous torque and force loading in a distributed, non-uniform manner). In this textbook, we will use simple cases to exemplify

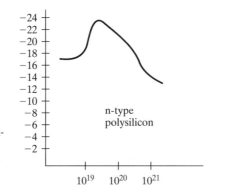
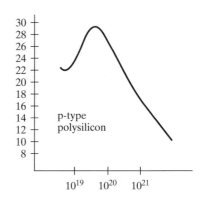

FIGURE 6.6

Gauge factor of polysilicon with different substrate doping types and concentrations of dopant [15].

the process. The general process for analyzing the internal distribution of stress uses the free-body isolation technique that was introduced in Chapter 3.

6.3.1 Stress in Flexural Cantilevers

In Chapter 3, we discussed the distribution of the longitudinal stress of beams under pure bending. Let us now consider a cantilever under a concentrated, transverse loading force applied at the free end. This situation is encountered very frequently in MEMS design. The transverse force loading introduces both longitudinal strains and shear strains. In this textbook, the shear stress components are ignored.

The distribution of longitudinal stress is first described qualitatively (Figure 6.7). Under a transverse loading of a concentrated force at the free end, the torque distribution through the length of the beam is non-uniform—it is zero at the free end and reaches a maxim at the fixed end. At any cross section, the signs of the longitudinal stresses change across the neural axis. The magnitude of stresses at any point on the cross section is linearly proportional with respect to the distance to the neutral axis.

The magnitude of the maximum stress associated with individual cross sections changes linearly with respect to the distance to the free end, reaching a section-wide maxim at the top and bottom surfaces. These are the reasons why piezoresistors are commonly found on the surface of a cantilever and near the fixed end.

Quantitatively, the magnitude of stresses at arbitrary locations along the length of the cantilever can be calculated by following a procedure similar to the one discussed in Chapter 3. The length of the cantilever is L. An x-axis starts at the free end and points towards the fixed end. The normal stress at any given cross section (located at x) and distance h to the neutral plane is denoted $\sigma(x, h)$. The total reactive torque associated with a cross section is simply the area integral of normal force acting on any given area dA, called $dF(x,h)$, multiplied by the arm distance between the force and the neutral plane; namely,

$$M = \iint_A dF(x,h)h = \int_w \int_{h=-\frac{t}{2}}^{\frac{t}{2}} (\sigma(x, h)\, dA)h. \tag{6.14}$$

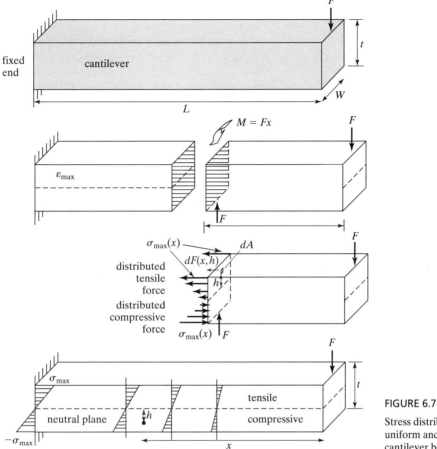

FIGURE 6.7

Stress distribution in a uniform and symmetric cantilever beam.

Under the assumption that the magnitude of stress is linearly related to h and is the greatest at the surface (denoted $\sigma_{max}(x)$) at any cross section, the torque balance equation at any given cross section yields

$$M = \int_w \int_{h=-\frac{t}{2}}^{\frac{t}{2}} \left(\sigma_{max}(x) \frac{h}{\left(\frac{t}{2}\right)} dA \right) h. \tag{6.15}$$

The maximum strain for the entire cantilever occurs at the fixed end, where $x = L$. In fact, in many routine design tasks, the sole interest is to find the magnitude of the maximum stress/strain

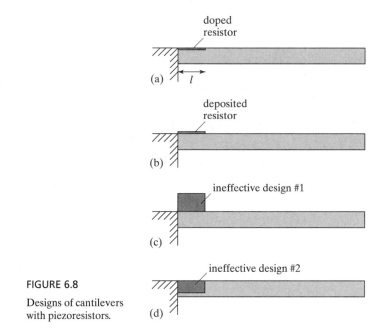

FIGURE 6.8

Designs of cantilevers with piezoresistors.

at the fixed end. The maximum strain is expressed as a function of the total torque $M(x)$:

$$s_{max} = \frac{M(x)t}{2EI} = \frac{FLt}{2EI}. \qquad (6.16)$$

In reality, piezoresistors always have a finite length and thickness. If the resistor is formed by doping a silicon beam, the piezoresistive element will lie below the surface (Figure 6.8a). On the other hand, if the resistor is formed from deposited polysilicon or metal layers, the piezoresistive elements will lie above the surface (Figure 6.8b). In both cases, the resistor covers a finite length starting at the base of the cantilever. (The stress throughout the resistor is in fact non-uniform with respect to length and thickness. This non-uniformity should be accounted for in rigorous analysis.) If we assume the piezoresistive resistor occupies a relatively thin layer near the top surface of the cantilever and is short with respect to the length of the cantilever, the stress at the resistor can be approximated by a single value, as shown in Equation (6.16).

Overly thick piezoresistors complicate designs and fabrication and are undesirable. Two scenarios are depicted in parts c and d of Figure 6.8. The approximation given by Equation (6.16) will not be true if the thickness of the piezoresistor (whether doped or deposited) is large compared with the thickness of the beam. In Figure 6.8d, doped piezoresistors in a cantilever beam extend below the surfaces by a significant depth. If the doped region should reach below the neutral axis, the portion of piezoresistor beyond the neutral axis actually reduces the sensitivity. In the extreme case, if the doped piezoresistor covered the entire thickness, the resistance change due to compressive and tensile regions would cancel out each other.

If a deposited piezoresistor, which is lying above the surface of a cantilever, is overly thick compared with the beam, the piezoresistor must be considered as an integrated part of the cross section and the internal stress analysis must be redone. The position of the neutral axis will change. The term EI in Equation (6.16) must consider both the piezoresistor and beam materials.

Example 6.2. Maximum Stress Points

Consider a fixed-free cantilever beam made of single-crystal silicon with its length pointing in the <100> crystal orientation. Ten points (labeled A through J) are identified on the cantilever. The length (l), thickness (t), and width (w) of the beam are 100 μm, 10 μm, and 6 μm, respectively. If a 1-mN force acts at the end of the cantilever, what is the magnitude of the maximum stress in the cantilever? At which point does the maximum stress occur in each case under various loading conditions?

Solution. For Case 1, a single axis loading force is applied. The reaction force at every cross section along the length of the cantilever is constant. Consequently, the stress at points A through J is exactly identical. The value of the stress is

$$\sigma_{case1} = \frac{F}{A} = \frac{0.001}{10 \times 10^{-6} \times 6 \times 10^{-6}} = 1.6 \times 10^7 \, \text{N/m}^2.$$

For Case 2, the stress at points A, B, C, and E is the greatest. Points A, B, and C are under tension while point E is under compression. The magnitude of the maximum stress is

$$\sigma_{max} = \frac{Mt}{2I} = \frac{Flt}{2I} = \frac{Flt}{2\frac{wt^3}{12}} = \frac{6Fl}{wt^2} = \frac{6 \times 0.001 \times 100 \times 10^{-6}}{10 \times 10^{-6} \times (6 \times 10^{-6})^2} = 1.6 \times 10^9 \, \text{N/m}^2.$$

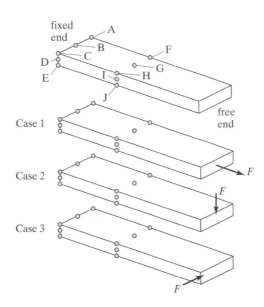

For Case 3, the stress at points C, D, and E is the greatest. The value of the stress at point A is the same as the stress values at points C, D, and E; however, the sign of the stress at point A is opposite of the stress at points C, D, and E:

$$\sigma_{max} = \frac{Mw}{2I} = \frac{Flw}{2I} = \frac{Flw}{2\frac{w^3 t}{12}} = \frac{6Fl}{tw^2} = \frac{6 \times 0.001 \times 100 \times 10^{-6}}{6 \times 10^{-6} \times (10 \times 10^{-6})^2} = 10^9 \, \text{N/m}^2.$$

Example 6.3. Resistance Change under Applied Force

A fixed–free cantilever is made of single-crystal silicon. The longitudinal axis of the cantilever points in the [100] crystal orientation. The resistor is made by diffusion doping, with a longitudinal gauge factor of 50. The length (l), width (w), and thickness (t) of the cantilever are 200 μm, 20 μm, and 5 μm, respectively. If a force $F = 100 \, \mu$N is applied at the end of the cantilever in the longitudinal direction, what would be the percentage change of resistance?

Solution. In this case, the stress level is constant across any particular cross section. The magnitude of the stress is given by

$$\sigma = \frac{F}{wt} = \frac{100 \times 10^{-6} \, \text{N}}{100 \times 10^{-12} \, \text{m}^2} = 1 \, \text{MPa}.$$

Young's modulus of silicon along the longitudinal direction of the resistor is 130 GPa. The strain is

$$s = \frac{\sigma}{E} = 0.00077\%.$$

The relative change of resistance is

$$\frac{\Delta R}{R} = Gs = 0.038\%.$$

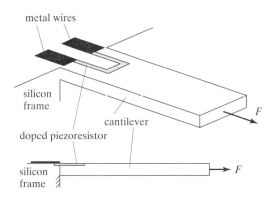

Example 6.4. Resistance Change under Applied Force

A fixed–free cantilever is made of single-crystal silicon. The resistor is made by diffusion doping, with a longitudinal gauge factor of 50. The depth of the doped region is less than 0.5 μm. The length, width, and thickness of the cantilever are 200 μm, 20 μm, and 5 μm, respectively. If a force $F = 100\ \mu$N is applied in the middle of the cantilever beam, what would be the percentage change of resistance?

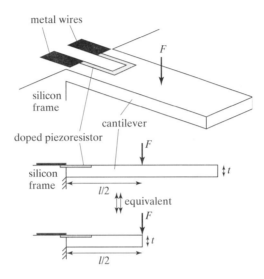

Solution. If the force is applied in the middle, the portion of the cantilever from the point of force to the free end does not bear the load or undergo deformation. Therefore, for the purpose of analyzing the maximum stress at the fixed end of the cantilever, an equivalent system is used. In the equivalent system, the length of the cantilever is half as long, and the force is applied at the distal end.

The resistors span a certain depth and length. The depth appears small compared with the thickness of the cantilever. The distribution of the stress in the cantilever is therefore not uniform. If we assume a uniform stress, equaling that at the top surface of the fixed end, then the maximum strain is expressed as

$$s_{max} = \frac{Mt}{2EI} = \frac{F\frac{l}{2}t}{2EI} = \frac{Flt}{4EI} = 0.019\%.$$

6.3.2 Stress in the Membrane

Membranes are often used in microsensors. The stress analysis of membranes is generally more complex than for beams because a membrane is two-dimensional in nature. In this section, we will review one of the simplest cases of membrane loading—by a uniformly distributed pressure on one side.

The governing equation for membrane displacement under a uniform pressure loading p is

$$\frac{\partial^4 w}{\partial x^4} + 2\frac{\partial^4 w}{\partial x^2 \partial y^2} + \frac{\partial^4 w}{\partial y^4} = \frac{p}{D}, \tag{6.17}$$

where w is the normal displacement for a point of the membrane at a location (x, y). The term D represents the rigidity of the membrane. It is related to Young's modulus (E), Poisson's ratio (ν), and the thickness of the material (t) according to

$$D = \frac{Et^3}{12(1 - \nu^2)}. \tag{6.18}$$

In the case of a square membrane with fixed boundaries, the two-dimensional distribution of membrane displacement and the magnitude of longitudinal stress along the x-axis are illustrated in Figure 6.9. Several important qualitative observations can be made:

1. The maximum displacement occurs at the center of the diaphragm.
2. The maximum stress occurs at the center points of two opposite edges and in the center of the membrane. The stresses along the edge and the center have different signs. These locations with high stress values are preferred for the placement of piezoresistive sensors for detecting membrane deformation.

In many application cases, only the maximum displacement and the maximum stress are of interest. These can be calculated using an empirical formula. The maximum displacement at

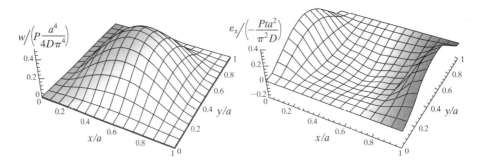

FIGURE 6.9

Normalized displacement (left) and stress in the x-axis (right).

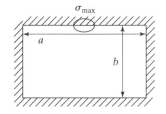
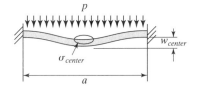

a/b	1.0	1.2	1.4	1.6	1.8	2.0	∞
β_1	0.3078	0.3078	0.3078	0.3078	0.3078	0.3078	0.3078
β_1	0.1386	0.1386	0.1386	0.1386	0.1386	0.1386	0.1386
α	0.0138	0.0138	0.0138	0.0138	0.0138	0.0138	0.0138

FIGURE 6.10
Bending of a rectangular plate under uniform stress.

the center (w_{center}) of a rectangular diaphragm (with the dimension of $a \times b$) under a uniform pressure of p is

$$w_{center} = \frac{\alpha p b^4}{E t^3}, \tag{6.19}$$

with the value of the proportional constant α determined by the ratio of a to b. The value of α can be found by using the look-up table in Figure 6.10. The maximum stress (at the center point of the long edge) and the stress in the center of the plate are

$$\sigma_{max} = \frac{\beta_1 p b^2}{t^2}, \tag{6.20}$$

$$\sigma_{center} = \frac{\beta_2 p b^2}{t^2}, \tag{6.21}$$

with the values of β_1 and β_2 listed in the table as well.

The displacement and stress analysis are considerably more complex if Poisson's ratios and intrinsic stress are present and need to be considered. The analytical expression for calculating the displacement of a square membrane under distributed pressure based on given Poisson's ratio and stress can be found in Reference [16]. Alternatively, one can use computer-aided finite element simulation to solve for displacement and stress distribution of membranes with custom geometries and/or under complex loading conditions.

6.4 APPLICATIONS OF PIEZORESISTIVE SENSORS

Over the years, piezoresistive sensing has been used for many categories of sensor applications. A few representative examples are discussed in the following; they illustrate the unique device designs, fabrication processes, and performance specifications achievable.

6.4.1 Inertia Sensors

Under an applied acceleration, a proof mass experiences an inertial force, which in turn deforms the mechanical support elements connected to the proof mass and introduces stress and strain. By measuring the magnitude of the stress, the value of the acceleration can be inferred. This is the basic principle of piezoresistive accelerometers.

We will examine two examples below. In one case (Case 6.1), the mechanical support is made by using wet silicon etching; another case uses dry silicon etching (Case 6.2). The piezoresistors are made from doped single-crystal silicon in both cases.

Case 6.1. Single-Crystal Silicon Piezoresistive Accelerometers

One of the earliest examples of a micromachined strain gauge accelerometer is the device made by Roylance and Angell in 1979 [17]. It was used for biomedical implants to measure heart wall accelerations. This application required sensitivity of approximately 0.01 g over a 100-Hz bandwidth and small sensor sizes.

The sensor consists of a cantilever with a rigid proof mass attached at its distal end (Fig. 6.11). Piezoresistors located at the base of the cantilever consist of p-type doped resistors with a sheet resistivity of 100 $\Omega/\square$. The cantilever points in the <110> crystal orientation.

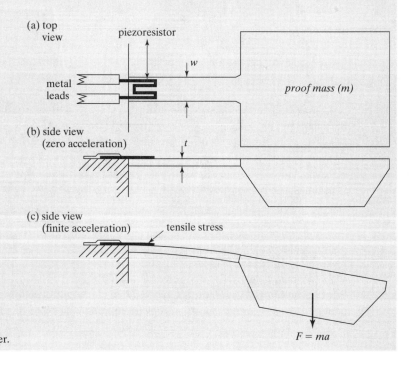

FIGURE 6.11
Piezoresistive accelerometer.

This particular alignment with crystal orientation is partially influenced by the need to define the cantilever and the proof mass using wet silicon etch. (The wet silicon etching technique will be reviewed in Chapter 10.)

Under a given acceleration a, the force experienced by the proof mass is

$$F = m \times a.$$

The force is experienced uniformly throughout the volume of the proof mass. However, for the purpose of calculating the moment produced by the force, we may assume the force is concentrated at the center of the proof mass. The moment experienced at the fixed end of the beam is expressed as

$$M = F\left(l + \frac{L}{2}\right).$$

(Here, the distributed inertia force acting on the cantilever is ignored relative to the contribution from the proof mass). The maximum magnitude of longitudinal strain occurs at the base of the cantilever, on the top surface. The magnitude is given by

$$s_{max} = \frac{Mt}{2EI} = \frac{F\left(l + \frac{L}{2}\right)t}{\frac{Ewt^3}{6}} = \frac{6F\left(l + \frac{L}{2}\right)t}{Ewt^3}.$$

The piezoresistor covers a certain distance and depth. The stress distribution in the resistor body is not uniform. If the resistor is relatively thin and short compared to the cantilever, we can safely assume that the stress on the resistor is uniform and equals ε_{max}.

Here, the strain ε_{max} is applied in the longitudinal direction of the resistor. Consequently, the relative change in resistance is

$$\frac{\Delta R}{R} = G \cdot s_{max} = \frac{6GF\left(l + \frac{L}{2}\right)}{Ewt^2} = \left(\frac{6Gm\left(l + \frac{L}{2}\right)}{Ewt^2}\right)a.$$

The accelerometers were fabricated from one silicon wafer. Two 7740 Pyrex glass wafers were anodically bonded to the silicon to form an enclosure. To allow for the excursion of the silicon proof mass, cavities were isotropically etched into the glass. The displacement of the proof mass was sensed using a diffused piezoresistor in the bending beam connecting the proof mass to the supporting rim of silicon. The overall dimensions of the devices were $2 \times 3 \times 0.6$ mm^3.

The fabrication process begins with an n-type (100) silicon wafer (Figure 6.12). The first step is to grow 1.5-μm-thick thermal oxide on both sides of the wafer (step b). The oxide on the front is photolithographically patterned to open a diffusion-doping window, through which the underlying silicon is exposed (step c). Exposed silicon in the open window region is doped for forming piezoresistors with 100 $\Omega/\square$ target sheet resistivity (step d). A second window is opened to allow doping for contact regions (to 10 $\Omega/\square$ sheet resistivity). The oxide on

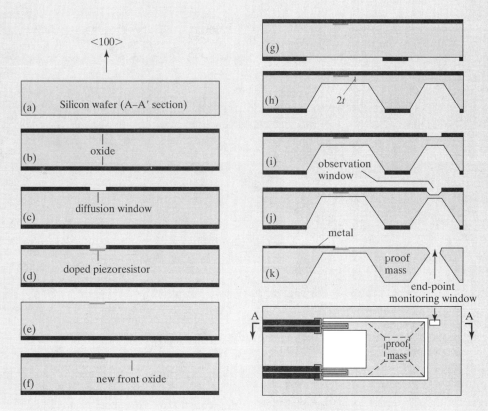

FIGURE 6.12

Fabrication of silicon proof mass with cantilever.

the front side is selectively removed, leaving the oxide on the back side intact (step e). To selectively remove oxide on one side in a wet bath (e.g., HF), the oxide on the front can be blanket coated with photoresist before the wet etching.

Oxide is grown on both sides of the wafer again (step f). This time, the front-side oxide is protected while the back-side oxide is photolithographically patterned (step g). The patterning produces the exposed regions of silicon. In anisotropic etching solutions, the bulk silicon is etched (step h). The etching is stopped until the remaining thickness of the wafer in the regions of the cantilever is approximately twice the desired thickness of the beam.

This step is controlled by carefully calibrating and observing the process. However, controlling the thickness is difficult given the non-uniform etch rate over a wafer and the time-varying nature of etch rates. Nonetheless, it is still easier and more tolerant of errors than if the desired thickness of the membrane is to be achieved by time etch alone.

A window is opened on the front side (step i), and the etching resumes (step j). The etch rate on the front and back will be roughly identical (step j). Hence, when the desired thickness is reached, the observation window is etched through (step k). This event can be reported by visual observation or using optical sensors while the wafer is immersed in wet etching solutions.

This device has been used to demonstrate the detection of accelerations down to 0.001 g, allowing cardiac accelerations to be measured directly. The authors reported a full scale range of +/− 200 g, a sensitivity of 50 $\mu V/(gV_{supply})$, an off-axis sensitivity of 10%, a piezoresistive effect temperature coefficient of −0.2 to −0.3%/°C, and a resonant frequency of 2330 Hz. The relative low resonant frequency is due to the large proof mass in use, which is necessary in order to increase the sensitivity.

Case 6.2. Bulk Micromachined Single-Crystal Silicon Acceletometer

Another bulk micromachined acceleration sensor with its sensitive axis in the plane of the wafer has been demonstrated [18]. The fabrication process takes advantage of the deep reactive ion etching technique to produce the proof mass and narrow cantilever beams with well-defined vertical walls. The deep reactive ion etching also circumvents fabrication issues associated with backside etching and wafer protection.

The schematic diagram of the device is shown in Figure 6.13. A fan-shaped proof mass is supported by a single high-aspect ratio flexure, which is implanted to form a piezoresistive sensor on its vertical surface. Metal conductive traces run on the top wafer surface and connect with resistors located on *vertical side walls*. The piezoresistors are formed by implanting at approximately 31° from vertical. Two proof masses are employed to form two active Wheatstone bridge elements, but only one is shown in Figure 6.13. These proof masses face opposite directions to cancel bridge responses to rotation. Primary temperature compensation is achieved by building the entire Wheatstone bridge with resistive elements of the same geometry and with matched temperature coefficients.

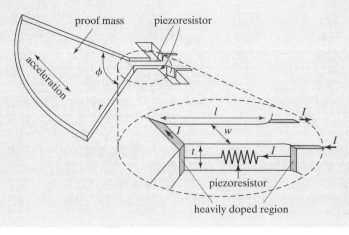

FIGURE 6.13

In-plane accelerometer design.

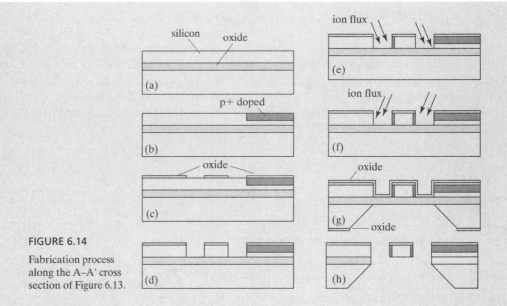

FIGURE 6.14

Fabrication process along the A–A' cross section of Figure 6.13.

According to the authors of Reference [18], the estimated magnitude of strain generated by acceleration with magnitude a is

$$\varepsilon = \frac{4\rho r^3 \sin(\phi/2)}{Ew^2} a, \quad (6.22)$$

where r is the radial length of the proof mass, ϕ the included angle of the proof mass, and w the width of the flexure.

The fabrication process starts with a silicon-on-insulator (SOI) substrate, which is built by oxidizing, bonding, and polishing standard wafers. The SOI substrate has two thin surface layers: a 1.2-μm-thick silicon oxide insulator layer (the "I" layer) under a 30-μm-thick phosphorus doped (100) n-type silicon (the "S" layer) with 0.5-Ωcm resistivity. First, heavily doped p-type regions are formed for contacts (Figure 6.14b). A low-temperature oxide (LTO) thin film is deposited and patterned photolithographically (Figure 6.14c), which serves as a mask for a deep reactive ion etching step (Figure 6.14d). The etching is selective over oxide, and it virtually stops at the buried oxide layer.

Conventionally, ion implantation is performed with the ion flux perpendicular to the substrate. The ions will only impinge on horizontal surfaces that are exposed to the line of sight of incoming ions. To implant on vertical surfaces, an oblique-angle ion implantation process is used, whereas the wafer is set at an angle to the entrant ion flux, allowing the ions to hit the vertical surfaces. The wafer must be implanted twice (Figure 6.14e and f) for both sides of the vertical surfaces to be doped. The sidewall doping results in a sheet resistivity of 2–10 kΩ/□. These dopants are restrained to the flexural region using a photoresist mask to avoid forming leakage paths around the sensor. The LTO film protects the top of the flexure from dopant atoms. The wafer is then coated with an LTO layer again (Figure 6.14g), which serves to protect the entire front side of the wafer during a subsequent wet silicon etching. The wet etching stops at the buried oxide layer due to its reduced etch rate (compared with that of silicon). The LTO and the buried oxide are then removed by using solutions of hydrofluoric acid, which does not attack silicon (Figure 6.14h).

The sensor showed a sensitivity of 3 mV/g and a resolution of 0.2 mg/$\sqrt{Hz}$ at 100 Hz. It compared favorably with commercial capacitive accelerometers based on capacitive and piezoresistive principles.

6.4.2 Pressure Sensors

The micromachined pressure sensor was one of the earliest demonstrations of micromachining technology. It is commercially very successful because of several important traits, including high sensitivity and uniformity.

Bulk microfabricated pressure sensors with thin deformable diaphragms made of single-crystal silicon are the earliest products and still dominate the market today. One example is shown in Figure 6.15 [3]. Piezoresistors are located in the center of four edges. The location of these piezoresistors corresponds to regions of maximum tensile stress when the diaphragm is bent by a uniformly applied pressure difference across the diaphragm.

Four resistors are connected in a full Wheatstone bridge configuration. The resistors were organized along the <110> crystalline direction. Each piezoresistor has a resistance of 2 kΩ. The sheet resistivity is 150 $\Omega/\square$. A fully functional on-chip signal-processing unit consists of two stage amplifiers, compensation circuitry, and two forms of output (frequency and voltage). In the Wheatstone bridge configuration, the temperature sensitivity of the piezoresistors cancels each other.

The diaphragm with embedded piezoresistors is made by using silicon bulk micromachining steps. Piezoresistors are made by selectively doping the silicon diaphragm.

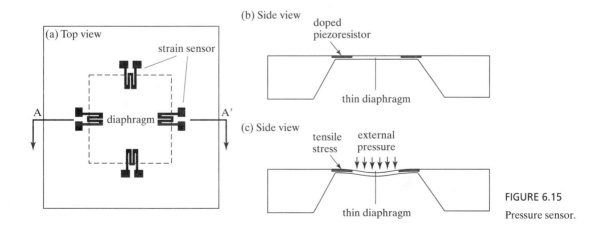

FIGURE 6.15
Pressure sensor.

Using microfabrication, the diaphragm thickness can be controlled precisely (at approximately 25 μm or below). The sensor chip provides a sensitivity of 4 mV/mm Hg, with the nonlinearity lower than 0.4% over the full scale. The temperature coefficient of the sensitivity is less than 0.06%/°C in the temperature range of −20 to 110°C.

The state-of-the-art microfabricated pressure sensors advance very rapidly. Both bulk and surface micromachining can be used. A representative example of a pressure sensor made using surface micromachining processes is discussed in Case 6.3.

Case 6.3. Surface Micromachined Piezoresistive Pressure Sensor

Surface micromachined pressure sensors have been developed using silicon nitride thin film as the membrane and using polycrystalline silicon as the strain sensors [19, 20]. Compared with the bulk micromachined pressure sensors discussed earlier, the thickness of the silicon nitride thin film is even smaller and more controllable. The reduced thickness results in greater stress for a given size membrane and pressure difference according to Equation (6.20).

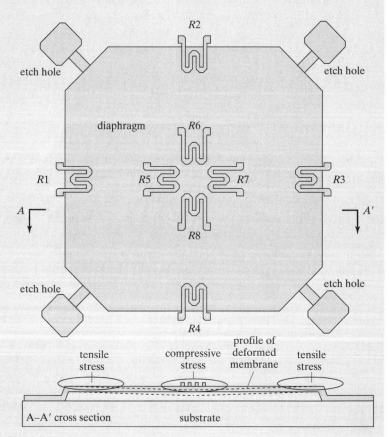

FIGURE 6.16
Top and cross-sectional view of a surface micromachined pressure sensor.

The sensor consists of a membrane suspended from the substrate by a small gap (Fig. 6.16). Eight piezoresistive sensors are involved in the design. Four resistors are located in the middle of each of the four membrane edges. These four sensors experience tension when the membrane is deflected downward. Alternatively, four other sensors can be located in the center of the diaphragm. They experience compression when the membrane is deflected downwards. For a given membrane bending, the resistances of the group of resistors on the edge (R_1 through R_4) and the one in the center (R_5 through R_8) change in opposite directions, further increasing the sensitivity if these two groups are connected in one of the branches of a Wheatstone bridge circuit.

The device fabrication process, shown in Figure 6.17, shares steps with the thermal-transfer based flow shear stress sensor discussed earlier (Case 5.7). The suspended membrane

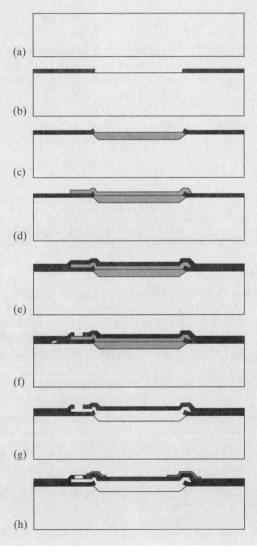

FIGURE 6.17

Fabrication process for the surface micromachined sensor.

is made of silicon nitride (Figure 6.17e). The sacrificial oxide is removed from the cavity through access channels (Figure 6.17g). The opening at the end of each access channel is sealed by a layer of chemical vapor deposition thin film, and a layer of polycrystalline silicon is deposited and patterned photolithographically (Figure 6.17h). This is followed by the deposition and patterning of metal thin film as wire leads.

The fabrication process does not involve time-consuming and chemically aggressive silicon wet etching. The surface micromachining process is therefore more compatible with integrated circuits. However, the gauge factor of polycrystalline silicon is smaller than that of single-crystal silicon. Surface micromachined sensors based on polymer have been made (Section 12.3 of Chapter 12).

6.4.3 Tactile Sensors

Tactile sensors are used to measure contact forces and to characterize surface profiles and roughness. Micromachined tactile sensors offer the potentials of high-density integration. We discuss a silicon micromachined tactile sensor with multiple sensitive axes in Case 6.4. Another piezoresistive tactile sensor, based on metal strain gauges and polymer materials, is discussed in Chapter 12.

Case 6.4. Multi-Axis Piezoresistive Tactile Sensor

Humans achieve dexterous manipulation tasks through the sophisticated tactile perception of objects. When a fingertip comes into contact with an object, a distributed contact stress profile is produced. The stress field at the point of contact consists of three components: a normal stress component and two in-plane shear stress components.

A tactile imager array with 4096 elements has been developed. The sensors' mechanical elements have been made using bulk micromachining technology [5]. Each element consists of a central shuttle plate suspended by four bridges over an etched pit, which allows a greater range of displacement and a larger dynamic range (Fig. 6.18). Embedded in each of the four bridges are polysilicon piezoresistors, labeled R_1 through R_4. Each of the piezoresistors acts as a variable leg of a resistive half-bridge circuitry. A direct measurement of the three stress components can be extracted simultaneously by measuring the changes of the resistances of R_1 through R_4.

The authors performed both analytical studies and the finite element simulation of mechanical sensing characteristics. The overall response of an individual piezoresistor is the direct sum of the responses induced by each stress component.

The response of the sensor structure to pure applied shear stress is determined by assuming that the resulting structural deformation would be primarily axial strain of the

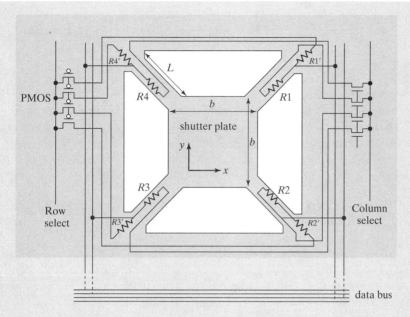

FIGURE 6.18

Piezoresistive tactile sensor.

bridge. For small lateral deflections of the shuttle plate, the shear stress (s_τ) induced in the bridge element by a lateral shear stress τ is described by

$$s_\tau = \frac{b^2}{2\sqrt{2}(EA)}\tau,$$

where b is the shuttle plate width, and E and A are Young's modulus and the cross section of the bridge. The term EA is the overall stiffness of the bridge, which is the sum of stiffness (EA) of each compositional layer. The bridge consists of silicon nitride, silicon oxide, and polycrystalline silicon layers. For the device, the cross-sectional areas are 10.2 μm², 12 μm², and 1.6 μm² for silicon nitride, silicon oxide, and polycrystalline silicon layers, respectively.

The application of normal stress induces the vertical displacement of the shutter plate. The magnitude of induced longitudinal strain in each bridge under a normal stress of σ_n is

$$s_n = \frac{b^2 L}{2(EA)\delta}\sigma_n,$$

where L is the length of the bridge, and δ the normal deflection of the plate.

Each piezoresistor (R_i, $i = 1, 4$) is connected to a reference resistor (R_i') with identical nominal value. Each resistor and its companion reference resistor form a half bridge and provide a voltage output that is fed into the data bus.

The arrayed sensor is built on a silicon wafer with existing complementary MOS circuitry, or CMOS. Two steps are added to the CMOS process sequence to control the state of internal stress in the final freestanding sensor structures. The first modified step happens at the beginning of the process. A 0.35-μm-thick tensile stressed (300 MPa) silicon nitride is deposited on the bare silicon wafer using the low-pressure chemical vapor deposition method (LPCVD). This layer forms the basis of the freestanding shuttle plate and bridges. The second modification comes at the end of the fabrication process, when a 0.6-μm-thick tensile stressed (200 MPa) silicon nitride is deposited using the plasma-enhanced chemical vapor deposition (PECVD) method.

After completion of the CMOS portion of the fabrication process, the sensing structures (shuttle plate and bridges) are released from the underlying silicon substrate using a bulk silicon wet etchant (tetramethyl ammonium hydroxide solution consisting of 5 wt% TMAH in H_2O with 16 g/l dissolved silicon) at 85°C. The TMAH with predissolved silicon is used instead of EDP or KOH because of its low etch rate on aluminum contact pads, which are standard in CMOS circuits. The die is packaged electronically and finally capped with a layer of elastomer rubber with an adhesive backing material.

To allow for the measurement of the mechanical characteristics of the array and individual elements, the chip is placed in a semiconductor microprobe station. Uniform stress over an array is provided by a pneumatic cavity sealed against the front polymer surface. The response of an individual sensor to a pure normal stress is determined by slowly pressurizing the cavity. The normal stress sensitivity was found to be 1.59 mV/kPa after amplification. The response of the individual element to an applied shear stress sensor was determined by the lateral displacement of a mechanical plate attached to the polymer. The shear stress sensitivity was 0.32 mV/kPa.

Cross talk between different measurement modes (normal vs. shear stress) was observed during calibration studies. Nonzero voltages were recorded in both shear stress axes with the application of a pure normal stress. The average value for this false shear stress response is 2.1% of the true normal shear stress. Similarly, a false normal stress results when pure shear stress is applied.

Hysteresis in the transduction mechanism, likely due to the residual compressing of the overlying elastomer, was calibrated. A sensor was loaded from zero load to a full-scale normal stress level of 78 kPa over the course of 20 s. The load was then removed. The average no-load error observed was 3.3 mV, corresponding to a 2.1-kPa normal stress.

The shuttle's resonant frequency changes as a result of adhesion with the polymer. The mechanical resonant frequency of the shuttle plate with the polymer backing is determined experimentally by applying a load and rapidly removing the loading probe. The response during the transitional period provides the mechanical resonant frequency 102 Hz.

Temperature sensitivity of the sensor was determined by altering the temperature of the probe substrate. The sensor output was recorded at several levels of temperature. The unamplified temperature coefficient of the normal stress signal is −0.83 mV/°C, corresponding to −0.52 kPa/°C in the normal stress measurement.

6.4.4 Flow Sensors

Microstructures can be used for flow sensing applications. Their small physical sizes reduce the impact on the flow field under test. Fluid flow around a microstructure can impart a lifting force [21, 22], drag force [23], or momentum transfer on a floating element [24, 25]. These forces can cause the microstructure to deform, producing minute changes of stress in the floating element or its supporting structures. Piezoresistors located strategically on these structures can therefore infer the bending by measuring the resistance.

We will review a direct shear stress sensor and a flow velocity sensor. Case 6.5 illustrates the design and fabrication of a flow shear stress sensor made using wafer bonding techniques. Case 6.6 is a momentum-transfer type flow rate sensor made using surface micromachining and three-dimensional assembly.

Case 6.5. Piezoresistive Flow Shear Stress Sensor

A floating-element shear stress sensor consists of a plate (120 μm wide and 140 μm deep) and four tethers (each 30 μm long and 10 μm wide) [26]. The tethers function as mechanical supports for both the plate and resistors. A flow over the floating element and parallel to the length of the tethers generates a shear stress on top of the suspended plate.

Assume the plate moves as a rigid body. The intended direction of flow is parallel with tethers. Shear stress introduces drag forces acting along the longitudinal directions of tethers. Two of the tethers experience a compressive stress and the other two a tensile stress. The changes in the resistance come from the piezoresistive properties of single-crystal silicon.

The force on the plate is equally divided across four tethers, with each tether experiencing a longitudinal stress of

$$\sigma = \frac{\tau A_p}{4 A_t}, \tag{6.23}$$

where A_p is the area of the plate and A_t is the cross-sectional area of the tether. The changes in the resistance of the tethers is therefore

$$\frac{\Delta R}{R} = Gs = G\frac{\tau A_p}{4 E A_t}. \tag{6.24}$$

The plate and the tethers are made from a 5-μm-thick lightly doped n-type silicon layer and are suspended 1.4 μm above another silicon surface. The fabrication of the sensor involves the processing of two wafers (wafers #1 and #2). The process begins with the growth of a 1.4-μm-thick silicon oxide on wafer #1 (handle wafer), with a background resistivity of 10–20 W/□ (Figure 6.19a). The oxide, which will lie underneath the floating element, is patterned and etched using a plasma etch (Figure 6.19c). The oxide determines the distance between the underside of the floating element and the substrate.

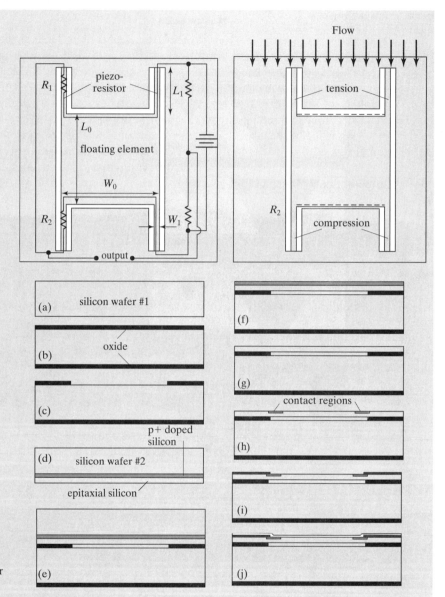

FIGURE 6.19

Piezoresistive shear stress sensor.

A device wafer, wafer #2 (Figure 6.19d), consists of a 5-μm lightly doped n-type epitaxial region (doped to approximately 10^{15} cm^{-3}) on top of a highly doped (10^{20} cm^{-3}) boron region (p$^+$ region). The two wafers are bonded (Figure 6.19e). The bonding sequence includes a preoxidation cleaning of the two wafers, hydration of the bonding surface using a 3:1 H$_2$SO$_4$:H$_2$O$_2$ solution for 10 minutes, a DI water rinse, spin dry, physical contact of the two bonding surfaces at room temperature, and a high temperature anneal at 1000°C in a dry oxygen environment for 70 minutes.

The device wafer is then thinned down all the way to the heavily doped layer by dissolving the silicon substrate in KOH anisotropic etching solutions (Figure 6.19f). The etch stops at the heavily doped region automatically because of the high selectivity. A different wet etchant, consisting of 8:3:1 mixture of $CH_3COOH:HNO_3:HF$ (HNA) solutions, is used to etch the heavily doped layer selectively until the epitaxial layer is reached (Figure 6.19g).

Diffusion doping is performed to provide a heavily doped silicon region (Figure 6.19h). A thin film metal layer is deposited and patterned to provide the electric contact. The metal and the heavily doped region form ohmic (non-rectifying) electrical contacts (Figure 6.19i). A layer of oxide is deposited on top of the device wafer to provide passivation against conducting or caustic environments (Figure 6.19j).

The sensor has been tested to show an overall sensitivity of the sensor being 13.7 μV/V-kPa. The devices were able to withstand high pressure (2200–6600 psi) and temperatures of 190–220°C in shear stresses of 1–100 kPa for up to 20 h.

Case 6.6. Metal Piezoresistive Flow-Rate Sensor

Haircells, a kind of mechanoreceptor, are commonly found across the animal kingdom. The haircell consists of a cilium attached to a neuron. Mechanical displacement of the cilium due to input stimulus causes the neuron to produce pulse output. This seemingly simple mechanical transduction principle is used by vertebrate animals (for hearing and balancing), fish (in lateral line flow sensors), and insects (for flow and vibration sensing). Artificial haircell sensors (AHC) mimicking the biological haircell can serve as a modular building block to perform different engineering sensor applications. A flow sensor based on this biological inspiration is discussed below.

The schematic of an AHC based on a polymer cilium is shown in Figure 6.20 [27]. The AHC is composed of a vertical beam (artificial cilium) rigidly attached to the substrate. Situated at the base of the beam, between the cilium and the substrate, is a strain gauge. The strain gauge is composed of a thin film nichrome (NiCr) resistor on a thicker polyimide backing that runs the length of the cilium. When an external force is applied to the vertical beam, either through direct contact with another object (functioning as a tactile sensor) or by the drag force from fluid flow (flow sensing), the beam will deflect and cause longitudinal strains in the vertical strain gauges.

There are two novel aspects of the device design and fabrication process. First, the vertical cilium is made of a polymeric material and is therefore mechanically robust. Second, the cilium is formed using an efficient 3D assembly and can be conducted on a wafer scale.

The fabrication comprises a series of metalization and polymer deposition steps (Figure 6.21). First, a 0.5-μm Al sacrificial layer is evaporated and patterned onto the substrate. Then, a 5.5-μm photo-definable polyimide (HD-4000 from HD Microsystems) is spun on and patterned photolithographically (Figure 6.21a). The polyimide is cured at 350°C in a 1

238 Chapter 6 Piezoresistive Sensors

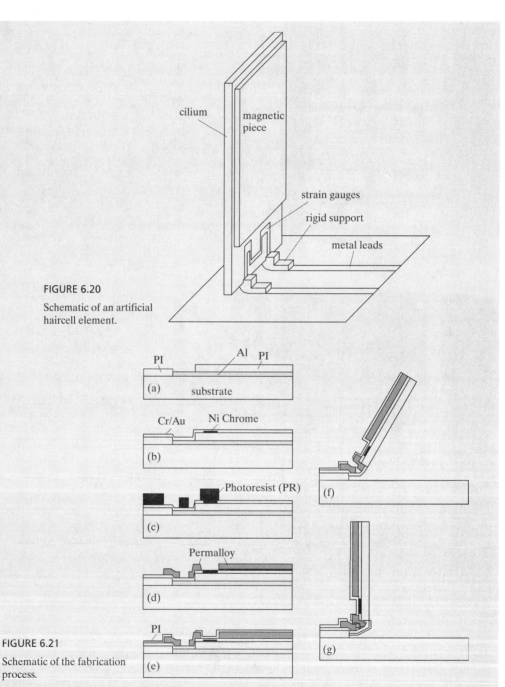

FIGURE 6.20
Schematic of an artificial haircell element.

FIGURE 6.21
Schematic of the fabrication process.

Torr N_2 vacuum for 2 hours. This is the highest temperature used in the process. Afterwards, a 750-Å-thick NiCr layer used for the strain gauge is deposited by electron beam evaporation. This is followed by a 0.5-μm-thick Au/Cr evaporation used for electrical leads and the bending hinge (Figure 6.21b). The authors built an electroplating mold with a patterned photoresist

FIGURE 6.22

SEM of a fabricated haircell array with different heights and widths. The fabricated device has a cilium length varying from 600 μm to 1.5 mm.

(Figure 6.21c). The Au/Cr layer is then used as a seed layer to electroplate approximately 5 μm of Permalloy (Figure 6.21d). The final surface micromachining step is another 2.7-μm polyimide film to serve as a protective coating for the Permalloy cilium and the NiCr strain gauge (Figure 6.21e).

The Al sacrificial layer is then etched in a TMAH solution for over a day to free the cilium structure. The sample is carefully rinsed and placed in the electroplating bath, where an external magnetic field is applied; this interacts with the Permalloy to raise the cilium out of the plane of the substrate (Figure 6.21f and g).

An array of AHC with different cilium and strain gauge geometry is shown in Figure 6.22. The array illustrates the parallel and efficient nature of the fabrication process. Overall, the fabrication method does not exceed temperatures over 350°C, allowing it to be completed on polymer substrates. Silicon, glass, and kapton film have all been used as a substrate for this process. The resistance of the strain gauges tested ranges from 1.2 kΩ to 3.2 kΩ. TCR measurement of the as-deposited NiCr film has a value of -25 ppm/°C, which is very small and should not contribute to anemometric effects during airflow testing.

SUMMARY

The following is a list of major concepts, facts, and skills associated with this chapter:

- The origin of piezoresistivity in silicon.
- The origin of piezoresistivity in metals.
- The influence of doping concentrations on silicon piezoresistors.
- The influence of wafer orientation on silicon piezoresistors.
- The definition of gauge factors for longitudinal and transverse loading cases for silicon.
- The influence of temperature on silicon piezoresistors and on the gauge factors.
- The difference of longitudinal and transverse piezoresistors.
- Analysis of Wheatstone bridge circuit output.
- Method for analyzing the magnitude of stress anywhere in a cantilever beam under simple transverse force loading conditions.

240 Chapter 6 Piezoresistive Sensors

- Identification of relative magnitude and signs of stress in different regions of a membrane under uniform pressure loading.
- Procedures for analyzing relative resistance changes under simple loading on a cantilever.

PROBLEMS

Problem 6.1: Design
A cantilever with a doped silicon piezoresistor is shown below. The width of the beam is w. Assume the entire resistor experiences a distributed stress. Find the analytical expression for the resistance chance, considering both the shear stress (reactive to F) and normal stresses. If Young's modulus is 120 GPa, the L and l are 400 μm and 40 μm, respectively, w and t are 20 and 10 μm, and $F = 1$ mN. Find the relative change of resistance due to normal stress and strain.

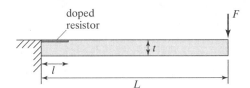

Problem 6.2: Design
Suppose a piezoresistor located on a single-crystal silicon beam is made by doping the entire cantilever thickness to a certain concentration (similar to the force loading condition of Problem 6.1). A vertical force is applied on the cantilever beam's end. Discuss the change of resistance in this case. The effect of the finite thickness of the metal wires should be neglected.

Problem 6.3: Design
For the case in Problem 6.1, discuss the advantages and disadvantages of increasing the length of the resistor (l) by four times.

Problem 6.4: Design
Consider the cantilever below with a width w. Find the analytical expression of the resistance change as a function of F.

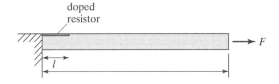

Problem 6.5: Design
A silicon nitride beam is 100 μm long, 5 μm wide, and 0.5 μm thick. How much force can be exerted at the free end of the beam before the beam fractures? (Assume the fracture strain of silicon nitride is 2% and Young's modulus is 385 GPa.) Show your analysis steps.

1. 16 MN
2. 16 mN
3. 16 μN

4. 0.16 mN

5. none of the above

Problem 6.6: Design

A single-crystal cantilever beam with a diffused piezoresistor is shown below. The longitudinal gauge factor is $G_l = 20$, while the transverse gauge factor is $G_t = 10$. Find the change in the resistance of the piezoresistor when the maximum longitudinal strain near the sensor is 1%.

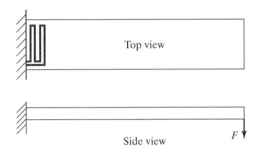

1. $\dfrac{\Delta R}{R} = 0.2$

2. $\dfrac{\Delta R}{R} = 0.1$

Problem 6.7: Design

A fixed–free silicon cantilever is 200 μm long (l), 5 μm wide, and 0.5 μm thick. A force of 1 μN is applied in the transverse direction at the center of the beam (half length). Find the maximum tensile stress at the fixed end and $x = l/4$ from the fixed end. Young's modulus of silicon along the longitudinal direction is 160 GPa.

Problem 6.8: Design

A silicon cantilever with the longitudinal direction pointing in the <110> crystal direction is under a transverse force loading. The cantilever is 200 μm long, 20 μm wide, and 1 μm thick. The resistor is 20 μm long. Calculate the percentage change of resistance when a 10 μN force is applied. The force is in the direction of beam thickness.

Problem 6.9: Design

Consider a silicon cantilever with two arms pointing in <100> direction. (Derive an analytical expression for the relative change of resistance under F.) The resistor is doped uniformly throughout the length. Each arm of the cantilever has a length of L, a width W, and a thickness t. The resistor along each arm has a length L, a width w, and a uniform doping depth of t_p. (Hint: The mechanical deformation and resistance associated with the horizontal bar connecting the two arms should be ignored.)

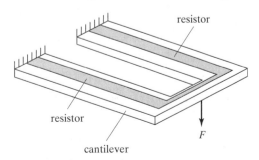

Problem 6.10: Design

A square membrane with four piezoresistors is diagrammed below. Resistors R_1 and R_4 are located on at the midpoints of two edges. Resistors R_2 and R_3 are located in the center of the membrane. The size of the membrane is b, and the thickness is t. A pressure difference p is applied across the membrane. Find the analytical expression of the output voltage under two Wheatstone bridge configurations depicted in (a) and (b).

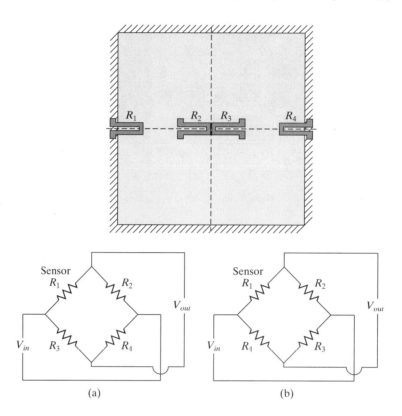

Problem 6.11: Review

The thickness, length, and width of the support beams were never provided in the paper for the acceleration sensor discussed in Case 6.1. Can you use the information provided in Case 1 and the reference within to identify all three design parameters? Explain the methodology for finding the dimensions of the supporting silicon cantilever beams.

Problem 6.12: Design

For the structure discussed in Case 6.2, derive the analytical expression that relate the resistance change to the input acceleration. Compare the results with the formula derived in the reference paper.

Problem 6.13: Fabrication

For the structure discussed in Case 6.2, what is the advantage and disadvantage if the beam width is reduced to 1/3 of its current value?

Problem 6.14: Design

For the pressure sensor discussed in Case 6.3, what is the advantage and disadvantage of increasing the length of the resistor on the edge to 1/4 of the total length of the membrane? What is the advantage and disadvantage of increasing the length of the resistor on the edge to 1/2 of the total length of the membrane?

Problem 6.15: Design

For the tactile sensor in Case 6.4, complete the analysis of the response of a single piezoresistor under shear stress and normal stress.

Problem 6.16: Fabrication

The polymer protective cover in Case 6.4 is provided by gluing an elastomer with adhesive backing. However, the adhesion of the polymer to individual shuttle plates is not controllable. Discuss an alternative design and process of a tactile sensor, with similar functions as in the case, but with an improved method for integrating the polymer with the silicon parts.

Problem 6.17: Review

For Case 6.4, discuss the reason for selecting TMAH as the silicon etchant rather than EDP or KOH. Identify and cite supporting evidence from literature.

Problem 6.18: Design

In Case 6.5, what is the sensitivity to in-plane cross flow (i.e., flow perpendicular to the longitudinal axis) and out-of-plane cross flow (i.e., flow impacting the flat plate in a normal direction)?

Problem 6.19: Challenge

The artificial haircell sensor discussed in Case 6.6 consists of a cilium with a rectangular cross-sectional area. Discuss a method to build high-aspect-ratio polymeric cilia in an efficient microfabrication process. Can you integrate a sensor with the element (either on it or connected to it) so the displacement of the artificial hair can be detected with high sensitivity?

REFERENCES

1. Smith, C.S., *Piezoresistance Effect in Germanium and Silicon*, Physics Review, 1954. **94**: p. 42–49.
2. Kanda, Y., *Piezoresistance Effect of Silicon*, Sensors and Actuators A: Physical, 1991. **28**: p. 83–91.
3. Sugiyama, S., M. Takigawa, and I. Igarashi, *Integrated Piezoresistive Pressure Sensor with Both Voltage and Frequency Output*, Sensors and Actuators A, 1983. **4**: p. 113–120.
4. Gretillat, F., M.-A. Gretillat, and N.F. de Rooij, *Improved Design of a Silicon Micromachined Gyroscope with Piezoresistive Detection and Electromagnetic Excitation*, Microelectromechanical Systems, Journal of, 1999. **8**: p. 243–250.
5. Kane, B.J., M.R. Cutkosky, and T.A. Kovacs, *A Traction Stress Sensor Array for Use in High-Resolution Robotic Tactile Imaging*, Microelectromechanical Systems, Journal of, 2000. **9**: p. 425–434.
6. Hautamaki, c., et al., *Experimental Evaluation of MEMS Strain Sensors Embedded in Composites*, Microelectromechanical Systems, Journal of, 1999. **8**: p. 272–279.
7. Geyling, F.T., and J.J. Forst, *Semiconductor Strain Transducers*, The Bell System Technical Journal, 1960. **39**: p. 705–731.
8. French, P.F., and A.G.R. Evans, *Piezoresistance in Polysilicon and Its Applications to Strain Gauges*, Solid-State Electronics, 1989. **32**: p. 1–10.
9. Engel, J., J. Chen, and C. Liu, *Polymer-Based MEMS Multi-Modal Sensory Array*. Presented at 226th National Meeting of the American Chemical Society (ACS). New York, NY: 2003.

10. Toriyama, T., and S. Sugiyama, *Analysis of Piezoresistance in p-type Silicon for Mechanical Sensors*, Microelectromechanical Systems, Journal of, 2002. **11**: p. 598–604.
11. Yamada, k., et al., *Nonlinearity of the Piezoresistance Effect of p-type Silicon Diffused Layers*, IEEE Transactions on Electron Devices, 1982. **ED-29**: p. 71–77.
12. Tufte, O.N., and E.L. Stelzer, *Piezoresistive Properties of Silicon Diffused Layers*, Journal of Applied Physics, 1963. **34**: p. 313–318.
13. Kerr, D.R., and A.G. Milnes, *Piezoresistance of Diffused Layers in Cubic Semiconductors*, Journal of Applied Physics, 1963. **34**: p. 727–731.
14. Senturia, S.D., *Microsystem Design*. Kluwer Academic Publishers, 2001.
15. French, P.J., and A.G.R. Evans, *Polycrystalline Silicon Strain Sensors*, Sensors and Actuators A, 1985. **8**: p. 219–225.
16. Maier-Schneider, D., J. Maibach, and E. Obermeier, *A New Analytical Solution for the Load-Deflection of Square Membranes*, Microelectromechanical Systems, Journal of, 1995. **4**: p. 238–241.
17. Angell, J.B., S.C. Terry, and P.W. Barth, *Silicon Micromechanical Devices*, Scientific American Journal, 1983. **248**: p. 44–55.
18. Partridge, A., et al., *A High-Performance Planar Piezoresistive Accelerometer*, Microelectromechanical Systems, Journal of, 2000. **9**: p. 58–66.
19. Liu, J., *Integrated Micro Devices for Small-Scale Gaseous Flow Study*. In Electrical Engineering. Pasadena, CA: California Institute of Technology, 1994, p. 169.
20. Lee, W.Y., M. Wong, and Y. Zohar, *Pressure Loss in Constriction Microchannels*, Microelectromechanical Systems, Journal of, 2002. **11**: p. 236–244.
21. Svedin, N., et al., *A New Silicon Gas-Flow Sensor Based on Lift Force*, Microelectromechanical Systems, Journal of, 1998. **7**: p. 303–308.
22. Svedin, N., E. Kalvesten, and G. Stemme, *A New Edge-Detected Lift Force Flow Sensor*, Microelectromechanical Systems, Journal of, 2003. **12**: p. 344–354.
23. Schmidt, M.A., *Wafer-to-Wafer Bonding for Microstructure Formation*, Proc. IEEE, 1998. **86**(8): p. 1575–1585.
24. Fan, Z., et al., *Design and Fabrication of Artificial Lateral-Line Flow Sensors*, Journal of Micromechanics and Microengineering, 2002. **12**: p. 655–661.
25. Svedin, N., E. Stemme, and G. Stemme, *A Static Turbine Flow Meter with a Micromachined Silicon Torque Sensor*, Microelectromechanical Systems, Journal of, 2003. **12**: p. 937–946.
26. Shajii, J., K.-Y. Ng, and M.A. Schmidt, *A Microfabricated Floating-Element Shear Stress Sensor Using Wafer-Bonding Technology*, Microelectromechanical Systems, Journal of, 1992. **1**: p. 89–94.
27. Chen, J., et al., *Towards Modular Integrated Sensors: The Development of Artificial Haircell Sensors Using Efficient Fabrication Methods*. Presented at IEEE/RSJ International Conference on Intelligent Robots and Systems. Las Vegas, NV: 2003.

CHAPTER 7

Piezoelectric Sensing and Actuation

7.0 PREVIEW

Piezoelectric materials are used for both sensing and actuation purposes. We will first review the principle of piezoelectricity and basic design methodology in Section 7.1. The piezoelectric properties of several representative materials are reviewed in Section 7.2. We will then discuss examples of sensors and actuators based on various mechanical and electrical configurations (Section 7.3).

7.1 INTRODUCTION

7.1.1 Background

The phenomenon of piezoelectricity was discovered in the late nineteenth century. It was observed that certain materials generate an electric charge (or voltage) when they are under mechanical stress. This is known as the **direct effect of piezoelectricity**. Alternately, the same materials would be able to produce a mechanical deformation (or force) when an electric field is applied to them. This is called the **inverse effect of piezoelectricity**. (Some literature refers to it as the converse effect of piezoelectricity.)

As an indicator of the magnitude of the piezoelectric effect, a field of 1000 V/cm applied between the ends of a quartz rod produces a strain of 10^{-7}. Conversely, a small strain can generate enormous fields.

In 1880, Pierre and Jacques Curie experimentally discovered the direct piezoelectric effect in various naturally occuring substances including Rochelle salt and quartz. In 1881, Hermann Hankel suggested using the term piezoelectricity, which is derived from the Greek "piezen" meaning "to press." In 1893, William Thomson (Lord Kelvin) published seminal papers on the theory of piezoelectricity. It was mathematically hypothesized and then experimentally proven that a material exhibiting the direct effect of piezoelectricity would also exhibit the inverse effect.

The piezoelectricity phenomenon was developed and applied in sonar and quartz oscillation crystals. In 1921, Walter Cady invented the quartz crystal-controlled oscillator and the narrow-band quartz crystal filter used in communication systems [1]. World War II spurred the growth of this field, especially with the urgent need by the military to detect submarines. Two important artificial piezoelectric crystals, barium titanate ($BaTiO_3$) and lead zirconate titanate ($PbZrTiO_3$-$PbTiO_3$, or PZT) were invented in the early 1950s. These materials are not naturally occurring piezoelectric materials. Rather, they are synthesized materials and must be electrically poled in order to exhibit significant piezoelectric effects. In 1958, synthetic quartz material became available.

Historically, well-known applications of piezoelectric sensors have included phonograph pickups; microphones; acoustic modems; and acoustic imaging for underwater, underground objects, and medical observation. A good book dedicated to piezoelectricity and applications is cited in Reference [2].

Now, piezoelectric materials are being used in MEMS sensors and actuators. Thin-film piezoelectric materials has been explored for use as on-chip acoustic transducers [3], pumps and valves for liquid and particles [4, 5], accelerometers [6, 7], speakers and microphones [8, 9], mirrors [10], and chemical sensors [11], among others.

Many important properties of piezoelectric materials stem from their crystalline structures. Piezoelectric crystals can be considered to be a mass of minute crystallites (domains). The macroscopic behavior of the crystal differs from that of individual crystallites, due to the orientation of such crystallites. The direction of polarization between neighboring crystal domains can differ by 90° or 180°. Owing to the random distribution of domains throughout the material, no overall polarization or piezoelectric effect is exhibited. A crystal can be made piezoelectric in any chosen direction by **poling**, which involves exposing it to a strong electric field at an elevated temperature. Under the action of this field, domains most nearly aligned with the field will grow at the expense of others. The material will also lengthen in the direction of the field. When the field is removed, the dipoles remain locked in an approximate alignment, giving the crystal a remnant polarization and a permanent deformation (albeit small).

The poling treatment is usually the final step of crystal manufacturing. Care must be taken in all subsequent handling and use to ensure that the crystal is not depolarized, since this will result in a partial or even total loss of its piezoelectric effect. A crystal may be depolarized mechanically, electrically, or thermally. Mechanisms for depolarization are further explained in the following paragraphs.

Exposure to a strong electric field of opposite polarity to the poling field will depolarize a piezoelectric element. The field strength required for marked depolarization depends on the material grade, the time the material is subjected to the depolarization field, and the temperature. For static fields, the threshold is typically between 200–500 V/mm. An alternating field will also have a depolarizing effect during the half cycles that it opposes the poling field.

Mechanical depolarization occurs when mechanical stress on a piezoelectric element becomes high enough to disturb the orientation of the domains and hence destroy the alignment of the dipoles. The safety limits for mechanical stress vary considerably with material grade.

If a piezoelectric element is heated to a certain threshold temperature, the crystal vibration may be so strong that domains become disordered and the element becomes completely depolarized. This critical temperature is called the **Curie point** or the **Curie temperature**. A safe operating temperature would normally be halfway between 0°C and the Curie point.

The properties of piezoelectric elements are time dependent. The stability as a function of time is of particular interest. Material characteristics may be degraded through aging effects due to the intrinsic process of spontaneous energy reduction. The speed of aging can be controlled through the addition of composite elements or through accelerated aging.

Many piezoelectric materials suffer from finite ion mobility. In other words, they do not provide long-term static holding power when used in actuators. The design of piezoelectric actuators operating in DC conditions must consider electric leakage.

7.1.2 Mathematical Description of Piezoelectric Effects

Piezoelectric materials are crystals. The microscopic origin of piezoelectricity is the displacement of ionic charges within a crystal, leading to the polarization and electric field. A stress (tensile or compressive) applied to a piezoelectric crystal will alter the spacing between centers of positive and negative charge sites in each domain cell; this leads to a net polarization manifested as open circuit voltages measurable at the crystal surface. Compressive and tensile stresses will generate electric fields and hence voltages of opposite polarity.

Inversely, an external electric field will exert a force between the centers of positive and negative charges, leading to an elastic strain and changes of dimensions depending on the field polarity.

Not all naturally occurring or synthesized crystals exhibit piezoelectricity. Crystals can be classified into 32 groups according to crystal symmetry. Centrosymmetric crystal structures are crystals that are symmetric along all axes through the center of the crystal. These crystals occupy 11 out of 32 possible groups and are non-piezoelectric materials because the positive and negative charge sites will not be spatially separated under stress. Out of 21 non-centrosymmetric groups, 20 are piezoelectric crystals.

Piezoelectric effects are strongly orientation dependent. The notation conventions for crystal orientations in the context of piezoelectric polarization are discussed first. A piezoelectric material needs to be poled in a particular direction to provide a strong piezoelectric effect, although some materials exhibit natural or spontaneous polarization. The direction of positive polarization is customarily chosen to coincide with the Z-axis of a rectangular system of crystallographic axes X, Y, and Z (Fig. 7.1). Alternatively, the normal stress components along axes X, Y, and Z are denoted by subscripts 1, 2, and 3, respectively. As such, the poling axis always coincides with axis-3. Shear stress and strain components about these axes are denoted by subscripts 4, 5, and 6, respectively.

In a piezoelectric crystal, the constitutive equation that relates electrical polarization (D) and applied mechanical stress (T) is

$$D = dT + \varepsilon E, \tag{7.1}$$

where d is the **piezoelectric coefficient matrix**, ε the electrical permittivity matrix, and E the electrical field. Here, an electric field is applied in conjunction with the mechanical stress to provide more generality. The electrical polarization is contributed by two parts—one stemming from electrical biasing and one from mechanical loading.

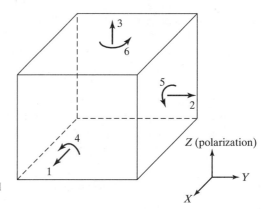

FIGURE 7.1

Schematic illustration of a piezoelectric crystal in a rectangular system.

If no electric field is present (i.e., $E = 0$), then the second term on the right-hand side of Equation (7.1) can be eliminated.

The general constitutive equation can be written in the full matrix form:

$$\begin{bmatrix} D_1 \\ D_2 \\ D_3 \end{bmatrix} = \begin{bmatrix} d_{11} & d_{12} & d_{13} & d_{14} & d_{15} & d_{16} \\ d_{21} & d_{22} & d_{23} & d_{24} & d_{25} & d_{26} \\ d_{31} & d_{32} & d_{33} & d_{34} & d_{35} & d_{36} \end{bmatrix} \begin{bmatrix} T_1 \\ T_2 \\ T_3 \\ T_4 \\ T_5 \\ T_6 \end{bmatrix} + \begin{bmatrix} \varepsilon_{11} & \varepsilon_{12} & \varepsilon_{13} \\ \varepsilon_{21} & \varepsilon_{22} & \varepsilon_{23} \\ \varepsilon_{31} & \varepsilon_{32} & \varepsilon_{33} \end{bmatrix} \begin{bmatrix} E_1 \\ E_2 \\ E_3 \end{bmatrix}. \qquad (7.2)$$

The terms T_1 through T_3 are normal stresses along axes 1, 2, and 3, whereas T_4 through T_6 are shear stresses. The units of electrical displacement (D_i), stress (T_j), permittivity (ε_i), and electrical field (E_j) are C/m², N/m², F/m, and V/m, respectively. The unit of the piezoelectric constant d_{ij} is the unit of electric displacement divided by the unit of the stress, namely,

$$[d_{ij}] = \frac{[D]}{[T]} = \frac{[\varepsilon][E]}{[T]} = \frac{\frac{F}{m}\frac{V}{m}}{\frac{N}{m^2}} = \frac{Columb}{N}. \qquad (7.3)$$

The inverse effect of piezoelectricity can be similarly described by a matrix-form constitutive equation. In this case, the total strain is related to both the applied electric field and any mechanical stress, according to

$$s = ST + dE, \qquad (7.4)$$

where s is the strain vector and S is the compliance matrix.

Equation (7.4) can be expanded to a full matrix form:

$$\begin{bmatrix} s_1 \\ s_2 \\ s_3 \\ s_4 \\ s_5 \\ s_6 \end{bmatrix} = \begin{bmatrix} S_{11} & S_{12} & S_{13} & S_{14} & S_{15} & S_{16} \\ S_{21} & S_{22} & S_{23} & S_{24} & S_{25} & S_{26} \\ S_{31} & S_{32} & S_{33} & S_{34} & S_{35} & S_{36} \\ S_{41} & S_{42} & S_{43} & S_{44} & S_{45} & S_{46} \\ S_{51} & S_{52} & S_{53} & S_{54} & S_{55} & S_{56} \\ S_{61} & S_{62} & S_{63} & S_{64} & S_{65} & S_{66} \end{bmatrix} \begin{bmatrix} T_1 \\ T_2 \\ T_3 \\ T_4 \\ T_5 \\ T_6 \end{bmatrix} + \begin{pmatrix} d_{11} & d_{21} & d_{31} \\ d_{12} & d_{22} & d_{32} \\ d_{13} & d_{23} & d_{33} \\ d_{14} & d_{24} & d_{34} \\ d_{15} & d_{25} & d_{35} \\ d_{16} & d_{26} & d_{36} \end{pmatrix} \begin{bmatrix} E_1 \\ E_2 \\ E_3 \end{bmatrix}. \quad (7.5)$$

If there is no mechanical stress present ($T_{i,\,i=1,6} = 0$), the strain is related to the electric field by

$$\begin{bmatrix} s_1 \\ s_2 \\ s_3 \\ s_4 \\ s_5 \\ s_6 \end{bmatrix} = \begin{pmatrix} d_{11} & d_{21} & d_{31} \\ d_{12} & d_{22} & d_{32} \\ d_{13} & d_{23} & d_{33} \\ d_{14} & d_{24} & d_{34} \\ d_{15} & d_{25} & d_{35} \\ d_{16} & d_{26} & d_{36} \end{pmatrix} \begin{bmatrix} E_1 \\ E_2 \\ E_3 \end{bmatrix}. \quad (7.6)$$

Note that, for any given piezoelectric material, the d_{ij} components connecting the strain and the applied field in the inverse effect are identical to the d_{ij} connecting the polarization and the stress in the direct effect. The unit of d_{ij} can be confirmed from Equation (7.6) as well. It is (m/m)/(V/m) = m/V = C/N.

The **electromechanical coupling coefficient** k is a measure of how much energy is transferred from electrical to mechanical energy, or vice versa, during the actuation process:

$$k^2 = \frac{energy\_converted}{input\_energy}. \quad (7.7)$$

This relation holds true for both mechanical-to-electrical and electrical-to-mechanical energy conversion. The magnitude of k is a function of not only the material, but also the geometries of the sample and its oscillation mode.

7.1.3 Cantilever Piezoelectric Actuator Model

Piezoelectric actuators are often used in conjunction with cantilevers or membranes for sensing and actuation purposes. General models for such piezoelectric actuators are rather complex. Accurate analysis often involves finite element modeling. For limited cases, such as a cantilever actuator with two layers, analytical models has been successfully developed. In this chapter, we will focus on the analysis of cantilevers with two layers of materials, at least one of them being a piezoelectric layer.

The deflection of a two-layer piezoelectric structure can be described by a compact formula. Consider a cantilever with two layers, one elastic and one piezoelectric, joined along one side (Fig. 7.2). These two layers share the same length. A compact model for calculating the curvature of bending has been made under the following assumptions:

1. The induced stress and strain are along axis-1, or the longitudinal axis of the cantilever;
2. Cross sections of the beam that were originally plane and perpendicular to the beam axis remain plane and perpendicular to the resulting curved axis;
3. The beam maintains a constant curvature throughout the beam;
4. Shear effects are negligible;
5. Beam curvature due to intrinsic stress may be ignored;
6. The beam thickness is much less than the piezoelectric-induced curvature;
7. Second-order effects (such as the influence of d_{33} and electrostriction) are ignored;
8. Poisson's ratio is isotropic for all films.

The beam bends into an arc when the piezoelectric layer is subjected to a longitudinal strain, s_{long} (Fig. 7.2). The radius of curvature can be found by

$$\frac{1}{r} = \frac{2|s_{long}|(t_p + t_e)(A_p E_p A_e E_e)}{4(E_p I_p + E_e I_e)(A_p E_p + A_e E_e) + (A_p E_p A_e E_e)(t_p + t_e)^2}, \tag{7.8}$$

where A_p and A_e are the cross-sectional areas of the piezoelectric and the elastic layer; E_p and E_e are Young's modulus of the piezoelectric layer and Young's modulus of the elastic layer; and t_p and t_e are the thickness of the piezoelectric layer and the elastic layer.

Once the radius of curvature is known, the vertical displacement at any location (x) of the cantilever can be estimated:

$$\delta(x) = r - r\cos(\phi) \approx \frac{x^2 d_{31} E_3 (t_p + t_e) A_e E_e A_p E_p}{4(A_e E_e + A_p E_p)(E_p I_p + E_e I_e) + (t_e + t_p)^2 A_e E_e A_p E_p}. \tag{7.9}$$

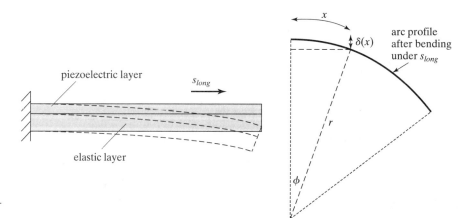

FIGURE 7.2

Bending of a piezoelectric bimorph.

The amount of force achievable at the free end of a piezoelectric bimorph actuator equals the force required to restore the tip of the actuator to its initial undeformed state. Since the displacement is linearly related to force according to

$$\delta(L) = F/k, \tag{7.10}$$

the expression of the force is

$$F = \delta(x = L)k. \tag{7.11}$$

Piezoelectric sensors and actuators with more than two layers are commonly encountered. Several general techniques can be found in References [12–16], both under simple or arbitrary loading.

Example 7.1. Bending of a Piezoelectric Beam

A 500-μm-long cantilever-type piezoelectric actuator is made of two layers: a ZnO layer and a polysilicon layer (Fig. 7.3). The width, thickness, and material properties of these two layers are listed in Table 7.1. Find the amount of vertical displacement at the end of cantilever and the transverse force at the end when the applied voltage is 10 V.

Solution. In this particular scenario, the polarization axis, axis-3, is perpendicular to the front surface of the cantilever. Axis-1 points in the longitudinal direction of the beam. The primary applied electric field is applied along axis-3; the intended direction of stress is in axis-1.

TABLE 7.1 Dimensions and Material Properties.

	ZnO	Polysilicon
Width (μm)	20	20
Thickness (μm)	1	2
Young's modulus (GPa)	160	160
Piezoelectric coefficient (pC/N)	5×10^{-12}	N/A

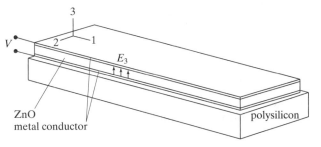

FIGURE 7.3 Bimorph piezoelectric actuator.

The longitudinal stress in the beam is denoted S_1. According to Equation (7.6), the longitudinal stress s_{long} is related to the electric field E_3 by

$$s_{long} = s_1 = d_{31}E_3. \tag{7.12}$$

Plugging in the expression for s_{long}, Equation (7.8) can be rewritten as

$$\frac{1}{r} = \frac{2d_{31}(t_p + t_e)(A_p E_p A_e E_e)E_3}{4(E_p I_p + E_e I_e)(A_p E_p + A_e E_e) + (A_p E_p A_e E_e)(t_p + t_e)^2}, \tag{7.13}$$

where E_3 is the electrical field in axis-3, and it is perpendicular to the cantilever substrate.

We find the maximum transverse displacement, occurring at the free end of the beam, by using

$$\delta(x=l) = \frac{l^2 d_{31} E_3 (t_p + t_e) A_e E_e A_p E_p}{4(A_e E_e + A_p E_p)(E_p I_p + E_e I_e) + (t_e + t_p)^2 A_e E_e A_p E_p} =$$

$$\frac{(500 \times 10^{-6})^2 \cdot 5 \times 10^{-12} \cdot \dfrac{10}{1 \times 10^{-6}} \cdot (3 \times 10^{-6}) \cdot A_e E_e A_p E_p}{4((20 \times 10^{-12}) \cdot 160 \times 10^9 + (40 \times 10^{-12}) \cdot 160 \times 10^9)(E_p I_p + E_e I_e) + (3 \times 10^{-6})^2 A_e E_e A_p E_p}.$$

Here,

$$(E_p I_p + E_e I_e) = 160 \times 10^9 \frac{20 \times 10^{-6} \cdot (10^{-6})^3}{12} + 160 \times 10^9 \frac{20 \times 10^{-6} \cdot (2 \times 10^{-6})^3}{12}$$

$$= 2.4 \times 10^{-12} \text{ Nm}^2$$

$$A_e E_e A_p E_p = (20 \times 10^{-12}) \cdot 160 \times 10^9 \cdot (40 \times 10^{-12}) \cdot 160 \times 10^9 = 20.48 \text{ N}^2.$$

Therefore,

$$\delta(x=l) =$$

$$\frac{(500 \times 10^{-6})^2 \cdot 5 \times 10^{-12} \cdot \dfrac{10}{1 \times 10^{-6}} \cdot (3 \times 10^{-6}) \cdot A_e E_e A_p E_p}{4((20 \times 10^{-12}) \cdot 160 \times 10^9 + (40 \times 10^{-12}) \cdot 160 \times 10^9)(E_p I_p + E_e I_e) + (3 \times 10^{-6})^2 A_e E_e A_p E_p}$$

$$= \frac{7.68 \times 10^{-16}}{9.216 \times 10^{-11} + 1.83 \times 10^{-10}} = 2.79 \times 10^{-6} \text{ m}.$$

7.2 PROPERTIES OF PIEZOELECTRIC MATERIALS

Since semiconductor materials are often used to make circuits and MEMS, it is of interest to discuss the piezoelectricity of important semiconductor materials. Elemental semiconductors such as silicon and germanium show centrosymmetric crystal structures and do not exhibit piezoelectric behavior. The III–V compound and II–VI compounds (such as GaAs and CdS), on the other hand, are held together by covalent and ionic bonding. They show acentric crystal symmetry and are thus piezoelectric. However, these are not ideal candidates because of the high cost and low piezoelectric coefficients.

Commonly employed piezoelectric materials and their properties are summarized in Table 7.2. Detailed information about the piezoelectric coefficients of selected materials are summarized in the subsections. Note that the properties of thin-film materials may be different from their bulk counterparts [17]. The progress of developing new piezoelectric materials is fast-paced in recent years [2, 18].

A number of common piezoelectric materials and their representative properties are described in the following paragraphs.

7.2.1 Quartz

The most familiar use of quartz crystal, a natural piezoelectric material, is as a resonator in watches. In a quartz-crystal oscillator, a small plate of quartz is provided with metal electrodes on its faces. Just as a bell rings when struck, the quartz plate also "rings"; but it rings at a very high frequency and produces an AC voltage between the electrodes at its mechanical resonant frequency. When such a crystal is used in an oscillator, positive feedback provides energy to the quartz crystal to keep it ringing, and the oscillator output frequency is precisely controlled by the quartz crystal. Quartz is not the only crystalline material that exhibits a piezoelectric effect, but it is used in this application because its oscillation frequency is quite insensitive to temperature changes. Quartz-crystal oscillators are able to produce output frequencies from about 10 kHz to more than 200 MHz; in carefully controlled environments, they can have a precision of one part in 100 billion, though one part in 10 million is more common.

TABLE 7.2 Properties of Selected Piezoelectric Materials.

Material	Relative permittivity (dielectric constant)	Young's modulus (GPa)	Density (kg/m^3)	Coupling factor (k)	Curie temperature (°C)
ZnO	8.5	210	5600	0.075	**
PZT-4 (PbZrTiO$_3$)	1300–1475	48–135	7500	0.6	365
PZT-5A (PbZrTiO$_3$)	1730	48–135	7750	0.66	365
Quartz (SiO$_2$)	4.52	107	2650	0.09	**
Lithium tantalate (LiTaO$_3$)	41	233	7640	0.51	350
Lithium niobate (LiNbO$_3$)	44	245	4640	**	**
PVDF	13	3	1880	0.2	80

The material properties are well characterized for quartz. The compliance matrix, piezoelectric coefficient matrix, and dielectric constants for quartz are summarized below:

$$s = \begin{bmatrix} 12.77 & -1.79 & -1.22 & -4.5 & 0 & 0 \\ -1.79 & 12.77 & -1.22 & 4.5 & 0 & 0 \\ -1.22 & -1.22 & 9.6 & 0 & 0 & 0 \\ -4.5 & 4.5 & 0 & 20.04 & 0 & 0 \\ 0 & 0 & 0 & 0 & 20.04 & -9 \\ 0 & 0 & 0 & 0 & -9 & 29.1 \end{bmatrix} \times 10^{-12}\, m^2/N \quad (7.14)$$

$$d = \begin{bmatrix} -2.3 & 2.3 & 0 & -0.67 & 0 & 0 \\ 0 & 0 & 0 & 0 & 0.67 & 4.6 \\ 0 & 0 & 0 & 0 & 0 & 0 \end{bmatrix} \times 10^{-12}\, C/N \quad (7.15)$$

$$\varepsilon_r = \begin{bmatrix} 4.52 & 0 & 0 \\ 0 & 4.52 & 0 \\ 0 & 0 & 4.52 \end{bmatrix}. \quad (7.16)$$

7.2.2 PZT

The lead zirconate titanate—$Pb(Zr_x, Ti_{1-x})O_3$ or PZT—system is widely used in polycrystalline (ceramic) form with very high piezoelectric coupling. The name PZT actually represents a *family* of piezoelectric materials. Depending on the formula of preparation, PZT materials may have different forms and properties. Manufacturers of PZT use proprietary formulas for their products. For example, PZT-4, PZT-5, PZT-6, and PZT-7 are doped by Fe, Nb, Cr, and La, respectively [19].

Techniques that are commonly used for preparing bulk PZT materials (such as PZT-4 and PZT-5A) are not suited for microfabrication. A number of techniques for preparing PZT films have been demonstrated, including sputtering, laser ablation, jet molding, and electrostatic spray deposition [20]. One of the most widely used methods to prepare thin-film PZT material for MEMS is sol-gel deposition. Using this method, relatively large thickness (e.g., 7 μm) can be reached easily [4, 6] using single- or multiple-layer deposition.

Using a processing technique called screen printing, even thicker PZT films can be reached in a single pass [3, 21, 22] with the highest piezoelectric coupling coefficient being 50 pC/N, which is significantly lower than what is achievable in bulk PZT. The screen printing ink consists of submicron PZT powders obtained commercially, as well as lithium carbonate and bismuth oxide as bonding agents. After screen printing, the deposited materials are dried and then fired at high temperatures for densification. The sol-gel deposition process is constantly being advanced. Pin-hole-free PZT films up to 12 μm thick have been realized with d_{33} in the 140–240 pC/N range [3], though a single-layer deposition thickness of 0.1 μm is more common.

7.2 Properties of Piezoelectric Materials

Properties of representative PZT materials are discussed below. The d matrix for $Pb(Zr_{0.40}, Ti_{0.60})TiO_3$ is given as

$$d_{ij} = \begin{pmatrix} 0 & 0 & 0 & 0 & 293 & 0 \\ 0 & 0 & 0 & 293 & 0 & 0 \\ -44.2 & -44.2 & 117 & 0 & 0 & 0 \end{pmatrix} pC/N. \quad (7.17)$$

The d matrix for $Pb(Zr_{0.52}, Ti_{0.48})TiO_3$ is

$$d_{ij} = \begin{pmatrix} 0 & 0 & 0 & 0 & 494 & 0 \\ 0 & 0 & 0 & 494 & 0 & 0 \\ -93.5 & -93.5 & 223 & 0 & 0 & 0 \end{pmatrix} pC/N. \quad (7.18)$$

For PZT-4, a special PZT material developed for underwater sonar applications, the compliance, piezoelectric coupling, and relative permittivity matrix are summarized below:

$$s = \begin{bmatrix} 12.3 & -4.05 & -5.31 & 0 & 0 & 0 \\ -4.05 & 12.3 & -5.31 & 0 & 0 & 0 \\ -5.31 & -5.31 & 15.5 & 0 & 0 & 0 \\ 0 & 0 & 0 & 39 & 0 & 0 \\ 0 & 0 & 0 & 0 & 39 & 0 \\ 0 & 0 & 0 & 0 & 0 & 32.7 \end{bmatrix} \times 10^{-12} \, m^2/N \quad (7.19)$$

$$d = \begin{bmatrix} 0 & 0 & 0 & 0 & 496 & 0 \\ 0 & 0 & 0 & 496 & 0 & 0 \\ -123 & -123 & 289 & 0 & 0 & 0 \end{bmatrix} \times 10^{-12} \, C/N \quad (7.20)$$

$$\varepsilon_r = \begin{bmatrix} 1475 & 0 & 0 \\ 0 & 1475 & 0 \\ 0 & 0 & 1300 \end{bmatrix}. \quad (7.21)$$

For PZT-5A, the compliance, piezoelectric coupling, and relative permittivity matrix are

$$s = \begin{bmatrix} 16.4 & -5.74 & -7.22 & 0 & 0 & 0 \\ -5.74 & 16.4 & -7.22 & 0 & 0 & 0 \\ -7.22 & -7.22 & 18.8 & 0 & 0 & 0 \\ 0 & 0 & 0 & 47.5 & 0 & 0 \\ 0 & 0 & 0 & 0 & 47.5 & 0 \\ 0 & 0 & 0 & 0 & 0 & 44.3 \end{bmatrix} \times 10^{-12} \, m^2/N \quad (7.22)$$

$$d = \begin{bmatrix} 0 & 0 & 0 & 0 & 584 & 0 \\ 0 & 0 & 0 & 584 & 0 & 0 \\ -171 & -171 & 374 & 0 & 0 & 0 \end{bmatrix} \times 10^{-12} \text{ C/N} \qquad (7.23)$$

$$\varepsilon_r = \begin{bmatrix} 1730 & 0 & 0 \\ 0 & 1730 & 0 \\ 0 & 0 & 1730 \end{bmatrix}. \qquad (7.24)$$

7.2.3 PVDF

The polyvinylidenfluoride (PVDF) is a synthetic fluoropolymer with monomer chains of $(-CH_2-CF_2-)_n$. It exhibits piezoelectric, pyroelectric, and ferroelectric properties; excellent stability to chemicals; mechanical flexibility; and biocompatibility [23]. The piezoelectric effect of PVDF has been investigated and modeled [24].

Thin-stretched PVDF films are flexible and easy to handle as ultrasonic transducers. The material is carbon based, and it is usually deposited as a spin cast film from a dilute solution in which the PVDF powder has been dissolved. As with most piezoelectric materials, process steps after deposition greatly affect the behavior of the film. For example, heating and stretching can increase or decrease the piezoelectric effect. PVDF and most other piezoelectric films require polarizing after deposition.

The d matrix of PVDF is

$$d = \begin{bmatrix} 0 & 0 & 0 & 0 & <1 & 0 \\ 0 & 0 & 0 & <1 & 0 & 0 \\ 20 & 2 & -30 & 0 & 0 & 0 \end{bmatrix} \text{pC/N}. \qquad (7.25)$$

7.2.4 ZnO

ZnO material can be grown using a number of methods, including RF or DC sputtering, ion plating, and chemical vapor deposition. In the MEMS field, ZnO is most commonly deposited by magnetron sputtering [25, 26] on various materials, with the c-axis (or Z-axis) close to the normal of a substrate. For ZnO, the c-axis is *spontaneously formed* without poling.

Strategies for reducing the intrinsic stress of ZnO have been explored, in order to realize a larger area and thicker films [27]. As-deposited ZnO films have significant compressive stress, ranging from 1 GPa to 135 MPa [7]. The stress can be reduced using thermal annealing (e.g., at 500°C for 5 minutes) in the 80 MPa to 100 MPa range.

A popular electrode material on top of the ZnO thin film is aluminum, which can be etched using a solution of KOH, $K_3Fe(CN)_6$, and water (1 g:10 g:100 ml). ZnO itself can be etched using wet etchants such as $CH_3COOH:H_3PO_4$:water (1 ml:1 ml:80 ml) at a fast rate [28].

The piezoelectric coefficient matrix for ZnO is

$$d = \begin{bmatrix} 0 & 0 & 0 & 0 & -11.34 & 0 \\ 0 & 0 & 0 & -11.34 & 0 & 0 \\ -5.43 & -5.43 & 11.37 & 0 & 0 & 0 \end{bmatrix} \text{pC/N}. \qquad (7.26)$$

7.2 Properties of Piezoelectric Materials

However, the exact values of the matrix depend on treatment conditions and crystallinity regularity (single crystal vs. polycrystal).

Example 7.2. ZnO Piezoelectric Force Sensor

A patch of ZnO thin film is located near the base of a cantilever beam, as shown in Fig. 7.4. The ZnO film is vertically sandwiched between two conducting films. The length of the entire beam is l. It consists of two segments: A and B. Segment A is overlapped with the piezoelectric material while segment B is not. The length of segments A and B are l_A and l_B, respectively. If the device is used as a force sensor, find the relationship between the applied force F and the induced voltage.

Solution. The c-axis (axis-3) of the deposited ZnO is generally normal to the front surface of the substrate on which it is deposited, in this case, the beam. A transverse force would produce a longitudinal tensile stress in the piezoelectric element (along axis-1), which in turn produces an electric field and output voltage along the c-axis. The shear stress components due to the force are ignored.

The stress along the length of the piezoresistor is actually not uniform and changes with the position. For simplicity, we assume the longitudinal stress is constant and equals the maximum stress value at the base. The maximum stress induced along the longitudinal direction of the cantilever is given by

$$\sigma_{1,max} = Mt/(2I) = Flt_{beam}/2I_{beam}.$$

The stress component $\sigma_{1,max}$ is parallel to axis 1.

According to Equation (7.2), the output electric polarization in the direction of axis-3 is

$$D_3 = d_{31}\sigma_{1,max}.$$

The overall output voltage is then

$$V = E_3 t_{piezo} = \frac{D_3 t_{piezo}}{\varepsilon} = \frac{Flt_{beam} t_{piezo}}{2\varepsilon I_{beam}},$$

with T_{piezo} being the thickness of the piezoelectric stack.

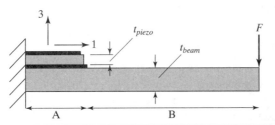

FIGURE 7.4 A piezoelectric force sensor.

Example 7.3. ZnO Piezoelectric Actuator

For the same cantilever as in Example 7.2, derive the vertical displacement at the end of the beam if it was used as an actuator. The applied voltage is V_3.

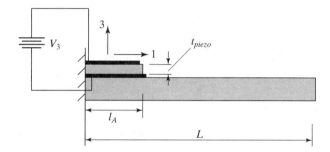

Solution. Under the applied voltage, the electrical field in axis-3 is

$$E_3 = \frac{V_3}{t_{piezo}}.$$

The applied electric field creates a longitudinal strain along axis-1, with the magnitude given by Equation 7.5 as

$$S_1 = E_3 d_{31}.$$

Segment A is curved into an arc. The radius of the curvature r due to the applied voltage can be found from Equation (7.13).

The displacement at the end of segment A, $\delta(x = l_A)$, can be found by following a procedure that is similar to the one used in Example 7.1. The angular displacement at the end of the piezoelectric patch is

$$\phi(x = l_A) = \frac{l_A}{r}.$$

The segment B does not curl and remains straight. The vertical displacement at the end of the beam is

$$\delta(x = l) = \delta(x = l_A) + l_B \sin[\phi(x = l_A)].$$

Example 7.4. ZnO Piezoelectric Actuator

A ZnO thin-film actuator on a cantilever is biased by coplanar electrodes. The geometry of the beams and the piezoelectric patches is identical as in Example 7.2. Find the output voltage under the applied force. If the structure is used as an actuator, what are the stress components when a voltage is applied across the electrodes?

7.2 Properties of Piezoelectric Materials 259

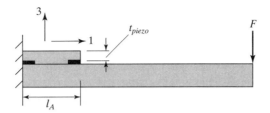

Solution. The applied force generates two stress components: normal stress T_1 and shear stress T_5. The output electric field is related to the stresses according to the formula for the direct effect of piezoelectricity:

$$\begin{bmatrix} D_1 \\ D_2 \\ D_3 \end{bmatrix} = \begin{bmatrix} d_{11} & d_{12} & d_{13} & d_{14} & d_{15} & d_{16} \\ d_{21} & d_{22} & d_{23} & d_{24} & d_{25} & d_{26} \\ d_{31} & d_{32} & d_{33} & d_{34} & d_{35} & d_{36} \end{bmatrix} \begin{bmatrix} T_1 \\ T_2 \\ T_3 \\ T_4 \\ T_5 \\ T_6 \end{bmatrix} + \begin{bmatrix} \varepsilon_{11} & \varepsilon_{12} & \varepsilon_{13} \\ \varepsilon_{21} & \varepsilon_{22} & \varepsilon_{23} \\ \varepsilon_{31} & \varepsilon_{32} & \varepsilon_{33} \end{bmatrix} \begin{bmatrix} E_1 \\ E_2 \\ E_3 \end{bmatrix}.$$

Because no external field is applied, the terms E_1, E_2, and E_3 on the right-hand side of the above equation are zero. The formula can be simplified to the form

$$\begin{bmatrix} D_1 \\ D_2 \\ D_3 \end{bmatrix} = \begin{bmatrix} 0 & 0 & 0 & 0 & -11.34 & 0 \\ 0 & 0 & 0 & -11.34 & 0 & 0 \\ -5.43 & -5.43 & 11.37 & 0 & 0 & 0 \end{bmatrix} \begin{bmatrix} T_1 \\ 0 \\ 0 \\ 0 \\ T_5 \\ 0 \end{bmatrix} \times 10^{-12}.$$

Therefore,

$$D_1 = -11.34 \times 10^{-12} \times T_5$$

$$D_3 = -5.43 \times 10^{-12} \times T_1.$$

The output voltage is related to the polarization in axis-1:

$$V_1 = \frac{D_1}{\varepsilon} \times l_A.$$

Let's find the output stress when the device is used as an actuator. Suppose a voltage V is applied across the longitudinal direction. Here, we assume the spacing between the two electrode is l_A; hence, the magnitude of the electric field is

$$E_1 = \frac{V}{l_A}.$$

The applied electric field creates a longitudinal strain along axis-1. The strain is found by

$$\begin{bmatrix} S_1 \\ S_2 \\ S_3 \\ S_4 \\ S_5 \\ S_6 \end{bmatrix} = \begin{bmatrix} S_{11} & S_{12} & S_{13} & S_{14} & S_{15} & S_{16} \\ S_{21} & S_{22} & S_{23} & S_{24} & S_{25} & S_{26} \\ S_{31} & S_{32} & S_{33} & S_{34} & S_{35} & S_{36} \\ S_{41} & S_{42} & S_{43} & S_{44} & S_{45} & S_{46} \\ S_{51} & S_{52} & S_{53} & S_{54} & S_{55} & S_{56} \\ S_{61} & S_{62} & S_{63} & S_{64} & S_{65} & S_{66} \end{bmatrix} \begin{bmatrix} T_1 \\ T_2 \\ T_3 \\ T_4 \\ T_5 \\ T_6 \end{bmatrix} + \begin{pmatrix} d_{11} & d_{21} & d_{31} \\ d_{12} & d_{22} & d_{32} \\ d_{13} & d_{23} & d_{33} \\ d_{14} & d_{24} & d_{34} \\ d_{15} & d_{25} & d_{35} \\ d_{16} & d_{26} & d_{36} \end{pmatrix} \begin{bmatrix} E_1 \\ E_2 \\ E_3 \end{bmatrix}.$$

Since no external stresses are applied, we set T_1 through T_6 as zero. The simplified formula for strain is

$$\begin{bmatrix} S_1 \\ S_2 \\ S_3 \\ S_4 \\ S_5 \\ S_6 \end{bmatrix} = \begin{pmatrix} 0 & 0 & -5.43 \\ 0 & 0 & -5.43 \\ 0 & 0 & 11.37 \\ 0 & -11.34 & 0 \\ -11.34 & 0 & 0 \\ 0 & 0 & 0 \end{pmatrix} \begin{bmatrix} E_1 \\ 0 \\ 0 \end{bmatrix} \times 10^{-12} = \begin{bmatrix} 0 \\ 0 \\ 0 \\ 0 \\ S_5 \\ 0 \end{bmatrix}.$$

No longitudinal strain components are generated in this manner.

Example 7.5. ZnO Piezoelectric Actuator

Derive the expression for the end displacement of a piezoelectric transducer that has a similar configuration to Example 7.4; the difference is that the electrodes are used to pole the ZnO material. In other words, axis-3 is now forced to lie in the longitudinal direction of the beam length. A voltage V is applied across two electrodes.

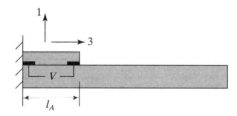

Solution. The electric field in the longitudinal axis is

$$E_3 = \frac{V}{l_A}.$$

The applied field induced a longitudinal strain (S_3) according to

$$\begin{bmatrix} s_1 \\ s_2 \\ s_3 \\ s_4 \\ s_5 \\ s_6 \end{bmatrix} = \begin{pmatrix} 0 & 0 & -5.43 \\ 0 & 0 & -5.43 \\ 0 & 0 & 11.37 \\ 0 & -11.34 & 0 \\ -11.34 & 0 & 0 \\ 0 & 0 & 0 \end{pmatrix} \begin{bmatrix} 0 \\ 0 \\ E_3 \end{bmatrix} \times 10^{-12}$$

or

$$s_3 = d_{33} E_3.$$

We should use s_3 to replace s_{long} in Equation (7.8). The subsequent analysis is similar to the one performed for Example 7.3.

7.2.5 Other Materials

Aluminum nitride (*AlN*) is another common thin-film piezoelectric material. However, it is not as popular as ZnO, as its piezoelectric coefficients are lower. The d matrix for aluminum nitride is

$$d_{ij} = \begin{pmatrix} 0 & 0 & 0 & 0 & 4 & 0 \\ 0 & 0 & 0 & 4 & 0 & 0 \\ -2 & -2 & 5 & 0 & 0 & 0 \end{pmatrix} \text{pC/N}. \tag{7.27}$$

Lithium niobate (LiNbO$_3$) and barium titanate are not as commonly used in the MEMS field, but they have found a wider use in acoustic areas. The coupling matrix for lithium niobate is

$$d_{ij} = \begin{pmatrix} 0 & 0 & 0 & 0 & 68 & -42 \\ -21 & 21 & 0 & 68 & 0 & 0 \\ -1 & -1 & 6 & 0 & 0 & 0 \end{pmatrix} \text{pC/N}. \tag{7.28}$$

The properties of barium titanate depend on the crystallinity configurations. For single-crystal bulk barium titanate, the d matrix is

$$d_{ij} = \begin{pmatrix} 0 & 0 & 0 & 0 & 392 & 0 \\ 0 & 0 & 0 & 392 & 0 & 0 \\ -34.5 & -34.5 & 85.6 & 0 & 0 & 0 \end{pmatrix} \text{pC/N}. \tag{7.29}$$

For polycrystalline bulk barium titanate, the d matrix is

$$d_{ij} = \begin{pmatrix} 0 & 0 & 0 & 0 & 270 & 0 \\ 0 & 0 & 0 & 270 & 0 & 0 \\ -79 & -79 & 191 & 0 & 0 & 0 \end{pmatrix} \text{pC/N}. \tag{7.30}$$

7.3 APPLICATIONS

Piezoelectric materials can be used in many microsensors and actuators. We will focus on the discussion of four types of sensors: inertia sensors, pressure sensors, tactile sensors, and flow sensors. Meanwhile, two examples of piezoelectric actuators will be reviewed. The case studies collectively will reveal design, materials, and fabrication issues specifically related to piezoelectric MEMS devices.

7.3.1 Inertia Sensors

Commercial MEMS accelerometers are primarily based on electrostatic or piezoresistive sensing. Piezoelectric sensors require more complex materials and fabrication processes. Nonetheless, piezoelectric acceleration sensors have been made in the past. In this section, we will review two such examples, one based on a cantilever proof mass (Case 7.1) and another based on a membrane proof mass (Case 7.2). The sensor in Case 7.1 is made using a surface micromachining process whereas the one in Case 7.2 was realized using a bulk micromachining process.

Integrating piezoelectric material in MEMS is not straightforward. First, controlling the microstructure of piezoelectric thin films requires careful calibration and dedicated equipment. Secondly, many piezoelectric thin films are not chemically inert. Care must be exercised to prevent damage to piezoelectric thin films during processing.

Case 7.1. Cantilever Piezoelectric Accelerometer

One of the applications of piezoelectric sensors is micromachined accelerometers. An exemplary device has been reported by a research group at the University of California at Berkeley [7]. The schematic diagram of the sensor is shown in Figure 7.5. A proof mass is attached to the end of a fixed–free cantilever beam. The cantilever beam consists of multiple layers of materials, with ZnO being the functional piezoelectric material. ZnO is used instead

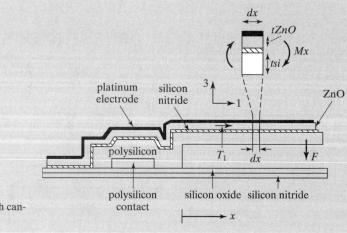

FIGURE 7.5
A piezoelectric accelerometer with cantilever design.

of PZT because, although the PZT material offers greater piezoelectric coefficient, it does so at the expense of a greater dielectric constant and hence a larger capacitance.

The piezoelectric layer is sandwiched between a top conducting layer (platinum) and a bottom conducting layer (polysilicon), which is similar to Example 7.2. A vertical acceleration will deflect the cantilever, producing a longitudinal stress in axis-1 along the length of the cantilever. (The ZnO film is polarized in the direction normal to the substrate.)

The simple analysis in Example 7.2 does not apply here, however. The analysis is relatively complex because the proof-mass weight and the piezoelectric layer are distributed in length and thickness. The distribution of stress is non-uniform along the length. However, it can be said that the output voltage is proportional to d_{31}, since the stress is applied in axis-1 and the electric polarization is measured in axis-3. The value of d_{31} is 2.3 pC/N in this case. Comprehensive modeling of the device performance can be found in Reference [13].

The fabrication process is briefly discussed below. A silicon oxide and a silicon nitride layer are deposited over the silicon wafer to serve as insulating layers. A phosphorus doped polycrystalline silicon film is deposited and patterned via reactive ion etching to define electrical contacts to the bottom electrodes of the accelerometer. Here, the authors used polysilicon instead of metal because polysilicon can withstand a much higher processing temperature and provide greater flexibility for downstream processing steps. Next, a 2-μm-thick layer of phosphosilicate glass is deposited by LPCVD, and it is patterned to define regions under the suspended cantilever. A second layer of 2-μm-thick silicon is deposited, covering bare silicon surfaces, the first polysilicon trace, and the PSG sacrificial layer. This layer is patterned by using a reactive ion etching (RIE) process with photoresist as the mask.

The RIE etch rate on silicon nitride and oxide (including PSG) is smaller, which reduces the damages of long-term etching. However, care must be exercised to prevent or minimize over-etching of the first polycrystalline layer. The authors then removed the photoresist layer with acetone.

A thin LPCVD silicon nitride layer is deposited over the wafer. This film serves as a stress-compensation layer for balancing a highly compressive stress in the ZnO film. The exact thickness of this layer depends on the actual stress and thickness of the ZnO layer.

Next, a ZnO layer on the order of 0.5-μm thick is deposited by RF-magnetron sputtering from a lithium-doped ZnO target. Finally, a 0.2-μm-thick Pt thin film is sputtered. The stress of the ZnO film is reduced by a rapid thermal annealing step. Afterwards, the three layers are defined using ion milling to produce precisely defined patterns. The underlying sacrificial layer is removed using HF solutions while the patterned ZnO patch is protected by photoresist. The ZnO protection is necessary because, although the film is covered by Pt on top, it is exposed on the side and through possible pinholes on the Pt films.

The device exhibited a sensitivity of 0.95 fC/g and a resonant frequency of 3.3 kHz.

Case 7.2. Membrane Piezoelectric Accelerometer

A second example of an accelerometer uses PZT instead of ZnO as the sensing material [6] because of its greater piezoelectric coefficient. The structure is also different from the previous example. The sensor consists of a silicon proof mass suspended by an annular diaphragm supporting a center proof mass (Figure 7.6). The annular ring design provides the desired mechanical characteristics, including a high-resonant frequency and insensitivity to transverse acceleration due to symmetry.

Three ring-shaped electrodes are involved in this device. Two concentric electrode rings are placed on the top of the membrane. A bottom electrode ring is placed underneath the PZT ring. The electric field between the bottom electrode and the top electrodes provides *in situ* poling. The output voltage is measured between two top electrodes.

The vertical displacement of the proof mass induced radial stress distribution. This radial stress element produces a radial electric field between the two concentric electrode rings. Vertical deformation of a proof mass causes a complex radial stress distribution. For example, if the membrane is vertically moved down, the stress is compressive within the radial vicinity of the proof mass but tensile within the radial vicinity of the outside frame.

The membrane of the annular ring vertically deforms due to the applied acceleration. The stress components in this annular ring are radially distributed. Unlike other examples in this chapter, in which a single coordinate system is used for the entire device, each cross section of the device is assigned to a coordinate system, with axis-1 pointing in the radial direction and axis-3 being perpendicular to the substrate.

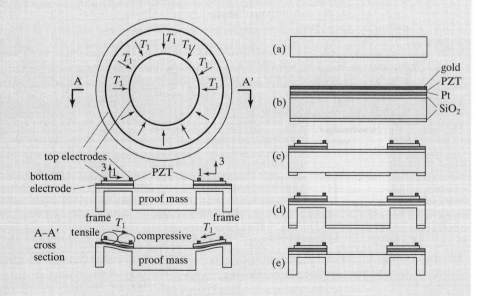

FIGURE 7.6
A piezoelectric accelerometer with membrane design.

Recall the general form of the piezoelectric coefficient matrix for PZT is

$$d_{ij} = \begin{pmatrix} 0 & 0 & 0 & 0 & d_{15} & 0 \\ 0 & 0 & 0 & d_{24} & 0 & 0 \\ d_{31} & d_{32} & d_{33} & 0 & 0 & 0 \end{pmatrix} \text{pC/N}.$$

The stress component along the radial direction, T_1, produces an electric polarization (output) in the direction of axis-3, with $D_3 = d_{31}T_1$.

Because the signs of radial stress change from the frame to the proof mass, the PZT materials in these two regions are poled with opposite polarization. This way, the output voltages from these two regions add up instead of canceling each other out. If the device relied on spontaneous poling, which would have a uniform direction throughout the entire material, the voltage output from differently stressed regions would cancel out each other.

The fabrication process starts with a silicon wafer (step a), which is coated with silicon oxide, platinum, PZT film, and then gold. Top electrodes (annular rings) are made in gold, whereas the bottom electrode is made in platinum (step b). The bottom electrode is used only for poling the PZT material *in situ*. Deep reactive ion etching is performed using patterned oxide on the back side as the mask (step d). The oxide on the front side is then etched to finalize the process (step e).

The MEMS sensor is tested by connecting the output of the sensor to a charge amplifier, which has a 10-pF feedback capacitor with an amplification of 10 mV/pC. Sensitivity ranging from 0.77 to 7.6 pC/g in the frequency range from 35 to 3.7 kHz were measured. This high sensitivity coupled with a broad frequency range were attributed to the FEM simulation and optimization. Unfortunately, it is difficult to model the behavior using a simple analytical formula because the annular ring design involves a range of spatial crystalline orientations, as well as Young's modulus.

7.3.2 Acoustic Sensors

There is growing interest in using micromachining technology to create microphones. MEMS-based microphones offer good dimensional control, miniaturization, direct integration with on-chip electronics, an arrayed format, and a potentially low cost due to batch processing. Piezoelectric microphones using diaphragms made of silicon nitride, silicon, and even organic thin film (e.g., Parylene [28]) have been made. On-chip integrated signal-conditioning circuitry has shown an unamplified sensitivity of 0.92 mV/Pa [8]. Let us review two cases. Case 7.3 focuses on a membrane-type acoustic sensor with PZT as the transduction material. Case 7.4 is a cantilever-type acoustic sensor using ZnO. Both sensors are made using bulk micromachining material.

Case 7.3. PZT Piezoelectric Acoustic Sensor

Bernstein and colleagues described the use of an array of piezoelectric transducers as an underwater acoustic imager [3]. The imager is akin to CCD images for optics. It consists of an 8 × 8 array of acoustic imaging sites. The cross-sectional view of each site is shown in

266 Chapter 7 Piezoelectric Sensing and Actuation

Figure 7.7. A layer of piezoelectric material—sol-gel deposited lead zirconate titanate (PZT)—lies on top of a silicon micromachined diaphragm. The size of each membrane varies from 0.2 to 2 mm.

Two electrodes sandwich a PZT thin film. The fabrication process is noteworthy because it involves the use of a sol-gel deposition of piezoelectric materials. After oxidizing the silicon wafers (Figure 7.8a) and patterning the oxide, a heavy boron dose was diffused in all but the oxide protected areas where through-wafer cavities would later be deposited (Figure 7.8b). All the remaining oxide is removed. A layer of low temperature oxide (LTO) is deposited using the LPCVD method. A 50-nm-thick Ti and 300-nm-thick Pt layer are deposited as the bottom electrode, with the Ti serving to increase the adhesion between the Pt and oxide (Figure 7.8c). This was followed by the deposition of PZT by spinning on a sol-gel mixture of lead acetate trihydrate, zirconium n-propoxide, and titanium isoproposide in a glacial acetic acid solvent. The sol-gel material is spun on, dried at 150°C to remove the solvent, heated at 400°C to remove residual organics, and preannealed at 600°C to densify the

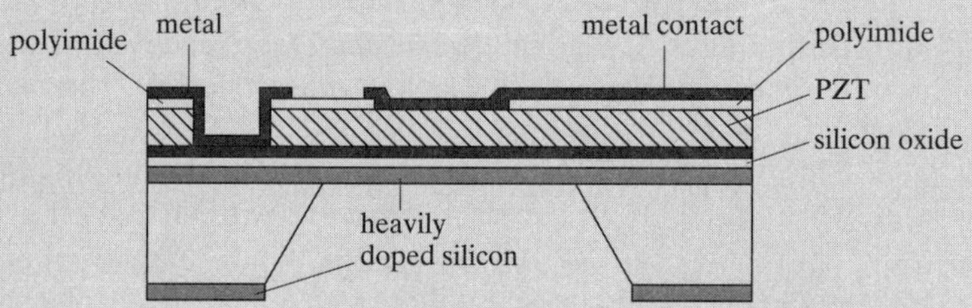

FIGURE 7.7
Schematic diagram of a piezoelectric microphone.

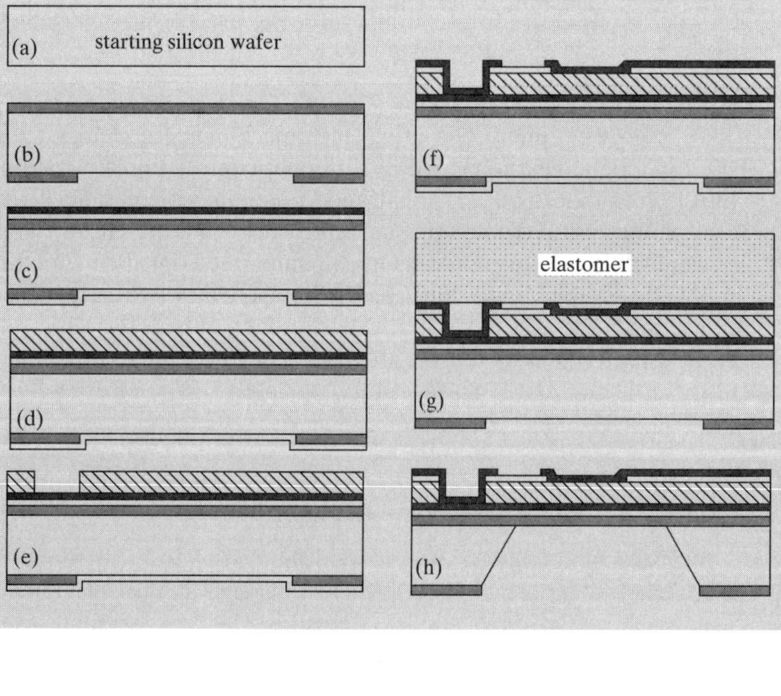

FIGURE 7.8
Fabrication process.

layer and prevent further shrinkage (Figure 7.8d). A detailed recipe for making the sol-gel solution can be found in the paper. The PZT material is poled at room temperature at a 36 V DC bias for 2 minutes to yield a resulting PZT film with a relative permittivity of 1400 and a d_{33} of 246 pC/N.

The PZT film is patterned by wet etching PZT in a solution containing buffered hydrogen fluoride (BHF) and hydrochloric acid (HCl) (Figure 7.8e). The BHF is a mixture of aqueous ammonium fluoride (NH_4F) and aqueous hydrogen fluoride (HF) [29]. The top electrode is separated from the ZnO by a 2-μm-thick polyimide dielectric layer. This is followed by the deposition of a top electrode and its patterning (Figure 7.8f). The back-side cavity is produced in an anisotropic etchant, with the front temporarily protected by a silicone elastomer cover (Figure 7.8g). Prior to silicon etching, the LTO oxide on the back is removed.

Case 7.4. PZT Piezoelectric Microphone

A second example is a piezoelectric cantilever microphone and microspeaker [9]. This example is selected because of its unique transduction principle, the choice of ZnO as the material, and the involvement of wet and dry silicon etching in one process flow.

Earlier work in Case 7.3 uses micromachined diaphragms that are clamped on all four sides. Cantilever microphones are chosen because of the compliance (Fig. 7.9). The cantilever is also free from any residue stress induced on a membrane. The authors stated that the use of a cantilever actually created a microphone with a great sensitivity (20 mV/μbar at 890 Hz, the resonant frequency). Conversely, acoustic output can be generated when the device is used as an actuator, with a sound pressure level of 75 dB at 890 Hz with 4 V (zero-to-peak) drive.

The size of the cantilever is 2 mm × 2 mm, and the overall thickness is 4.5 μm. A ZnO thin film is located on the cantilever. The fabrication begins with a <100>-oriented silicon wafer. A 0.2-μm-thick oxide is grown by thermal oxidation, followed by the LPCVD deposition of a 0.5-μm-thick silicon nitride. The nitride is deposited at 835°C and 300 mTorr deposition pressures from a vapor with a 6:1 ratio of dichlorosilane (DCS, SiH_2Cl_2) to ammonia. An anisotropic silicon etch is performed on the back of the wafer until the silicon oxide is reached. The membrane is rather large and thin. Care must be exercised when conducting further processing on the front of the wafer. To provide sufficient strength to the membrane to survive the remainder of the chemical processes, a second 0.5-μm-thick LPCVD nitride layer is deposited on both sides of the silicon wafer with a reaction gas ratio of 4:1.

A 0.2-μm-thick LPCVD polysilicon electrode is deposited on the wafer. The front of the wafer is coated with spin-on photoresist and patterned. The photoresist serves as a mask in a plasma-etching process to etch the polysilicon. Because the etch rate on silicon nitride is finite, care must be taken to prevent excessive over etch of the silicon nitride. Fortunately, two coatings of silicon nitride (with a total thickness of 1 μm) are present on the front. The front is then coated with an LPCVD low temperature oxide (LTO), which is an

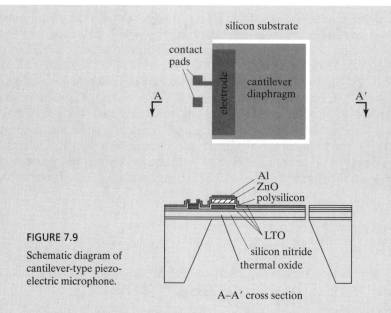

FIGURE 7.9
Schematic diagram of cantilever-type piezoelectric microphone.

insulating layer. The wafer is annealed for 25 minutes at 950°C to relieve stress and activate dopants in the polysilicon. A 0.5-μm-thick layer of ZnO is deposited using RF-magnetron sputtering. The ZnO is then encapsulated with another layer of LTO (0.3-μm thick). Contact windows are made at the contact pad sites.

Next, a layer of aluminum is deposited by sputtering and is photolithographically patterned. The wafer is then diced into individual dies. Because the dicing process involves a vacuum on the back for holding the wafer and running water on the front for removing particles and heat, the membranes are subjected to rather harsh conditions. Optionally, a 0.5-mm-thick aluminum thin film can be deposited on the back to further strengthen the membrane during this step. At the die level, the LTO, nitride, and aluminum on the membrane are patterned using HF, plasma etch, and a wet etchant containing $K_3Fe(CN)_6$ and KOH, respectively.

7.3.3 Tactile Sensors

The thrust of the tactile sensor research is to quantitatively measure contact forces (or pressure) that mimics a human's spatial resolution and sensitivity, has a large bandwidth, and has a wide dynamic range. A piezoelectric tactile sensor is discussed in Case 7.5.

Case 7.5. Polymer Piezoelectric Tactile Sensor

To reduce electrical noise and impedance mismatch effects, a two-dimensional matrix of high-input impedance metal-oxide-semiconductor field-effect transistor (MOSFET) amplifiers have been directly gate-contact coupled to the lower surface of a piezoelectric PVDF polymer film [30]. This MOSFET amplifier arrangement provides a separate, but identical, high-input impedance (10^{12} Ω) voltage measurement capability for each taxel (tactile pixel).

To realize a tactile sensor no larger than an adult's fingertip, a silicon IC with the peripheral dimensions of 9.2 mm by 7.9 mm is designed and fabricated. The prototyping was performed by MOSIS (Metal-oxide-semiconductor Implementation System) foundry service. A portion of the IC's area was reserved for the MOSFET amplification and output interface circuitry. The 8 × 8 taxel matrix was allocated to an area of 5.3 mm × 5.3 mm.

A schematic diagram of the sensor array with its cross section revealed is shown in Figure 7.10. A continuous, poled PVDF film is attached to the front surface of the silicon chip. A 6-μm-thick urethane conformal coating layer was deposited on the PVDF film by spin coating. Individual taxel electrodes (400 × 400 μm) are separated from their nearest neighbors by 300 μm.

Although this work focuses on measuring the normal stress components, tactile sensors with polymer piezoelectric materials have been developed with component-selective response characteristics [31, 32].

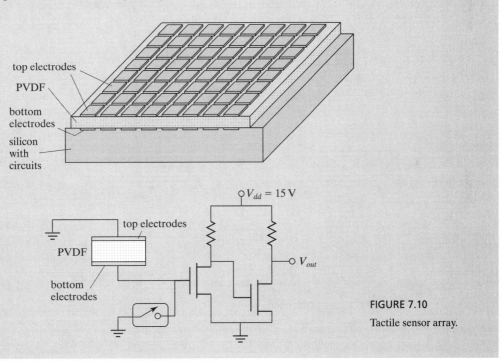

FIGURE 7.10
Tactile sensor array.

7.3.4 Flow Sensors

Flow sensors can be built using piezoelectric principles in a similar fashion as piezoresistive flow sensors, although the material deposition and optimization will require more efforts in general. For example, floating element shear stress sensors have been made using piezoelectric bimorph sensors [33]. We focus on discussing a flow-rate sensor based on piezoelectric principles (Case 7.6).

Case 7.6. Piezoelectric Flow-Rate Sensor

A volumetric flow-rate sensor based on piezoelectric sensing is discussed in Reference [34]. This example is selected because it involves integrating piezoelectric thin films on polymer materials and with fluid channels.

The sensor consists of two pressure sensors with piezoelectric readouts connected to a hydraulic restriction channel at two locations (up- and downstream). The pressure difference measured using the piezoelectric sensors provides information about the flow rate, since the pressure difference along a channel and the volumetric flow rate are linked by Bernoulli's equation. The device has been designed to measure flow rates from 30 μl/h to 300 μl/h. The restriction has a hydraulic resistance of $R_h = 60$ mbar/(ml/h) with a channel length of 10 mm and a hydraulic diameter of 67 μm.

Each pressure sensor consists of a membrane (made of polyimide) carrying an annular ring made of ZnO. The diameter of the membrane is 1 mm with the thickness being 25 μm. The strain distribution is not uniform throughout the ring. It was found that the average

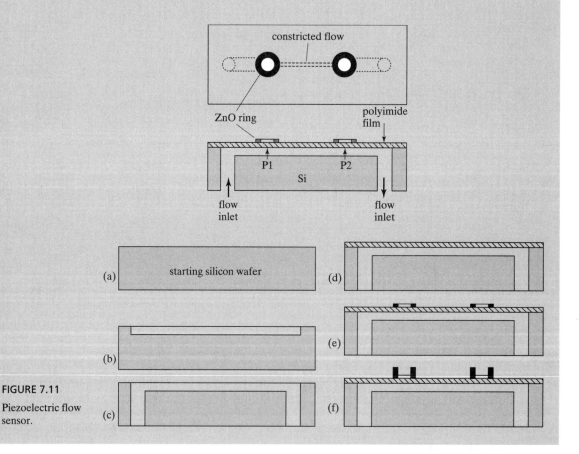

FIGURE 7.11 Piezoelectric flow sensor.

strain is 6.8×10^{-5} for a pressure of 100 mbar. An annular ring is used rather than a continuous circular membrane. The ZnO material is located in the region with an identical sign of radial stress when the membrane is deformed.

The device is fabricated starting from a silicon wafer (Figure 7.11). First, 50-μm deep channels are etched by reactive ion etching into the wafer surface (step b). Fluid interconnection holes are drilled using an ultrasonic drilling tool (step c). Next, a commercially available heat bonding type polyimide sheet (UPILEX@ VT) is bonded to the etched silicon wafer at a bonding pressure of 50–100 bars and a temperature of 300°C (step d). On the sealed wafer, a gold electrode is evaporated and patterned on the planar polyimide membrane (step e), followed by a silicon dioxide insulation layer deposited by plasma-enhanced chemical vapor deposition (PECVD). Here, the authors did not use LPCVD oxide because the deposition temperature is too high for the polyimide. A 1-μm-thick ZnO film is then sputtered using RF magnetron sputtering from a sintered ZnO target and is then coated with a second insulation layer. Finally, an aluminum electrode is deposited using a lift-off process (step f). The lift-off process does not involve wet etching for patterning, which may attack underlying layers including ZnO.

The average pressure sensor sensitivity is 8 mV/mbar. A flow volume of 1 to 10 nl has been measured.

7.3.5 Surface Elastic Waves

Piezoelectric materials, under proper electrical bias, can launch elastic waves in bulk or thin films. Two of the most commonly encountered elastic waves are the surface acoustic wave (SAW) and the flexural plate wave (or Lamb wave) (Fig. 7.12). The SAW occurs on samples of appreciable depth, whereas Lamb waves occur in thin plates of materials.

Surface elastic waves can be launched using comb-drive electrodes. The principle of launching a SAW wave on a bulk piezoelectric material (e.g., PZT) is illustrated in Figure 7.13. Electrodes are arranged in an interdigitated fashion. AC voltage between electrode fingers creates electric field lines between neighboring conductors. For the launcher, the electric field lines are parallel to axis-1. Under the influence of E_1, the mechanical stress is

$$\begin{bmatrix} s_1 \\ s_2 \\ s_3 \\ s_4 \\ s_5 \\ s_6 \end{bmatrix} = \begin{pmatrix} d_{11} & d_{21} & d_{31} \\ d_{12} & d_{22} & d_{32} \\ d_{13} & d_{23} & d_{33} \\ d_{14} & d_{24} & d_{34} \\ d_{15} & d_{25} & d_{35} \\ d_{16} & d_{26} & d_{36} \end{pmatrix} \begin{bmatrix} E_1 \\ E_2 \\ E_3 \end{bmatrix} = \begin{pmatrix} 0 & 0 & d_{31} \\ 0 & 0 & d_{32} \\ 0 & 0 & d_{33} \\ 0 & d_{24} & 0 \\ d_{15} & 0 & 0 \\ 0 & 0 & 0 \end{pmatrix} \begin{bmatrix} E_1 \\ 0 \\ 0 \end{bmatrix} = \begin{bmatrix} 0 \\ 0 \\ 0 \\ 0 \\ s_5 \\ 0 \end{bmatrix}, \quad (7.31)$$

which corresponds to a moment acting along axis-2. The disturbance to the bulk lattice dissipates by propagating as an elastic wave.

272 Chapter 7 Piezoelectric Sensing and Actuation

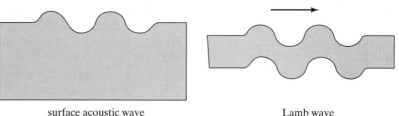

FIGURE 7.12

Surface acoustic wave and flexural plate wave.

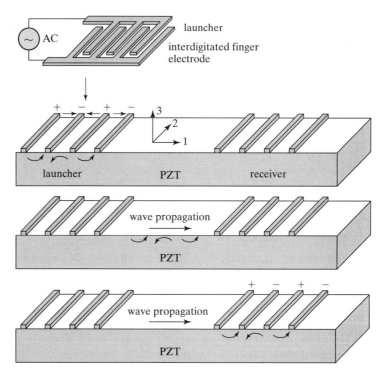

FIGURE 7.13

Launch of a SAW wave.

The elastic wave propagates along axis-1 towards a set of receiver electrodes. As the wave propagates, it interacts with the solid interior as well as with the surface. Once the wave arrives at the receiver electrode, it is converted back into electrical polarization in axis-1, according to this formula:

$$\begin{bmatrix} d_1 \\ d_2 \\ d_3 \end{bmatrix} = \begin{bmatrix} 0 & 0 & 0 & 0 & d_{15} & 0 \\ 0 & 0 & 0 & d_{24} & 0 & 0 \\ d_{31} & d_{32} & d_{33} & 0 & 0 & 0 \end{bmatrix} \begin{bmatrix} 0 \\ 0 \\ 0 \\ 0 \\ T_5 \\ 0 \end{bmatrix} = \begin{bmatrix} d_{15}T_5 \\ 0 \\ 0 \end{bmatrix}. \quad (7.32)$$

The surface elastic wave has a wide variety of functions [35], ranging from chemical sensing, to environmental monitoring, to electrical circuitry, to the transportation of fluid in contact with the surface [4]. The possibility of pumping liquid with a flow rate up to 0.255 μl/min has been demonstrated. Propagation characteristics, such as amplitude and frequency, are influenced by the density, viscosity, and molecular weight of particles or solutions in contact with the surface of the bulk. As such, the surface elastic wave can be used to broadly characterize the physical and chemical phenomena occurring in a region between the launcher and receiver electrodes [26].

SUMMARY

This chapter addresses the governing equations of piezoelectricity, materials, and device designs. The topic of piezoelectric sensing and actuation is quite broad and cannot possibly be covered in one chapter. This chapter is intended to provide a starting point for interested readers to explore further.

At the end of this chapter, a reader should understand the following concepts and facts:

- The origin of the direct and inverse effects of piezoelectricity.
- Essential crystal properties of materials with piezoelectricity.
- Governing equations of the direct effect of piezoelectricity.
- Governing equations of the inverse effect of piezoelectricity.
- Commonly used piezoelectric materials and their major properties.
- Quantitative analyses of piezoelectric sensors based on a cantilever beam configuration.
- Analysis of piezoelectric actuators based on a cantilever beam configuration.
- Qualitative understanding about the design issues of piezoelectric sensors and actuators based on suspended membranes.
- The functional principle of surface elastic wave devices.

PROBLEMS

Problem 7.1: Review
Prove that the unit of the piezoelectric coefficient is C/N from the governing equation of the inverse effect of piezoelectricity, as shown in Equation 6.

Problem 7.2: Design
For PZT material, how many different ways are there to generate a pure torque? Assume axis-3 is aligned normal to the substrate. Draw perspective view diagrams for the position of electrodes.

Problem 7.3: Design
A piezoelectric force sensor has the configuration shown in the following figure. If a force is applied in the direction shown, what will be the expression of the output voltage?

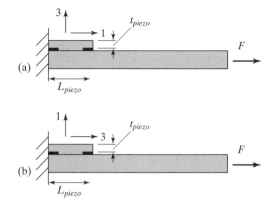

Problem 7.4: Design

A piezoelectric actuator has the configuration shown in the following figure. If a voltage is applied across the two electrodes, what is the expression of the resultant linear displacement at the end of the cantilever?

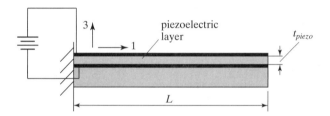

Problem 7.5: Design

Consider the piezoelectric actuator diagrammed in Problem 7.4. The applied voltage is fixed. If the thickness of the piezoelectric layer can be continuously changed within the range of 5% to 100% of the thickness of the elastic layer, qualitatively plot the output force as a function of the piezoelectric layer thickness. Qualitatively plot the output displacement as a function of the piezoelectric layer thickness.

Problem 7.6: Design

Derive the expression for the vertical displacement at the end of the cantilever.

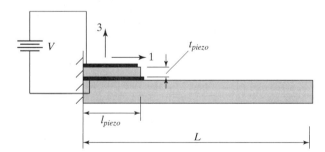

Problem 7.7: Analysis

Assume that two layers of the piezoelectric bimorph actuator have the same width; reduce Equation (7.9) to a simpler form.

Problem 7.8: Fabrication

Draw cross-sectional fabrication sequence associated with Case 7.1. Include details of lithography steps and consider the process on the backside of the wafer as well. For each step that involves etching or material removal, comment on the selectivity of the etching agents on all materials exposed to it during the particular step.

Problem 7.9: Challenge

Complete the design of a floating element fluid shear stress sensor using a similar design principle as the piezoresistive shear stress sensor discussed in Section 5.5.4 [36]. Develop a fabrication process and draw a detailed process flow in a representative cross section. The cross section must include piezoelectric elements. Include full details of lithographic steps. For each statement that involves etching of material removal, comment on the selectivity of the etching agents on all materials exposed to it.

Problem 7.10: Review

Draw a diagram illustrating the signs of stress and the direction of the electrical polarization at one arbitrary cross section in Case 7.2, when the proof mass is bent down. Repeat the process for when the diaphragm is bent upward.

Problem 7.11: Fabrication

For Case 7.2, if the membrane were to be made of LPCVD silicon nitride instead of single-crystal silicon, draw an alternative fabrication process. Include full details of lithographic steps. For each statement that involves etching of material removal, comment on the selectivity of the etching agents on all materials exposed to it.

Problem 7.12: Design

For the acoustic sensor example in Case 7.3, derive an analytical expression for the output response of the sensor with respect to a uniformly applied pressure with a magnitude of p. Is the sandwich electrode configuration optimal for high sensitivity? Consider at least one alternative configuration and companion design, performance, and fabrication steps. (Hint: the acoustic sensor is a circular membrane with fixed boundary conditions under uniform pressure.)

Problem 7.13: Fabrication

The example in Case 7.3 used doped silicon as the membrane layer. To increase the performance, one can use thinner film, such as silicon nitride, as the membrane. Develop a process for realizing a similar device with a 200-nm-thick silicon nitride thin film. Skip description of detailed lithography steps.

Problem 7.14: Review

The example in Case 7.5 uses a continuous sheet of commercial PVDF material. This introduces cross-sensitivity among pixels. In Case 7.5, estimate the cross-sensitivity of the vertical pixels. For example, if a normal force is applied on one pixel, what is the output of its nearby pixel?

Problem 7.15: Design

For the example in Case 7.5, what is the expression for the sensitivity to shear loading?

Problem 7.16: Challenge

The cross-sensitivity will be significantly reduced if the PVDF pixels can be mechanically separated. Find the material and processing technique that may be used to provide separate high-resolution PVDF pixel arrays on CMOS circuitry.

Problem 7.17: Design

The capability of large-scale deformable optics devices is limited by the great stiffness in typical macro- or mini-mirrors caused by the appreciable thickness of the actuator and mirror layers. Large changes in focal length can be achieved if this stiffness is reduced. Electrostatically actuated devices suffer from the pull-in effect and limited displacement range due to gap sizes.

A piezoelectric actuator has been used for moving (deflecting) light beams [10]. The basic structure consists of a circular membrane released by a back-side through-wafer etch. The top surface of the membrane has an iris-shaped piezoelectric actuation layer, which is composed of a PZT layer as the active piezoelectric material. An insulating layer separates the ZnO and electrodes from the bulk substrate. An in-plane poling scheme makes use of the d_{33} rather than other piezoelectric coefficients. The amplitude of d_{33} is roughly twice as large as the amplitude of d_{31}. The mirror achieved maximum deflection at the center of approximately 7 μm at a bias voltage of 700 V.

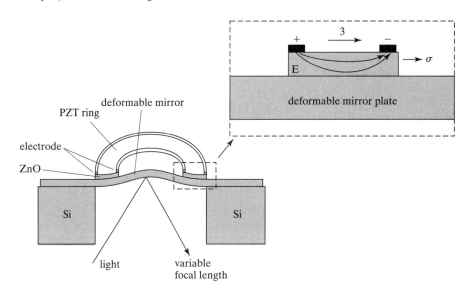

Problem 7.18: Review

Compare the design of piezoelectric loaded membranes in Cases 7.2 and 7.6. Discuss the major differences in the mechanical design. Discuss the pros and cons in terms of the difficulty of material preparation and processing.

Problem 7.19: Challenge

Most piezoelectric actuators created piston motion or cantilever deflection normal to the substrate. Find a design by which a microstructure can be moved in a plane parallel to the substrate surface using piezoelectric actuation. The design should be accompanied by a practical fabrication method.

REFERENCES

1. Cady, W.G., *Piezoelectricity: An Introduction to the Theory and Applications of Electromechanical Phenomena in Crystals*, 2d ed. New York: Dover, 1964.
2. Ikeda, I., *Fundamentals of Piezoelectricity*. Oxford University Press, 1996.
3. Bernstein, J.J., et al., *Micromachined High Frequency Ferroelectric Sonar Transducers*, Ultrasonics, Ferroelectrics and Frequency Control, IEEE Transactions, 1997. **44**: p. 960–969.

4. Luginbuhl, P., et al., *Microfabricated Lamb Wave Device Based on PZT Sol-Gel Thin Film for Mechanical Transport of Solid Particles and Liquids*, Microelectromechanical Systems, Journal of, 1997. **6**: p. 337–346.
5. Li, H.Q., et al., *Fabrication of a High Frequency Piezoelectric Microvalve*, Sensors and Actuators A: Physical, 2004. **111**: p. 51–56.
6. Wang, L.-P., et al., *Design, Fabrication, and Measurement of High-Sensitivity Piezoelectric Microelectromechanical Systems Accelerometers*, Microelectromechanical Systems, Journal of, 2003. **12**: p. 433–439.
7. DeVoe, D.L., and A.P. Pisano, *Surface Micromachined Piezoelectric Accelerometers (PiXLs)*, Microelectromechanical Systems, Journal of, 2001. **10**: p. 180–186.
8. Ried, R.P., et al., *Piezoelectric Microphone with On-Chip CMOS Circuits*, Microelectromechanical Systems, Journal of, 1993. **2**: p. 111–120.
9. Lee, S.S., R.P. Ried, and R.M. White, *Piezoelectric Cantilever Microphone and Microspeaker*, Microelectromechanical Systems, Journal of, 1996. **5**: p. 238–242.
10. Mescher, M.J., M.L. Vladimer, and J.J. Bernstein, *A Novel High-Speed Piezoelectric Deformable Varifocal Mirror for Optical Applications*. Presented at Micro Electro Mechanical Systems, 2002. The Fifteenth IEEE International Conference. 2002.
11. Vig, J.R., R.L. Filler, and Y. Kim, *Chemical Sensor Based on Quartz Microresonators*, Microelectromechanical Systems, Journal of, 1996. **5**: p. 138–140.
12. Tadmor, E.B., and G. Kosa, *Electromechanical Coupling Correction for Piezoelectric Layered Beams*, Microelectromechanical Systems, Journal of, 2003. **12**: p. 899–906.
13. DeVoe, D.L., and A.P. Pisano, *Modeling and Optimal Design of Piezoelectric Cantilever Microactuators*, Microelectromechanical Systems, Journal of, 1997. **6**: p. 266–270.
14. Smits, J.G., and W. Choi, *The Constituent Equations of Piezoelectric Heterogeneous Bimorphs*, Ultrasonics, Ferroelectrics and Frequency Control, IEEE Transactions. 1991. **38**: p. 256–270.
15. Elka, E., D. Elata, and H. Abramovich, *The Electromechanical Response of Multilayered Piezoelectric Structures*, Microelectromechanical Systems, Journal of, 2004. **13**: p. 332–341.
16. Weinberg, M.S., *Working Equations for Piezoelectric Actuators and Sensors*, Microelectromechanical Systems, Journal of, 1999. **8**: p. 529–533.
17. von Preissig, F.J., H. Zeng, and E.S. Kim, *Measurement of Piezoelectric Strength of ZnO Thin Films for MEMS Applications*, Smart Materials and Structures, 1998. **7**: p. 396–403.
18. Muralt, P., *Ferroelectric Thin Films for Microsensors and Actuators: A Review*, Journal of Micromechanics and Microengineering, 2000. **10**: p. 136–146.
19. Tressler, J.F., S. Alkoy, and R.E. Newnham, *Piezoelectric Sensors and Sensor Materials*, Journal of Electroceramics, 1998. **2**: p. 257–272.
20. Lu, J., et al., *Microstructure and Electrical Properties of Pb(Zr, Ti)O3 Thick Film Prepared by Electrostatic Spray Deposition*, Sensors and Actuators A: Physical, 2003. **108**: p. 2–6.
21. Chen, H.D., et al., *Development and Electrical Characterization of Lead Zirconate Titanate Thick Films on Silicon Substrates*. Presented at Applications of Ferroelectrics, ISAF '94, proceedings of the Ninth IEEE International Symposium. 1994.
22. Walter, V., et al., *A Piezo-Mechanical Characterization of PZT Thick Films Screen-Printed on Alumina Substrate*, Sensors and Actuators A: Physical, 2002. **96**: p. 157–166.
23. Manohara, M., et al., *Transfer by Direct Photo Etching of Poly(vinylidene flouride) Using X-rays*, Microelectromechanical Systems, Journal of, 1999. **8**: p. 417–422.
24. Brei, D.E., and J. Blechschmidt, *Design and Static Modeling of a Semiconductor Polymeric Piezoelectric Microactuator*, Microelectromechanical Systems, Journal of, 1992. **1**: p. 106–115.

25. Yamamoto, T., T. Shiosaki, and A. Kawabata, *Characterization of ZnO Piezoelectric Films Prepared by RF Planar Magnetron Sputtering*, Journal of Applied Physics, 1980. **51**: p. 3113–3120.
26. Wenzel, S.W., and R.M. White, *A Multisensor Employing an Ultrasonic Lamb-Wave Oscillator*, Electron Devices, IEEE Transactions. 1988. **35**: p. 735–743.
27. Zesch, J.C., et al., *Deposition of Highly Oriented Low-Stress ZnO Films*. Presented at Ultrasonics Symposium, 1991. Proceedings, IEEE. 1991.
28. Niu, M.-N., and E.S. Kim, *Piezoelectric Bimorph Microphone Built on Micromachined Parylene Diaphragm*, Microelectromechanical Systems, Journal of, 2003. **12**: p. 892–898.
29. Inagaki, Y., and M. Shimizu, *Resource Conservation of Buffered HF in Semiconductor Manufacturing*, Semiconductor Manufacturing, IEEE Transactions. 2002. **15**: p. 434–437.
30. Kolesar, Jr., E.S., and C.S. Dyson, *Object Imaging with a Piezoelectric Robotic Tactile Sensor*, Microelectromechanical Systems, Journal of, 1995. **4**: p. 87–96.
31. Domenici, C., and D. De Rossi, *A Stress-Component-Selective Tactile Sensor Array*, Sensors and Actuators A: Physical, 1992. **31**: p. 97–100.
32. Domenici, C., et al., *Shear Stress Detection in an Elastic Layer by a Piezoelectric Polymer Tactile Sensor*, Electrical Insulation, IEEE Transactions [see also Dielectrics and Electrical Insulation, IEEE Transactions], 1989. **24**: p. 1077–1081.
33. Roche, D., et al., *Piezoelectric Bimorph Bending Sensor for Shear-Stress Measurement in Fluid Flow*, Sensors and Actuators A: Physical, 1996. **55**: p. 157–162.
34. Kuoni, A., et al., *Polyimide Membrane with ZnO Piezoelectric Thin Film Pressure Transducers as a Differential Pressure Liquid Flow Sensor*, Journal of Micromechanics and Microengineering, 2003. **13**: p. S103–S107.
35. White, R.M., *Surface Elastic Waves*, Proceedings of the IEEE, 1970. **58**: p. 1238–1276.
36. Shajii, J., K.-Y. Ng, and M.A. Schmidt, *A Microfabricated Floating-Element Shear Stress Sensor Using Wafer-Bonding Technology*, Microelectromechanical Systems, Journal of, 1992. **1**: p. 89–94.

CHAPTER 8

Magnetic Actuation

8.0 PREVIEW

Magnetic sensing and actuation is one of the most widely used transduction principles found in our daily lives. This chapter focuses on the designs and fabrication methods for *micro*magnetic actuators, which often involve permanent magnets and electromagnetic coils. In Section 8.1, basic principles pertaining to microscale magnetic actuation are reviewed. We will discuss representative fabrication processes for various elements of an on-chip electromagnetic system in Section 8.2. Six cases are discussed in Section 8.3 to exemplify opportunities and methodologies.

Magnetic sensing and actuation are closely related. Indeed, many materials, components, and fabrication methods for microactuators originated in the magnetic sensing and disk storage industry. An excellent review of magnetic sensing can be found in Reference [1].

8.1 ESSENTIAL CONCEPTS AND PRINCIPLES

8.1.1 Magnetization and Nomenclature

A magnetic field may cause the internal magnetic polarization of a piece of magnetic material within the field. This phenomenon is called **magnetization**.

A piece of magnetic material is made of magnetic domains. Each magnetic domain is said to consist of a magnetic dipole. The strength of the internal magnetization of the bulk magnetic material depends on the extent of the ordering of these domains. These domains contribute to a net internal magnetic field within the magnetic material itself, if they are somewhat aligned.

Magnetism has been studied for hundreds of years. Researchers and practitioners utilize a mixture of concepts and units; some are acquired historically and are based on the CGS unit system while some conform to the SI unit system. One of the purposes of this review is to clarify the units for various variables.

Magnetic field intensity (symbol H) represents the driving magnetic influence external to a magnetic material. Its convenient SI unit is A/m. The conventional unit in CGS unit system is oersted (1 A/m = $4\pi/10^3 Oe$).

Another term, called **magnetic field density** (symbol B), represents the induced total magnetic field inside a piece of magnetic material. The total magnetic field accounts for the influence of the induction field and the internal magnetization. The term B is often referred to as magnetic induction or magnetic flux density as well. The magnitude of B can be expressed in units within the SI unit system—Tesla, or Wb/m^2—or within the CGS unit system—Gauss (1 T = 10^4 Gauss). The convenient SI unit for Weber is V · s. The convenient SI unit for Tesla is therefore V · s/m^2.

The magnetic field densities of commonly encountered magnetic objects are

common refrigerator magnet: 100–1000 Gauss;

rare earth magnet used in Magnetic Resonance Imaging: 1–2 T;

magnetic storage media: 10 mT or 100 Gauss;

earth magnetic field (near equator): 1 Gauss.

The relationship between B and H can be described using the following equation:

$$B = \mu_0 H + M = \mu_0(H + \chi H) = \mu_r \mu_0 H, \tag{8.1}$$

where μ_0 is the magnetic permeability of space (SI unit: Henry/meter or Wb/(A · m)), μ_r the relative permeability of the magnetic material, and M the internal magnetization. The magnetic susceptibility χ is defined as $\mu_r - 1$. A magnetic material with a weak and positive χ is called **paramagnetic**; one with a weak and negative χ is **diamagnetic**. For paramagnetic and diamagnetic materials, the relative permeability is very close to 1.

For **ferromagnetic materials** (e.g., iron, nickel, cobalt, and some rare earths), the values for relative permeability are very large. A ferromagnetic material is so named because iron is the most common example of this group. Ferromagnetic materials are often used in MEMS actuation applications. We will focus on the magnetization of ferromagnets in the next few paragraphs.

The linear relationship between B and H is only valid within a certain range of H. The full magnetization curve for a ferromagnetic material is illustrated in Figure 8.1. There are a number of important features to note:

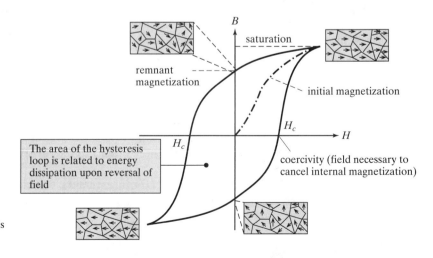

FIGURE 8.1

Magnetization hysteresis curve.

1. After the external induction field reaches a certain level, the magnetization will reach a saturation point, called **saturation magnetization**. The saturation represents a situation when all available domains within a piece of magnetic material have been aligned to one another.
2. A ferromagnetic material will lose a portion of its magnetization upon the removal of the external magnetic field. The fraction of the saturation magnetization which is retained after H is removed is called the **remnance** of the material (or remnant magnetization).
3. The coercivity is a measure of the reverse field needed to drive the magnetization to zero after having reached saturation at least once.
4. The area enclosed by the hysteresis curve indicates the amount of magnetic energy stored in a magnetic material.

There are two important classes of ferromagnets—**hard magnets** and **soft magnets**. The word "hard" means that the magnet retains certain magnetic polarization even under zero external magnetic driving fields. The alternative, a "soft" ferromagnetic material, has very low remnance and exhibits internal magnetization only when it is subjected to a biasing, external magnetic field.

Their differences are easily explained using the B–H hysteresis curves. The curve to the left in Figure 8.2 shows the hysteresis curve of a hard magnetic material, such as a

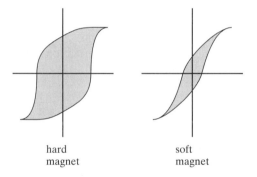

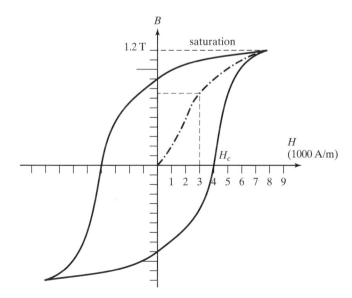

FIGURE 8.2

Representative hysteresis curves for hard and soft magnets.

permanent magnet. Permanent magnetic materials not only exhibit a large remnant field. They also require a larger reverse field and energy input in order to switch or destroy the built-in magnetic field. The curve to the right shows the hysteresis curve of a soft magnetic material. The core of a transformer, for example, should be made of a soft magnet. The small stored energy allows high efficiency, low power consumption, and rapid transition.

Example 8.1. Magnetization Hysteresis Curve

Find the internal magnetization ($\chi\mu_0 H$) of a ferromagnetic material (with the hysteresis curve provided) at two induction field strengths—(a) 3000 A/m (first time magnetization); (b) 10,000 A/m—after several full magnetization cycles. Calculate the induction field created by this magnet, measured at its surface.

Solution. When the induction field is 3000 A/m, the total magnetization is 0.75 Tesla. The magnetic field density associated with the driving field is

$$\mu_0 \times 3000 \text{ A/m} = 0.00377 \frac{Wb}{A \cdot m} \frac{A}{m} = 0.00377 \text{ Tesla}.$$

The internal magnetization $\chi\mu_0 H$ is $0.75 - 0.00377 = 0.746$ Tesla.

When the induction field is 10,000 A/m, the internal magnetization is 1.2 Tesla. The magnetic field density associated with the driving field is

$$\mu_0 \times 10{,}000 \text{ A/m} = 0.00377 \frac{Wb}{A \cdot m} \frac{A}{m} = 0.01257 \text{ Tesla}.$$

Therefore, the internal magnetization $\chi\mu_0 H$ is $1.2 - 0.01257 = 1.187$ Tesla.

When the driving magnetic field is removed, a remnant magnetic field of 0.9 Tesla is present at the surface of the material. Since the magnetic field lines are continuous at the interface of a magnetic material and the surrounding media, we can assume that the magnetic flux density in air is 0.9 T as well. In air, the magnetization force is

$$H = \frac{B}{\mu_0} = \frac{0.9}{1.257 \times 10^{-6}} = 716 \times 10^3 \text{ A/m}.$$

This value is much greater than the external field of 9000 A/m necessary to magnetize the ferromagnet in the first place.

8.1.3 Selected Principles of Micromagnetic Actuators

A magnetic field can be used to produce force, torque, or displacement of microstructures, according to several important magnetic actuation principles [2]. A driving magnetic field can act on a number of elements, including current-carrying wires, inductor coils, pieces of magnetic material, or magnetostrictive materials.

In this section, we will discuss the formula for estimating the magnetic interaction of a current-carrying wire and magnetized magnetic pieces.

A Lorentz force actuator exploits the interaction between a current-carrying conductor and an external magnetic field (Fig. 8.3). The Lorentz force acting on a single moving charge q is given by

$$\vec{F} = q\vec{v} \times \vec{B}, \tag{8.2}$$

where $\vec{v}$ is the velocity of the charge. The magnitude of the force is

$$F = qvB \sin \theta, \tag{8.3}$$

with θ being the angle between the velocity and the magnetic field ($\theta < 180°$). The direction of the resulting force can be easily determined by a simple mnemonic procedure. Extend your right hand. Point your thumb in the direction of the velocity of a positive charge, and point your fingers in the magnetic field direction. The palm faces the direction of the force on charge. The force is perpendicular to the velocity of the charge and the magnetic field.

Example 8.2. Lorentz Force on a Current-Carrying Wire

Calculate the force acting on a 100-μm long metal wire carrying a current of 10 mA when it is placed inside a uniform magnetic field of 1 T with the field lines transverse to the direction of the conducting wire.

Solution. A current-carrying wire hosts a large number of moving charged particles at any given time. The force on a single charged particle can be easily found. To find the force on the entire wire, we need to find the number of particles present in a wire (n) at a given moment and the velocity of the particles (v) in order to calculate the total force.

The current is a measure of the number of positive charges passing through the cross section of a wire per second. In a time period of l/v, which equals the average time for a charge carrier to traverse through the entire length of the wire, all electrons (n) in the wire would pass through the end terminal of the wire. As far as the end section is concerned, a total number of nq charges pass through within l/v seconds. Therefore, the expression of the current is

$$I = \frac{nq}{\left(\dfrac{l}{v}\right)}. \tag{8.4}$$

We can rearrange it to obtain

$$nqv = Il. \tag{8.5}$$

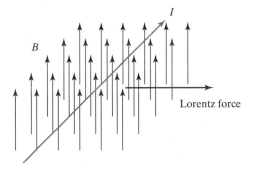

FIGURE 8.3

Lorentz force on a current-carrying wire inside a magnetic field.

The total Lorentz force is the number of carriers multiplied by the force on each carrier:

$$F = n(qvB \sin \theta) = IlB \sin \theta = 1\mu N \qquad (8.6)$$

Example 8.3. Unit Analysis

Check the consistency of the unit of the Lorentz force expression derived in Example 8.2 in the SI system.

Solution. In the SI unit system, the unit of F is Newton. The unit of the term on the right-hand side of Equation (8.6) is

$$[IlB \sin \theta] = A \cdot m \cdot Tesla = A \cdot m \cdot \frac{Wb}{m^2} = A \cdot m \cdot \frac{V \cdot s}{m^2} = \frac{A \cdot V \cdot s}{s}.$$

Since the product of A and V gives the unit of power (W), and the product of W and s gives the unit of work, $N \cdot m$, the equivalent unit of the terms to the right is *Newton*.

Magnetic actuators can occur as a result of the interaction between a permanent magnet and an external DC magnetic field. A classic example is the familiar magnetic compass (Fig. 8.4). The permanent magnet used in the magnetic compass is a hard ferromagnetic material. If the internal and external magnetic field lines are aligned, no force or torque will be exerted on the compass needle. The compass needle will experience a torque (called magnetic torque) when the direction of the internal magnetization is not aligned with the local earth magnetic field lines. The torque causes the needle to rotate until the internal magnetic field is lined with the external field lines. This principle of interaction can be extended to microscale sensors and actuators. Indeed, micromachined magnetic actuators using manually attached [3] or integrated [4] permanent magnets have been developed.

Irrespective of the method by which it is generated, the external magnetic field can be classified into two broad categories: a spatially **uniform magnetic field** and a **non-uniform magnetic field** with a gradient. Depending on the type of magnet (hard or soft) and the initial orientation in the field (aligned or misaligned with field lines), net forces and/or torques can be produced.

The interaction of hard and soft magnetic pieces in these two types of field is illustrated in Figure 8.5. Two pieces of hard (permanent) magnet (pieces 1 and 2) and two soft magnets (3 and 4) are used as examples. Their initial orientations are different. Pieces 1 and 3 are oriented such that

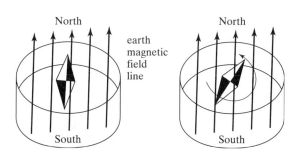

FIGURE 8.4

Magnetostatic actuation of a magnetic compass.

8.1 Essential Concepts and Principles 285

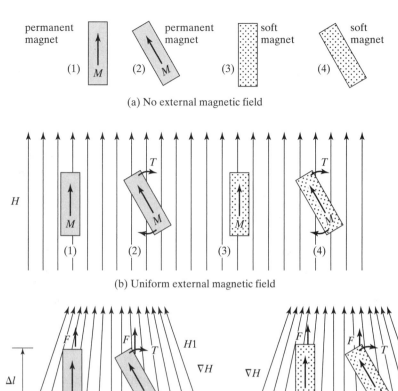

FIGURE 8.5

Magnetization states and force states in magnetic fields. Net magnetic forces (F) and moment (T) are indicated for each piece under each biasing case.

their internal magnetization is parallel to the local external magnetic field lines. Pieces 2 and 4 are intentionally misaligned.

In the situation depicted in part (a), zero external magnetic bias is present. The permanent magnets are polarized, whereas the soft magnets are not. Neither pieces experience any net force or moment due to the lack of magnetic driving force (H).

In the situation depicted in part (b), a spatially uniform magnetic field is applied. The magnetic field polarizes the soft magnet pieces (3 and 4), so that they become magnetized. The hard magnets—pieces 1 and 2—are already magnetized. We assume the strength of their internal magnetization is not changed by the external magnetic field, although in reality it will be changed slightly.

When the internal magnetization is parallel to the external magnetic field, no force or moment is generated (pieces 1 and 3). However, if the internal magnetization is placed at an angle with respect to the external field lines, the magnetic pieces will experience torques (pieces 2 and 4), but not net forces.

Note that the internal magnetic fields of pieces 2 and 4 are aligned along their longitudinal axes, rather than being parallel to external field lines. **Shape anisotropy** is said to play an

TABLE 8.1   Presence of Magnetic Force and Torque.

	Case a				Case b				Case c			
Pieces	1	2	3	4	1	2	3	4	1	2	3	4
Net force	0	0	0	0	0	0	0	0	F	F	F	F
Net torque	0	0	0	0	0	T	0	T	0	T	0	T

important role in determining the direction of magnetization or remnant magnetization. For example, if a piece of ferromagnetic material is shaped into a long-aspect ratio rod, the internal magnetization will usually point in the longitudinal direction of the rod, irrespective of the direction of the induction field relative to the longitudinal direction of the rod. (Similarly, a thin flat piece of a ferromagnetic material may exhibit strong magnetization in the plane of the plate rather than vertical to the plate surface, even when the induction field is primarily lined to the normal direction of the plate surface.) This phenomenon has to do with the energy required to align magnetic domains. It simply takes less energy to magnetize along the longitudinal direction or the in-plane direction of these magnetic pieces.

In the situation of part (c), a non-uniform magnetic field is presented. Again, both soft magnets (pieces 3 and 4) are magnetized. If the internal magnetization is parallel to the external field lines, a net force materializes on the magnetic piece (pieces 1 and 3). On the other hand, if the internal magnetization is placed at an angle with respect to the external field lines, the magnetic piece experiences a net force *and* a torque (pieces 2 and 4).

A summary of the force and torque for each piece under these three situations (depicted in parts a, b, and c) is provided in Table 8.1.

Next, let us review a systematic approach for analyzing the net force and the moment acting on a piece of magnetic material inside an arbitrary external magnetic field.

For practical purposes, a magnetized magnetic material with an internal magnetization M can be considered as a charged magnetic dipole. For simplicity, we assume a magnetized magnetic piece contains two monopoles of opposite polarity. A concentrated force acting on each monopole can be expressed as

$$F = M(wt)H, \tag{8.7}$$

where w and t are the width and thickness of the cross section, and H is the local strength of the biasing magnetic field. The force is proportional to the magnitude of the internal magnetization and the external driving magnetic field, as well as the cross-sectional dimensions of the piece.

The net force on a piece of magnetized material is the vector sum of the forces concentrated on two monopoles.

In part (b), the external magnetic field is uniform and the field lines are straight. The magnitude of H is identical everywhere, hence no magnetic pieces experience the net force.

In the case of a spatially non-uniform magnetic field (part c), the forces on two poles are not identical because the local magnitudes of H are different. A net force is therefore produced (e.g., for all pieces in part c). The net force is

$$F = M(wt)\Delta H, \tag{8.8}$$

where ΔH is the difference of the magnetic driving field at two monopoles. The magnitude of ΔH equals the distance between two poles along the field line directions and the gradient of the magnetic field ($\partial H/\partial l$), i.e.,

$$\Delta H = \Delta l \frac{\partial H}{\partial l}. \tag{8.9}$$

Rotational torques (T) can be developed between an external magnetic field and a magnetized magnetic piece, provided that the internal magnetization and the external field lines are not aligned. In fact, as long as the forces developed on two monopoles of a magnetic piece are not rested along the same line, a torque would be produced. For examples, pieces 2 and 4 in case b experience torque even though the net forces are zero. The magnitude of the torque equals the magnitude of the force multiplied by the distance between two lines of force. If the magnetized pieces are free to rotate, angular displacement can result. Large angular displacement angles (180°) have been achieved in 2.25-μm-thick polysilicon flexural beams with soft magnetic pieces attached to them [5].

In earlier chapters, we have discussed electrostatic, thermal, and piezoelectric actuation. What are the reasons for using magnetic actuation?

One of the advantages of magnetic actuation is the prospect of eliminating electrical wires, which is unavoidable in electrostatic, thermal, and piezoelectric actuators. This can drastically reduce the complexity of packaging and use. Magnetic MEMS actuators are capable of performing truly nontethered operations. For example, a set of MEMS-based microwings adorned with magnetic materials have been demonstrated. It is capable of generating lifting forces (capable of lifting the 165 μg wing) without any wires attached to it, where power is provided by a rotating magnetic field (500 Hz) [6]. Micromagnetic stir bars integrated with microfluid channels can provide mixing and pumping without wires attached for providing voltage or current [7].

Secondly, relatively large magnetic fields can be present in the free space without harm or damage to humans or environment. In contrast, a large electric field in the free space or dielectrics would cause problems such as dielectric breakdown or electrocution.

Further, external magnetic fields with sufficient strength to generate appreciable force and torques for microscale devices can be provided by passive, permanent magnets. Such magnets are extremely low cost and do not consume power at all during operation.

8.2 FABRICATION OF MICROMAGNETIC COMPONENTS

Magnetic actuators involve unique materials and unique structures (e.g., solenoids). The material preparation and fabrication techniques for representative components of a micromagnetic system are discussed in the following section.

8.2.1 Deposition of Magnetic Materials

Although it is possible to attach small pieces of magnetic materials to micromechanical structures for realizing sensors and actuators [3], this process is generally very inefficient. Monolithic integration of magnetic materials is more accurate and widely practiced.

The most common technique for depositing ferromagnetic materials for microdevice applications is **electroplating**. A chemical solution consisting of constituent ions of the desired magnetic material is used as a wafer bath. The work piece for the metal deposition (wafer) is biased negatively with respect to a counter electrode, which is placed in the bath during the electroplating session.

Since the magnetic force is related to the cross section of a magnetic element according to Equation (8.7), large thicknesses of ferromagnets are generally needed for the generation of large forces or torques. The electroplating process is often desirable over other thin-film deposition methods (such as sputtering) because it is relatively easy to reach appreciable thickness (e.g., 5 μm and above). The electroplating rate can be controlled by the current density supplied. The greater the current density, the faster the electroplating. Of course, there is a practical limit to how high the current can be due to concerns of heating and the tendency of increased surface roughness under high current densities.

In many cases, the wafer is not conductive on its own. Under these circumstances, the surface of a wafer is first coated with a thin-film metal layer for providing negative electric biasing. This thin film layer is called the **seedlayer**. Common seedlayer materials are copper, aluminum, or gold. Thin metal layers of Cr or Ti are often used to enhance adhesion between the seed layer metal and the substrate.

A typical electroplating process flow using a seedlayer is shown in Figure 8.6. A substrate covered with a metal seedlayer is prepared. In order to produce patterned ferromagnetic thin film, a mold electroplating method is often used (Figure 8.6a). The mold, made of a thin-film insulating layer (e.g., patterned photoresist), is deposited and patterned (Figure 8.6b). The wafer is immersed in an electroplating solution (Figure 8.6c). Electroplated metal grows in the open windows, where the seedlayer is exposed to the electroplating bath (Figure 8.6e). The electroplating mold is then selectively removed. The electroplating process may result in a thickness less than the height of mold (Figure 8.6d), or greater (Figure 8.6e), depending on the duration of the plating step. When the thickness of the electroplated metal reaches beyond the height of the mold, it tends to grow laterally. This property can be used to create metallic structures with unique shapes, such as the mushroom in Figure 8.7.

Many magnetic materials and processes have been developed for the magnetic data storage industry. The types of materials that can be deposited by electroplating are wide ranging. The nickel iron alloy, called Permalloy, is used widely since it has high permeability (500–1000), soft magnetic properties (coercivity being 1–5 Oe), high magnetoresistivity, and low magnetorestriction (which is changed to material dimensions under the applied magnetic field) [8]. Other materials (e.g., soft magnetic materials [9, 10], permanent magnets, and polymer-based magnets) and properties (e.g., greater permeability and coercivity) are needed for certain applications. The constitutions of electroplating bath and pertinent processing parameters for two representative materials, including NiFe and CoNiMnP, are summarized in Table 8.2.

Besides electroplating, other fabrication processes for realizing thick magnetic materials are also available. One such process relies on commercial polyimide mixed with ferrite magnetic powers at different concentration levels [15]. The magnetic polymer composite, consisting of particles of magnetic materials suspended in non-magnetic media, offers the ability to incorporate the magnetic materials of arbitrary characteristics available at the bulk scale to micromachining applications. Patterned polymer magnetic film can be produced using either a screen printing technique or, if the polymer matrix is photo-definable, spin coating followed by photolithography.

8.2.2 Design and Fabrication of Magnetic Coil

On-chip integrated solenoids are of great interest. They can be used for electromagnetic sources and coil actuators, as well as inductors, telemetry coils, and transformers for integrated

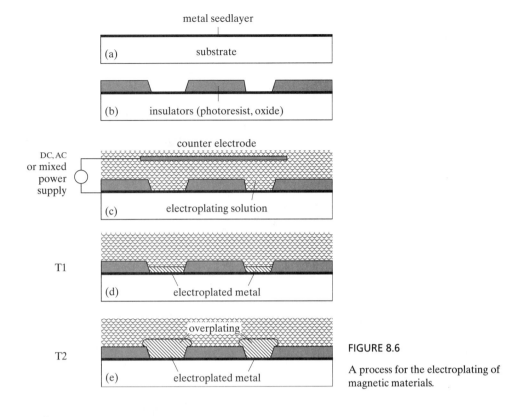

FIGURE 8.6

A process for the electroplating of magnetic materials.

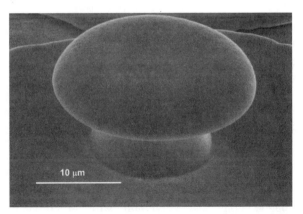

FIGURE 8.7

Mushroom-shaped electroplated metal, revealed after the electroplating mold is removed.

circuits. Solenoids by conventional machining involve winding conducting wires around a ferromagnetic core (Figure 8.8). In microfabricated devices, however, this practice is prohibitively difficult due to the small scale and lack of automated tools. Instead, the most prevalent and manufacturable form of the electromagnet is a single-layer, planar coil with an air core (Figure 8.9a). Such a coil is not capable of generating a strong magnetic flux density because of the lack of a magnetic core and the lateral spreading of wires away from the center of the coil.

290 Chapter 8 Magnetic Actuation

TABLE 8.2 Electroplating Bath Constitution for Representative Magnetic Materials.

Magnetic material	Electroplating solution composition	Amount (gram/liter)
$Ni_{80}Fe_{20}$, Permalloy (other alternative recipes have also been suggested for achieving different processes and functional characteristics [9, 11, 12])	$NiCl_2 \cdot 6H_2O$, Nickel (II) chloride hexahydrate $NiSO_4 \cdot 6H_2O$, nickel sulfate H_3BO_4, boric-acid powder (e.g., Fisher A74-500) Sodium saccharin, for stress adjustment NaCl, sodium chloride $FeSO_4 \cdot 7H_2O$, ferrous sulfate crystal *Note:* The solution should have a pH value between 2.7 and 2.8. To lower pH, drop in a small amount of diluted HCl. Ideal current density: 8–12 mA/cm^2.	39.0 16.3 25.0 1.5 25.0 1.4
CoNiMnP permanent magnet [13, 14]	$CoCl \cdot 6H_2O$ $NaCl \cdot 6H_2O$ $MnSO_4 \cdot H_2O$ NaCl $B(OH)_3$ $NaH_2PO_2 \cdot xH_2O$ Sodium lauryl sulfate Schchain *Note:* Cobalt anodes were used to avoid hypophosphite oxidation. Current density is 10–20 mA/cm^2.	24 24 3.4 23.4 24 4.4 0.2 1.0

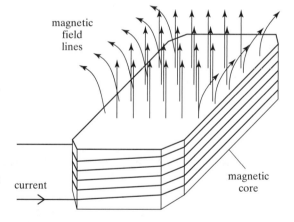

FIGURE 8.8

Magnetic field line distribution of a solenoid magnet.

More efficient electromagnet coils have been built with integrated cores and wraparound coils. Such coils can be classified into two categories according to the orientation of the magnetic core: those with the magnetic flux normal to the substrate plane or those with the flux lying parallel to the substrate plane. A good review on fabrication methods and materials can be found in Reference [10, 16].

A number of techniques for realizing microsolenoids with integrated ferromagnetic cores are shown in Figure 8.9. The simple scheme of electroplating a high-permeability magnetic material with a planar coil improves upon the performance of a single-layer magnetic coil

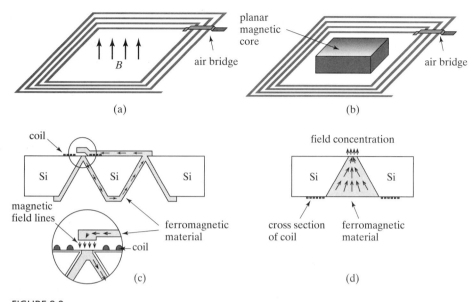

FIGURE 8.9

A planar magnetic inductor.

(Figure 8.9b). However, the improvement is not as significant as the issue of the strong dispersion of magnetic field persists. To contain and even concentrate the magnetic field lines, innovative structures and fabrication processes are involved. For example, through-wafer magnetic cores can be built by taking advantage of the sloped surfaces created using anisotropic etching (Figure 8.9c). The equivalent of a horseshoe magnet can be made by processing magnetic core materials on both sides of a wafer [17]. It has been shown that 200 mN of actuation force can be generated with an 80 mA actuation current and 320 mW of power. The magnetic flux density can be increased by reducing the cross-sectional area of the core (Figure 8.9d).

All solenoids in Figure 8.9 involve single-layered conducting wires. Coils with multiple turns can be made by stacking (Figure 8.10). The process involves depositing a metal coil layer, covering it with a dielectric material, planarizing the dielectric material, and repeating the cycle. In theory, this process can be repeated an infinite number of times. In reality, there is a practical limit to how many layers can be stacked due to finite process time, and to the potential degradation of registration quality and surface roughness as the number of layers increases.

Three-dimensional coils have been fabricated on various substrate surfaces, including non-planar surfaces, using techniques ranging from microcontact printing on a cylinder [18], three-dimensional assembly [19], laser-direct lithography [20], fluid self-assembly [21], and even conductive loops formed by bonding wires.

Magnetic coils with in-plane magnetization can be built as well. The process flow for a typical multi-turn coil is shown in Figure 8.11. The basic process consists of three major steps: deposit and pattern a bottom conductive layer (Figure 8.11a), electroplate the magnetic core as well as vertical conductive posts (Figure 8.11b and c), and deposit and pattern a top conductive layer (Figure 8.11d).

FIGURE 8.10

Process flow for a multiple-layer planar coil.

8.3 CASE STUDIES OF MEMS MAGNETIC ACTUATORS

Micromagnetic actuators can be categorized according to the types of magnetic sources and the microstructures involved.

The source of the magnetic field can be a permanent magnet, an integrated electromagnetic coil (with or without the core), or an external solenoid. Multiple types of sources may be used in a hybrid manner.

Force-generating microstructures, located on a chip, can be one of the following types: a permanent magnet (hard ferromagnet), a soft ferromagnet, or an integrated electromagnetic coil (with or without a core).

Overall, there are 12 possible permutations. In this chapter, we will not present an exhaustive overview of the cases for each possible category. Rather, six represented cases will be reviewed to illustrate typical challenges and solutions. The relation of these cases according to the classification scheme discussed previously is summarized in Table 8.3.

Magnetic actuators may also be combined with other modes of actuation (e.g., electrostatic, thermal, or piezoelectric actuation) or integrated with sensors for position control (e.g., see Reference [22]).

8.3 Case Studies of MEMS Magnetic Actuators   293

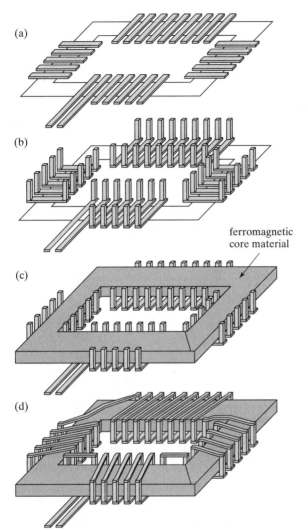

FIGURE 8.11

Fabrication of a three-dimensional microcoil.

TABLE 8.3 Types of Magnetic Actuators and Corresponding Case Numbers.

On-chip microstructure	Source of field			
	Stationary permanent magnet source	External solenoid source	Integrated electromagnetic coil source	Hybrid source
Permanent magnet		Case 8.5		
Soft ferromagnetic material		Case 8.2	Case 8.1	Case 8.6
Integrated coil	Case 8.3 and Case 8.4			

Case 8.1. Magnetic Motor

The first example is a planar variable-reluctance magnetic micromotor with a fully integrated stator and coils [23] (Figure 8.12). The stators are made of integrated electromagnets, whereas the rotor is made of a soft magnetic material. The motor has two sets of salient poles, one set in the stator (which usually has excitation coils wrapped around the magnetic poles) and another set on the rotor.

When a phase coil is excited, the rotor poles located closest to the excited stator poles are attracted to the stator pole (Figure 8.12a, b). Due to the rotation of the rotor, the rotor poles will align with the stator poles. The excited phase coil is turned off, and the next phase is excited for continuous motion. In this design, the wound poles of all phases are arranged in pairs of opposite polarity to achieve adjacent pole paths of short lengths. The stator coils arranged in one or more sets and phases are excited in sequence to produce a continuous rotor rotation.

The rotor is 40 μm thick and 500 μm in diameter. It is microassembled onto the chip containing the stators or fabricated in an integrated fashion by electroplating. When a 500-mA current is applied to each stator, 12° of rotation (one incremental stroke) is produced. By applying three-phase 200-mA current pulses to the stators, the rotation of the rotor was observed with its speed and direction adjusted by the frequency and phase firing order of the power supply. Continuous rotor motion at speeds up to 500 rpm has been observed. The predicted torque for the motor at 500-mA drive current is calculated as 3.3 nN-m.

A toroidal-meander type integrated inductive component is used in the motor for flux generation [12, 23]. Multilevel magnetic cores are "wrapped" around planar meander conductors (Figure 8.12c). This configuration can be thought of as the result of interchanging the roles of the conductor wire and the magnetic core in a conventional inductor (Figure 8.12d). The fabrication process begins with an oxidized silicon wafer (Figure 8.13). A 200-nm-thick titanium thin film is deposited as the electroplating seedlayer (step b). Polyimide (Dupont PI-2611) was spin coated on the wafer to build electroplating molds for the bottom layer of the magnetic core. Four coats were accomplished to obtain a thick polyimide film with an after-curing thickness of 12 μm (step c). This polyimide is coated with an aluminum metal thin film and again with a photoresist layer, which is photolithographically patterned. The photoresist serves as a mask for etching the aluminum (in a wet etching solution), which then serves as a mask for etching the polyimide (in oxygen plasma) (step d). Electroplating of nickel-iron Permalloy is grown to fill the openings in the polyimide (step f). Detailed processing parameters are described in Reference [12].

Another layer of polyimide is spin coated to insulate the bottom magnetic core (step g). A 7-μm-thick metal film (either aluminum or copper) was deposited and patterned on top of the polyimide insulator (step i). More polyimide is spin coated on the patterned metal to planarize the wafer and insulate the meander conductor (step j). The polyimide is patterned using the same procedure that was done previously (step k). Via holes are opened all the way to the bottom magnetic core, and an electroplating process is conducted to produce the top magnetic core (step l).

8.3 Case Studies of MEMS Magnetic Actuators 295

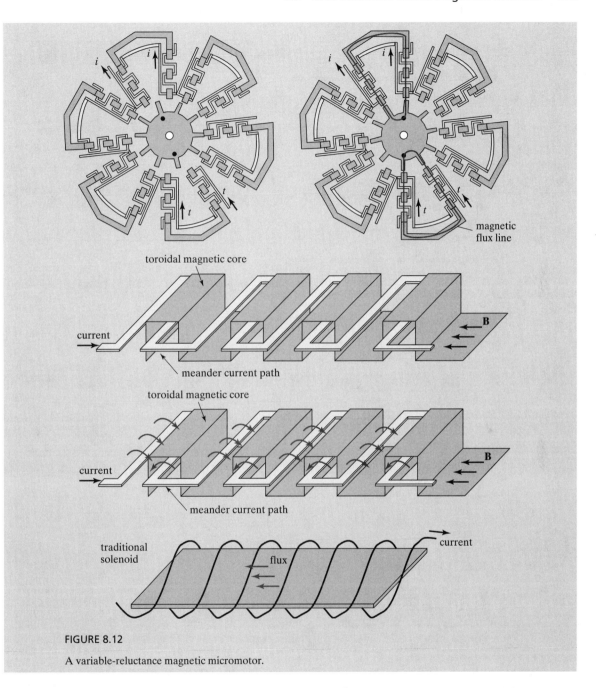

FIGURE 8.12

A variable-reluctance magnetic micromotor.

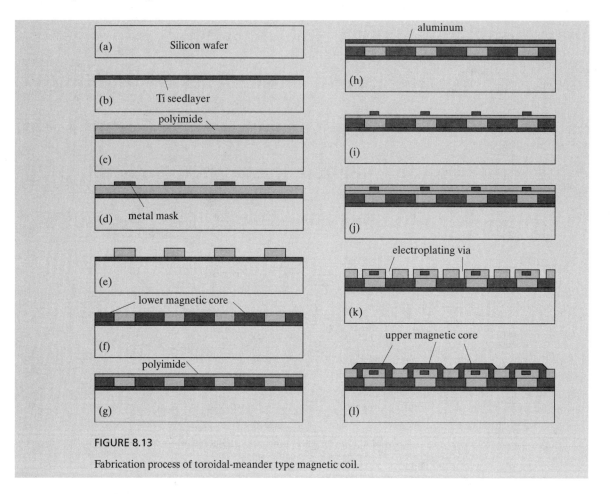

FIGURE 8.13

Fabrication process of toroidal-meander type magnetic coil.

Case 8.2. Magnetic Beam Actuation

One of the unique performance characteristics of magnetic actuation over electrostatic, thermal, and piezoelectric actuators is the ability to generate *torque* and achieve a large angular displacement. For example, large angular displacement—and hence, vertical displacement on the order of 90° and several millimeters—can be achieved under a magnetic field of 10 kA/m [5].

One actuator has been developed for dynamic aerodynamic control [24]. The actuator consists of a rigid flap supported on one side by two fixed–free cantilevers. Each flap is made of polycrystalline silicon (as the support structure) and electroplated ferromagnetic materials (Permalloy in this case). The cross-sectional side view is shown in Figure 8.14a.

8.3 Case Studies of MEMS Magnetic Actuators

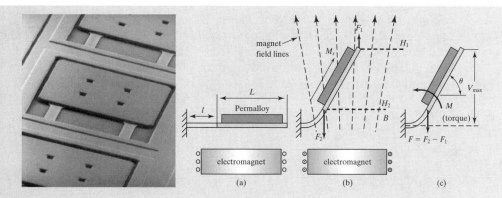

FIGURE 8.14

Flap magnetic actuator.

When an external magnetic field is present, an internal magnetization is developed in the ferromagnetic piece. The situation is similar to the one experienced by piece 4, part (c) of Figure 8.5. Experimental characterization of a single flap has been conducted to show that up to a 65° angular displacement can be achieved.

The magnitude of the internal magnetization equals the saturation magnetization, being approximately 1.5 T for the prepared material. The direction of the internal magnetization vector lies within the plane of the flap according to shape anisotropy (Figure 8.14b).

In a non-uniform magnetic field, a torque and a force are developed on the microflap. (The force element is ignored in the analysis.) The magnetic torque and the angular displacement are intricately related. The magnitude of the magnetic torque at a given displacement angle θ is approximated as

$$M_{mag} = FL \cos(\theta) = M_s(WTH_1) L \cos(\theta) = M_s V_{magnet} H_1 \cos(\theta). \quad (8.10)$$

According to the formula for flexural bending beams under a pure torque (see Appendix B), the angular displacement is related to the magnetic torque according to

$$\theta_{max} = \frac{M_{mag} l}{EI}. \quad (8.11)$$

By solving these two equations simultaneously, the magnitude of the magnetic torque and the angular displacement can be found. In turn, the maximum vertical displacement at the end of the support cantilever (but not the end of the entire plate) is

$$y_{max} = \frac{M_{mag} l^2}{2EI}. \quad (8.12)$$

The fabrication process of the magnetic actuator is shown in Figure 8.15. The process is carried out on a silicon wafer because of the high temperature associated with the structural layer (LPCVD polycrystalline silicon) and sacrificial layer (LPCVD oxide). First, LPCVD oxide is deposited and patterned, followed by deposition and patterning of polycrystalline silicon (step a). A metal seedlayer is deposited, followed by spin coating and the patterning of photoresist as the electroplating mold (step b). Electroplating occurs in regions not covered by the photoresist, to a height decided by the thickness of the spin-coated photoresist (step c). The photoresist is removed (step d). Subsequently, the sacrificial layer (oxide) is removed in an HF acid bath (step e).

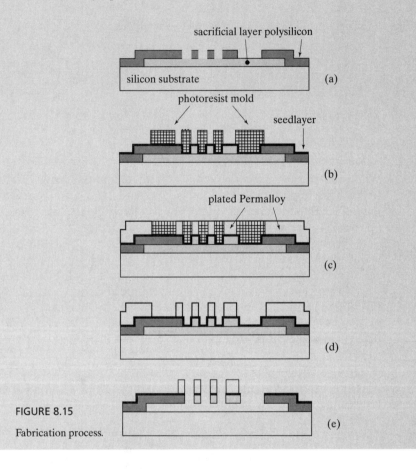

FIGURE 8.15
Fabrication process.

Case 8.3. Plate Torsion with Lorentz Force Actuation

A moving-coil electromagnetic optical scanning mirror capable of one-axis rotation is discussed here. A mirror plate is supported by torsional hinge structures consisting of multilayered polyimide films with aluminum lead wires in between (Fig. 8.16). The mirror consists of a planar microplate, with a smooth side for optical reflection and the opposite side hosting a planar coil. Two permanent magnets are placed on the side of the mirror, such that the magnetic field lines are parallel to the plane of the mirror and orthogonal to the torsional hinges. When current passes through the coil, Lorentz forces will develop and cause rotational torque on the mirror. The direction of the torque depends on the direction of the current input.

The magnitude of torque acting on the actuator is described as

$$T = iB_1 l_1 l_2 n, \tag{8.13}$$

where the term i represents the current, B_1 the field created by the permanent magnet, l_1 the average length of the coil parallel to the edge of the magnet, l_2 the average distance between wires in the direction perpendicular to the hinge, and n the number of coil turns.

The mirror hinges are made of polyimide material to increase the shock tolerance, up to 2500 g [25]. The dimensions of the hinge determine the frequency response range. A fast scanning mirror and a slow one were made using different geometric specifications. Driven with a solenoid current of 20 mA, the scanning mirrors rotate to an angle of 1° at the resonant frequency of 1.7 kHz, or 60° at a resonant frequency of 72 Hz. A lifetime of at least 13,000 hours has been proven.

A mirror array using a similar principle (but with more elaborate configuration) has been reported in Reference [26].

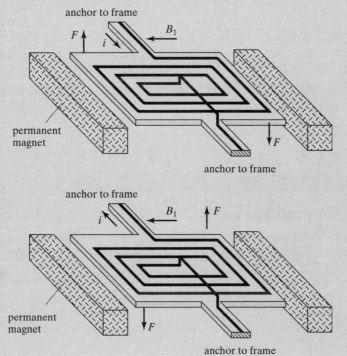

FIGURE 8.16
Diagram of a micromirror.

Case 8.4. Multi-Axis Plate Torsion Using On-Chip Inductors

Another micromirror capable of rotation along two axes is discussed. A mirror is suspended in a rotational gimbal that provides rotation degrees-of-freedom in two axes, as shown in Figure 8.17a. Four planar electroplated electromagnetic coils are located at the back of the mirror—one occupying each quadrant. Current passing through each coil generates a magnetic dipole. A strong permanent magnet based on rare-earth materials is located underneath the mirror plate. Based on the polarity of the dipole with the external field, an

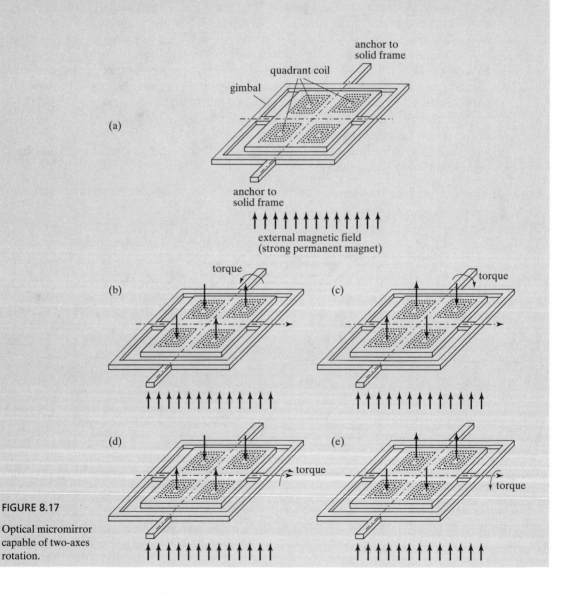

FIGURE 8.17
Optical micromirror capable of two-axes rotation.

attractive or repulsive force will be acted on each coil under the interaction of the induced magnetic dipole and the permanent magnet.

A rotational torque can be generated if two neighboring coils (joined by at least one side) are biased in the same direction. There are four distinct possibilities for generating torques (Figure 8.17b–e). The selective activation of these coils will induce coupled angular displacement along two axes in a highly selective manner. The motions are controlled electrically under a strong biasing electromagnetic field provided by a stationary permanent magnet. The permanent magnet eliminates any power consumption for biasing, and it can operate over a relatively large distance and large displacement angles. The large magnetic field H compensates for the fact that the electromagnetic field created by each planar coil is relatively weak.

Case 8.5. Bidirectional Magnetic Beam Actuator

A bidirectional cantilever-type magnetic actuator is discussed here (Fig. 8.18) [4]. At the tip of a silicon cantilever beam, permanent magnet arrays are electroplated so as to achieve a vertical magnetic actuator by taking advantage of the vertical magnetic anisotropy of the magnetic arrays. The adoption of array shapes in the design of permanent magnets (multiple vertical posts as opposed to a large sheet) allows suppression of the residual stress between the electroplated CoNiMnP films and the establishment of preferred internal magnetization, which is normal to the plane of the cantilever.

A commercial inductor is used as the electromagnet to drive the bidirectional actuator. The magnetic force acting on the cantilever is

$$F = V \cdot M_s \cdot \frac{\partial H}{\partial z}, \tag{8.14}$$

where V is the volume of the magnetic piece, and M_s is the magnetization of the magnet. To generate a large force, it is advantageous to use a spatially dispersed magnetic field to increase the gradient.

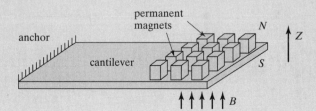

FIGURE 8.18

A bidirectional magnetic actuator.

The force acting on the surface of the electroplated film is 50 μN at the input current of 100 mA, which results in an 88-μm deflection. The cantilever beam is 6 mm long, 1 mm wide, and 13 μm thick. The maximum deflection was 80 μm at a 142-mW power consumption.

The permanent magnets are electroplated. During the electroplating process, a biasing magnetic field is provided by permanent magnets with a 3900-Gauss flux density measured at the surface. The bias improves the magnetic coercivity from 30–40 kA/m to 87.6 kA/m. The remnant magnetization improves from 50–80 mT to 170–190 mT.

Case 8.6. Hybrid Magnetic Actuator with Position Holding

Magnetic actuation with a hybrid magnetic source has been used to achieve a latchable, bistable electrical switch [27].

MEMS technology has been used to realize switches and relays. A variety of techniques have been investigated. For example, electrostatic actuators must require a constant voltage bias to hold the switch in the ON or OFF position. A switch with a bistable latching action is important because it only consumes power during the transition but does not hold the switch in either the ON or OFF position. This makes them susceptible to power interruption.

The structure and principle of a bistable magnetic switch is illustrated in Figure 8.19. A cantilever is elevated above the substrate surface by a torsion support. The cantilever consists of soft ferromagnetic material (Permalloy, in this case) on top and a layer of high-conductivity metal (gold) on the bottom (for electrical contact purposes). The biasing magnetic field is contributed by two sources—a planar coil (on-chip) and a permanent magnet (off-chip). A planar coil is embedded beneath the cantilever. A permanent magnet located on the back side of the silicon provides a constant background magnetic field, H_0.

The length of the cantilever is much greater than its width and thickness. When it is magnetized inside an external magnetic field, the internal magnetization (labeled M) is always in the longitudinal direction due to shape anisotropy. The interaction between the internal magnetization (M) and the external magnetic field creates a torque. However, the internal magnetization has two stable directions, due to the initial alignment between the cantilever and the external magnetic field. Depending on the direction of M, the torque can be either clockwise or counterclockwise. Both angular positions, corresponding to OFF and ON states, are stable.

The unique design of this switch is the fact that the bidirectional magnetization can be momentarily reversed by using a second magnetic field. This allows the torque and the position of the switch to be switched by supplying a small current. Towards this end, a planar coil situated between the cantilever and the external magnet is used to generate a magnetic field to compensate for the field created by the external magnet. The permanent magnet holds the cantilever in that position until the next switching event is applied.

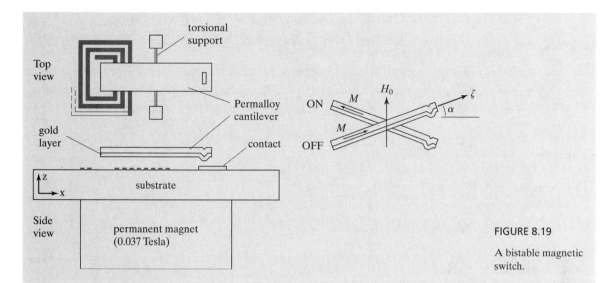

FIGURE 8.19

A bistable magnetic switch.

The magnitude of the applied torque is expressed as

$$\tau = M \times B_0 = \mu_0 M \times H_0. \tag{8.15}$$

For a torsion beam with a length of 600 μm, width of 10 μm, thickness of 1 μm, and Young's modulus of 200 GPa, the magnetic torque at $\alpha = 1.6°$ of bending is 8.4×10^{-10} N·m; this is 4.7 times greater than the elastic restoring force created by the torsion hinges.

How can the planar magnet create a sufficient magnetic field to toggle the orientations of the M? The planar coil generates magnetic field lines pointing in both the X and Z-axes. The magnetic field in the X-axis is mainly responsible for the switching. Using computational and analytical approaches, the magnetic field in the X-axis is estimated for both ON and OFF cantilever positions. The average field strength is 0.001 and 0.002 Tesla, greater than the X-axis component of the external field created by the permanent magnet ($\mu_0 H_0 \sin \alpha$).

The maximum DC current is greater than 500 mA. Lifetime tests show that a switch operated for 4.8 million cycles in ambient conditions did not have any observable damage on the contact region with an ON current of 200 μA.

SUMMARY

At the end of this chapter, the reader should understand the following concepts and facts:

- The relation between internal magnetization and the external magnetic field.
- The magnetization hysteresis curve for hard and soft magnetic materials.
- The properties and preparation methods of commonly used magnetic materials in MEMS.

- The wafer preparation procedure and equipment setup for electroplating.
- The analysis of the magnetic force on a piece of magnetized magnetic material that is placed within a magnetic field.
- The analysis of magnetic torque on a piece of magnetic material inside a magnetic field.
- The analysis of the angular bending of a cantilever carrying a piece of magnetic material.
- Design and fabrication methods for magnetic coils with single and multiple turns.

PROBLEMS

Problem 8.1: Review
Prove the consistency of units (in the SI system) in Equation (8.1).

Problem 8.2: Review
Prove the unit consistency of Equation (8.8).

Problem 8.3: Review
Prove that the unit of Tesla is $kg/(s^s A)$.

Problem 8.4: Review
Prove the consistency of the unit for Equation (8.14).

Problem 8.5: Design
A piece of permanent magnetic material (with a saturation magnetization of 1 Tesla) supported by torsional bars is placed inside an external magnetic field. At rest, the magnetic piece rests in a horizontal plane. The magnetic field lines point in the vertical direction. The magnetic piece is displaced at an angle with respect to the magnetic field. The bar is 400 μm long, 30 μm thick, and 30 μm wide. Two torsional bars are connected to the middle of the length of the bar. Each bar is made of polycrystalline silicon. The dimensions of each bar are 300 μm (length), 5 μm (width), and 1 μm (thickness). Find the magnitude of the external magnetic field (H) for two values of the bending angle θ ($\theta < 90°$): 20° and 45°. We assume shear modulus of polycrystalline silicon from the MUMPS process is 69 GPa.

Problem 8.6: Design
A 1-mm-long polysilicon wire with a width and thickness of 2 μm is fixed on both ends. Find the maximum displacement occurring on the wire when the applied transverse magnetic field is 0.1 T. A DC current of 1 mA is applied to the wire.

Problem 8.7: Design
Determine the magnetic field generated by a solenoid with multiple wire winding. The number of windings is 200, while the core has a cross-sectional dimension of 200 μm by 50 μm. The current is 10 mA. The diameter of the wire is 10 μm.

Problem 8.8: Design
Determine the magnetic field generated by a 10-turn planar inductor with a planar magnetic core. The magnetic core is 100 μm in diameter and 100 μm tall. The current is 10 mA. The wire winding begins at a diameter of 110 μm, with the increment between windings being 20 μm. The width and thickness of the wire are 16 μm and 10 μm, respectively. State your assumptions clearly.

Problem 8.9: Fabrication

Develop the fabrication process of a magnetic flap actuator similar to the example in Case 8.2, but use polyimide as the structural layer material. The cantilever may consist of multiple layers of polyimide in order to cancel out any intrinsic bending.

Problem 8.10: Design

Derive the analytical formula for calculating the magnitude of the angular displacement of the mirror based on known current and B in Case 8.3.

Problem 8.11: Design

Design an actuator device using inspiration from two cases—use processes discussed in Case 8.3 but magnetic actuator configuration of Case 8.4. Draw a schematic front and cross-sectional view of the new device, illustrating pertinent parts and materials. Explain the principle of operation. Identify major processing steps using a simplified process flow.

Problem 8.12: Design

Develop an analytical expression for the torque generated by the actuator of Case 8.4 when two adjacent coils are biased in the same direction of H and the other two are unbiased. Under a magnetic field of 0.3 T and a current of 100 mA, the desired angular displacement is 30°.

Problem 8.13: Design

Based on the calculated force acting on the cantilever (59 μN) of Case 8.5, find the amount of vertical displacement.

Problem 8.14: Design

Calculate the magnitude of the magnetic field gradient encountered in Case 8.5.

Problem 8.15: Fabrication

Discuss the detailed fabrication process for the bidirectional magnetic actuator of Case 8.5. Include details of lithography steps.

Problem 8.16: Fabrication

Draw and explain detailed fabrication process for the switch discussed in Case 8.6. Indicate all materials involved in the steps. Include details of photolithography steps.

REFERENCES

1. Ripka, P., *Magnetic Sensors and Magnetometers*. Norwood, MA: Artech House, 2001.
2. Niarchos, D., *Magnetic MEMS: Key Issues and Some Applications*, Sensors and Actuators A: Physical, 2003. **109**: p. 166–173.
3. Wagner, B., M. Kreutzer, and W. Benecke, *Permanent Magnet Micromotors on Silicon Substrates*, Microelectromechanical Systems, Journal of, 1993. **2**: p. 23–29.
4. Cho, H.J., and C.H. Ahn, *A Bidirectional Magnetic Microactuator Using Electroplated Permanent Magnet Arrays*, Microelectromechanical Systems, Journal of, 2002. **11**: p. 78–84.
5. Judy, J.W., R.S. Muller, and H.H. Zappe, *Magnetic Microactuation of Polysilicon Flexure Structures*, Microelectromechanical Systems, Journal of, 1995. **4**: p. 162–169.
6. Miki, N., and I. Shimoyama, *Soft-Magnetic Rotational Microwings in an Alternating Magnetic Field Applicable to Microflight Mechanisms*, Microelectromechanical Systems, Journal of, 2003. **12**: p. 221–227.

7. Lu, L.-H., K.S. Ryu, and C. Liu, *A Magnetic Microstirrer and Array for Microfluidic Mixing*, Microelectromechanical Systems, Journal of, 2002. **11**: p. 462–469.
8. Mallinson, J., *Magnetoresistive Heads—Fundamentals and Applications*. New York: Academic Press, 1996.
9. Taylor, W.P., et al., *Electroplated Soft Magnetic Materials for Microsensors and Microactuators*. Presented at International Conference on Solid State Sensors and Actuators, Transducers '97. Chicago: 1997.
10. Park, J.Y., and M.G. Allen, *A Comparison of Micromachined Inductors with Different Magnetic Core Materials*. Presented at Electronic Components and Technology Conference, 46th proceedings 1996.
11. Leith, S.D., and D. T. Schwartz, *High-Rate Through-Mold Electrodeposition of Thick 200 μm NiFe MEMS Components with Uniform Composition*, Microelectromechanical Systems, Journal of, 1999. **8**: p. 384–392.
12. Ahn, C.H., and M.G. Allen, *A New Toroidal-Meander Type Integrated Inductor with a Multilevel Meander Magnetic Core*, Magnetics, IEEE Transactions, 1994. **30**: p. 73–79.
13. Liakopoulos, T.M., W. Zhang, and C.H. Ahn, *Micromachined Thick Permanent Magnet Arrays on Silicon Wafers*, Magnetics, IEEE Transactions, 1996. **32**: p. 5154–5156.
14. Cho, H.J., S. Bhansali, and C.H. Ahn, *Electroplated Thick Permanent Magnet Arrays with Controlled Direction of Magnetization for MEMS Application*, The Electro-Chemical Society (ECS), 2000.
15. Lagorce, L.K., O. Brand, and M.G. Allen, *Magnetic Microactuators Based on Polymer Magnets*, Microelectromechanical Systems, Journal of, 1999. **8**: p. 2–9.
16. Ahn, C.H., and M.G. Allen, *Micromachined Planar Inductors on Silicon Wafers for MEMS Applications*, Industrial Electronics, IEEE Transactions, 1998. **45**: p. 866–876.
17. Wright, J.A., Y.-C. Tai, and S.-C. Chang, *A Large-Force, Fully-Integrated MEMS Magnetic Actuator*. Presented at International Conference on Solid State Sensors and Actuators, Transducers '97. Chicago: 1997.
18. Rogers, J.A., R.J. Jackman, and G.M. Whitesides, *Constructing Single- and Multiple-Helical Microcoils and Characterizing Their Performance as Components of Microinductors and Microelectromagnets*, Microelectromechanical Systems, Journal of, 1997. **6**: p. 184–192.
19. Zou, J., et al., *Development of Three-Dimensional Inductors Using Plastic Deformation Magnetic Assembly (PDMS)*, IEEE Transactions on Microwave Theory and Techniques, 2003. **51**: p. 1067–1075.
20. Young, D.J., et al., *A Low-Noise RF Voltage-Controlled Oscillator Using On-Chip High-Q Three-Dimensional Coil Inductor and Micromachined Variable Capacitor*. In Technical Digest, IEEE Solid-State Sensor and Actuator Workshop. Hilton Head, SC: 1998.
21. Scott, K.L., et al., *High-Performance Inductors Using Capillary Based Fluidic Self-Assembly*, Microelectromechanical Systems, Journal of, 2004. **13**: p. 300–309.
22. Bhansali, S., et al., *Prototype Feedback-Controlled Bidirectional Actuation System for MEMS Applications*, Microelectromechanical Systems, Journal of, 2000. **9**: p. 245–251.
23. Ahn, C.H., Y.J. Kim, and M.G. Allen, *A Planar Variable Reluctance Magnetic Micromotor with Fully Integrated Stator and Coils*, Microelectromechanical Systems, Journal of, 1993. **2**: p. 165–173.
24. Liu, C., et al., *Out-of-Plane Magnetic Actuators with Electroplated Permalloy for Fluid Dynamics Control*, Sensors and Actuators A: Physical, 1999. **78**: p. 190–197.
25. Miyajima, H., et al., *A Durable, Shock-Resistant Electromagnetic Optical Scanner with Polyimide-Based Hinges*, Microelectromechanical Systems, Journal of, 2001. **10**: p. 418–424.
26. Bernstein, J.J., et al., *Electromagnetically Actuated Mirror Arrays for Use in 3-D Optical Switching Applications*, Microelectromechanical Systems, Journal of, 2004. **13**: p. 526–535.
27. Ruan, M., J. Shen, and C.B. Wheeler, *Latching Micromagnetic Relays*, Microelectromechanical Systems, Journal of, 2001. **10**: p. 511–517.

CHAPTER 9

Summary of Sensing and Actuation

9.0 PREVIEW

The developer of microsensors and actuators faces many choices and obstacles. The selection of the sensing and actuation method is dependent on many factors, including performance, stability, reliability, energy consumption, cost and complexity of instrumentation, cost of development, and cost of ownership. It is rare that a sensing or actuation mechanism proves to be advantageous for all selection criteria. Often, relative minor concerns during the research phase, such as the temperature stability of performance, become major concerns in the commercialization phase. When selecting a transduction method or material, its ability to meet all primary and secondary criteria must be considered. The ability to select the correct principles and materials grows with the reader's experiences. The selection is not an exact science and must be dealt with on an individual basis. In Section 8.1, a first-order comparison of various major sensing and actuation methods is presented along several common criteria.

The sensing methods outlined in Chapters 3 through 6 are the most frequently encountered. However, many other transduction principles are used for unique applications. These transduction principles can enhance certain performance specifications and/or reduce instrumentation complexity. In this chapter, we will review a number of other sensing principles in Sections 9.1 through 9.4. The use of these sensing principles has been demonstrated by at least a few research groups and industrial laboratories, or they have been used in commercial devices. Interested readers are encouraged to explore literature pertaining to these topics. The discussion of alternative actuation methods is beyond the scope of this book. Interested readers should refer to Chapter 1.

9.1 COMPARISON OF MAJOR SENSING AND ACTUATION METHODS

The relative advantages and disadvantages of electrostatic sensing, thermal sensing, piezoresistive sensing, and piezoelectric sensing are summarized in Table 9.1. Often, the choice of a sensing principle is not just based on sensitivity. The sensitivity, which is dependent on noise, is just one of many points to consider.

The relative advantages and disadvantages of electrostatic actuation, thermal actuation, piezoelectric actuation, and magnetic actuation are summarized in Table 9.2.

TABLE 9.1 Comparison of Various Sensing Methods.

	Advantages	Disadvantages
Electrostatic sensing	• Simplicity of materials • Low-voltage, low-current operation • Rapid response	• Large footprint of device necessary to provide sufficient capacitance • Complexity of readout electronics • Sensitive to particles and humidity
Thermal sensing	• Simplicity of materials • Elimination of moving parts	• Relatively large power consumption • Generally slower response than electrostatic sensing
Piezoresistive sensing	• High sensitivity achievable • Simplicity of materials (metal strain gauge)	• Requires doping of silicon to achieve high-performance piezoresistors • Sensitive to environmental temperature changes
Piezoelectric sensing	• Self generating—no power necessary	• Complex material growth and process flow • Cannot sustain high-temperature operations

TABLE 9.2 Comparison of Various Actuation Methods.

	Advantages	Disadvantages
Electrostatic actuation	• Simplicity of materials • Fast actuation response	• Trade-off between magnitude of force and displacement • Susceptible to pull-in limitation
Thermal actuation	• Capable of achieving large displacement (angular or linear) • Moderately fast actuation response	• Relatively large power consumption • Sensitivity to environmental temperature changes
Piezoelectric actuation	• Fast response possible • Capable of achieving moderately large displacement	• Requires complex material preparation • Degraded performance at low frequencies
Magnetic actuation	• Capable of generating large angular displacement • Possibility of using very strong magnetic force as bias	• Moderately complex processes • Difficulty to form on-chip, high-efficiency solenoids

9.2 TUNNELING SENSING

Electron tunneling has been studied as an important displacement transducer because of its high sensitivity [1]. Under normal circumstances, electric current would not pass through insulators such as air or dielectrics. Electrons simply must acquire sufficient energy before they can cross the energy barrier that corresponds to the work function difference between the insulator and the conductors (Figure 9.1). However, when the distance between two electrodes reaches nanoscopic scale (e.g., 1 nm), electrons can move across the energy barrier by a quantum mechanical principle called tunneling.

The tunneling current between a tip and an opposing surface is generally represented as

$$I \propto V \exp^{(-\beta\sqrt{\phi}z)}, \qquad (9.1)$$

where V is a bias voltage, β a conversion factor with a typical value of $10.25(eV)^{-1/2}$/nm, ϕ the tunnel barrier height with the unit being electron volts (eV), and z the separation between conducting surfaces (typically on the order of 1 nm). The magnitude of the current is on the order of nA. For gold electrodes, one study has found the barrier height to be between 0.05 and 0.5 eV, depending on the cleanliness of the electrodes [2, 3].

For typical values of Φ and z, the current varies by an order of magnitude for each Å change in electrode separation. A 1% fluctuation in the tunneling current corresponds to the displacement of 0.003Å. If the detection were limited by shot noise in the tunnel current, the minimum detectable deflection would be 1.2×10^{-5} Å/$\sqrt{Hz}$. This sensitivity is independent of the lateral dimensions of the electrodes since tunneling only requires one metal atom on the surface of each side of the gap.

The tunneling phenomenon has been used to characterize surfaces with atomic resolution. The invention of the first scanning tunneling microscope was made by Heinrich Rohrer and Gerd Karl Binnig of IBM. The STM can image atomic details as tiny as 1/25th the diameter of a typical atom, which corresponds to a resolution several orders of magnitude better than the best electron microscope. The STM's significance was quickly recognized, and it has been used in fields as diverse as semiconductor science, metallurgy, electrochemistry, and molecular biology. More discussions on the topic of STM and the Scanning Probe Microscope (SPM) instrument can be found in Chapter 14.

Tunneling-based sensors have been developed for a number of applications. With a very large current-to-distance gain, they offer high resolution under a small device area. The tunneling phenomenon has been used to sense the distance between two conductors, and in turn, any physical phenomenon that may cause the distance to change. Published works on tunneling-based sensors have included force sensors [3], infrared sensors [4, 5], magnetometers [1], accelerometers [2, 3, 6], and pressure sensors [1]. High sensitivities are achieved for displacement (2×10^{-11} Å/$\sqrt{Hz}$ at 1 kHz, [2]), force (10^{-11} N/$\sqrt{Hz}$, [3]), infrared absorption (3×10^{-10}

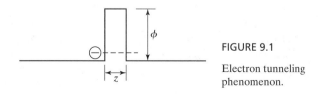

FIGURE 9.1

Electron tunneling phenomenon.

W/$\sqrt{Hz}$ at 25 Hz, [4]), and acceleration. High-sensitivity motion sensors have been developed for monitoring seismic activities. Acceleration sensors based on the tunneling principle have been demonstrated with sensitivity on the order of 1×10^{-7} g/$\sqrt{Hz}$ at a frequency of 10 kHz [2], which corresponds to a displacement sensitivity of 2×10^{-4} Å/$\sqrt{Hz}$. The sensitivity is three orders of magnitude better than that achievable with capacitive sensing. At low frequencies (below 1 kHz), the *1/f* noise dominates. At high frequencies, the shot noise and Johnson noise dominate. One example is reviewed in Case 9.1.

Although the tunneling sensing is extremely sensitive, it presents three major challenges to experimentalists—noise, cross sensitivity, and instrumentation complexity. Tunneling-based sensors are so sensitive that, in early experiments, they picked up unexpected interferences such as the vibration from the central air conditioning systems of large buildings or people walking several floors away. The high sensitivity comes at the price of the complexity of fabrication, packaging, and circuitry.

Long-term stability of tunneling-based sensors is also a major concern. For example, the tunnel current theoretically occurs between two metal atoms that, over time, may rearrange their positions due to Brownian motion or the chemical reaction with the environment. The migration of adsorbed molecules, such as water molecules, may also affect the tunneling characteristics in unexpected ways. Other possible sources of long-term drift include the variation of the mechanical characteristics of the mechanical microstructure (e.g., sagging or thermal expansion) and of the electrical properties of circuits. Early studies strongly suggest that the temperature coefficients of the mechanical properties of the sensor or the package contribute 95% of the noise at low frequency (below 0.1 Hz) [3].

For the best performance, tunneling-based devices should operate in a closed-loop mode. There are several reasons: (1) Due to the large current-to-displacement gain, the measurement range is limited if the device operates in the open-loop mode. (2) The height of the tunneling barrier may vary by one order of magnitude over time in air, thus affecting the open-loop sensitivity. The closed-loop control is usually accomplished by converting the tunneling current into voltage and applying correction signals to an actuator. A number of circuit strategies have been identified for implementing closed-loop control with integrated circuits [3, 6].

The characteristics of the electromechanical actuator used to control the separation between the tunneling electrodes often impose the dominant limitations on the performance of the tunneling sensor system. The actuator must have adequate bandwidth, dynamic range, and precision.

Case 9.1. Tunneling Accelerometer

Figure 9.2 shows a tunneling accelerometer with active capacitive actuators to control the gap in a closed-loop fashion [7]. The device incorporates two levels of silicon and one of glass. The first silicon piece, labeled silicon 2 in Figure 9.2, contains one proof mass with an integrated tunneling tip. The mass is supported by beams. The second silicon piece, labeled silicon 1 in Figure 9.2, is fixed and forms a parallel-plate capacitor with the top surface of the

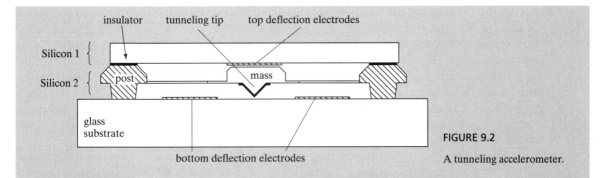

FIGURE 9.2
A tunneling accelerometer.

proof mass in the first silicon. The capacitor is used to generate acceleration for self-testing and for closed-loop gap control.

The active area of the proof mass is 400×400 μm^2. A CMOS circuit is used with the tunnel device regarded as a non-linear variable resistor. The resistor varies with the tunneling distance. Three variations of circuits were developed. These circuits provide different levels of power consumption and noise floor [6].

The accelerometer has a sensitivity of 125 mV/g, a bandwidth of 2.5 kHz, and a dynamic range of 30 g. The 1/f noise equivalent acceleration was 1 mg/$\sqrt{Hz}$. The drift over a 1-month operation was less than 0.5%.

9.3 OPTICAL SENSING

Optical sensing turns angular or translational displacement into changes of optical intensity or phase. Optical sensing offers a number of advantages for MEMS sensors applications:

1. The optical interrogation of the movement of microstructures eliminates the need for conductive wires associated with electrical biasing and sensing. This point is especially important if optics are used to address a large array of devices. The savings due to the reduced packaging complexity is dramatic in selected cases.
2. As we will show later, optical sensing also offers uniquely high sensitivity and versatility in many cases. Very high-sensitivity optical detectors can be obtained off the shelf and at low cost.

Certainly, optical sensing cannot be applied or involved in all cases. Certain applications preclude the use of external or internal light sources, for example, due to finite package size. Optical position sensing can be achieved in several configurations. A few major configurations are reviewed in Sections 9.3.1 through 9.3.3.

9.3.1 Sensing with Waveguides

Light can travel in man-made waveguides. An optical fiber is a form of a low-cost, highly efficient optical waveguide. A fiber generally consists of an inner core and an outer sheath. The refractive index of the core and sheath are different, allowing a total reflection of light if it shines at the interface from the core side. Hence, a light beam can be physically confined within the core region, and it can travel with little loss even when the fiber is curved.

Optical fibers form the basis of many sensors [8]. Fiber-based sensing takes advantage of the fact that the phase and intensity of light in fiber is a function of fiber bending, mechanical stress on fiber, temperature [9], surface optical properties, and interaction with chemistry and biological entities [10, 11]. For example, a segment of fiber is shown in Figure 9.3 to illustrate the principle of sensing fiber bending. If the fiber is straight (unstressed state), light will go through a certain optical path length, which is different from the physical length of the fiber. However, if the fiber is mechanically bent, the new effective optical length causes the phase and intensity of light at the output end of the fiber to change. This principle has been used for a variety of sensor applications [12, 13].

Optical fibers are formed by pulling glass materials at elevated temperatures. Advanced features, such as longitudinal microchannels [14] and non-circular cores [15], can be incorporated to introduce new ways to interact with light.

Sensing using on-chip integrated waveguides is also possible. An accelerometer with an integrated optical waveguide is made of mixed silicon oxide and silicon nitride thin films patterned into a linear guide structure. Part of the guide is located on a proof mass while the rest is on the frame. Acceleration-induced displacements change the optical light coupling coefficient by shifting the waveguide on the proof mass with respect to the frame [16]. The detection limit demonstrated was 0.17 μm, which corresponds to 1.7 dB/g for positive acceleration and 2.3 dB/g for negative acceleration.

9.3.2 Sensing with Free-Space Light Beams

Free-space light beams can be used to detect positions of objects and sense a phenomenon that causes that position change. The most straightforward configuration involves a beam of light bouncing off the reflective surface of a microstructure of interest such as the back side of a cantilever (Figure 9.4). The reflected beam is directed into a photodiode or onto a projection screen. If angular bending of the cantilever occurs, the reflected spot will move. The displacement of the reflected spot is proportional to the degree of cantilever bending. Specifically, the distance by which the spot moves equals the product of the angular displacement and the distance between the cantilever and the screen. The displacement of the reflected spot is

$$d = 2\theta \cdot L, \tag{9.2}$$

where L is the distance between the device under test and the photodetector. Since the angular displacement is amplified by the distance L, this principle is commonly referred to as an "optical lever."

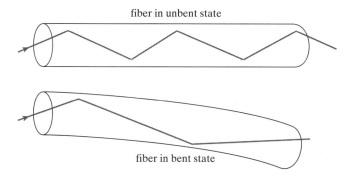

FIGURE 9.3

Optical-fiber sensing.

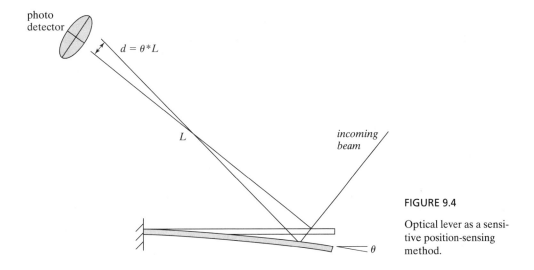

FIGURE 9.4

Optical lever as a sensitive position-sensing method.

9.3.3 Position Sensing with Optical Interferometry

One of the most sensitive optical techniques for measuring the displacement of a cantilever is the interferometer. Optical interference provides a highly accurate measurement of relative movement between a reference body and a moving body. Interferometric measurement can take many forms, including a Michelson interferometer [17], a Febry–Perot cavity [18], or interdigitated fingers as diffraction gratings [19, 20]. A displacement resolution at 0.01 Å or lower can be achieved. Optical interference transducers are capable of achieving a position resolution equivalent to electron tunneling transducers, with all practical noise sources considered.

Optical measurement requires an external optical source and receivers. In many cases, they are ideal for the precision characterization of micromachined devices, regardless of the actual sensing method. The comparison of the displacement sensing by electrostatic and interferometric methods has been conducted [21, 22]. In another example, the optical interferometry has been used to characterize the responses of a gyro for automotive applications [23].

A few examples of sensors based on optical interferometry are discussed next. They rely on gratings to produce diffraction. These applications include accelerometers (Case 9.2), cantilever displacement sensors (Case 9.3), and pressure sensors (Case 9.4).

Case 9.2. Interferometric Accelerometer

An acceleration sensor with nano-g resolution consists of a bulk silicon micromachined proof mass with interdigitated fingers that are alternatively connected to the proof mass and the substrate [20]. The geometry of the fingers forms a phase-sensitive optical diffraction grating, which reflects the incident coherent optical beam into several orders with an intensity that depends on the relative displacement between the two sets of fingers (Figure 9.5). In the equilibrium position, where the relative deflection of moving fingers is zero, the intensities of even-numbered orders are at their maximum. The spatial separation of the second-order component from the central, zeroth-order component is $\lambda D f_g$, where f_g is the spatial frequency of the grating, D the observation distance, and λ the wavelength of the incident beam.

When the moving fingers are displaced by $\lambda/4$, the central beam spot vanishes and the energy is divided between the two first-order components and other odd-numbered components.

The mechanical deflection can be determined by measuring the intensity of the zeroth-mode spot, first-mode spot, or the difference between the two modes. The intensity of the diffracted modes depends on the out-of-plane offset between the two sets of fingers (d) and is given by

$$I(d) = I_0 \sin^2\left(\frac{2\pi d}{\lambda}\right) \tag{9.3}$$

with λ being the illumination wavelength. Using an interferometer, the deflection of the cantilever can be resolved to 0.003 Å at 10 Hz.

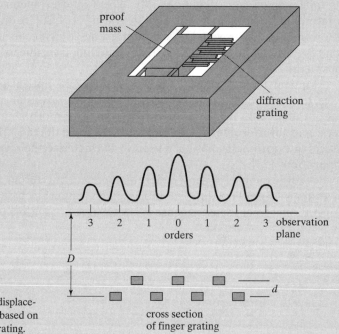

FIGURE 9.5

Principle of displacement sensor based on diffraction grating.

In the design, the 50 fingers (each 175 μm long, 6 μm wide, and 20 μm in thickness) are separated by gaps of 3 μm. The fingers on the proof mass overlap with the fingers on the support substrate for 125 μm, giving an area of 450 μm × 125 μm in which to focus the laser.

The optical source is a 670-nm wavelength, 5-mW laser diode with a focusing lens. The 80-Hz resonant proof mass has a noise equivalent acceleration of 40 ng/$\sqrt{Hz}$ and a dynamic range of 85 dB at 40 Hz. This is at least identical to, if not better than, the ones obtained using tunneling sensing.

The main sources of noise are the shot noise of the photodetector, the thermal mechanical noise of the cantilever, the laser intensity noise, the laser phase noise, the laser 1/f noise, the resistor Johnson noise, the electronic noise of the detection electronics, and the mechanical vibration of the whole system. Elemental noise components have been discussed and characterized in Reference [24]. The measured noise of the device (0.02 Å) is large compared with the predicted thermal mechanical noise [25] and shot noise. A significant contribution of noise may be the intensity fluctuations of the laser. The noise floor can be further lowered if the laser intensity fluctuation is detected and subtracted.

Case 9.3. Cantilever with Optical Interference Position Sensing

An interferometer integrated with a cantilever is demonstrated to report the displacement of atomic force microscope probes with approximately 0.02 Å noise (rms) in the 10 Hz–1 kHz bandwidth [19]. A modified two-dimensional grating can also be used to measure the minute relative displacement of two neighboring beams [26] with applications in biological and chemical sensing.

A part of the cantilever is micromachined into the shape of interdigitated fingers (Fig. 9.6). One set of fingers is attached to the movable cantilever whereas a matching set is fixed to the device frame. When the cantilever is illuminated, the fingers form a phase-sensitive diffraction grating, and the tip displacement is determined by measuring the intensity of the diffracted modes. When a force acts on the cantilever, only the alternating fingers that are connected to the outer portion of the cantilever are vertically displaced. The remaining set of fingers, or reference fingers, are attached to the inner portion of the cantilever and remain fixed.

An illumination source (a laser diode with the wavelength = 670 nm) and a standard photodiode are used with the light passing through the fingers. The dominant reflected mode from the grating is the zeroth mode. As the tip is displaced, the interference between the light reflecting off the reference fingers and the moving fingers causes the zeroth mode intensity to decrease while a first mode is enhanced. When the cantilever is deflected by an amount of λ/4, the zeroth mode is minimized and the first mode is maximized. The cantilever deflection can be measured from the intensity of the zeroth mode, first mode, or the difference between the modes.

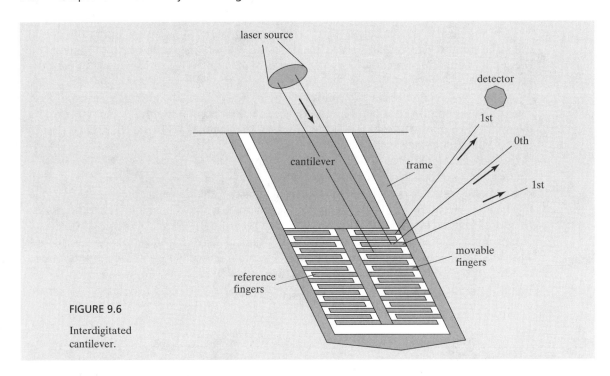

FIGURE 9.6

Interdigitated cantilever.

Case 9.4. Membrane Displacement Sensor with Interferometric Sensing

Acoustic sensors can be based on many principles, as we have illustrated in previous chapters. Here, we compare two surface micromachined capacitors: one is based on capacitive sensing and one is based on acoustic sensing.

Capacitive micromachined acoustic sensors (CMUT) have been demonstrated as an alternative to piezoelectric sensors [27]. The CMUT device (shown in Figure 9.7) consists of a metallized silicon nitride membrane. In its receiving mode, acoustic waves impinging on the membrane cause displacement of the membrane. The capacitance between the membrane and the underlying counter electrode changes as a result. The output current i of the CMUT in response to a membrane displacement Δx is given by

$$i = w_a V_{bias} C \frac{\Delta x}{d_0}, \qquad (9.4)$$

where w_a is the angular frequency of the acoustic wave, V_{bias} the DC bias voltage, C the CMUT capacitance, and d_0 the gap between the membrane and the substrate. The output current is proportional to the frequency, resulting in a relatively poor performance in the low-frequency range. The fabrication process involves the surface micromachining of suspended conductive membranes. Parasitic capacitances will affect the measurement.

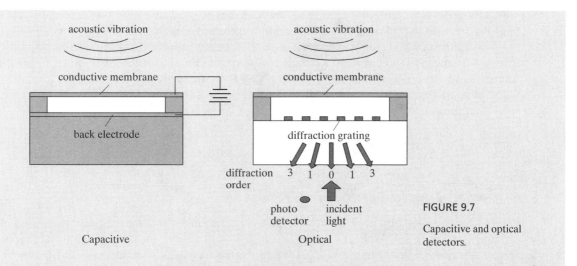

FIGURE 9.7

Capacitive and optical detectors.

Alternatively, the displacement of a membrane can be detected using optical interferometry. An optical diffraction grating has been integrated with a surface micromachined membrane to detect membrane displacement [28]. The device is schematically illustrated in Figure 7. The substrate is made of a transparent material (quartz) to allow optical transmission. The reflective membrane and the grated electrode form a phase-sensitive diffraction grating. When the grating under the membrane is illuminated through the transparent substrate by a coherent light source, the reflected field will split into odd diffraction orders in addition to specular reflection (or zeroth order). Some of the incident light passes through the electrode grating and reflects back from the membrane. This produces interference from the diffracted and reflected light, and it changes the intensity of the diffracted orders. The output intensity for a small displacement of Δx is obtained as

$$i = RI_{in}\frac{4\pi}{\lambda_0}\Delta x, \tag{9.5}$$

where R is the responsivity of the optical detector, and i_{in} is the intensity of the incoming light.

Using this method, optical detection can be used to achieve a very high sensitivity $(2 \times 10^{-4}\, \text{Å}/\sqrt{\text{Hz}})$ in the DC to 2 MHz range. In order to produce a response down to DC, the cavity must be sealed under a vacuum.

9.4 FIELD-EFFECT TRANSISTORS

Many integrated circuit components can be used for sensing applications. For example, it is well known that the device characteristics of solid-state devices such as transistors, diodes, and resistors exhibit temperature sensitivity. Indirectly, they can serve as temperature sensors. Characteristics of field-effect devices such as the field-effect transistor (FET) are also influenced by the strength of the electric field, which is a function of the distance between two moving electrodes with a fixed potential difference. As such, the FET can be used for measuring acceleration [29], pressure (sensitivity = 0.1 mA/bar) [30], and acoustic waves (sensitivity = 0.1–1 mV/Pa) [31].

In addition, the electrical characteristics of many integrated circuit elements can be affected by mechanical stress. Stress and strain applied in active device regions change the energy band structure; this is similar to the way that stress and strain cause the piezoresistivity of semiconductor materials. Here, let us review an accelerometer based on the FET transduction in Case 9.5.

Case 9.5. Displacement Using the Gate of FET

We first consider how an FET transistor works. The cross section diagram of an FET is shown in Figure 9.8. An FET consists of three electrical terminals: source, drain, and gate. Current flowing between the source and the drain is controlled by the potential applied to the gate. In the case of an n-type substrate, the majority of the charge carriers are electrons. When no voltage is applied to the polysilicon gate, current flow between the source and the drain regions is minimal as there are two back-to-back diodes involved. Regardless of the bias between the source and the drain, one of the diodes will invariably be reverse biased and limit the current flow capacity. A strong negative bias to the gate will locally invert the polarity of the major charge carriers in the region immediately under the gate region. In a simplistic view, holes are attracted to the surface by the bias voltage. This region of inverted polarity is called the channel; it helps the flow of the current between the source and drain.

The current between the source and the drain under a certain favorable voltage bias to the source, drain, and gate is

$$I_D = \frac{\overline{\mu} Z C_i}{L} \left[(V_G - V_T)V_D - \frac{1}{2} V_D^2 \right], \tag{9.6}$$

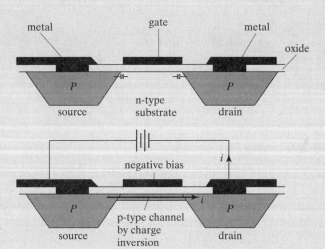

FIGURE 9.8

Operation principle of a regular transistor.

where $\bar{\mu}$ is the mobility of charge carriers, Z and L the width and length of the channel, C_i the capacitor per unit area of the dielectrics (including gate oxide and air media), V_G the voltage difference between gate and source, V_D the voltage difference between the gate and drain, and V_T the threshold voltage of the FET.

It is of interest to explore the mechanical sensing capabilities of such electronic components because they allow mechanical sensing to be directly coupled to electronic signal processing and logic circuits with minimal parasitics involved. Another advantage lies in the fact that the fabrication of the sensor is compatible with integrated circuits and can be carried in an IC foundry.

An acceleration sensor based on the field effector transistor is demonstrated in Reference [29]. The seismic mass of the accelerometer is the gate of an FET; consequently, the distance between the gate and the channel changes as a function of the applied acceleration (Fig. 9.9). This distance is equivalent to the typical oxide dielectric layer for the CMOS FET. As a consequence, the value of the threshold voltage V_T of the transistor changes.

Both the C_i and the V_T terms are functions of the air gap between the gate and the channel. The relation between these two terms and gap a is

$$C_i = \frac{1}{\frac{1}{C_{dielec}} + \frac{1}{C_{media}}} = \frac{1}{\frac{t_{dielec}}{\varepsilon_{diele}} + \frac{a}{\varepsilon_{media}}}, \tag{9.7}$$

$$V_T = \Phi_{ms} + 2\Phi_F - \frac{Q_i}{C_i} - \frac{Q_d}{C_i}, \tag{9.8}$$

where the subscripts *dielec* and *media* represent values assigned to the dielectric (e.g., the gate oxide) and the media (e.g., vacuum or air between the floating gate and the dielectric). The terms ε and t are the dielectric constant and the thickness of layers, Φ_{ms} is the difference of the work function between the gate and the semiconductor, Φ_F the flat-band voltage, Q_i the trapped charge (per unit area) in the dielectric material, and Q_d the accumulated charges in the channel.

The acceleration sensor uses a floating mass with four cantilever spring supports, where the thickness is 2 μm. The length of the support ranges from 290 μm to 350 μm, where the width is 5 μm in both cases. The size of the center plate ranges from 350 × 350 μm^2 to 300 × 300 μm^2. The mechanical sensitivity of the designs ranges from 11.5 nm/g to 4.6 nm/g. The air gap between the mass and the substrate is 1 μm. At zero displacement, the threshold voltage is 10 V. The channels are 5 μm long and 10 μm wide.

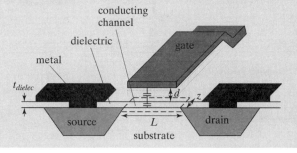

FIGURE 9.9

Position sensing principle.

9.5 RADIO FREQUENCY RESONANCE SENSING

In many sensing applications, the signals are represented by the magnitude of voltage or current. However, signals can also be represented by the frequency of output signals in a scheme called frequency modulation. For example, one can realize piezoresistive sensors with either voltage or frequency output [32]. One of the most important benefits of frequency modulation is the high noise immunity. The disadvantage of this scheme lies in the fact that the signal process electronics are more complex in order to decipher signals represented in the frequency domain. In Case 9.6, we discuss a pressure sensor whose output is encoded by the frequency of signals.

Case 9.6. Resonance Mode Pressure Sensor

The following example illustrates a pressure sensor based on passive wireless resonant telemetry [33]. A pressure-sensitive membrane made of low-temperature cofireable ceramics (LTCC) is covered with a spiral planar inductor. The center contact pad of the inductor is intentionally made very large in order to form an appreciable capacitor with an opposing electrode surface. If pressure changes, the membrane displaces, changing the relative capacitance value. The inductance value may also change due to the curving of

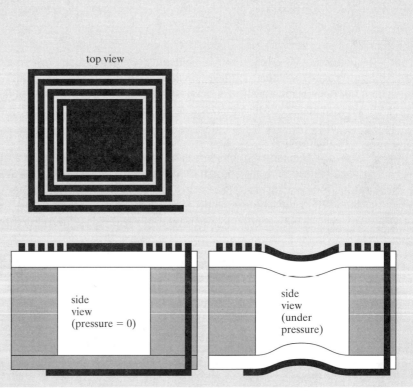

FIGURE 9.10

Principle of a pressure sensor using RF resonance sensing.

the membrane. The resonance frequency of the resonant circuit is related to pressure P, as the capacitance is given by

$$C_s(P) = C_0 \sum_{i=0}^{\infty} \frac{1}{2i+1} \left(\frac{2d_0}{t_g + 2t_m \varepsilon_r^{-1}} \right)^i, \tag{9.9}$$

where C_0 is the capacitance at zero pressure, d_0 the displacement of the membrane under pressure P, t_g and t_m the size of the gap and the thickness of the membrane, and ε_r the relative dielectric constant of the membrane. The authors state that the magnitude of d_0 is related to pressure by

$$\frac{d_0}{t_m} + 0.488 \left(\frac{d_0}{t_m} \right)^3 = \frac{3P(1 - \nu^2)}{16E} \left(\frac{a}{t_m} \right)^4, \tag{9.10}$$

where a is the radius of the membrane, E Young's modulus, and ν Poisson's ratio.

The measurement sensitivity and accuracy of a typical sensor is -141 kHz/bar and 24 mbar, respectively.

SUMMARY

The purpose of this chapter is twofold. First, this chapter provides a comparison among sensing principles discussed in earlier chapters. The same sensing task can be achieved using a number of transduction principles. The selection of a successful sensing method is case dependent. Major advantages and disadvantages of each method are presented.

The second purpose of this chapter is to highlight a few promising sensing principles. These principles showed a considerable performance advantage over electrostatic, piezoelectric, and piezoresistive sensing in important aspects. Often, they provide simplicity and low cost that is not available in other methods. These methods have been used in a number of applications but have not been used as widely as their performance warrants.

At the end of this chapter, the reader should understand the following concepts and facts:

- Major advantages and disadvantages associated with capacitive sensing, piezoresistive sensing, and piezoelectric sensing.
- Major advantages and disadvantages associated with capacitive actuation, piezoelectric actuation, and thermal actuation.
- The physical principle behind tunneling-based sensing.
- The optics principle behind optical interference position sensing.
- Qualitative behavior analysis of field-effect transistors.

PROBLEMS

Problem 9.1: Fabrication

Draw detailed fabrication process for the three individual layers in Case 9.1. Clearly mark all materials involved in each step.

Problem 9.2: Review

Review all acceleration sensors reviewed in case studies in Chapters 4, 5, 6, 7, and 9. Rank these sensors in terms of (1) sensitivity; (2) simplicity of fabrication; and (3) sensitivity to temperature. For each design, identify at least one major advantage and at least one major drawback.

First, identify all devices investigated for this work. Identify them by the case number and chapter number. Then label them numerically. Rank them according to the three criteria given. Explain your selection process for the best and worst in each category.

Problem 9.3: Review

Repeat Problem 9.3 for one or more of the following sensor types: membrane pressure sensors, flow speed sensors, and cantilever displacement sensors.

Problem 9.4: Design

An optical level measures the vertical displacement of a cantilever that is 100 μm long. The vertical displacement t corresponds to an angular displacement q. The length of the optical level is 1 mm. A disk-shaped quadrant optical detector is used with a radius of 1 cm. Find the range of the displacement that can be measured with this configuration.

Problem 9.5: Design

In Case 9.2, derive the analytical relationship between the intensity of the diffracted beam and the applied acceleration. Identify the design strategies for increasing the sensitivity to acceleration.

Problem 9.6: Review

In Case 9.2, identify and explain the simple design strategy that was used to minimize the cross-sensitivity to acceleration applied in other axes. (*Hint*: Refer to Figure 9.5.)

Problem 9.7: Design

Derive the analytical expression of the sensitivity as percentage change per unit displacement at the free-end of the structure in Case 9.3, between the optical sensing case, a hypothetical piezoresistive sensing case, and a hypothetical capacitive sensing case. The gap size for the capacitive sensing is d_0.

Problem 9.8: Fabrication

Draw the fabrication process for the membrane acoustic sensor (Case 9.4) based on capacitive sensing and optical sensing. The details of photolithography steps may be skipped. Discuss at least one major aspect in which the optical sensing have simplified the process and materials compared to the capacitive sensing method.

Problem 9.9: Challenge

Consider both the acoustic sensor in Case 9.4 and the resonance sensing structure in Case 9.6. Discuss the design and fabrication process of a surface micromachined pressure sensor using similar electronics sensing architecture as in Case 9.6 and a mechanical microstructure as in Case 9.4. Discuss the materials compatibility for each step.

Problem 9.10: Challenge

Design an actuator, based on any principle of actuation that has been discussed in the previous few chapters or outside of this textbook, that provide the largest possible displacement normal to the chip surface. A singular actuator is to occupy a chip space of no more than 50 μm $\times$ 50 μm, excluding conductive wire leads that may be extended from this region. The actuator must be based on practical materials that are suitable for microfabrication. There is an upper limit of allowable voltage, current, and power input, whichever is invoked first. The limits for voltage, current and power are: 100 V, 0.2 A, and 300 mW

Problem 9.11: Challenge

Repeat Problem 9.10, targeting designs that produce the largest possible angular displacement.

Problem 9.12: Challenge

Repeat Problem 9.10 targeting designs that produce the largest possible force under normal translational movement.

Problem 9.13: Challenge

Form a group of three to four students and complete Problem 9.10, 9.11, or 9.12. Participate in a class-wide competition.

REFERENCES

1. Grade, J., et al., *Progress in Tunnel Sensors*. Presented at Technical Digest, Solid-State Sensor and Actuators workshop. Hilton Head, SC: 1996.
2. Kenny, T.W., et al., *Micromachined Silicon Tunnel Sensor for Motion Detection*, Applied Physics Letters, 1991. **58**: p. 100–102.
3. Kenny, T.W., et al., *Wide-Bandwidth Electromechanical Actuators for Tunneling Displacement Transducers*, Microelectromechanical Systems, Journal of, 1994. **3**: p. 97–104.
4. Kenny, T.W., et al., *Micromachined Infrared Sensors Using Tunneling Displacement Transducers*, Review of Scientific Instruments, 1996. **67**: p. 112–128.
5. Grade, J., et al., *Wafer-Scale Processing, Assembly, and Testing of Tunneling Infrared Detectors*. Presented at International Conference on Solid State Sensors and Actuators, Transducers '97. Chicago: 1997.
6. Yeh, C., and K. Najafi, *CMOS Interface Circuitry for a Low-Voltage Micromachined Tunneling Accelerometer*, Microelectromechanical Systems, Journal of, 1998. **7**: p. 6–15.
7. Yeh, C., and K. Najafi, *Micromachined Tunneling Accelerometer with a Low-Voltage CMOS Interference Circuit*. Presented at 1997 International Conference on Solid-State Sensors and Actuators, Chicago, IL: 1997.
8. E. Udd, *Fiber-Optic Sensors*. In Wiley Series in Pure and Applied Optics. S.S. Ballard and J.W. Goodman, Eds. John Wiley and Sons, 1991.
9. Maurice, E., et al., *High Dynamic Range Temperature Point Sensor Using Green Fluorescence Intensity Ratio in Erbium-Doped Silica Fiber*, Lightwave Technology, Journal of, 1995. **13**: p. 1349–1353.

10. Mignani, A.G., and F. Baldini, *In-vivo Biomedical Monitoring by Fiber-Optic Systems*, Lightwave Technology, Journal of, 1995. **13**: p. 1396–1406.
11. Michie, W.C., et al., *Distributed pH and Water Detection Using Fiber-Optic Sensors and Hydrogels*, Lightwave Technology, Journal of, 1995. **13**: p. 1415–1420.
12. Knowles, S.F., et al., *Multiple Microbending Optical-Fiber Sensors for Measurement of Fuel Quantity in Aircraft Fuel Tanks*, Sensors and Actuators A: Physical, 1998. **68**: p. 320–323.
13. Luo, F., et al., *A Fiber-Optic Microbend Sensor for Distributed Sensing Application in the Structural Strain Monitoring*, Sensors and Actuators A: Physical, 1999. **75**: p. 41–44.
14. Mach, P., et al., *Tunable Microfluidic Optical Fiber*, Applied Physics Letters, 2002. **80**: p. 4294–4296.
15. Kopp, V.I., et al., *Chiral Fiber Gratings*, Science, 2004. **305**: p. 74–75.
16. Plaza, J.A., et al., *BESOI-Based Integrated Optical Silicon Accelerometer*, Microelectromechanical Systems, Journal of, 2004. **13**: p. 355–364.
17. Rugar, D., H.J. Mamin, and P. Guethner, *Improved Fiber-Optic Interferometer for Atomic Force Microscopy*, Applied Physics Letters, 1989. **55**: p. 2588–2590.
18. Stephens, M., *A Sensitive Interferometric Accelerometer*, Review of Scientific Instruments, 1993. **64**: p. 2612–2614.
19. Manalis, S.R., et al., *Interdigital Cantilevers for Atomic Force Microscopy*, Applied Physics Letters, 1996. **69**: p. 3944–3946.
20. Loh, N.C., M.A. Schmidt, and S.R. Manalis, *Sub-10 cm^3 Interferometric Accelerometer with Nano-g Resolution*, Microelectromechanical Systems, Journal of, 2002. **11**: p. 182–187.
21. Annovazzi-Lodi, V., S. Merlo, and M. Norgia, *Comparison of Capacitive and Feedback-Interferometric Measurements on MEMS*, Microelectromechanical Systems, Journal of, 2001. **10**: p. 327–335.
22. Jensen, B.D., et al., *Interferometry of Actuated Microcantilevers to Determine Material Properties and Test Structure Nonidealities in MEMS*, Microelectromechanical Systems, Journal of, 2001. **10**: p. 336–346.
23. Annovazzi-Lodi, V., et al., *Optical Detection of the Coriolis Force on a Silicon Micromachined Gyroscope*, Microelectromechanical Systems, Journal of, 2003. **12**: p. 540–549.
24. Yaralioglu, G.G., et al., *Analysis and Design of an Interdigital Cantilever as a Displacement Sensor*, Journal of Applied Physics, 1998. **83**: p. 7405–7415.
25. Gabrielson, T.B., *Mechanical-Thermal Noise in Micromachined Acoustic and Vibration Sensors*, Electron Devices, IEEE Transactions on, 1993. **40**: p. 903–909.
26. Savran, C.A., et al., *Fabrication and Characterization of a Micromechanical Sensor for Differential Detection of Nanoscale Motions*, Microelectromechanical Systems, Journal of, 2002. **11**: p. 703–708.
27. Jin, X., et al., *Fabrication and Characterization of Surface Micromachined Capacitive Ultrasonic Immersion Transducers*, Microelectromechanical Systems, Journal of, 1999. **8**: p. 100–114.
28. Hall, N.A., and F.L. Degertekin, *Integrated Optical Interferometric Detection Method for Micromachined Capacitive Acoustic Transducers*, Applied Physics Letters, 2002. **80**: p. 3859–3861.
29. Plaza, J.A., et al., *New FET Accelerometer Based on Surface Micromachining*, Sensors and Actuators A: Physical, 1997. **61**: p. 342–345.
30. Svensson, L., et al., *Surface Micromachining Technology Applied to the Fabrication of an FET Pressure Sensor*, Journal of Micromechanics and Microengineering, 1996. **6**: p. 80–83.

31. Kuhnel, W., *Silicon Condenser Microphone with Integrated Field-Effect Transistor*, Sensors and Actuators A: Physical, 1991. **26**: p. 521–525.
32. Sugiyama, S., M. Takigawa, and I. Igarashi, *Integrated Piezoresistive Pressure Sensor with Both Voltage and Frequency Output*, Sensors and Actuators A, 1983. **4**: p. 113–120.
33. Fonseca, M.A., et al., *Wireless Micromachined Ceramic Pressure Sensor for High-Temperature Applications*, Microelectromechanical Systems, Journal of, 2002. **11**: p. 337–343.

CHAPTER 10

Bulk Micromachining and Silicon Anisotropic Etching

10.0 PREVIEW

In this chapter, we will focus on discussing bulk micromachining technology. Section 10.1 outlines major terminology. In Section 10.2, we will discuss the anisotropic wet etching of silicon. We begin by reviewing the simplest case, and then we gradually increase the level of complexity. In Sections 10.3 through 10.6, we will review plasma etching, deep reactive ion etching, isotropic wet etching, and gas-phase etching. Native oxide is a commonly occurring material form on silicon and affects micromachining processes. We will discuss the behavior and processing techniques related to the native oxide in Section 10.7. Wafer bonding is a technique often used in conjunction with bulk etching to create more sophisticated microstructures. Such hybrid processes are discussed in Section 10.8. In Section 10.9, we will systematically review processing techniques for realizing two types of structures: suspended beams and plates, and suspended membranes.

10.1 INTRODUCTION

Bulk micromachining is an important class of the MEMS process [1]. In bulk micromachining processes, a portion of the substrate (bulk) is removed in order to create freestanding mechanical structures (such as beams and membranes) or unique three-dimensional features (such as cavities, through-wafer holes, and mesas). Bulk micromachining can be applied to silicon, glass, gallium arsenide, and other materials of interest. In this chapter, we focus on discussing bulk micromachining of *silicon* substrates.

There are two major categories of processes for bulk silicon etching according to the medium of the etchant: wet etching and dry etching. Wet silicon etching processes use liquid chemical solutions in contact with silicon. Dry etching processes use plasma (high-energy gas containing ionized radicals) or vapor-phase etchants to remove materials.

Bulk silicon etching can be classified according to the three-dimensional distribution of etch rates and the profiles of resultant microstructures. The etch rate of wet bulk etching may

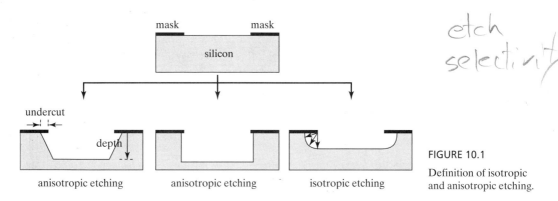

FIGURE 10.1 Definition of isotropic and anisotropic etching.

depend on crystal orientation. The etch rate of dry bulk etching may depend on the directions in a wafer. If the etch rates in all directions are identical, the etching is said to be **isotropic**. If the etch rate is orientation dependent, the etching is said to be **anisotropic**. Representative cross-sectional profiles of cavities created by isotropic and anisotropic bulk etching are shown in Figure 10.1.

As shown in Figure 10.1, etching action is not confined to the region exposed through the open areas of a mask. Significant material removal may occur underneath masked areas. The technical term for this lateral material removal is **undercut**. Undercut is desired in order to create suspended mechanical structures. For certain applications, undercut may be undesirable and should be minimized by proper mask design and careful process control.

The following is a list of important issues to consider when selecting a bulk micromachining process and etchant.

1. **Etch rate.** The speed of the material removal is important. A higher etch rate translates into shorter etching time and greater manufacturing throughput.
2. **Etch rate selectivity.** Selectivity is defined as the ratio between the etch rate of the targeted material and that of non-intended materials, such as mask layers. The selectivity ratio should be as large as possible. For example, an etchant should not attack the mask at all in the ideal situation.
3. **Processing temperature.** The temperatures of the bulk and of the etching medium are highly relevant. Process steps with a high temperature limit the selection of materials. The etch rate of many systems is temperature dependent.
4. **Etch uniformity across a wafer.** More often than not, the etch rate is non-uniform across a wafer surface. This complicates process control, especially when wafer sizes are large.
5. **Sensitivity to over-time etch.** Because of non-uniform etch rate over a wafer, some structures on a wafer are finished earlier than others. These structures that are finished early must endure over-time etch to allow all structures on a given wafer to reach a desired end point. Over-time etch is unavoidable. A robust process that is insensitive to over etch is always desirable.
6. **Safety and cost of etchants.** Certain etchants are hazardous when they are inhaled or contact skin. It is important to know the material safety issues associated with each etchant. The material safety data sheet (MSDS) of a given product should always be consulted and strictly followed.

TABLE 10.1 Properties of Bulk Etching Solutions and Methods.

	EDP	Alkali-OH	TMAH	Gas phase	Plasma etch	HNA
Dry/wet	Wet	Wet	Wet	Dry	Dry	Wet
Isotropic/Anisotropic	Anisotropic	Anisotropic	Anisotropic	Isotropic	Anisotropic or isotropic	Isotropic
Etch rate on <100> Si (μm/min)	0.3–1.25	0.5–1	0.3–1	1–10 per pulse/cycle	0.5–2.5	0.5–1
Etch rate on silicon nitride (nm/min)	Very low	Very low	1–10	Low	100–400	Very low
Etch rate on doped silicon	Low on highly doped Si	Low on highly doped Si	Low on highly doped Si	Not sensitive to doping	Not sensitive to doping	Selectivity dependent on mixing ratio
Etch rate on silicon oxide	Low	Low, but higher than that of EDP	Low	Very low	Low	Moderate
Cost	Moderate	Low	Low to medium	Moderate to high	Moderate to high	Low

7. **Surface finish and defects.** Different etching methods and materials result in varying degrees of smoothness of surfaces and crack densities. The smoothness is important for many device and material performance aspects. For example, the fracture strength of single-crystal silicon microstructures is found to be affected by the etchants used to form them [2]. In practice, it is important to know the degree of smoothness associated with various etching methods and conditions, as well as ways to improve the smoothness or artificially roughen a surface if necessary.

Several relevant properties of common bulk etchants and etching methods are summarized in Table 10.1.

10.2 ANISOTROPIC WET ETCHING

10.2.1 Introduction

The technology for silicon anisotropic wet etching has been developed over the past 20 years [3–5]. It is a versatile process and can be used to create recessed structures (e.g., cavities with or without membranes), protruding structures (e.g., pyramidal tips and mesas), and suspended mechanical structures. The process has been used to successfully fabricate many commercial products, including silicon pressure sensors, accelerometers, neuron recording probes, and probes for scanning probe microscopes.

Today, new processes are being developed to supplement and replace silicon anisotropic etching. The deep reactive ion etching process, for example, offers a high vertical aspect ratio, unique vertical wall profiles, and better material selectivity. New materials, such as polymer substrates, are being used more frequently.

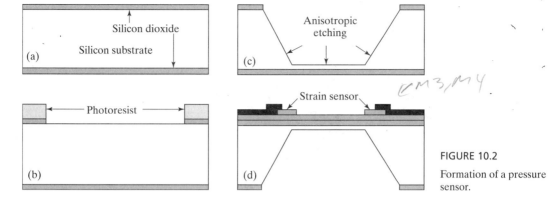

FIGURE 10.2

Formation of a pressure sensor.

Despite new materials and processes, silicon wet anisotropic etching remains a major force in MEMS fabrication. Certain features (such as pyramidal protrusions and inverted pyramidal cavities) can only be formed in silicon using wet anisotropic etching. The etching method offers unique three-dimensional profiles and properties (e.g., ultra smooth {111} sloped surfaces). The cost for a wet etching system is also significantly lower than that needed for dry etching. Hence, we dedicate a significant portion of this chapter to this topic.

Let us first review a few examples of structures created using processes that prominently feature bulk etching steps. The wet anisotropic etch process has been used to realize a membrane-based silicon pressure sensor, one of the earliest MEMS commercial products [6]. The fabrication process of this device consists of several major steps (Figure 10.2). The process begins with a silicon wafer with the front surface of {100}. Thermal oxide is grown on both sides of the wafer next (Figure 10.2a). A layer of photosensitive polymer (photoresist) is spin coated on the front side of the wafer. The photoresist is then photolithographically patterned and developed. The pattern in the photoresist is subsequently transferred into the oxide using hydrofluoric acid solutions (Figure 10.2b). The photoresist is selectively removed using acetone, which does not attack oxide or silicon. The oxide serves as a mask during a wet anisotropic etching. The cross-sectional profile of the wafer after a certain etching time (Figure 10.2c) shows a cavity that consists of {111} sloped sidewalls and a bottom {100} plane. The etching is stopped with a desired membrane thickness left. Strain sensors are formed at strategic locations of the membrane by doping the single-crystal silicon membrane (as in the case of Reference [6]) or by depositing and patterning doped polycrystalline silicon (as shown in Figure 10.2d). Metal thin films are deposited and patterned to provide wire leads.

In another example of an application, the silicon bulk micromachining process is used to create the suspended cantilever and protruded tip of an atomic force microscope probe shown in Figure 10.3. The detailed fabrication process of this device is further discussed in Chapter 14.

Anisotropic etching produces three-dimensional shapes from two-dimensional mask features. The three-dimensional profile is made of various crystal planes and may evolve with time. How are these three-dimensional microstructures with unique geometric characteristics formed? We will seek answers to this question by reviewing the geometric transformation rules in Sections 10.2.2 through 10.2.5.

FIGURE 10.3

Cantilever beams formed by anisotropic etching.

10.2.2 Rules of Anisotropic Etching—Simplest Case

Let us first examine the simplest case—the profile of an etched pit in a <100>-oriented silicon substrate when the mask contains a rectangle or a square open window with edges aligned to the <110> direction (Figure 10.4a).

When a silicon wafer is immersed in a wet silicon etching solution, the atomic layer that is exposed through the open window is etched first. Wet anisotropic etchants exhibit drastically different etch rates along different crystal planes (see Section 9.3.5 for details). In general, the etch rate along the <111> direction is the slowest among all crystal orientations.

The reason for the crystalline-dependent etch rate is not yet fully understood though conjecture exists. It is believed that since wet anisotropic etchants remove silicon by first oxidizing a silicon surface (using the oxidizing constituents within the solution) and then removing the oxide (using oxide-etching constituents), the difference of the etch rate is attributed to the difference of the oxidation rates. However, there seems to be no reported direct evidence correlating experimental data and reaction rate analysis.

Microscopically, the etch rate difference is attributed to the atomic bonding energy of silicon atoms on various surfaces. Atoms on different crystalline surfaces are associated with a different number of neighbors, bonding energy, and degrees of difficulty for removal.

At this stage, it is sufficient for the reader to realize that the etch rates are different along crystal orientations, and that the etch rates along the <100> and <110> directions are much greater than that along the <111> direction. Etch rates along certain high-index crystalline directions are greater.

With the knowledge of the etch rate distribution in a crystal lattice, let's now examine the progression of the etched profile. We shall focus on the cross section of the silicon wafer cut through the middle of a window (Figure 10.4). Atoms on the top layer are closely packed and all face the <100> direction. A representative atom, atom A in Figure 10.4b, can be removed according to the etch rate in the <100> direction.

As the first layer of the atom is removed, atoms along the edges of the mask opening are exposed in <100> *and* other directions (Figure 10.4c). An atom in the middle of the etched floor (e.g., atom B) can only be attacked from the <100> direction as it is flanked by other <100>-facing atoms on the same plane. An atom on the edge (e.g., atom C) has a greater bond strength than atom B. One intuitive way to understand the higher bond strength associated with atom C is that in the A–A' cross section, atom C has a greater than 180° solid angle on the solid side. In contrast, atom B is only associated with a solid angle of 180°.

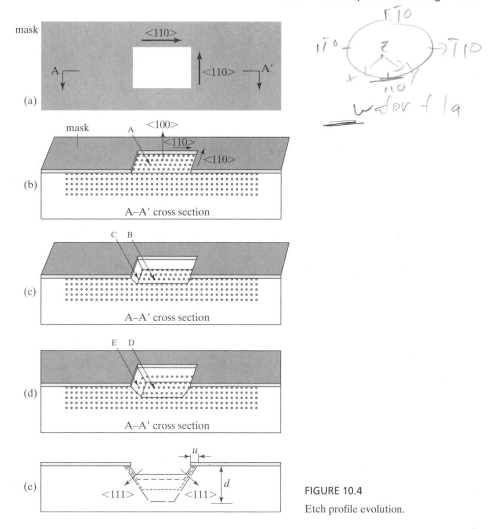

FIGURE 10.4
Etch profile evolution.

After the second layer of atoms is removed (Figure 10.4d), a representative atom on the slope (atom E) is now flanked by atoms facing the <111> direction. The energy it takes to remove atom E is much greater than that needed to remove atom D, for example. Consequently, the etch rate of the {111} slope is much slower than that of the bottom surface.

As the etching progresses, the cross-sectional profile of the cavity changes according to Figure 10.4e. Dashed lines represent the progression of etch profiles. The depth of the cavity increases with the etching time according to the etch rate in the <100> direction. Originating from the edges, sloped {111} surfaces are formed. Atoms lying in the middle of this {111} plane are etched very slowly as it is flanked by other <111>-facing atoms.

A computer simulation program is used to graphically illustrate the three-dimensional etched profile (Figure 10.5), with the mask shown in Figure 10.5a. The top view of the wafer after a few layers of silicon atoms are removed is shown in Figure 10.5b. A perspective view of the cavity with a magnified view of one of the corners is shown in Figure 10.5c. The corner is defined by two {111} planes and a bottom {100} plane.

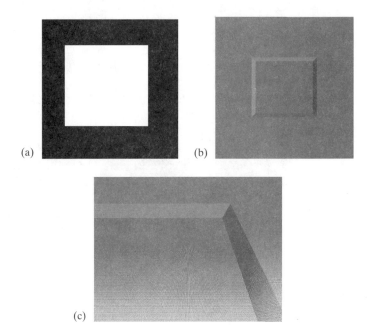

FIGURE 10.5

A sequence of images created using simulation software.

If we assume the wafer thickness is much greater than the size of the window opening and that the etching process continues for a long time, the {100} surface would eventually disappear and the four {111} planes would meet.

In the case of a square window, the cavity would end in a perfect point. In the case of a rectangular opening, a cavity would end in a knife-edge at the bottom. In practice, since masks are not ideal, a perfect point is almost never encountered.

A schematic view of the cavity obtained from a square window opening is shown in Figure 10.6. There is no undercut if we assume the etch rate on the {111} surfaces is negligible. The angle between the {111} and the {100} surfaces is 54.7°. In this case, the width of the window (w) and the final etching depth are related, according to $d = \dfrac{w}{2} \tan 54.7°$.

The scanning electron micrograph of an inverted cavity with four {111} slopes is shown in Figure 10.7.

Note that once the {100} bottom surface disappears, the profile of the cavity won't change significantly over time. Indeed, if the etch rate of {111} planes is completely ignored in a particular case, the profile will *never* change no matter how much longer the chemical etching is conducted. A three-dimensional profile bound by slow-etching {111} planes is called a **self-limiting stable profile (SLSP)**. Before the self-limiting profile is reached, the profile is called a **transitional profile**, and it is subjected to changes over time (e.g., Figure 10.4c). Transitional profiles can be further classified according to how fast they will change—**unstable transitional profiles** change with time in a rapid and complex manner, whereas **stable transitional profiles** change slowly and with better predictability.

These three categories of profiles are summarized in Table 10.2.

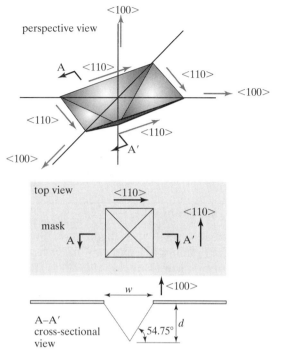

FIGURE 10.6

Cross-section profile of an etched cavity.

FIGURE 10.7

SEM micrograph of an inverted cavity.

The SLSP profiles are of particular interest in production and manufacturing—they are the most insensitive to over-time etching.

In practice, wafer thickness may be comparable with or even smaller than the width of the window opening. Depending on the width of the mask opening, wafer thickness (t), and etch time (T), four stable profiles (of STP or SLSP types) may result. These profiles can be classified in two categories: through-wafer holes or blind cavities. A blind cavity may end at a point, a line, or a {100} plane. The conditions and outcomes are summarized in Table 10.3. The etch rate in the <100> direction is denoted $r_{<100>}$.

TABLE 10.2 Types of Three-Dimensional Profiles for the Anisotropic Wet Etching of Silicon.

Type of profile		Definition
Transitional profile	Unstable transitional profile (**UTP**)	A three-dimensional etching profile that consists of fast-etching, high-index planes (e.g., {211}, {411}, etc.)
	Stable transitional profile (**STP**)	A three-dimensional profile that consists of only {100}, {110}, and {111} planes
Self-limiting stable profile (**SLSP**)		A three-dimensional profile that consists of only {111} planes

Example 10.1. Silicon Anisotropic Etching

A square mask opening is used to create a cavity in a silicon wafer. Suppose the thickness of the wafer is 500 μm (t) and the window opening is 1 mm (w) on each side. The etch rate on the {100} surface is 2 μm/min (T). Ignoring the etch rate on the {111} surface, calculate the size of the opening on the back side of the wafer if the etching time is longer than $t/r_{<100>}$.

What would be the thickness of the wafer if we desire to form a blind cavity with an inverted point instead?

Solution. Let us first check whether an SLSP-type through-wafer window will be formed. First, we find

$$2t/\tan(54.7°) = 0.714 \text{ mm.}$$

Since the width of the window is greater than $2t/\tan(54.47°)$, the etch will result in a through-wafer hole instead of a blind cavity. The width of the window opening on the back of the wafer is

$$w - 2t/\tan(54.7°) = 1 - 0.714 = 0.286 \text{ mm.}$$

If a blind cavity with an inverted point is desired, the following relation must be satisfied:

$$w < 2t/\tan(54.7°).$$

Rearranging the terms, we have

$$t > \frac{w \tan(54.7°)}{w} = 0.7 \text{ mm.}$$

The previous analysis ignores any etching in the <111> direction. In reality, however, {111} surfaces do experience a finite etch rate. The profile of an inverted pyramidal cavity with a finite undercut is shown in Figure 10.8. For a total etching time of T, the lateral undercut distance u is found as

$$u = \frac{r_{<111>} \times T}{\sin(54.7°)}. \tag{10.1}$$

In the case of an inverted SLSP-type blind cavity, the undercut may modify its depth over time.

10.2 Anisotropic Wet Etching

TABLE 10.3 Etching Profiles.

If $T \geq \dfrac{t}{r_{<100>}}$	If $w > 2t/\tan(54.7°)$ The SLSP profile is a through-wafer hole.	Profile at any $\infty > T \geq \dfrac{t}{r_{100}}$
	If $w \leq 2t/\tan(54.7°)$ The SLSP profile is a blind cavity ending at a point (if the mask opening is a square) or a line (if the opening is a rectangle).	Profile at any $\infty > T \geq \dfrac{t}{r_{100}}$
If $T < \dfrac{t}{r_{<100>}}$	If $w > 2T \cdot r_{<100>}/\tan(54.7°)$ This STP profile is a blind cavity ending at a {100} plane. This profile can be used to produce a thin silicon membrane with controlled thickness.	Profile at any $T < \dfrac{t}{r_{<100>}}$. The term d is the vertical etching depth. In this case, $d < Tr_{<100>}$.
	If $w \leq 2T \cdot r_{<100>}/\tan(54.7°)$ The SLSP shape is a blind cavity ending at a point or a line. The self-limiting depth is $d' = \dfrac{w \cdot \tan(54.7°)}{2}$.	Profile at any $\dfrac{t}{r_{<100>}} T \geq \dfrac{w \cdot \tan(54.7°)}{2r_{<100>}}$.

10.2.3 Rules of Anisotropic Etching—Complex Structures

In the previous section, we reviewed a very simple and commonly encountered case: masks with rectangular or square openings that have edges that are aligned to the <110> direction lying in a {100} silicon surface. We now build up the complexity of the analysis by relaxing the limitation on mask shapes. First, we eliminate the restriction that the edges of a window must

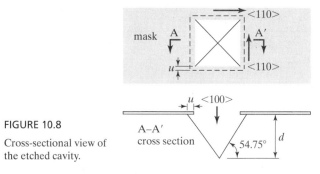

FIGURE 10.8

Cross-sectional view of the etched cavity.

be a non-interrupted straight line. Next, we eliminate the restriction that the edges must be parallel to the <110> direction.

Instead of first reviewing the text of rules, let us observe illustrated examples of actual etching cases as predicted by a computer simulation software [7]. The progression of etch profiles from two independent mask patterns (labeled A and B) is illustrated in Figure 10.9. These two window openings in the mask are very different. However, after a sufficiently long etch time (e.g., 170 minutes), they result in very similar three-dimensional profiles: blind STP-type cavities with edges aligned to <110> (Figure 10.9h).

Note that etching occurs underneath certain regions that are originally masked. For example, the material underneath the extended beam in B is gradually removed starting at its distal end.

The etching rules can be discussed easily by tracking how the corners of the mask patterns are transformed. There are two types of corners: convex corners and concave corners [8]. At a **convex corner**, the solid angle of the masked region is less than 180°. At a **concave corner**, the solid angle of the masked region is greater than 180°. Solid arrows in Figure 10.9a point to concave corners, while convex corners are indicated by arrows with dashed lines.

For convex corners, the fastest etching planes dominate the three-dimensional shapes evolving underneath the masked region. For concave corners, the slowest etching planes dominate the three-dimensional shape. In other words, convex corners tend to be undercut rapidly, exposing fast etching planes such as {211} and {411}. At concave corners, slow etching planes such as {111} tend to develop and eventually prevail.

The undercut can be used to create suspended micromechanical structures. For example, the mask opening B will result in a suspended cantilever (Figure 10.10), provided that the mask is chemically and mechanically sturdy enough to survive the entire wet etching sequence. The material of the cantilever is that of the mask layer. The cantilever can be made of a variety of materials, including silicon nitride, silicon oxide, and heavily doped silicon. Other materials are possible as long as they exhibit good etching selectivity. A comprehensive survey of etch rates by various etchants on silicon substrate and important mask materials can be found in References [9, 10].

Example 10.2. Prediction of Etching Profile

In the previous example, the protruding cantilever is parallel to the <110> direction. However, cantilevers do not have to be pointing in the <110> direction. The shape of the cantilever can be different from the rectangle as well. Determine the stable etching profile resulting from the mask below.

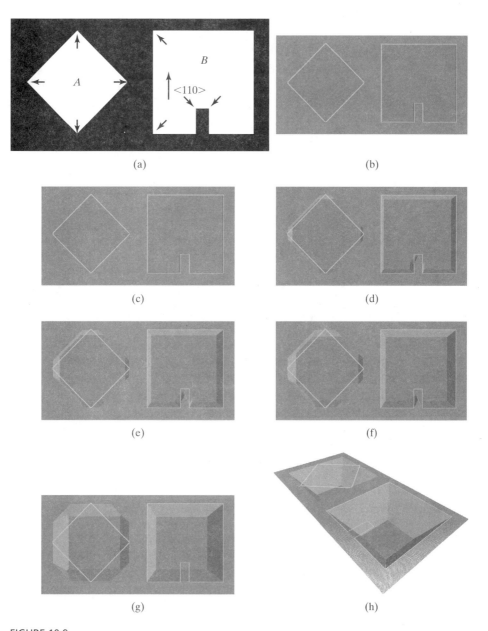

FIGURE 10.9

Progression of an etch profile. Frame (a) is the top view of mask. Frames (b) through (h) are the top or perspective views of the cavities at 5, 10, 20, 40, 50, 80, and 170 minutes of etching.

Solution. The predicted etch profiles at different etching time intervals are shown next (Fig. 10.11). Materials underneath extended beams are undercut starting from the convex corners. Frames (a) through (g) depict UTP cavities. After a sufficiently long etch time (e.g., at 180 minutes for frames (h) and (i)), a blind STP-type cavity bound by four {111}-sloped planes and a {100} bottom is formed.

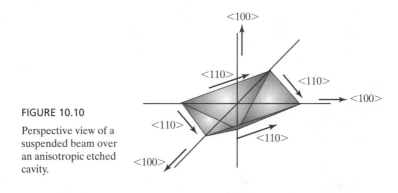

FIGURE 10.10

Perspective view of a suspended beam over an anisotropic etched cavity.

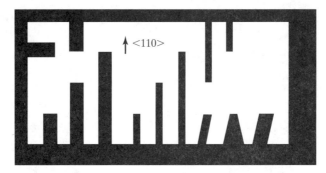

In certain cases, the mask opening may consist of curved boundaries, making it difficult to identify distinct concave and convex corners. It is rather difficult, and often unnecessary, to predict UTP profiles precisely under these circumstances. On the other hand, the STP and SLSP profiles originating from such mask patterns can be determined readily. Let's first observe the progression of the etch profile for a mask shown below (Figure 10.12), which contains five opening patterns, including two with curved boundaries.

The profiles of etching at various time intervals are shown in Figure 10.13. Frames (a) through (j) depict cavity profiles at 5, 10, 20, 40, 50, 70, 105, 125, 230, and 310 minutes. Frame (k) is a perspective view of the wafer with STP and SLSP cavities at 310 minutes. Frame (l) is similar to (k) but reveals a cross section. After a sufficiently long etching time, the two openings with curved boundaries result in SLSP cavities bound by {111} planes.

The self-limiting stable profile produced from a given opening window can be found in a straightforward manner without going through the analysis of UTP profiles.

To simplify future analysis, we define a **footprint of self-limiting cavity**, or FSLC, as the intersection of an SLSP cavity with the front surface of a wafer. If the front surface of the wafer is the {100} plane, the procedure to determine FSLC is as follows:

1. Find the left-most, right-most, top-most, and bottom-most points (or lines) associated with the continuous periphery of a mask opening.
2. Draw the vertical <110> lines passing through the left- and right-most points.
3. Draw the horizontal <110> lines passing through the top- and bottom-most points.
4. The FSLC is the area bound by the four lines identified in Steps 2 and 3.

10.2 Anisotropic Wet Etching **339**

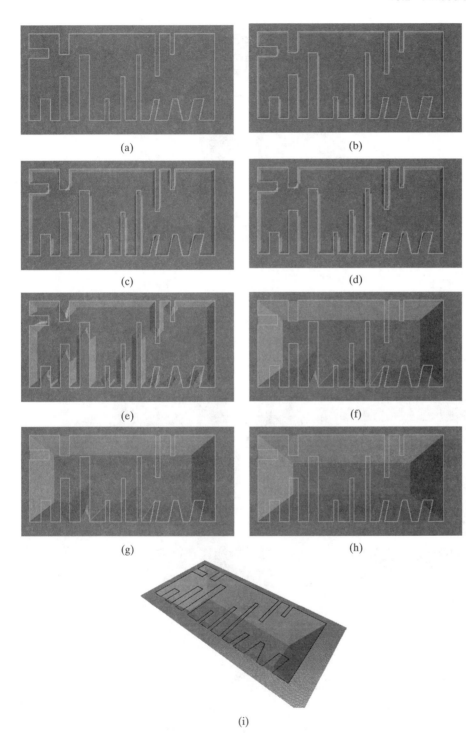

FIGURE 10.11

Progression of the etch profile. Frames (a) through (h) depict etching profiles at 3, 9, 15, 24, 36, 60, 120, and 180 minutes of etching time. Frame (i) depicts the perspective view of the etching profile at 180 minutes.

340 Chapter 10 Bulk Micromachining and Silicon Anisotropic Etching

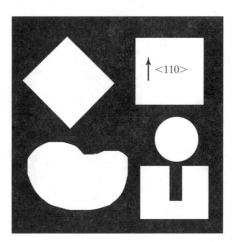

FIGURE 10.12
Mask.

Example 10.3. Recognition of Cavity Types

A mask is given below, together with perspective view of the front surface of the silicon wafer after a certain etching time. Identify the types of etched cavities at this particular moment.

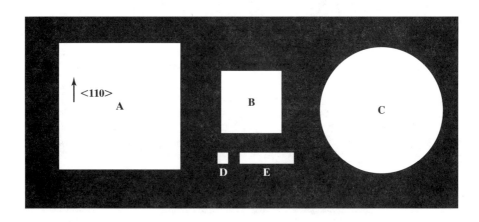

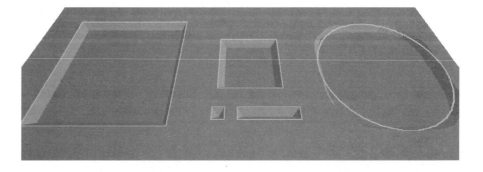

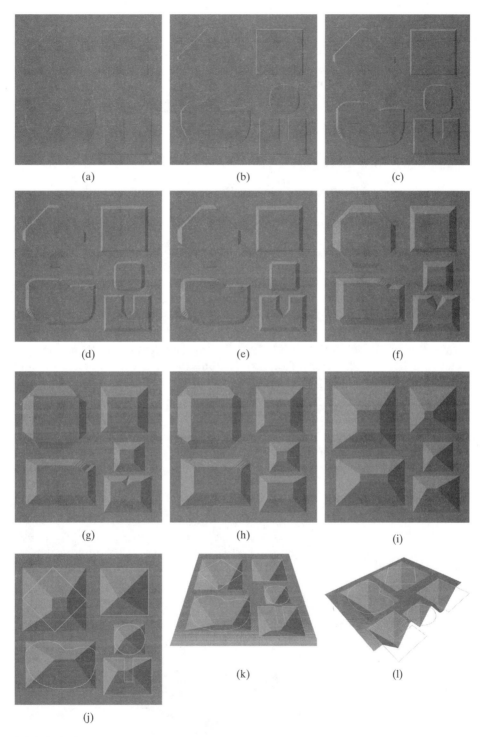

FIGURE 10.13
Etch profile at various time intervals.

Answers:
Opening A: Stable transitional cavity (STP).
Opening B: Stable transitional cavity (STP).
Opening C: Unstable transitional cavity (UTP) with fast-etching and high-index planes.
Opening D: Self-limiting stable cavity (SLSP) ending at a point.
Opening E: Self-limiting stable cavity (SLSP) ending at a line.

Example 10.4. Analysis of FSLC

Predict the FSLC of windows in the mask pattern depicted in Figure 10.14.

Solution. This mask consists of many opening windows. The etching profile progression is shown in Figure 10.15.

So far, we have discussed the evolution of three-dimensional etch profiles from masks with open (transparent) windows. What if a mask consists of an opaque patch on a large transparent background? In this case, the etching profile will always be unstable and transitional. Fast etching planes dominate the profile. Let us examine how etching profiles evolve in this case from an exemplary mask that contains two independent features: a square and a circle (Figure 10.16).

We draw the conclusion that if the mask used is a patch rather than a hole, a protruding structure will be formed. The protrusion will be unstable transitional, and it will ultimately disappear if a sufficiently long etch time is allowed.

Example 10.5. Corner Compensation

Predict the UTP etching profiles for the following patterns (A and B) and given wafer orientation.

Solution. The top view of the wafer at various time intervals is in Fig. 10.17. Mesas with rounded corners are formed. For a given depth of etching and etch time (e.g., 39 minutes), the corner rounding under pattern B is less severe than that under pattern A. Beams extending from the corners of the square of pattern B are used to delay corner rounding and create sharp, instead of rounded, corners with the desired depth. This general technique is called **corner compensation**.

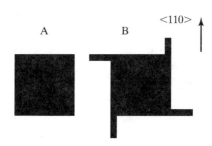

10.2 Anisotropic Wet Etching 343

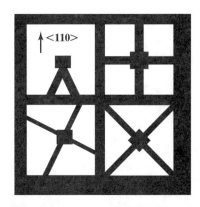

FIGURE 10.14
Mask.

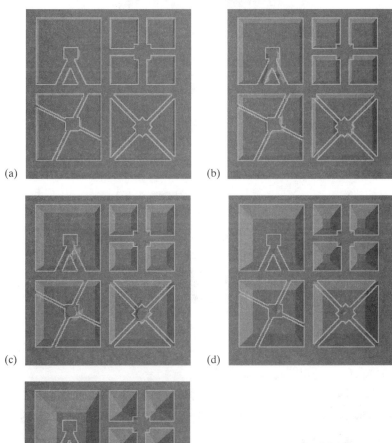

FIGURE 10.15
Progression of an etch profile. Frames (a) through (e) show etch profiles at 5, 20, 30, 50, and 70 minutes. Frame (f) is the perspective view at 70 minutes.

344 Chapter 10 Bulk Micromachining and Silicon Anisotropic Etching

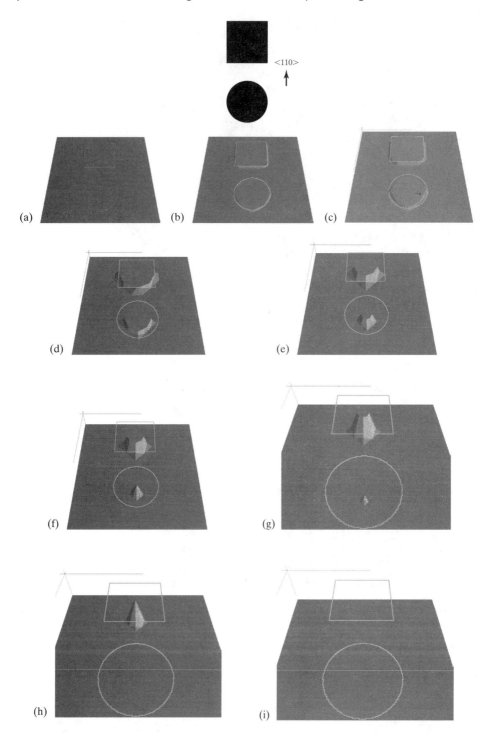

FIGURE 10.16

Progress of an etch profile. Frames (a) through (i) depict etching profile at 5, 15, 25, 50, 90, 110, 120, 150, and 200 minutes into the etching process.

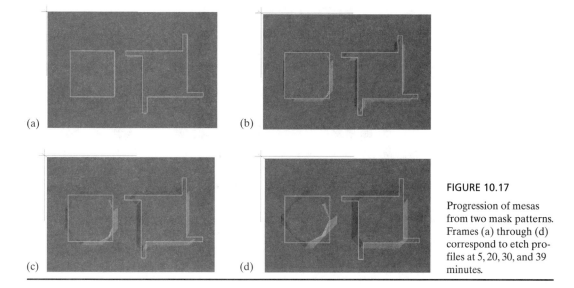

FIGURE 10.17

Progression of mesas from two mask patterns. Frames (a) through (d) correspond to etch profiles at 5, 20, 30, and 39 minutes.

Although <100>-oriented silicon wafers are used predominantly, other wafer orientations are possible. For example, the wet etching of <111>-oriented silicon allows useful and novel geometries [11]. Wafers with slanted crystallographic cuts have been used as well [12].

10.2.4 Interaction of Etching Profiles from Isolated Patterns

In previous sections, we have generally placed independent openings on masks far apart from each other so that their etching profiles won't interfere with each other. Such structures are considered isolated and their evolution is independent. Let's look at two isolated polygons placed side by side. Their FSLC envelopes do not overlap. The progression of the etched cavity is shown in Figure 10.18.

For comparison, let's consider the case where these two polygons are brought closer to each other, such that their FSLC envelopes now overlap. The progression of the etched profile is drastically different, as illustrated in Figure 10.19. At 50 minutes, the edges of the two cavities begin to merge. Over time, the cavities that originated from these two isolated windows merge to form a connected SLSP entity.

10.2.5 Summary of Design Methodology

The anisotropic etching process can be used to create unique three-dimensional features and suspended mechanical elements. We discussed rules for predicting the UTP, STP, and SLSP profiles originating from a mask pattern on a {100} silicon wafer. The profiles at a given time depend on the mask orientation, wafer thickness, window opening sizes, and etch rates in various crystal directions.

Depending on the mask shape and etch time, three types of three-dimensional structures may form: *unstable transitional*, *stable transitional*, and *self-limiting stable*.

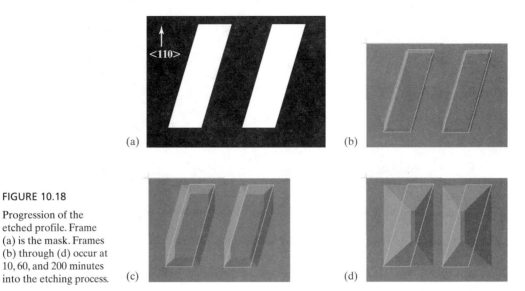

FIGURE 10.18

Progression of the etched profile. Frame (a) is the mask. Frames (b) through (d) occur at 10, 60, and 200 minutes into the etching process.

Unstable transitional profiles can be used to create unique three-dimensional shapes, but the process control is difficult [8].

Stable transitional profiles and self-limiting stable profiles are desirable for manufacturing because they are tolerant of over-time etching and they are robust against uncertainties and the variation of process parameters. The SLSP profiles are the most robust.

10.2.6 Chemicals for Wet Anisotropic Etching

EDP (ethylene diamine pyrocatechol) is a frequently used anisotropic etchant. It is sometimes referred to as EPW (ethylenediamine pyrocatechol and water). The etch rate ratio for <100> and <111> directions reaches 35:1 or higher. The etch rate on the <100> silicon orientation is between 0.5 to 1.5 μm/min. Etch rates for silicon nitride (LPCVD) and silicon oxide are almost negligible. The etch rate on silicon dioxide is 1–2 Å/min and on silicon nitride is <1 Å/min. On one hand, it is highly selective, allowing thermally grown silicon oxide as a mask. On the other hand, this means that even very thin natively grown oxide becomes a barrier to etching if it is not properly removed.

Etching is often conducted in a chemical fume hood because of the toxicity of the EDP vapor. The chemical solutions are often heated to 90–100°C during the etching process. Escape of vapor may alter the concentration of etching chemicals and the etch rate. To maintain a constant concentration, the escaped water vapor is condensed on a water-cooled top and allowed to drip back into the flask. This system, called a reflux system, is illustrated in Figure 10.20.

KOH (potassium hydroxide) is a low-cost alternative to EDP. The KOH etchant can be easily made in a laboratory by mixing solid KOH with water in a slightly exothermic reaction. The etching characteristics depend on the concentration of KOH and the temperature of solutions. The most commonly used concentration is 20–40 wt %. The etch rate ratio for <100> and <111> is also very high (even higher than that of EDP). The etch rate on silicon nitride (LPCVD) is negligible. However, the etch rate on silicon oxide (thermally grown) is 14 Å/min, which is much greater than the etch rate on silicon oxide by EDP.

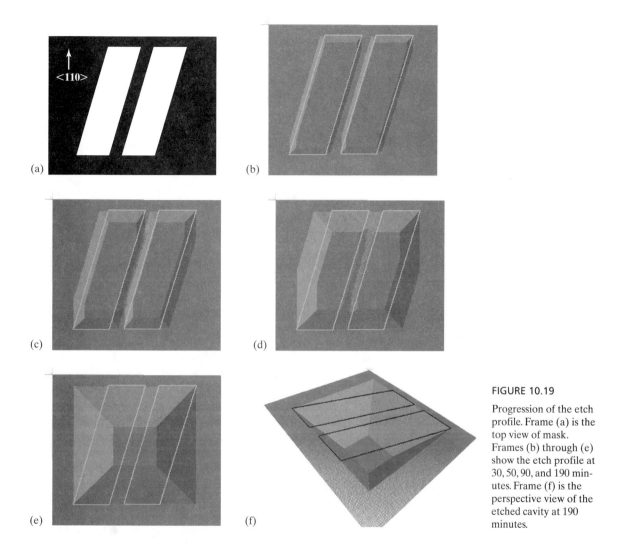

FIGURE 10.19

Progression of the etch profile. Frame (a) is the top view of mask. Frames (b) through (e) show the etch profile at 30, 50, 90, and 190 minutes. Frame (f) is the perspective view of the etched cavity at 190 minutes.

For both EDP and KOH, the etch rate for silicon is greatly reduced when the silicon is heavily doped with boron or phosphors. The degree of etch rate reduction as a function of doping concentration is different for these two chemicals [5, 13]. When the EDP is greater than 7×10^{18} cm^{-3}, boron reduces the etch rate by 50 times. Therefore, it is possible to use heavily doped silicon as an **etch stop layer** against over-time etch.

Normally, the wet etching is conducted without electrical bias to the wafer and under normal ambient light conditions. Active electrical and optical activation, however, can be used to introduce new etching behavior. For example, pulsed potential annodization exploits the etching behavior of a p–n junction [14]. The photovoltaic electrochemical etch stop technique (PHET) is based on the electrochemical growth of a passivation silicon oxide. This method can be used to provide external, flexible control of etch stop and does not require a doping process [15].

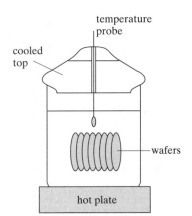

FIGURE 10.20

Wet anisotropic etching reflux system.

Tetramethyl ammonium hydroxide (TMAH) is an alternative silicon etchant. It was discovered after EDP and KOH [16]. It has a very slow etch rate on aluminum, a common IC metallization material. As such, TMAH can be used to etch wafers with aluminum wire leads. The etch rate on aluminum can be further reduced if silicon is predissolved into the solution; however, the dissolved silicon is known to create more surface roughness. TMAH also exhibits a reduced etch rate for heavily doped silicon: a 10× reduction at 10^{20} cm^{-3} and a 40× reduction at 2×10^{20} cm^{-3}. The etch rate on silicon oxide is low (0.05–0.25 nm/min). The disadvantages of TMAH are (1) the etch rate ratio of the <100> and <111> directions are not as high as KOH and EDP, and (2) the roughness of the finished surface is generally worse than the cases of EDP.

The etch rate distribution with respect to crystal orientation can be determined using experimental means. Instead of preparing samples of different front-surface crystal orientations and determining the etch rate by etching each piece individually, the etch rate along all directions in a given plane can be determined using special patterns, such as the wagon wheel mask pattern shown in Figure 10.21. The pattern consists of slit openings with angular offsets being θ covering the entire 360°C span.

Etching in a chemical solution would cause the slits to widen due to undercut. The amount of widening for each slit is proportional to the etch rate in the direction normal to the longitudinal edges of slits. The bottom part of Figure 10.21 illustrates the widening of a slit, which in turn translates into the shortening of radial lines (i.e., changes of d_r).

The computer-simulated progression of the etching profile from a wagon wheel pattern is shown in Figure 10.22. Actual results of the etched wheel are illustrated in Figure 10.23.

The wagon wheel pattern provides a qualitative and direct graphical representation of the etch rate but the results lack quantitative accuracy. The angular resolution is limited. The minimal angular increment depends on the mask resolution and size of the wagon wheel pattern.

An alternative methodology for obtaining the etch rate diagram with a higher angular resolution has been developed; it uses a mask that contains a set of independent rectangular holes with incremental angular offset (1° in Reference [7]) (Figure 10.24a). Etching for a relatively short period of time (5–10 minutes), certain edges of the UTP undercut cavity are parallel with the long edge of the rectangle. Therefore, the etch rate associated with a particular orientation can be obtained by measuring the amount of lateral undercuts along the long

FIGURE 10.21

Wagon wheel mask pattern.

edges (Figure 10.24b). A single rectangle opening and the underlying cavity are shown in a magnified view (Figure 10.24c). A polar etch rate diagram with 1° resolution is obtained for silicon in EDP (Fig. 10.25).

10.3 DRY ETCHING OF SILICON—PLASMA ETCHING

Gas-phase etching of substrates and thin-film material can be achieved using plasma etching techniques. Plasma is a discharge in which the ionization and fragmentation of gases takes place and produces chemically active species, namely oxidizers and/or reducing agents. A high-voltage, radio frequency AC signal applied between two parallel electrodes is often used for generating plasma, though DC bias is also feasible. The high electrical field causes the gas molecules to ionize, which produces active species. Different gas contents and mixing ratios lead to unique etch characteristics and profiles. Some commonly used etching gases and corresponding materials are summarized in Table 10.4.

Two forms of reactions may occur at the wafer-gas interface: chemical etching and physical etching. Chemically, active ion species can react with silicon or thin-film materials to form a gaseous compound that is removed through gas circulation. Physically ionized species are accelerated toward the wafer. These species physically bombard the substrate and remove atoms.

FIGURE 10.22

A sequence of images created by simulation software.

In general, physical etching is more directional and anisotropic, whereas chemical etching is isotropic and material selective.

TABLE 10.4 Common Gas-Phase Etching Gases and Byproducts.

Materials to be etched	Candidate gases	Result products
Si, SiO_2, Si_3N_4	CF_4, SF_6, NF_3	SiF_4
Si	Cl_2	$SiCl_2$, $SiCl_4$
Al	BCl_3, CCl_4, $SiCl_4$, Cl_2	$AlCl_3$, Al_2Cl_6
Organic solids (photoresist)	Oxygen, $O_2 + CF_4$	CO, CO_2, H_2O
Refractory metals	CF_4	WF_6

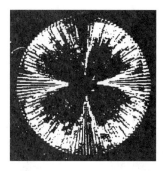

FIGURE 10.23

Etch result from a wagon wheel pattern.

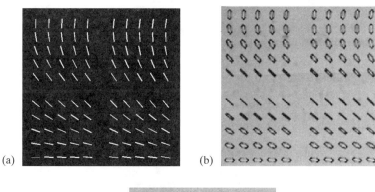

FIGURE 10.24

The etching experiment and simulation for an EDP etch rate diagram.

If the electrode holding the wafer is grounded, the etching is called **plasma etching**. If the wafer is fixed to an electrode on which AC bias is applied, the etching is called **reactive ion etching**. Compared with plasma etching, reactive ion etching is more physical in nature and its etch rate distribution is more anisotropic.

Gas-phase plasma etching is a complementary alternative to wet silicon anisotropic etching. Plasma etch does not require wafers to be in contact with wet chemical solutions; thus it simplifies cleaning steps. Plasma etch offers different options of masking material. The process temperature is also lower. Generally speaking, wafers are not heated although prolonged plasma etching may cause the wafer temperatures to increase.

The sidewall profile is different from what is achievable with web anisotropic etching. In fact, the anisotropy of the vertical wall can be controlled, within a certain range, by adjusting several process parameters including the process pressure, DC bias voltage, power input, and the chemical species used.

10.4 DEEP REACTIVE ION ETCHING (DRIE)

Deep reactive ion etching (DRIE) is a special kind of RIE process. It is capable of producing deep and high-aspect ratio features with near vertical sidewalls. They are based on patents currently owned by Robert Bosch GmbH and Texas Instruments [17, 18]. The key to achieving continuous deep etching with a high aspect ratio is to conduct the etching process in small depth increments. This approach utilizes repeated cycles that contain an etching step followed by a passivation step. A layer of inhibition film is deposited during the inhibition step on all walls of a cavity. During the next etching step, the inhibition film is preferentially removed from the bottom of the trenches due to ion bombardment, while preventing the etching of the sidewalls. Process parameters play an important role in determining the surface morphology and mechanical performance of devices [19].

DRIE silicon etching is rapidly gaining popularity despite the high cost of processing equipment. The DRIE process offers a fast etch rate, vertical sidewalls, a room temperature process capability, and the ability to easily correlate the mask with the resultant three-dimensional structures. The etch selectivity is a major benefit. The DRIE process can use photoresist, silicon oxide, and metal as a mask. In addition, the etching uniformity on a wafer scale is generally much improved over those of regular RIE etch.

Though powerful, the DRIE technique alone is capable of creating nothing more than passive three-dimensional parts. For this reason, deep reactive ion etching must be combined with other processing styles and steps to increase the electrical and mechanical complexity of microstructures. The DRIE technique can be combined with many processing styles, including anisotropic etching, to create a rich variety of innovative geometries (e.g., microneedles and comb fingers [20, 21, 22]).

It is possible to have deep reactive ion etching of materials other than silicon, including piezoelectric materials (i.e., quartz, PZT) [23], Pyrex glass [24], and commercial PMMA [25].

Example 10.6. HARPSS Process

One example of such a process is a high aspect-ratio combined poly- and single-crystal silicon (HARPSS) MEMS technology [26], which is capable of creating high Q resonators, accelerometers, and gyros. Later, a related process called trench-refill polysilicon technology (TRiPs) was invented [27]. Key steps of the HARPSS process are illustrated in Figure 10.25. Silicon nitride is first deposited on a silicon wafer and then photolithographically patterned. The nitride effectively provides electrical insulation. Photoresist or oxide is used as a mask for deep reactive ion etching to create vertical grooves that are 100 μm deep and 6 μm wide (step a). An LPCVD oxide film is deposited over the entire silicon wafer (step b). The deposition is relatively conformal in open trenches, although the thickness of oxide on the bottom of the trenches is lower than at the open surface. The wafer is spin coated with photoresist. The photoresist is lithographically patterned and used as a mask in the subsequent etch step (step c). Anchor windows are opened over the silicon nitride structures to allow subsequent polysilicon films to firmly adhere to the silicon nitride film. The photoresist is then removed. Here, care should be exercised to completely remove the photoresist from the bottom of trenches, because the next step involves a high-temperature process. Any remaining photoresist could contaminate the oxidation tube and degrade device performance.

A boron-doped LPCVD polysilicon layer is deposited (step d). The polysilicon is patterned using photoresist as a mask (step e). Another layer of photoresist is spin coated and patterned,

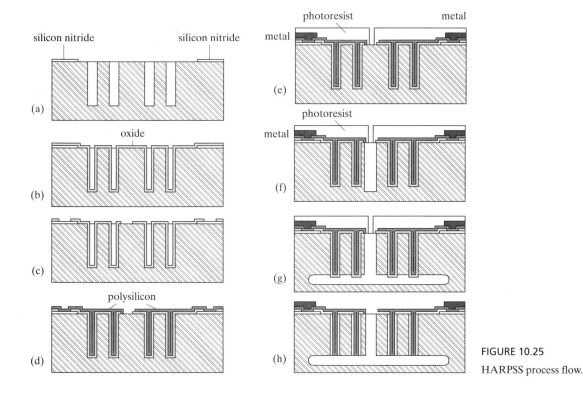

FIGURE 10.25
HARPSS process flow.

creating windows in which silicon would be removed by a subsequent DRIE etching step (step f). After the target depth has been reached, an isotropic plasma etching is performed using SF_6 reactive gas. The plasma etching is undercut laterally by 25 μm in each direction (step g). The oxide film, whose thickness defines the gap between the polysilicon vertical electrodes and counter electrodes (bulk silicon), is removed using HF solutions (step h).

10.5 ISOTROPIC WET ETCHING

The most commonly used isotropic silicon etchant is HNA, a mixed chemical solution consisting of at least three constituent acids—the letter H stands for hydrofluoric acid, N for nitric acid, and A for acetic acid [28]. Various mixing ratios of these three major ingredients are possible, resulting in different etch rates on silicon and etch selectivity against masking layers. Because the solution contains HF, the etch rate on oxide is relatively quick, at approximately 300 Å/min.

10.6 GAS-PHASE ETCHANTS

Gas-phase etching of silicon using XeF_2 or BrF_3 is a complementary alternative to wet etching and plasma etching [29, 30]. At room temperature, XeF_2 exists in solid phase and BrF_3 exists in liquid phase. XeF_2 sublimates and BrF_3 vaporizes to form active gas when subjected to pressure below 100 mtorr. These gases react with silicon (in single-crystal or polycrystalline forms) strongly, even under room temperature, resulting in isotropic etch profiles. On the other hand, the gas-phase etchants have a negligible etch rate on common mask materials, including silicon dioxide, metal, and even photoresist [10].

Care must be taken because the reaction byproducts contain HF vapor that, if released into air accidentally, may become absorbed into water vapor and constitute airborne drops of highly concentrated HF acid.

The bulk silicon etch rate is very high. The etch rate is generally not characterized as an etching depth per unit time. Rather, since etching is administered by injecting a fixed amount of etching agents in burst cycles, the etch rate is given as μm/cycle. The effective etch rate can be as high as 20–50 μm/cycle.

10.7 NATIVE OXIDE

A thin layer of silicon oxide would inevitably form on silicon naturally when the silicon is placed or stored under room temperature and environmental oxygen and moisture levels. This layer of material is very thin, as the oxide growth temperature is low. However, this oxide may present a significant etching barrier if the silicon etchant has a small etch rate on silicon dioxide. It is a common practice to remove native oxide before wet anisotropic etch or plasma etch steps to ensure that the reaction with silicon starts predictably.

The thickness of the native oxide is so small that it often eludes detection by thin-film thickness measurement instruments. A practical laboratory technique to determine whether a native oxide layer is present is the water-beading test. A pure silicon surface is hydrophobic while an oxidized silicon surface is hydrophilic. The hydrophobic/hydrophilic nature of a surface can be quickly and inexpensively determined by dispensing drops of water on the substrate surface and observing the shape of the water drop (Figure 10.26). Water beads easily on hydrophobic silicon surfaces and rolls off the substrate when the substrate is tilted. On the other hand, water would spread uniformly on a hydrophilic oxide surface.

The native oxide layer can be removed easily by dipping the wafer in a low-concentration hydrofluoric acid solution (e.g., 5%) for a few seconds.

10.8 WAFER BONDING

Wafer-to-wafer bonding is a versatile technique that allows wafers with disparate materials, surface profiles, and functional characteristics to be joined to form unique structures [31] [32].

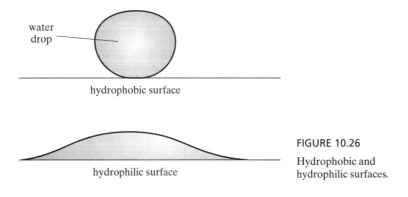

FIGURE 10.26

Hydrophobic and hydrophilic surfaces.

Bonding processes can be categorized according to the temperature of the operation: room-temperature bonding, low-temperature bonding (<100°C) and high-temperature bonding (>100°C). Wafer bonding can be *direct* (i.e., without involving any intermediate adhesive layer), or *indirect* (where an adhesion layer is used). Bonding can be initiated by mechanical contact force, molecular attractive force, or electrostatic force. Several major types of bonding techniques are summarized in Table 10.5.

Wafers can be chemically or mechanically modified after bonding. For example, they can be thinned to a desired thickness by mechanical polishing or chemical etching. Wafer-to-wafer transfer has been demonstrated to achieve surface planarization [44] and to produce devices such as mirrors [45] and membranes [46, 47], even ones with large sizes [48].

Let us examine one specific example of a MEMS sensor made using the wafer bonding and transfer process. In the early days of MEMS, bonding technology was successfully used by Lucas NovaSensor in 1985 to make a silicon-based pressure sensor with a sealed cavity (acting as a pressure reference). The fabrication process is illustrated in Figure 10.27. A {100} silicon wafer (wafer #2 in the figure) is etched in anisotropic etchant to form a cavity bound by {111} surfaces. Two possible SLSP-type profiles are shown side by side—an inverted-pyramidal cavity and a through-wafer hole. Another wafer, wafer #1, is then bonded to the top surface. Wafer #1 is polished to a desired thickness, which defines the thickness of pressure sensing membranes. Selective doping is performed to form piezoresistors for sensing membrane displacement. The pressure sensor using the blind cavity is an absolute pressure sensor, whereas the one using the through-wafer hold is a relative pressure sensor, measuring the difference of pressure across the membrane. It is impossible to form large, ultra-thin (e.g., less than 1 μm), flat membranes with single-crystal silicon using any other means.

TABLE 10.5 Representative Bonding Techniques.

	Representative materials	Comments
Anodic bonding [33, 34]	Glass to silicon; oxide to silicon	Performed under 400°C with electric field present (e.g., 1.2 kV) Works well in atmosphere or vacuum
Fusion bonding [35–37]	Silicon to silicon	Very sensitive to surface defects and particles
Low-temperature adhesive bonding [38]	Many possibilities of substrates and materials	Representative adhesive layers include photoresist, polymer adhesive, spin-on glass adhesive
Eutectic bonding [36]	Gold to silicon	Processing temperature is between 450–550°C
Low-temperature silicon direct bonding [39]	Silicon to silicon	Temperature of bonding is below 110°C It may be followed by long-term storage or high-temperature treatment, during which the bonding energy increases
Solder bonding	Many possibilities of substrates	It uses low melting-temperature metals such as indium [40], aluminum [41], or others
Mechanical bonding	Many possibilities	Examples include microriveting [42] and micromechanical Velcro [43]

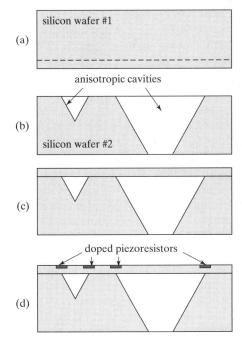

FIGURE 10.27

A fabrication process to realize absolute and differential pressure-sensing membranes using silicon bulk etching and bonding.

Most bonding operations are conducted at the wafer scale. However, bonding can be performed at die level or device level [36, 41].

10.9 CASE STUDIES

In this section, we will discuss a number of representative fabrication processes for two important classes of structures—suspended beams and membranes. It is important to compare different design and fabrication options under each class. The comparison is not meant to serve as an exhaustive survey. Rather, it is a beginning introduction about the rich possibilities of microfabrication.

10.9.1 Suspended Beams and Plates

Suspended beams and plates can be made of many different materials and under various geometries. A part list of representative suspension structures (up to two levels) and substrate structures is drawn in Figure 10.28. A microstructure is constructed by selecting one beam/plate piece and one substrate piece. The materials for the plate structure could be single-crystal silicon (with various doping levels), silicon nitride, silicon oxide, polysilicon, polymers, and metals.

A number of generic fabrication processes are available depending on the materials, the substrate, the vertical profile in silicon, and the composition of layers. A few representative fabrication process classes are shown in Figure 10.29. These methods can be classified into three major categories: undercutting a suspension structure, etch back, and processes involving wafer bonding and structure transfer.

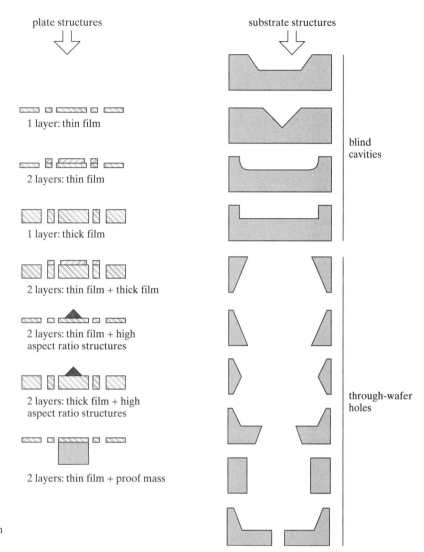

FIGURE 10.28

List of possible suspension and substrate structures.

10.9.2 Suspended Membranes

Suspended membranes are often used in MEMS. Applications include pressure sensors, acoustic sensors, acoustic actuators, and optical mirrors, to name a few.

There are many fabrication techniques for making membranes of various materials, sizes, and thicknesses. Several generic methods are diagrammed in Figure 10.30. The techniques can be classified into several categories—wet silicon etching, dry plasma etching, wafer bonding and transfer, and membrane bonding. There are several implementation strategies within each category.

Why do we need so many different options for achieving similar structures? Let us evaluate a few specific processes to discuss their relative merits.

In the first example, an ultra-thin membrane (less than 1 μm) is desired. The film does not have to be intrinsically piezoresistive. A preferred process uses LPCVD silicon nitride. The process,

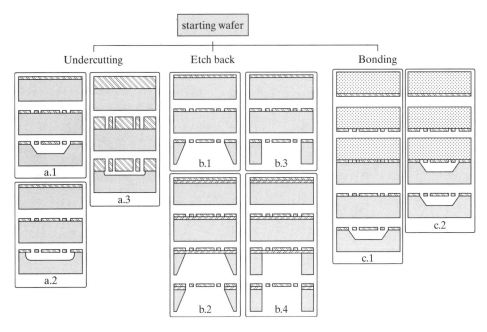

FIGURE 10.29

Representative methods for forming micromachined plates by bulk micromachining.

in line with option a.1 of Figure 10.30, starts with a <100>-oriented silicon wafer (Figure 10.31). A layer of silicon nitride is grown using a low-pressure chemical vapor deposition process. The silicon nitride film has tensile intrinsic stress. Thickness of the silicon nitride film is generally less than 1.5 μm. Greater film thickness is generally not used because the time to grow LPCVD nitride will be excessively long. The silicon nitride film on the back side of the wafer is photolithographically patterned by using reactive ion etching with photoresist as mask. An opening is produced in the silicon nitride, with the four sides parallel to <110> directions. The wafer is immersed in a silicon etchant until the cavity reaches the other side of the wafer.

Silicon membranes may be desired as opposed to silicon nitride membranes for a number of reasons. This is true if the desired membrane thickness is in excess of 1.5 μm, or if piezoresistors made of doped silicon are to be embedded in the membrane itself.

A representative process for making a silicon membrane starts with a silicon wafer with a <100>-oriented front surface (Figure 10.32). It is also in line with option a.1 of Figure 10.30. Silicon dioxide is grown on the wafer. The oxide thin film on the back of the wafer is photolithographically patterned and etched with HF solutions. EDP or KOH is used to etch the silicon wafer. The etch rate is calibrated and the exact thickness of the wafer is measured accurately. By controlling the timing of the process, a silicon membrane with a target thickness can be formed. The wafer is then immersed in HF to remove oxide.

This process, however, makes use of STP cavities and is not robust to over-time etch and inevitable uncertainties of process parameters. Using this process, it is difficult to achieve a thickness of less than 1 μm.

How can one make silicon membranes with more process robustness? Yet another process, again in line with option a.1 of Figure 10.30, makes use of the fact that heavily doped silicon

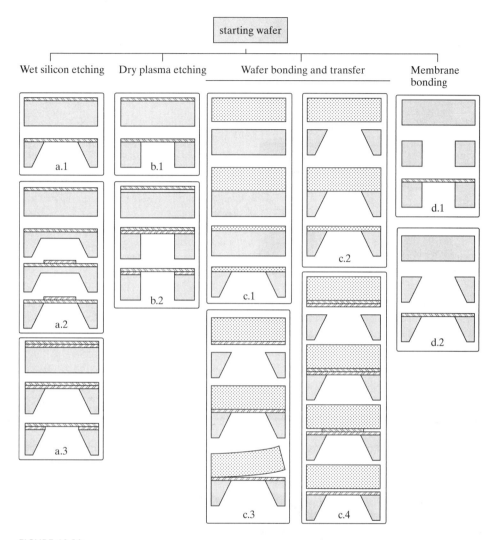

FIGURE 10.30

General strategies for fabricating a membrane.

slows down anisotropic wet etching (Figure 10.33). A silicon wafer is doped on the front side using diffusion doping or ion implantation. The wafer is cured at a high temperature to allow dopant atoms to be incorporated into the silicon lattice. This creates a layer of highly doped silicon with a doping concentration in the range of 10^{20} cm^{-3}. The thickness of the layer can be controlled accurately. At this concentration, the etch rate by anisotropic etchants is reduced by a factor of 35 or more. The wafer is then oxidized on both sides, or it is passivated using silicon nitride films. The oxide or nitride film is patterned on the back side to open windows, in which silicon is exposed. Anisotropic through-wafer etching is conducted until the highly doped silicon on the front side is exposed. If oxide is used, the film can be removed using HF. If silicon nitride is used, the film can be selectively removed using a H_3PO_4 solution at 180°C.

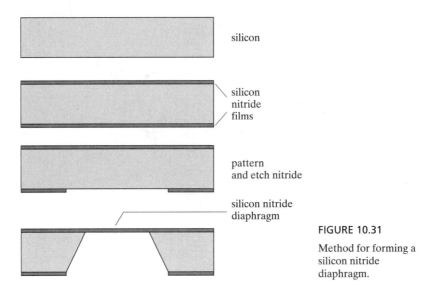

FIGURE 10.31

Method for forming a silicon nitride diaphragm.

This last process takes advantage of high selectivity to improve process robustness and yield. However, since the membrane is heavily doped, it is impossible to host effective piezoresistors.

SUMMARY

This chapter is dedicated to the bulk microfabrication technology. The anisotropic wet etching of silicon is the most unique and frequently used process in the MEMS field. Section 10.3 discusses the rules for shape transformation from a two-dimensional mask to a three-dimensional etched profile with an increasing degree of mask complexity. Other etching techniques are introduced in this chapter as well.

At the end of this chapter, the reader should understand the following concepts and facts:

- The definition of anisotropic and isotropic etching.
- The etching profiles associated with common wet and dry etching agents including wet silicon anisotropic etch, plasma etch, deep reactive ion etch, isotropic wet etch, and gas-phase etch.
- Rules for mask transformation when using wet silicon anisotropic etching.
- Types of etched profiles at a given time during etch.

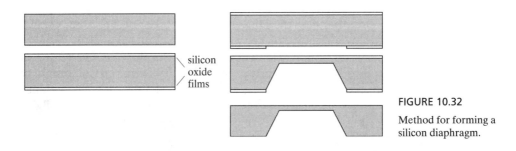

FIGURE 10.32

Method for forming a silicon diaphragm.

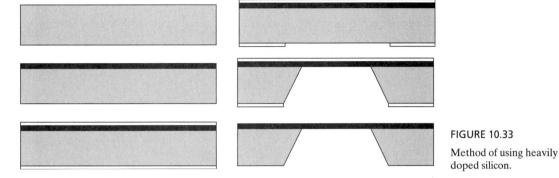

FIGURE 10.33
Method of using heavily doped silicon.

- Methods for identifying FSLC associated with a mask shape.
- Common fabrication methods for realizing suspended beams.
- Common fabrication methods for realizing suspended membranes.

PROBLEMS

Problem 10.1: Design
A silicon wafer is 500 μm thick. The front surface is {100}. A mask consists of a rectangular window of unknown size. The sides of the window are parallel to <110>. After through-wafer etch, a window (50 μm by 80 μm) is formed on the other side of the wafer. Find the size of the mask window. The undercut rate is negligible.

Problem 10.2: Design
Repeat Problem 10.1 if the etch rate of {111} surfaces is 1/100 that of the {100} surfaces.

Problem 10.3: Design
A through-wafer hole is shown in the following figure. A 10-μm opening is desired at the front side of the wafer. It is fabricated based on anisotropic etching of an <100>-oriented wafer. Determine the designed size of the window W at the back side of the wafer. The etch rate on the <111> surface is negligible.

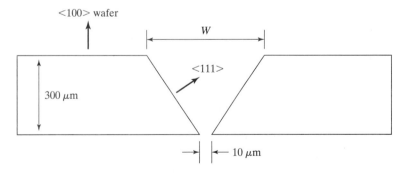

1. 222 μm
2. 435 μm
3. 745 μm
4. 377 μm

Problem 10.4: Fabrication

Prove whether the silicon underneath the center plate of the mask in Figure 10.34 will be completely undercut.

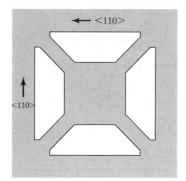

FIGURE 10.34

Etch mask for piezoresistive tactile sensor.

Problem 10.5: Design

Anisotropic silicon etching is used to create an inverted pyramid to serve as a microliter fluid reservoir. The mask layer is silicon nitride. The anisotropic etchant is EDP. Which of the following masks will yield the largest reservoir volume? Assume all the diagrams are drawn with same scale.

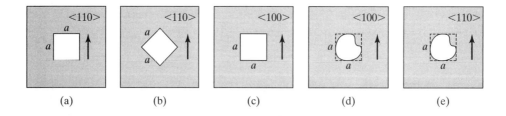

1. a
2. b and c
3. e
4. b and d
5. a and c
6. a, c, and d

Problem 10.6: Design

The mask shown below is used to create an anisotropically etched cavity in silicon. The top surface of the wafer is {100}. What is the size of the opening at the front surface of the wafer after one hour of etching time, taking into account the lateral undercut underneath the original mask? Assume that the etch rate in the <111> direction is finite at 0.05 μm/min, whereas the etch rate in <100> direction is 1 μm/min.

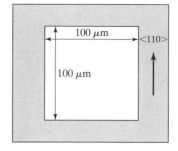

1. 100 μm × 100 μm
2. 103.6 μm × 103.6 μm
3. 103 μm × 103 μm
4. 101.7 μm × 101.7 μm
5. 102 μm × 102 μm
6. 107.3 μm × 107.3 μm

Problem 10.7: Design

Draw the top and cross-sectional view of the SLSP profile. Assume the etch rate in the <111> direction is zero.

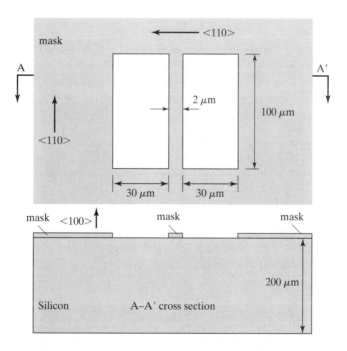

Problem 10.8: Design

Draw the top and cross-sectional view of the SLSP etched profile. Assume the etch rate in the <111> direction is 1/100 that in the <100> direction.

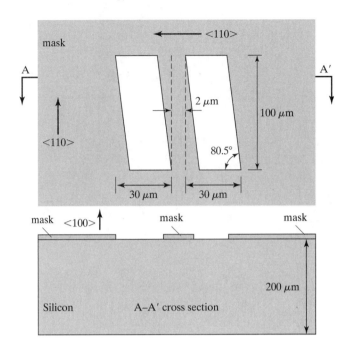

Problem 10.9: Design

Draw the top and cross-sectional view of the SLSP etched profile. Assume the etch rate in the <111> direction is zero.

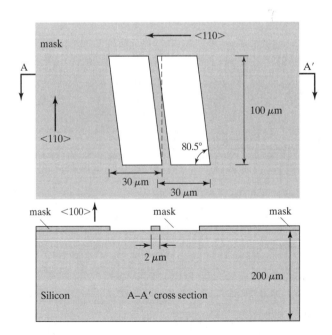

Problem 10.10: Design

Draw the top and cross-sectional view of the SLSP etched profile. Assume the etch rate in the <111> direction is zero.

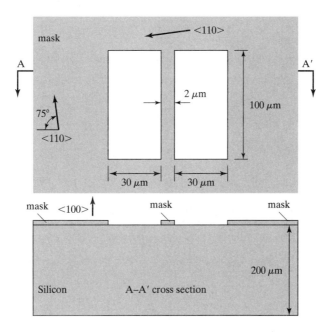

Problem 10.11: Design

Using computer-aided simulation tools to draw the mask and simulate the SLSP profile, assume the thickness of the wafer is much greater than the window opening. Assume $L = 100\ \mu m$.

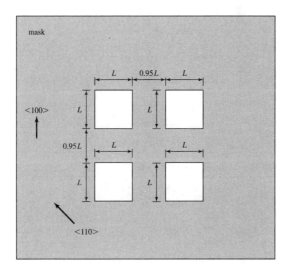

Problem 10.12: Design

Misalignment during the fabrication process can influence the outcome of anisotropic silicon etching. A square opening with length of each side being 100 mm is supposed to be aligned to the <110> direction. Angular misalignment can occur due to instrumental error or mistakes in correctly identifying the true crystallographic directions during the wafer manufacturing process. Suppose the error is 5°. What is the size of undercut at each side?

Problem 10.13: Design

Draw the top and cross-sectional view of etched profiles of the SLSP etched profile. Assume the etch rate in the <111> direction is zero.

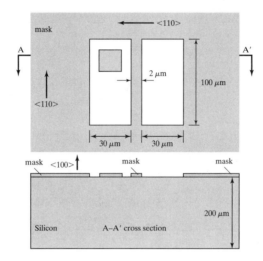

Problem 10.14: Design

Draw the top and cross-sectional view of the SLSP etch profile. Assume the etch rate in the <111> direction is zero.

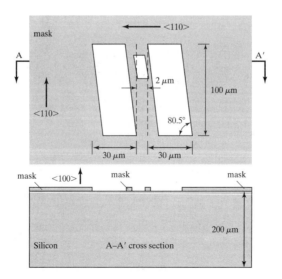

Problem 10.15: Fabrication

Determine the mask and three-dimensional profile evolution for generating an array of protrusion features as illustrated in Figure 10.35. The sloped surfaces consists of {111} planes and higher-index planes.

FIGURE 10.35
Results of tip etching.

Problem 10.16: Design

What is the terminal etching profile of Example 10.2 if the edge of the mask is aligned to the <100> instead of the <110> direction? Make line drawings of the mask and cavity (top view).

Problem 10.17: Fabrication

How many methods according to Figure 10.31 are suitable for making 200-nm thick silicon nitride membranes with a size of 1×1 mm, with a polycrystalline silicon piezoresistor located on the top surface of the silicon nitrite membrane? The polysilicon resistor, which is entirely located on the membrane and covers a fraction of the membrane area, is connected to metal wire leads. Out of all possible fabrication methods, discuss their advantages and disadvantages in terms of cost, process yield, and restrictions.

Problem 10.18: Fabrication

How many methods according to Figure 10.31 are suitable for making 200-nm thick silicon nitride membranes. Discuss the reason why unsuccessful methods fail. Out of all possible fabrication methods, discuss their advantages and disadvantages in terms of cost, process yield, and restrictions.

Problem 10.19: Fabrication

How many methods according to Figure 10.31 are suitable for making $1 \mu m$ thick Parylene membranes with a size of 1×1 mm? Discuss the reason why unsuccessful methods fail. Out of all possible fabrication methods, discuss their advantages and disadvantages in terms of cost, process yield, and restrictions.

REFERENCES

1. Kovacs, G.T.A., N.I. Maluf, and K.E. Petersen, *Bulk Micromachining of Silicon*. Proceedings of the IEEE, 1998. **86**(8): p. 1536–1551.
2. Yi, T., L. Li, and C.-J. Kim, *Microscale Material Testing of Single Crystalline Silicon: Process Effects on Surface Morphology and Tensile Strength*. Sensors and Actuators A: Physical, 2000. **83**(1–3): p. 172–178.
3. Bean, K.E., *Anisotropic Etching of Silicon*. IEEE Transaction on Electron Devices, 1978. **ED 25**: p. 1185–1193.
4. Seidel, H., et al., *Anisotropic Etching of Crystalline Silicon in Alkaline Solutions, I. Orientation Dependence and Behavior of Passivation Layers*. Journal of Electrochemical Society, 1990. **137**(11): p. 3612–3626.
5. Seidel, H., et al., *Anisotropic Etching of Crystalline Silicon in Alkaline Solutions II. Influence of Dopants*. Journal of Electrochemical Society, 1990. **137**(11): p. 3626–3632.
6. Sugiyama, S., M. Takigawa, and I. Igarashi, *Integrated Piezoresistive Pressure Sensor with Both Voltage and Frequency Output*. Sensors and Actuators A, 1983. **4**: p. 113–120.
7. Zhu, Z., and C. Liu, *Micromachining Process Simulation Using a Continuous Cellular Method*. IEEE/ASME Journal of Microelectromechanical Systems (JMEMS), 2000. **9**(2): p. 252–261.
8. Hubbard, T.J., and E.K. Antonsson, *Emergent Faces in Crystal Etching*. Microelectromechanical Systems, Journal of, 1994. **3**(1): p. 19–28.
9. Williams, K.R., and R.S. Muller, *Etch Rates for Micromachining Processing*. Microelectromechanical Systems, Journal of, 1996. **5**(4): p. 256–269.
10. Williams, K.R., K. Gupta, and M. Wasilik, *Etch Rates for Micromachining Processing—Part II*. Microelectromechanical Systems, Journal of, 2003. **12**(6): p. 761–778.
11. Oosterbroek, R.E., et al., *Etching Methodologies in <111>-Oriented Silicon Wafers*. Microelectromechanical Systems, Journal of, 2000. **9**(3): p. 390–398.
12. Strandman, C., et al., *Fabrication of 45-deg Mirrors Together with Well-Defined v-Grooves Using Wet Anisotropic Etching of Silicon*. Microelectromechanical Systems, Journal of, 1995. **4**(4): p. 213–219.
13. Raley, N.F., Y. Sugiyama, and T.V. Duzer, *(100)-Silicon Etch-Rate Dependence on Boron Concentration in Ethylenediamine-Pyrocatechol-Water Solutions*. Journal of Electrochemical Society, 1984. **131**(1): p. 161–171.
14. Wang, S.S., V.M. McNeil, and M.A. Schmidt, *An Etch-Stop Utilizing Selective Etching of N-type Silicon by Pulsed Potential Anodization*. Microelectromechanical Systems, Journal of, 1992: p. 187–192.
15. Peeters, E., et al., *PHET, An Electrodeless Photovoltaic Electrochemical Etchstop Technique*. Microelectromechanical Systems, Journal of, 1994. **3**(3): p. 113–123.
16. Landsberger, L.M., et al., *On Hillocks Generated During Anisotropic Etching of Si in TMAH*. Microelectromechanical Systems, Journal of, 1996. **5**(2): p. 106–116.
17. Douglas, M.A., *Trench Etch Process for a Single-Wafer RIE Dry Etch Reactor*. 1989, Texas Instruments Incorporated: US patent # 4,855,017.
18. Laermer, F., and A. Schilp, *Method for Anisotropic Plasma Etching of Substrates*, in *US PTO*. 1996, Robert Bosch GmbH: US patent 5,498,312.
19. Chen, K.-S., et al., *Effect of Process Parameters on the Surface Morphology and Mechanical Performance of Silicon Structures After Deep Reactive Ion Etching (DRIE)*. Microelectromechanical Systems, Journal of, 2002. **11**(3): p. 264–275.
20. Gardeniers, H.J.G.E., et al., *Silicon Micromachined Hollow Microneedles for Transdermal Liquid Transport*. Microelectromechanical Systems, Journal of, 2003. **12**(6): p. 855–862.

21. Lee, S., et al., *Surface/Bulk Micromachined Single-Crystalline-Silicon Micro-Gyroscope.* Microelectromechanical Systems, Journal of, 2000. **9**(4): p. 557–567.
22. Lee, S., S. Park, and D.-I. Cho, *The Surface/Bulk Micromachining (SBM) Process: A New Method for Fabricating Released MEMS in Single Crystal Silicon.* Microelectromechanical Systems, Journal of, 1999. **8**(4): p. 409–416.
23. Wang, S.N., et al., *Deep Reactive Ion Etching of Lead Zirconate Titanate Using Sulfur Hexafluoride Gas.* Journal of Americal Ceramic Society, 1999. **82**(5): p. 1339–1341.
24. Li, X., T. Abe, and M. Esashi, *Deep Reactive Ion Etching of Pyrex Glass Using SF6 Plasma.* Sensors and Actuators A: Physical, 2001. **87**(3): p. 139–145.
25. Zhang, C., C. Yang, and D. Ding, *Deep Reactive Ion Etching of Commercial PMMA in O_2/CHF_3, and O_2/Ar-based Discharges.* Journal of Micromechanics and Microengineering, 2004. **14**(5): p. 663–666.
26. Ayazi, F., and K. Najafi, *High Aspect-Ratio Combined Poly and Single-Crystal Silicon (HARPSS) MEMS Technology.* Microelectromechanical Systems, Journal of, 2000. **9**(3): p. 288–294.
27. Selvakumar, A., and K. Najafi, *Vertical Comb Array Microactuators.* Microelectromechanical Systems, Journal of, 2003. **12**(4): p. 440–449.
28. Petersen, K.E., *Silicon as a Mechanical Material.* Proceedings of the IEEE, 1982. **70**(5): p. 420–457.
29. Warneke, B., and K.S.J. Pister, *In situ Characterization of CMOS Post-Process Micromachining.* Sensors and Actuators A: Physical, 2001. **89**(1–2): p. 142–151.
30. Yao, T.-J., X. Yang, and Y.-C. Tai, *BrF3 Dry Release Technology for Large Freestanding Parylene Microstructures and Electrostatic Actuators.* Sensors and Actuators A: Physical, 2002. **97–98**: p. 771–775.
31. Schmidt, M.A., *Wafer-to-Wafer Bonding for Microstructure Formation.* Proc. IEEE, 1998. **86**(8): p. 1575–1585.
32. Tsau, C.H., S.M. Spearing, and M.A. Schmidt, *Fabrication of Wafer-Level Thermocompression Bonds.* Microelectromechanical Systems, Journal of, 2002. **11**(6): p. 641–647.
33. Albaugh, K.B., P.E. Cade, and D.H. Rasmussen. *Mechanisms of Anodic Bonding of Silicon to Pyrex Glass.* In *Solid-State Sensor and Actuator Workshop, 1988. Technical Digest, IEEE.* 1988.
34. Chavan, A.V., and K.D. Wise, *Batch-Processed Vacuum-Sealed Capacitive Pressure Sensors.* Microelectromechanical Systems, Journal of, 2001. **10**(4): p. 580–588.
35. Shimbo, M., et al., *Silicon-to-Silicon Direct Bonding Method.* Journal of Applied Physics, 1986. **60**(8): p. 2987–2989.
36. Cheng, Y.T., L. Lin, and K. Najafi, *Localized Silicon Fusion and Eutectic Bonding for MEMS Fabrication and Packaging.* Microelectromechanical Systems, Journal of, 2000. **9**(1): p. 3–8.
37. Mehra, A., et al., *Microfabrication of High-Temperature Silicon Devices Using Wafer Bonding and Deep Reactive Ion Etching.* Microelectromechanical Systems, Journal of, 1999. **8**(2): p. 152–160.
38. Field, L.A., and R.S. Muller, *Fusing Silicon Wafers with Low Melting Temperature Glass.* Sensors and Actuators A: Physical, 1990. **23**(1–3): p. 935–938.
39. Tong, Q.-Y., et al., *Low Temperature Wafer Direct Bonding.* Microelectromechanical Systems, Journal of, 1994. **3**(1): p. 29–35.
40. Singh, A., et al., *Batch Transfer of Microstructures Using Flip-Chip Solder Bonding.* Microelectromechanical Systems, Journal of, 1999. **8**(1): p. 27–33.
41. Cheng, Y.-T., L. Lin, and K. Najafi, *A Hermetic Glass-Silicon Package Formed Using Localized Aluminum/Silicon-Glass Bonding.* Microelectromechanical Systems, Journal of, 2001. **10**(3): p. 392–399.
42. Shivkumar, B., and C.-J. Kim, *Microrivets for MEMS Packaging: Concept, Fabrication, and Strength Testing.* Microelectromechanical Systems, Journal of, 1997. **6**(3): p. 217–225.
43. Han, H., L.E. Weiss, and M.L. Reed, *Micromechanical Velcro.* Microelectromechanical Systems, Journal of, 1992. **1**(1): p. 37–43.

44. Spiering, V.L., et al., *Sacrificial Wafer Bonding for Planarization After Very Deep Etching*. Microelectromechanical Systems, Journal of, 1995. **4**(3): p. 151–157.
45. Niklaus, F., S. Haasl, and G. Stemme, *Arrays of Monocrystalline Silicon Micromirrors Fabricated Using CMOS Compatible Transfer Bonding*. Microelectromechanical Systems, Journal of, 2003. **12**(4): p. 465–469.
46. Bang, C.A., et al., *Thermal Isolation of High-Temperature Superconducting Thin Films Using Silicon Wafer Bonding and Micromachining*. Microelectromechanical Systems, Journal of, 1993. **2**(4): p. 160–164.
47. Yun, C.-H., and N.W. Cheung, *Fabrication of Silicon and Oxide Membranes Over Cavities Using Ion-Cut Layer Transfer*. Microelectromechanical Systems, Journal of, 2000. **9**(4): p. 474–477.
48. Yang, E.-H., and D.V. Wiberg, *A Wafer-Scale Membrane Transfer Process for the Fabrication of Optical Quality, Large Continuous Membranes*. Microelectromechanical Systems, Journal of, 2003. **12**(6): p. 804–815.

CHAPTER 11

Surface Micromachining

11.0 PREVIEW

Surface micromachining methods are used widely in MEMS. As the name suggests, surface micromachining processes are responsible for creating microstructures that reside near the surfaces of a substrate. Unlike bulk micromachining, surface machining does not involve removal or etching of bulk substrate materials.

A key technique for making suspended microstructures is the sacrificial etching. We will discuss general sacrificial etching techniques in Section 11.1 using polysilicon-based electrostatic micromotors as examples. Three generations of micromotors with increasing complexity are presented.

There are many candidate materials for surface micromachining. In Section 11.2, we will discuss criteria for selecting structural and sacrificial materials.

Sacrificial layer etching is often performed in liquid solutions. Many suspended, compliant micromechanical structures cannot survive the drying process without special designs and procedures. Drying-related failures and correction methods are reviewed in Section 11.3.

Surface micromachined structures can be used to form strong three-dimensional devices by employing post-processing assembly. Representative assembly techniques are discussed in Section 11.5.

Surface micromachining processes are more closely related to, and compatible with, integrated circuit fabrication than bulk micromachining. In recent years, commercial MEMS devices are increasingly being made in foundries. In Section 11.6, we will outline pertinent issues related to the use of foundry processes.

11.1 BASIC SURFACE MICROMACHINING PROCESSES

11.1.1 Sacrificial Etching Process

A simple, two-layer sacrificial etching process for forming a suspended cantilever beam is illustrated in Figure 11.1. Here, the desired freestanding structure is a fixed–free cantilever anchored at one end to the substrate. The fabrication process starts with the deposition of a sacrificial

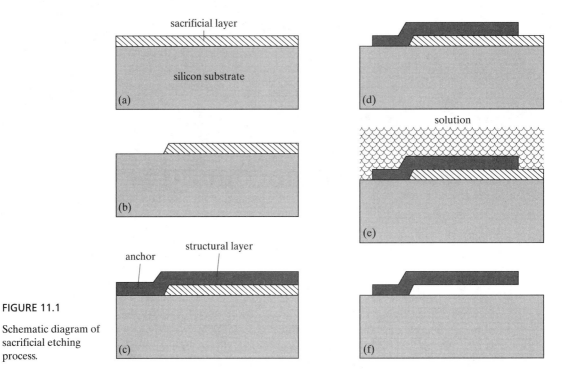

FIGURE 11.1

Schematic diagram of sacrificial etching process.

material on the silicon wafer (Figure 11.1a). The sacrificial material is defined by photolithography and patterned (Figure 11.1b), then it is followed by the deposition and patterning of a structural thin film (Figure 11.1c and d). The sacrificial material should ideally provide a mechanically rigid and chemically reliable support for the structural layer. In other words, the sacrificial layer needs to behave as a reliable *placeholder* during the process. The sacrificial material is later removed selectively to free the overlying structural layer (Figure 11.1e). If the sacrificial layer etching is conducted in a wet chemical solution, the liquid must then be removed to produce the final structure (Figure 11.1f).

11.1.2 Micromotor Fabrication Process—A First Pass

In Sections 11.1.2 through 11.1.4, we will review three complete surface micromachining processes for making electrostatic motors. For details about the operational principles of the micromotor, refer to Chapter 4. The fabrication involves the deposition, patterning, and etching of *multiple* structural and sacrificial layers. Major performance objectives include friction reduction and wear resistance.

We will discuss the processes for realizing micromotors in three incremental passes. In this section, we discuss a generic process. In Section 1.3, a process that allows a reduced contact area between the rotor and the substrate is discussed. In Section 1.4, we review a more complex design that reduces sidewall friction and wear by employing low-friction side bearing.

The basic motor fabrication process is shown in Figure 11.2. The process begins with a silicon wafer. Since no bulk machining is involved, the orientation of the wafer is not relevant. The

11.1 Basic Surface Micromachining Processes

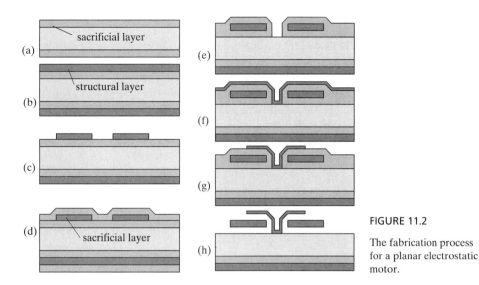

FIGURE 11.2

The fabrication process for a planar electrostatic motor.

wafer is first covered by a sacrificial thin film, such as silicon oxide deposited by using the LPCVD method. (In this case, the thin film covers both sides of the wafer. However, only the materials on the front side are relevant to the outcome of fabrication.)

A structural material, polycrystalline silicon, is deposited on the wafer next (Figure 11.2b). In this case, the structural layer forms the rotor.

In Figure 11.2c, a photoresist thin film covers the front surface of the wafer. The photoresist, after development and curing, serves as a mask in a subsequent reactive ion etching step, which transfers the pattern in the photoresist to the polysilicon structural layer. The photoresist is then removed, either using oxygen plasma (dry etching) or organic solvents (wet etching).

The wafer is covered with yet another layer of oxide sacrificial material (Figure 11.2d). The deposited film conformally covers both horizontal surfaces and vertical sidewalls. The material of the second sacrificial layer may be different from the previous one, although in this case, LPCVD silicon oxide is a convenient choice. It offers satisfactory step coverage and temperature compatibility.

A center hub that restricts the lateral translation of the rotor is made next. The hub also prevents the rotor from escaping from the substrate. In order to produce the hub, an anchor window reaching to the substrate is made (Figure 11.2e). It is accomplished by photolithography and wet chemical etching through two sacrificial layers. A second structural layer (Figure 11.2f) is deposited. It is firmly attached to the substrate via the open window.

A photoresist film covers the second structural layer and then undergoes patterning and development. It is used to define the second structural layer (Figure 11.2g). Finally, both sacrificial layers are removed by immersing the wafer into HF etch solutions.

Though the rotor is shown to be at an elevated position in the drawing of the finished device (Figure 11.2h), it could easily drop to the substrate under gravity and form contact on its broad side. The rotor may also be stuck to the substrate if the wafer is not properly dried.

11.1.3 Micromotor Fabrication Process—A Second Pass

The fabrication process discussed earlier is capable of generating a basic micromotor. However, its structure and performance is far from optimal. One major drawback associated with the design shown in Figure 11.2 is the lack of friction control. One way for mitigating the friction is to reduce the contact area between the rotor and the substrate. A process for realizing such a feature is discussed in the following and diagrammed in Figure 11.3.

Starting with a silicon wafer, a dielectric insulating thin film is deposited first (Figure 11.3a). Notice that only the deposited material on the front side is shown for brevity. (Readers should

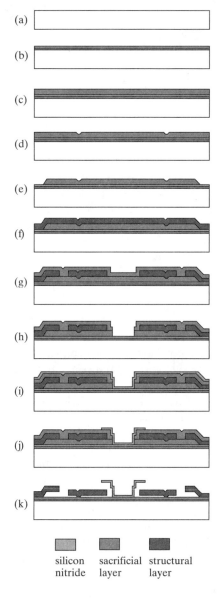

FIGURE 11.3

Micromotor fabrication.

silicon nitride | sacrificial layer | structural layer

keep in mind that materials are also deposited on the back side during some steps.) In Figure 11.3b, a sacrificial layer is deposited. Two sets of photolithography and etching steps are performed to produce the cross-sectional profile illustrated in Figure 11.3e. The first step involves coating the sacrificial layer with a photoresist, patterning and developing the photoresist, and using the photoresist as a mask to remove a *portion* of the sacrificial layer thickness (Figure 11.3d). The photoresist layer is stripped and then replaced by a new one. A new round of processing steps—including photoresist coating, patterning, and development of photoresist—is performed again. This time, the exposed sacrificial layer is etched through its entire thickness (Figure 11.3e).

Process steps from Figure 11.3d through Figure 11.3k parallel the process discussed in Section 11.1.2. The only difference is that the second structural layer (for the hub) is silicon nitride instead of polycrystalline silicon. In Figure 11.3, the sacrificial layer is completely removed to free the rotor. Notice that the rotor may contact the substrate only at tiny bumps defined in Figure 11.3c. The effective contact area is much reduced compared to the previous case—this feature reduces friction and the chance of the rotor being stuck to the substrate.

The rotor and the hub are not in tight contact at all times. The rotor and the hub must be separated by an in-plane gap whose size is determined by the thickness of the second sacrificial layer applied. During high-speed rotation, the rotor may wobble within the constraint of the hub and create alternating contact points, much like a hula ring moves around the waist of a dancer.

11.1.4 Micromotor Fabrication Process—Third Pass

The contact between the rotor and the hub creates additional friction and wear. One solution for this problem is to reduce the friction coefficient between the rotor and the hub by making contact surfaces out of silicon nitride.

A process for integrating a sidewall silicon nitride bearing structure is illustrated in Figure 11.4. A silicon wafer is first coated with a dielectric barrier consisting of silicon oxide and silicon nitride (Figure 11.4a). A polysilicon grounding plane is deposited (Figure 11.4b), followed by the deposition and patterning of a first sacrificial layer (Figure 11.4c). A second polysilicon structural layer (Figure 11.4d) is conformally deposited. The oxide sacrificial layer and the polysilicon layer are patterned photolithographically and etched. A layer of silicon nitride is deposited conformally, covering horizontal and vertical surfaces (Figure 11.4f).

A globally reactive ion etching follows. The process parameter is designed such that the etch rate in the vertical direction is greater than that in the horizontal direction due to physical etching. While silicon nitride on front-facing surfaces (such as the hub bottom and rotor front surface) is removed, the silicon nitride on vertical surfaces is retained (Figure 11.4g).

A timed wet etch is performed to create controlled lateral undercut in the first sacrificial layer (Figure 11.4h). Following this step, a layer of sacrificial material is deposited conformally. The sacrificial layer covers all surfaces, including those below the overhanging structures (Figure 11.4i). A photoresist layer is spin coated and patterned, exposing the bottom of the hub region. Sacrificial material in the open window is etched to gain access to the underlying polycrystalline silicon (Figure 11.4j). After the deposition and patterning of another layer of silicon nitride as the hub (Figure 11.4k), a global sacrificial etch is performed to release the rotor.

The finished motor, with the cross section shown in Figure 11.4l, contains friction-control films on the sidewall of the hub. Silicon nitride deposited in the undercut region created in Figure 11.4h prevents the rotor from dropping to the substrate surface.

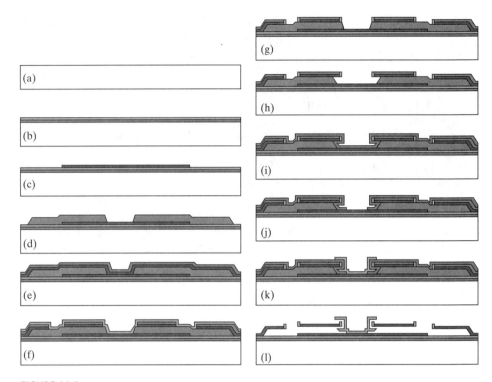

FIGURE 11.4

A motor with sidewall anti-friction coating.

11.2 STRUCTURAL AND SACRIFICIAL MATERIALS

There is a large selection of candidate structural and sacrificial materials. The choice of materials depends on the specific needs of each individual application. In this section, we will first evaluate the general criteria for identifying successful candidate materials for structural and sacrificial layers. The major focus is on the thin films of polysilicon, oxide, and silicon nitride. Other material systems, including polymers and metals, are also discussed in Section 11.2.3.

11.2.1 Material Selection Criteria

Some essential prerequisites for selecting structural and sacrificial layer materials in a basic two-layer process include the following:

1. The deposition of the structural material on top of the sacrificial material must not cause the sacrificial layer to melt, dissolve, crack, disintegrate, or become unstable or destroyed in any way;
2. The method used for patterning the structural layer must not attack the sacrificial layer or any existing layers on a substrate;

3. The method used for removing the sacrificial layer must not attack, dissolve, or destroy the structural layer or the substrate. The sacrificial undercut etch rate should be adequately high.

The most established material pair is the polycrystalline silicon (as a structural layer) and the phosphosilicate glass (or PSG, as the sacrificial layer). This material system satisfies the three criteria mentioned earlier. The PSG can withstand the deposition of the polysilicon layer. The polysilicon can be patterned by using plasma etching, which etches PSG and oxide at reduced rates. The PSG and oxide can be removed by using wet HF acid solution, which exhibits a very high undercut etch rate on PSG but a minimal etch rate on silicon.

Many surface micromachining processes involve multiple structural and sacrificial layers. A number of general factors should be considered when selecting the structural and sacrificial layers. These include (1) etch rates and etching selectivity; (2) achievable layer thickness; (3) temperature of material deposition; (4) intrinsic stress of the structural layer; (5) surface smoothness; and (6) cost of materials and processes. Several of these issues are elaborated in greater detail below.

Etch rate and etch selectivity. During the sacrificial etching, both the structural layer and the sacrificial layer are exposed to etchants. The etch rates of the etchant on the sacrificial layer and the structural layer are labeled r_{sa} and r_{st}, respectively. The etch selectivity, defined as the ratio of r_{sa} and r_{st}, should be as high as possible.

For devices involving multiple structural layers and sacrificial layers, the successful choice of materials, etchants, and etching conditions must strive to maintain high etch selectivity at each and every step of the process.

It takes time and effort to build a comprehensive and reliable etch rate table. A good starting point for beginners is published etch rate data. Papers published by Williams et al. [1, 2] archived the cross reactivity between 53 materials and 35 etching methods. However, it should be noted that etch rates can be influenced by a large number of factors.

If two competing process options have identical selectivity, the one with the higher r_{sa} is preferred. If a process has a lower etch rate on the sacrificial material but a higher etch selectivity over another, the choice should be made by considering other additional factors.

Achievable layer thickness. Materials that can cover a wide range of thickness are always preferred. Many applications call for films with large thicknesses. However, the practical thickness of materials that can be realized for a thin-film structure depends on several factors. Often, the thickness of a material is limited by one of the following reasons: finite deposition rate and excessive processing time, inherent physical and chemical limitations, or unacceptable stress associated with thick films. For example, the time it takes to grow a 10-μm-thick polysilicon film would be around 10–15 hours, which is feasible in theory but expensive in practice. For some materials, the stress and stress-induced bending may reach an unacceptable level when the thickness is large. For example, it is rare to grow LPCVD silicon nitride beyond a thickness of 2 μm. Long processing time aside, a suspended silicon nitride membrane may crack under intrinsic tensile stress on its own if the thickness reaches a critical value.

Temperature of material deposition. High-quality materials that can be grown at low temperatures are always desirable. High temperatures associated with the deposition or modification of a certain layer may limit the choice of materials for underlying layers. A specific concern with regard to the deposition temperature arises when the micromechanical components are integrated with on-chip signal-processing circuitry. Exposure to high temperature over long durations may cause the dopants to out-diffuse and thereby damage the transistors in an irreversible manner.

Surface roughness. The roughness of the front and bottom surfaces of the micromachined structure layers is a pertinent concern in many cases. Rough surfaces may be desired for reducing adhesion, whereas smooth surfaces may be used to increase optical reflectivity, reduce loss of radio frequency signals, or increase resonant quality factors.

11.2.2 Thin Films by Low-Pressure Chemical Vapor Deposition

A widely practiced method for depositing structural and sacrificial materials is the **chemical vapor deposition** (CVD). In a CVD deposition chamber, solid thin films are formed on wafers by the condensation of vapor or the adhesion of solid-phase reaction byproducts. The material growth is typically conducted in a sealed chamber to prevent the introduction of particles and the leakage of reaction gases.

The reaction energy is provided by heat or by plasma power. If the energy is provided by heat alone and conducted under low processing pressure (i.e., a few hundred mtorr), the process is called **low-pressure chemical vapor deposition (LPCVD)**. If the energy is provided by plasma power, the process is called **plasma-enhanced chemical vapor deposition (PECVD)**.

In an LPCVD process, wafers are placed in the chamber filled with flowing process gases (Figure 11.5). The chamber is typically made of quartz to sustain high deposition temperatures. The typical range of pressure is 200–400 mtorr (1 torr = 1/760 of 1 atmosphere pressure).

There are three major LPCVD materials used in MEMS: polycrystalline silicon, silicon nitride, and silicon dioxide. LPCVD polysilicon and oxide materials were established in the integrated circuit industry. Naturally, these two materials were the first to be used in MEMS. Professors Howe and Muller at the University of California at Berkeley, among others, were pioneers at establishing their mechanical properties and process guidelines. The LPCVD polycrystalline silicon (polysilicon) material is used extensively in integrated circuit manufacturing. Doped polysilicon typically serves as the transistor gates, conductors, and resistors. Undoped LPCVD oxide is often deposited on a finished chip as an encapsulation.

LPCVD Polycrystalline Silicon Polysilicon is deposited within a temperature range of 580–620°C by decomposing silane gas (SiH_4) [3]. The reaction formula is $SiH_4 = Si + 2H_2$. Process parameters such as pressure, gas flow rates, and temperature determine the microstructure; hence, they determine the electrical and mechanical properties of the polysilicon films [4, 5]. For

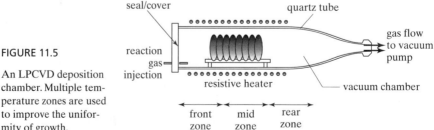

FIGURE 11.5

An LPCVD deposition chamber. Multiple temperature zones are used to improve the uniformity of growth.

example, it is known that at a deposition temperature below 580°C, amorphous silicon will form in a horizontal LPCVD system [6].

Doping affects not just electrical characteristics, but mechanical ones as well (e.g., stress of polysilicon [7]). The doping of polysilicon can be achieved using two methods. First, the dopant atoms can be incorporated into undoped polysilicon by diffusion or ion implantation. Second, the dopant atoms can be incorporated *in situ* during the LPCVD growth by introducing additive gas components containing dopant atoms.

LPCVD Silicon Nitride The LPCVD silicon nitride is an insulating dielectric, and it generally exhibits tensile intrinsic stress. It is deposited at around 800°C by reacting silane gas (SiH_4) or dichlorosilane ($SiCl_2H_2$) with ammonia (NH_3). The reaction formula in the silane case is $3SiH_4 + 4NH_3 = Si_3N_4 + 24H$. The chemical formula for stoichiometric silicon nitride is Si_3N_4. However, if the concentrations of gases deviate from the stoichiometric condition, a nitride film with the chemical formula Si_xN_y will result.

LPCVD Silicon Dioxide LPCVD oxide is deposited under relatively a low temperature, e.g., 500°C, by reacting silane with oxygen. The reaction formula is $SiH_4 + O_2 -> SiO_2 + 2H_2$. If no additive gas is introduced during the reaction, the oxide is undoped and called **low temperature oxide** (LTO). Phosphorus atoms may be introduced into LTO when an additive gas, phosphine (PH_3), is used in conjuction with silane and oxygen. The specific form of P-doped oxide is called **phosphosilicate glass** (PSG). The incorporation of P atoms increases the etch rate of oxide in HF acid solutions. The etch rate on PSG with 4 wt % phosphors can reach more than 1 μm/min in concentrated HF etchants (40%). The PSG material is also known to soften and reflow under high temperature (e.g., above 900°C), which creates smooth, rounded step edges.

The oxide sacrificial layer is often removed with liquid etchant. However, the dry etching of oxide by HF vapor has also been reported [8].

Process Parameter Control Process parameters (including the gas mixture ratio, flow rate, pressure, and temperature) influence the electrical, mechanical, and thermal properties of LPCVD films.

Process parameters affect the thermal properties of LPCVD structural films. For example, the thermal conductivity of LPCVD thin films have been studied [9–11]. The thermal conductivity and other thermal characteristics of low-stress silicon nitride has been investigated in Reference [12]. One comprehensive investigation [13] showed that the grain size, the concentration, and the type of dopants affect the thermal conductivity of polysilicon.

Process parameters also affect the intrinsic stress of LPCVD films. The stress levels of LPCVD materials may be tensile or compressive; the magnitude maybe determined by a variety of factors, including processing temperature, temperature profile during ramping, gas composition, and film thickness. Silicon nitride films generally exhibit tensile stress. Lowering of stress can be achieved by varying processing parameters and gas compositions [12]. Stress in polysilicon thin films is low (in terms of magnitude) compared with oxide and silicon nitride. The stress can be further reduced by regular thermal annealing [3, 14, 15] or rapid thermal annealing [16, 17].

Alternative Methods Silicon nitride, polysilicon, and silicon oxide can be deposited using methods other than LPCVD. The previously discussed LPCVD deposition processes result in wafer scale and a global deposition of materials. Polycrystalline or amorphous silicon have been grown using local chemical vapor deposition by resistively heating up microstructures [18]. In such cases, CVD growth only occurs on the heated element.

There is practical limit to the thickness of the sacrificial oxide grown by CVD methods due to finite deposition rates. If thicker oxide is needed, other forms of oxide can be used. For example, spin-on glass (SOG) has been explored as a sacrificial material [19]. (It has also been used as a structural material [20].)

It is also possible to deposit silicon, oxide, and silicon nitride using PECVD methods. Compared with LPCVD, the PECVD method is conducted at a lower processing temperature. Unfortunately, low temperature also means a lower material density and a reduced resistance to etching.

Alternatively, thin-film materials such as polysilicon may be formed by sputtering off commercial targets. The sputtering process can be performed at room temperature.

11.2.3 Other Surface Micromachining Materials and Processes

Other LPCVD materials, such as silicon germanium (SiGe) and polycrystalline germanium, are emerging as viable structural materials for MEMS in recent years. SiGe is being actively pursued in MEMS applications because its low processing temperature (450°C compared with 580°C for polysilicon) allows the processing of MEMS on a wider variety of substrate materials and enhances compatibility with CMOS [21–23]. The growth rate is also greater compared with polysilicon. Polycrystalline germanium offers a low deposition temperature ($<350°C$) and an excellent etch selectivity to thin-film materials commonly used in silicon MEMS [24].

Besides semiconductor materials and related thin films, other classes of materials are also used as structural and sacrificial layers. These materials include polymers and metal thin films. Compared with LPCVD materials, polymers and metals can be deposited and processed under a much lower temperature and using simpler equipment.

Polymers that are often used in MEMS today are reviewed in Chapter 12. Polymer materials can be deposited using a variety of means, including spin coating, vapor coating, spray coating, and electroplating. Polymers can serve as structural layers and provide unique mechanical, electrical, and chemical characteristics not available in semiconductor films.

As a sacrificial layer, polymer thin films can be removed by dry etching (e.g., oxygen plasma etching) or by using strong organic solvents such as acetone. Nonconventional fabrication processes are available for various polymers. For example, one method for removing the Parylene sacrificial layer is to first turn Parylene into carbon and then remove the carbon by reacting it with oxygen to turn solid carbon into carbon dioxide gas [25]. This process circumvents the stiction problem associated with liquid phase etching and subsequent drying.

Elemental metals (including gold, copper, nickel, aluminum, and metal alloys) can be used as sacrificial layers or structural layers [26–31]. Thin metal films (e.g., less than 1 μm) can be deposited by evaporation or sputtering. Thick metal layers (e.g., greater than 2 μm) can be made by using electroplating.

Even silicon—either in single-crystal or polycrystalline forms—can be used as a sacrificial material. Silicon sacrificial etching has been demonstrated using gas phase etchants, such as XeF_2 or BrF_3 [32, 33].

11.3 Acceleration of Sacrificial Etch

TABLE 11.1 Possible Combination of Sacrificial Layer (Columns) and Structural Layer (Rows). "No" Indicates Generally Impossible Combinations.

	Sacrificial layer			
Structural layer	CVD PSG or thermal oxide	Photoresist	Parylene	Metal
LPCVD polysilicon	OK	No, deposition temperature too high for resist	No, deposition temperature too high for Parylene	No, many metals cannot sustain the high temperature of LPCVD polysilicon
LPCVD silicon nitride	OK	No, deposition temperature too high for resist	No, deposition temperature too high for resist	No, deposition temperature too high for resist
Metal	OK[1]	OK[2]	OK[3]	OK (if different metals)
Photoresist	No, HF etching solution may attack resist	No, structural layer and sacrificial layer are etched simultaneously	No, all methods for etching Parylene (including dry etching) attack the resist structural layer	OK
Parylene	OK	OK, organic solvents may attack resist but not Parylene	N/A	OK

[1] Certain oxide etchants (such as concentrated HF) may attack certain metal.
[2] Evaporated metal may increase the temperature of wafer and cause polymer to locally melt. Carefully processing control is required.
[3] The Parylene (as sacrificial layer) must be removed using oxygen plasma, which may oxidize certain metals.

Example 11.1. Material Selection of a Two-Layer Surface Micromachining Process

Discuss the general compatibility between four sacrificial materials (CVD oxide, photoresist, Parylene, and metal) and five structural materials (CVD polysilicon, CVD silicon nitride, metal, photoresist, and Parylene). How many pairs are viable structural–sacrificial material combinations in a two layer process?

Answer. The compatibilities between the listed structural and sacrificial materials are summarized in Table 11.1. Overall, there are ten feasible material pairs.

11.3 ACCELERATION OF SACRIFICIAL ETCH

The sacrificial release of large-area plates or the removal of sacrificial material inside long or blind channels can be very time consuming. Due to small dimensions (thickness in the case of large-area plate, and cross section in the case of channels), the transport rate of fresh etchants and byproducts decreases as the etching progresses.

The chemical transport characteristics for the wet chemical etching of sacrificial materials in long blind channels has been studied for PSG (with silicon nitride as sacrificial layer) [34] and photoresist (with Parylene as sacrificial layer) [35].

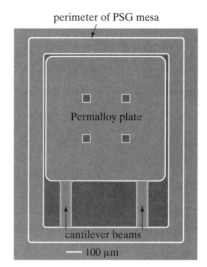

FIGURE 11.6

Surface micromachined magnetic actuator.

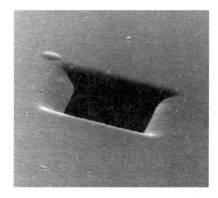

FIGURE 11.7

SEM micrograph of an etch hole.

One strategy for decreasing the overall time for sacrificially undercutting large-area plates is to deploy small openings called **etch holes**. A scanning electron micrograph of suspended micromachined flaps, each supported by two cantilever beams, is shown in Figure 11.6. The area of the flap is 200 μm $\times$ 200 μm. Four etch holes on each flap initiate sacrificial undercut from within the interior of the flap, as well as from the edges. It can significantly shorten the required time of etching. A micrograph of an etch hole in polysilicon is shown in Figure 11.7. In most cases, the etch holes are small so as not to adversely influence performances.

In certain applications, however, etch holes may exert a finite influence on the performance of devices. For example, etch holes on an optical reflector cause diffraction as well as reduction of the reflectance [36]. In cases where etch holes are not preferred or allowed, alternative etching methods and materials become necessary. A number of possible techniques have been demonstrated in the past. The following materials have been reported as having an extremely fast etch rate and high selectivity: (1) Dendritic polymer, such as hyper branched polymers (HBPs) [37], and

(2) ZnO thin films. The etch rate on ZnO by 2% HCl exceeds 1000 Å/s with no bubble formation. If the sacrificial material is metal, the undercut may be accelerated by an electrolytic dissolution with an active electrical bias [38]. Furthermore, self-assembled monolayers can be used to release large-area microdevices without wet chemical etching [39].

Researchers have also developed porous silicon (with through-film pores measuring 10–50 nm in diameter) for use as structural layers. The pores allow underlying sacrificial layers to be removed rapidly without resorting to macroscopically defined etch holes [40, 41]. Because the holes are nanoscopic in sizes, they have a minimal impact on performances.

11.4 STICTION AND ANTISTICTION METHODS

Sacrificial material removal is often accomplished using wet chemical solutions because of the high speed of etching, simplicity of setup, and generally good selectivity. It is necessary to dry the wafer and chips afterwards by natural or forced evaporation. However, the drying process is not always straightforward. The situation is explained in Figure 11.8.

As liquid is gradually removed through evaporation, the top surface of a microstructure is exposed to air. Liquid trapped underneath the suspended microstructure will take much longer to escape. A surface tension force develops at the interface of trapped liquid and air acting in the direction that is tangent to the liquid–air interface. For macroscopic devices, the surface tension force is negligible and does not cause any significant deformation. However, since microscale devices often use compliant structures and involve small gap spacing, the surface tension forces can cause appreciable deformation of the surface microstructures. Often, contacts with the substrate are made.

Detailed studies have been conducted to characterize the surface energies of the bonding of selective material systems [42]. The contact between the suspended structure and the substrate can lead to irreversible damage. Upon contact, strong molecular forces (e.g., van der Waals forces) are incurred to reinforce the attraction. Further, solid bridging is very likely due to the presence of fresh reaction byproducts.

This failure mode of microstructures is referred to as **stiction**; this is a hybrid word combining *sticking* and *friction*.

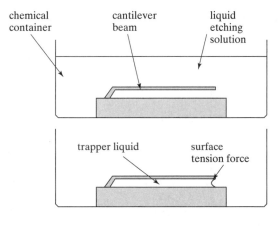

FIGURE 11.8

Schematic diagram of drying process.

Many practical methods have been developed to address the stiction issue. These methods take one of four approaches:

1. *Eliminate the capillary attractive force* by modifying the solid–liquid interface chemistry [43];
2. *Prevent the excessive bonding force* from occurring, for example, by elevating the solution temperature [44] or reducing the surface contact areas [45];
3. *Release structures that are stuck to the substrate* using various energetic input forms, locally or globally [46, 47];
4. *Provide counteractive force* to mechanical structures to prevent contact, for example, by taking advantage of bending stemming from intrinsic stress [48].

Two representative methods under these approaches are discussed in more detail in the following paragraphs.

Adverse surface tension associated with the liquid–air interface can be circumvented by using the supercritical fluid drying method. This technique was initially developed in the biology community to dehydrate delicate biological tissues for morphology studies while avoiding structure collapses and deformation. Let's examine the process by observing a generic phase diagram of a solvent material (Figure 11.9). Three phases—solid, liquid, and air—are the most familiar to readers. In addition, a *supercritical phase* occurs at high pressure and temperature.

A typical drying process utilizing the supercritical phase is described below. The chip containing released microstructures is immersed in a liquid and placed under a moderate pressure (instead of the atmosphere pressure) and room temperature. This starting condition is indicated by point 1 on the phase diagram. The temperature of the liquid is increased while the pressure constant is maintained. The solvent makes a transition from the liquid phase to the supercritical phase (point 2). The pressure of the supercritical fluid is then dropped, causing the supercritical fluid to turn into a vapor (point 3). The transformation from liquid to supercritical fluid and then from supercritical fluid to vapor involves virtually zero surface tension.

The stiction can be reduced by using hydrophobic coatings on the microstructure and the substrate to reduce bonding energy. Surfaces can be terminated with long-chain molecules that self-assemble to form a hydrophobic self-assembled monolayer (SAM) [43, 49]. Alternatively, fluorocarbon materials (similar to Teflon) deposited by plasma methods can be used [50].

Hydrophobic surface treatments find applications beyond antistiction coating. For example, the use of hydrophobic patterned regions reduces surface bonding energy and allows lifting of wafer-scale devices from a mold without using sacrificial wet etching [39]. Local

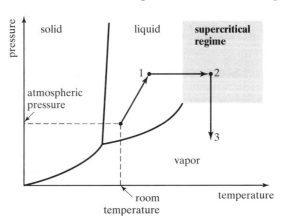

FIGURE 11.9

Carbon dioxide supercritical drying.

hydrophobicity modification also allows the automated assembly of small parts in an self-organized manner [51, 52].

11.5 ASSEMBLY OF 3D MEMS

Surface micromachining is very efficient compared with bulk micromachining. However, surface microstructures generally reside very close to the substrate. Many interesting solutions for realizing strongly three-dimensional structures out of two-dimensional surface micromachined structures have been proposed and investigated in the past decade.

Microstructures connected to rotational hinges have been realized using surface micromachined technology (Figure 11.10). The hinges allow in-plane microstructures to be erected with respect to the substrate plane [53].

The fabrication process of the hinge using polysilicon as the structural layer and oxide as the sacrificial layer is discussed below. First, a layer of sacrificial material, sacrificial layer 1, is deposited using LPCVD (Figure 11.11a). Another thin film, structural layer 1, is deposited (Figure 11.11b) and photolithographically patterned (Figure 11.11c). Yet another sacrificial layer, sacrificial layer 2, is deposited again, covering the structural layer conformally (Figure 11.11d). A thin-film photoresist is coated and patterned. It serves as a mask for realizing substrate access holes through the two sacrificial layers (Figure 11.11e). Another structural layer, structural layer 2, is deposited, filling the holes created in the previous step (Figure 11.11f). The wafer is covered with a spin-on photoresist and then photolithographically patterned (Figure 11.11g). The remaining photoresist serves as a mask to define the hinge housing (Figure 11.11i). A global sacrificial layer etch is performed to release the structure layers, which can now rotate freely with respect to the hinge housing.

This general process can be modified to create more complex devices. For example, a vertical micro windmill has been made in the past [54].

Devices connected by hinges are generally very small, and they are beyond the capability of efficient manual handling. Hinged structures can be moved out of plane using a variety of means and forces, including (1) parallel assembly using fluid disturbance during rinsing; (2) serial assembly using a micromanipulator [55]; (3) parallel assembly using built-in stress in layers for prying microstructures out of plane upon release; (4) surface tension forces created by the phase change of polymer or solder materials [56–58]; (5) parallel assembly using external magnetic field [59], or thermal kinetic forces resulting from thermally generated bubbles (at elevated temperatures of 100°C and a low pressure of 10 torr) [60]. Assembly using electrostatic or thermal actuators has also been demonstrated (e.g., Reference [61]), but they require more a complex setup and larger surface areas.

Any method of accomplishing off-plane angular displacement must have high yield and high efficiency (parallel actuation). It is also critical that any on-chip mechanism for achieving actuation must occupy minimal chip area. Two such methods are discussed below. Both use magnetic torque as the driving force for assembly. In Case 11.1, the rotation is enabled by a surface micromachined hinge. In Case 11.2, the rotation is produced by plastically bending a cantilever made of ductile metal.

FIGURE 11.10

A hinged microstructure rotated off the substrate.

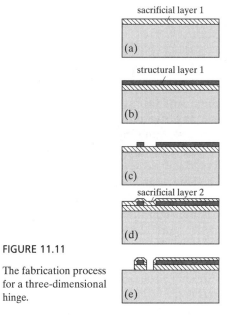

FIGURE 11.11

The fabrication process for a three-dimensional hinge.

Case 11.1. Magnetic Assisted Three-Dimensional Assembly

We will now discuss a method for deploying (rotating) hinged flaps using magnetic actuation [59]. A piece of electroplated Permalloy material is attached to a hinged flap. After the sacrificial etch and drying, the flap rests in the substrate plane. When an external magnetic field is applied, the Permalloy develops a magnetization M, which interacts with the external field to produce a torque. The induced torque causes the flap to rotate out of plane about its base.

The assembly of hinged structures using magnetic forces has several advantages, including high efficiency in chip-scale or wafer-scale parallel actuation and a minimum dedicated chip area for active actuation. A global magnetic field is capable of addressing a large number of devices in parallel. Since the Permalloy piece overlaps the area of the microflap, the actuation mechanism occupies minimal additional chip area.

FIGURE 11.12

Assembled 3D microstructures.

The magnitude of the mechanical torque can be expressed as

$$T_m = M w_m t_m l_m H_{ext} \cos\theta, \qquad (11.1)$$

where l_m, w_m, and t_m are the length, width, and thickness of the Permalloy piece, respectively, and θ represents the displacement angle between the flap and the substrate. The magnitude of the magnetization also depends on θ. When the displacement is small, the Permalloy material is partially magnetized; as θ increases, M gradually approaches a saturation magnetization M_s. In this application work, the regime of large displacement is the major focus. In a simplified analysis, the magnetic material can be treated as possessing a constant magnetization, with $M = M_s$.

In order to realize three-dimensional mechanically stable assemblies (such as the structure shown in Figure 11.12), it is important that flaps within a close vicinity must be able to achieve asynchronous actuation under a global magnetic field. The design objective of controlling the angular displacement of a flap has been accomplished using two techniques. In the first method, the volume of the magnetic piece determines the magnitude of T_m and, subsequently, the displacement of the device under a given H_{ext}. In the second method, a cantilever-beam loading mechanism is used to provide a resistive force to the otherwise free flap as it rotates out of plane. The stiffness (or the spring constant) of the cantilever beam will determine the angular displacement of the flap. A stiffer spring can provide greater resistive forces to the microstructure; consequently, a greater H_{ext} is needed to reach a desired angular displacement. By using these two methods, individual hinged components can be lifted in a prescribed sequence as the magnitude of H_{ext} is gradually increased from zero to a maximum.

Case 11.2. Plastic Deformation Magnetic Assembly

The assembly method using hinges can be improved in several aspects. It requires multiple structural and sacrificial layers. Meanwhile, it is difficult to create a good electrical connection between the vertical flaps and the substrate. A process, called plastic deformation magnetic assembly (PDMA), has been used to address these two issues.

A micromachined planar flap with a ductile bending region is used as an example to describe the essential steps (Figure 11.13). A basic PDMA process consists of three steps [62]. First, the planar flap is fabricated and a piece of magnetic material is deposited on the flap (Figure 11.13a). The planar flap is then released from the substrate (e.g., by using sacrificial layer etching).

The magnetic material piece on the planar flap will be magnetized in the external magnetic field. As a result, a torque will be generated in the magnetic material piece to bend the planar flap off the substrate (Figure 11.13b). The bending angle of the planar flap (θ) increases with H_{ext}.

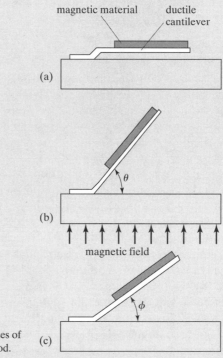

FIGURE 11.13

Three major phases of the PDMA method.

If the flexible region is made of a ductile material, it can be bent into a plastic deformation regime so that the planar flap maintains its position at a certain rest angle (ϕ) (e.g., 90°) above the substrate even after H_{ext} is removed. The magnitude of ϕ is usually smaller than the maximum value of θ due to the bending relaxation caused by the release of the elastic energy stored during the actuation.

The process is very efficient and can yield arrayed microdevices (Figure 11.14). This technique has been used to make out-of-plane flow sensors [63], among other applications. Details can be found in Section 5.5.2 of Chapter 5.

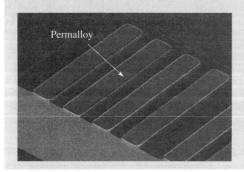
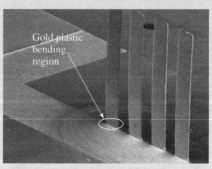

FIGURE 11.14

Micrograph of a microstructure before and after the assembly.

11.6 FOUNDRY PROCESS

Microfabrication processes used to be exclusively performed in custom research and development facilities. However, this is a rather expensive approach. In recent years, with increased industrialization activities in the MEMS area, the presence and capabilities of foundry services are rapidly growing. Some foundries allow users to share the cost of one run.

In such a foundry process, a wafer area is divided into dies, where each die represents a device design from a distinct user. Naturally, all dies on one wafer must go through the same process flow. Certainly one loses the ability to devise process sequences and structures of arbitrary materials and complexity. However, the process cost and the process time are significantly reduced.

MEMS devices with monolithically integrated mechanical and electronics elements can be made in three types of facilities: (1) IC-centric foundries; (2) newer, MEMS-centric foundries; and (3) dedicated production lines and research laboratories with vast possibilities of customization.

IC-centric foundries follow established circuit process flow; alternation is technically difficult and prohibitively expensive. Generally, the integration of micromechanical elements must conform strictly to the established foundry process flow. To develop MEMS products using IC-centric foundries may present some restrictions on materials and processes.

A second type of foundry provides dedicated MEMS services. They cater to MEMS-specific materials and processes, and therefore provide more flexibility and variations. The Multi-User MEMS Processes (or MUMPS) are a representative and early example of MEMS-specific foundries. The MUMPS process is a multiuser foundry process that allows many users to share the fabrication cost of a single processing batch.

The standard MUMPS process use three polysilicon layers: two oxide sacrificial layers and a metal layer. The layers and their thickness are summarized in Table 11.2 in the order they occur in the process flow. Young's modulus and the fracture strength of polycrystalline silicon grown by the MUMPs process have been studied in detail [64].

Using a similar process flow, various structures (ranging from accelerometers, micromotors, microfluid channels, tunable mirrors, capacitors, and inductors) can be made. The layered structure and two possible applications are illustrated in Figure 11.15.

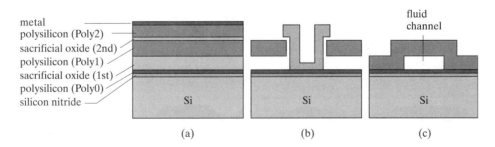

FIGURE 11.15

MUMPS process. (a) General stacking of layers without lithography and before any etching. (b) A motor. (c) A microchannel.

TABLE 11.2 Materials Involved in the MUMPS Foundry Process.

Material name	Deposition temperature °C	Thickness (μm)
Silicon nitride	800	0.6
Poly 0	600	0.5
1st oxide sacrificial	500	2
Poly 1	600	2
2nd oxide sacrificial	500	0.75
Poly 2	600	1.5
Metal (Au)	Room temperature	0.5

Since a foundry process is not run by the designer, it is important for designers to be able to communicate the design intent to foundry engineers. The way to communicate the design intent is to use standard mask design templates. Certain software programs have embedded support for selective foundry services, standardizing not only the names of layers but also the color scheme and appearances.

Microfabrication processes are never 100% accurate in terms of linewidth definition, registration, deposition thickness, and etching depth. Device failure may stem from such imperfections, especially if the designers and processors are physically separated. Seemingly minor process imperfections can cause drastic device failures. For example, multiple structural layers cannot always be aligned with perfect registration. Figure 11.16 illustrates a possible scenario in which a "minor" misalignment error can cause a device, a closed cavity, to fail completely. Due to crucial misalignment, the intended closed cavity may become an open cavity.

To prevent these failures, foundry services provide **design rules** to users to avoid commonly known failure modes. These rules provide safety margins based on statistical and historical data gathered by the foundries.

SUMMARY

Surface micromachining technology is the central focus of this chapter. Surface micromachining and bulk micromachining, which were discussed in the previous chapter, are often used together in a process. This chapter discusses the sacrificial layer technology based on the most familiar material

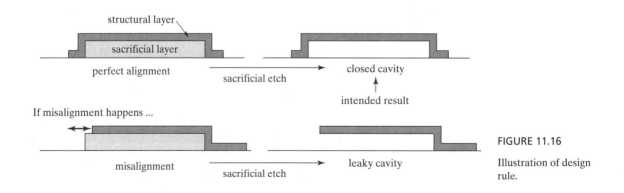

FIGURE 11.16

Illustration of design rule.

system—polysilicon as the structural layer and oxide as the sacrificial layer. Three processes for realizing micromotors are discussed to exemplify the technical complexity achievable. Other alternative materials are briefly discussed. We then discussed major fabrication-related issues, including stiction and the acceleration of sacrificial undercut.

The following is a list of major concepts, facts, and analytical skills introduced in this chapter:

- The basic, two-layer surface micromachining process.
- The use of the micromachining process for realizing micromotors.
- The etch rate selectivity of common etchants for polycrystalline silicon and oxide.
- General material selection criteria for structural and sacrificial layers, with up to three layers of structural layers in a process.
- The deposition conditions and etching methods of polysilicon, silicon nitride, oxide, photoresist, and metals (e.g., gold).
- Commonly practiced antistiction methods.
- Ability to synthesize a successful surface micromachining process flow by selecting correct structural and sacrificial layers and etchants, with maximized selectivity.

PROBLEMS

Problem 11.1: Design
A suspended cantilever made of gold is located on a silicon substrate with a silicon nitride passivation coating. The silicon nitride layer was deposited using the low-pressure chemical vapor deposition method. The sacrificial layer is silicon oxide. Find as many ways as possible to etch the sacrificial layer without damaging the structural layer and the substrate, using the materials listed in [2]. Explain your answers.

Problem 11.2: Design
A surface micromachining process uses LPCVD polycrystalline silicon as the structural layer on top of a plain silicon wafer. The sacrificial layer needs to be 5 μm thick. List all criteria that this sacrificial layer material must satisfy. Discuss whether there is a material that can be used for this purpose. (Consider all possible candidate sacrificial layers (out of the list of materials in [2]) that satisfy process temperature compatibility, etch chemistry compatibility, acceptable deposition time (less than 6 hours total), and acceptable etching time. It is acceptable to list materials whose temperature compatibility with poly silicon is questionable.).

Problem 11.3: Fabrication
A generic surface micromachining process is diagramed below, using two layers of structural and two layers of sacrificial materials. In this case, the substrate is silicon, the structural layer #1 is polycrystalline silicon, and the structural layer # 2 must be Parylene.

(a) Identify a set of possible candidate materials for sacrificial layer #1 and #2 out of a list of materials below: LPCVD silicon nitride, LPCVD silicon oxide, photoresist, and evaporated gold thin film. Briefly state the reasoning behind the decision for each item in the list. (Hint: it is possible that no materials can fulfill the requirements).

(b) Find the possible combination of other materials that provide process compatibility by using the etch compatibility table in [2]. (Limit your choice to those discussed in [2]).

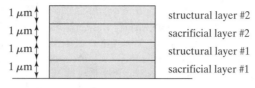

Problem 11.4: Fabrication

A generic surface micromachining process is diagrammed below, using two layers of structural and two layers of sacrificial materials. In this case, the substrate is silicon, sacrificial layer #1 is gold and the structural layer #1 is Parylene.

Part a:

Identify a set of possible candidate materials for sacrificial layer #2 and structural layer #2 out of a list of materials below:

Sacrificial layer choices: LPCVD silicon oxide, photoresist, and evaporated gold thin film.

Structural layer choices: LPCVD silicon nitride, LPCVD silicon oxide, evaporated gold thin film.

Briefly state the reasoning behind the decision for each item in the list. (Hint: it is possible that no materials can fulfill the requirements).

Part b:

Find the possible combination of other materials that provide process compatibility by using the etch compatibility table. (Limit your choice to those discussed in [2]). Describe the complete process flow for building the stack, including details of photoresist spinning and development. (Hint: Find the deposition temperature of various materials from on-line sources and summarize them in a table).

Problem 11.5: Fabrication

A generic surface micromachining process is diagrammed below, using two layers of structural and two layers of sacrificial materials. In this case, the substrate is silicon, structural layer #1 is gold, and structural layer # 2 is Parylene.

Identify a set of possible candidate materials for sacrificial layer #1 and 2 out of a list of materials below: LPCVD silicon nitride, LPCVD silicon oxide, and photoresist.

Briefly state the reasoning behind the decision for each item in the list. (Hint: it is possible that no materials can fulfill the requirements).

Problem 11.6: Fabrication

A generic surface micromachining process is diagramed below, using two layers of structural and two layers of sacrificial materials. In this case, the substrate is glass, structural layer 1 is polycrystalline silicon, and structural layer # 2 is gold.

Identify a set of possible candidate materials for sacrificial layer #1 and #2 out of a list of materials below: LPCVD silicon nitride, LPCVD silicon oxide, photoresist, and evaporated gold thin film. Briefly state the reasoning behind the decision for each item in the list. (Hint: it is possible that no materials can fulfill the requirements).

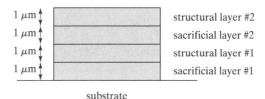

Problem 11.7: Fabrication

Repeat Problem 11.3 for the diagram shown below. Here, we apply an arbitrary criteria that no deposition step should take more than 3 hours. (Hint: find the growth rate data from literatures and on-line sources.)

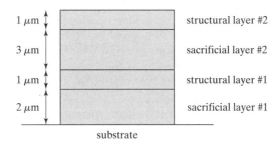

Problem 11.8: Design

Configure a mask layout software so that the layer names corresponds to the names of MUMPS process layers (SIN (silicon nitride), Poly0 (1st polysilicon layer), PSG1 (1st sacrificial layer), Poly 1 (2nd polysilicon layer), PSG2 (2nd sacrificial layer), Poly2 (3rd polysilicon layer), and Metal). (Hint for instructor: you may define a standard template file and distribute to the class for consistency of color and fill patters for each layer. There are a number of choices, including free software such as Xkic). Draw mask layout of micromotor according to Section 11.1.2. Here, assume the diameter of the rotor is 100 μm. Obtain other dimensions by approximate scaling from Fig. 4.1.

Problem 11.9: Design

Use the software and template created in Problem 8 to draw a mask for making a microfluid channel. The channel is 500 μm long and 10 μm wide. Calculate the flow resistance associated with the channel for water flow. The height of the channel should correspond to the thickness of one sacrificial layer.

Problem 11.10: Fabrication

Develop a process based on the MUMPS service to realize a microfluid channel with polysilicon walls around. The fluid should not be exposed to materials other than silicon when flowing in the channel. Draw each step of the fabrication process. Determine the height of the channel.

Problem 11.11: Fabrication

Develop a process to realize a rotational micromirror with mechanical structures resembling the Texas Instruments digital light processor using the MUMPS process. Draw the complete mask for realizing the micromirror using the template developed in Problem 8. Draw the cross-section of the finished mirror device.

Problem 11.12: Fabrication

Can one use the MUMPS process to create a crude microfluid ink jet nozzle array for ink jet printing? If so, explain and draw the design and processes. Draw the cross-section of the finished nozzle device.

If not, explain the reasons. (Hint: do not consider the incorporation of ink reservoir in this problem).

REFERENCES

1. Williams, K.R., and R.S. Muller, *Etch Rates for Micromachining Processing*. Microelectromechanical Systems, Journal of, 1996. **5**(4): p. 256–269.
2. Williams, K.R., K. Gupta, and M. Wasilik, *Etch Rates for Micromachining Processing—Part II*. Microelectromechanical Systems, Journal of, 2003. **12**(6): p. 761–778.
3. French, P.J., et al., *The Development of a Low-Stress Polysilicon Process Compatible with Standard Device Processing*. Microelectromechanical Systems, Journal of, 1996. **5**(3): p. 187–196.
4. Kamins, T.I., and T.R. Cass, *Structure of Chemically Deposited Polycrystalline-Silicon Films*. Thin Solid Films, 1973. **16**(2): p. 147–165.
5. Kamins, T.I., *Structure and Propreties of LPCVD Silicon Films*. Journal of Electrochemical Society, 1980. **127**: p. 686–690.
6. Kamins, T., *Polycrystalline Silicon for Integrated Circuits and Displays*. 2d ed. Kluwer Academic Publishers, 1998.
7. Ylonen, M., A. Torkkeli, and H. Kattelus, *In Situ Boron-Doped LPCVD Polysilicon with Low Tensile Stress for MEMS Applications*. Sensors and Actuators A: Physical, 2003. **109**(1–2): p. 79–87.
8. Lee, Y.-I., et al., *Dry Release for Surface Micromachining with HF Vapor-Phase Etching*. Microelectromechanical Systems, Journal of, 1997. **6**(3): p. 226–233.
9. Volklein, F., and H. Balles, *A Microstructure for Measurement of Thermal Conductivity of Polysilicon Thin Films*. Microelectromechanical Systems, Journal of, 1992: p. 193–196.
10. Paul, O.M., J. Korvink, and H. Baltes, *Determination of the Thermal Conductivity of CMOS IC Polysilicon*. Sensors and Actuators A: Physical, 1994. **41**(1–3): p. 161–164.
11. Paul, O., and H. Baltes, *Thermal Conductivity of CMOS Materials for the Optimization of Microsensors*. Journal of Micromechanics and Microengineering, 1993. **3**: p. 110–112.
12. Mastrangelo, C.H., Y.-C. Tai, and R.S. Muller, *Thermophysical Properties of Low-Residual Stress, Silicon-Rich, LPCVD Silicon Nitride Films*. Sensors and Actuators A: Physical, 1990. **23**(1–3): p. 856–860.
13. McConnell, A.D., S. Uma, and K.E. Goodson, *Thermal Conductivity of Doped Polysilicon Layers*. Microelectromechanical Systems, Journal of, 2001. **10**(3): p. 360–369.
14. Guckel, H., et al., *Fine-Grained Polysilicon Films with Built-In Tensile Strain*. Electron Devices, IEEE Transactions on, 1988. **35**(6): p. 800–801.
15. Gianchandani, Y.B., M. Shinn, and K. Najafi, *Impact of High-Thermal Budget Anneals on Polysilicon as a Micromechanical Material*. Microelectromechanical Systems, Journal of, 1998. **7**(1): p. 102–105.
16. Ristic, L., et al., *Properties of Polysilicon Films Annealed by a Rapid Thermal Annealing Process*. Thin Solid Films, 1992. **220**(1–2): p. 106–110.
17. Zhang, X., et al., *Rapid Thermal Annealing of Polysilicon Thin Films*. Microelectromechanical Systems, Journal of, 1998. **7**(4): p. 356–364.
18. Joachim, D., and L. Lin, *Characterization of Selective Polysilicon Deposition for MEMS Resonator Tuning*. Journal of Microelectromechanical Systems, 2003. **12**(2): p. 193–200.
19. Azzam Yasseen, A., J.D. Cawley, and M. Mehregany, *Thick Glass Film Technology for Polysilicon Surface Micromachining*. Microelectromechanical Systems, Journal of, 1999. **8**(2): p. 172–179.
20. Liu, R.H., M.J. Vasile, and D.J. Beebe, *The Fabrication of Nonplanar Spin-On Glass Microstructures*. Microelectromechanical Systems, Journal of, 1999. **8**(2): p. 146–151.

21. Sedky, S., et al., *Structural and Mechanical Properties of Polycrystalline Silicon Germanium for Micromachining Applications.* Microelectromechanical Systems, Journal of, 1998. **7**(4): p. 365–372.
22. Franke, A.E., et al., *Polycrystalline Silicon-Germanium Films for Integrated Microsystems.* Microelectromechanical Systems, Journal of, 2003. **12**(2): p. 160–171.
23. Rusu, C., et al., *New Low-Stress PECVD Poly-SiGe Layers for MEMS.* Microelectromechanical Systems, Journal of, 2003. **12**(6): p. 816–825.
24. Li, B., et al., *Germanium as a Versatile Material for Low-Temperature Micromachining.* Microelectromechanical Systems, Journal of, 1999. **8**(4): p. 366–372.
25. Hui, E.E., C.G. Keller, and R.T. Howe. *Carbonized Parylene as a Conformal Sacrificial Layer.* In Technical Digest, Solid-State Sensor and Actuator Workshop. Hilton Head Island, SC: 1998.
26. Storment, C.W., et al., *Flexible, Dry-Released Process for Aluminum Electrostatic Actuators.* Microelectromechanical Systems, Journal of, 1994. **3**(3): p. 90–96.
27. Zavracky, P.M., S. Majumder, and N.E. McGruer, *Micromechanical Switches Fabricated Using Nickel Surface Micromachining.* Microelectromechanical Systems, Journal of, 1997. **6**(1): p. 3–9.
28. Frazier, A.B., and M.G. Allen, *Uses of Electroplated Aluminum for the Development of Microstructures and Micromachining Processes.* Microelectromechanical Systems, Journal of, 1997. **6**(2): p. 91–98.
29. Zou, J., C. Liu, and J. Schutt-aine, *Development of a Wide-Tuning-Range Two-Parallel-Plate Tunable Capacitor for Integrated Wireless Communication Systems.* International Journal of RF and Microwave CAE, 2001. **11**: p. 322–329.
30. Buhler, J., et al., *Electrostatic Aluminum Micromirrors Using Double-Pass Metallization.* Microelectromechanical Systems, Journal of, 1997. **6**(2): p. 126–135.
31. Kim, Y.W., and M.G. Allen, *Single- and Multi-Layer Surface-Micromachined Platforms Using Electroplated Sacrificial Layers *1.* Sensors and Actuators A: Physical, 1992. **35**(1): p. 61–68.
32. Tea, N.H., et al., *Hybrid Postprocessing Etching for CMOS-Compatible MEMS.* Microelectromechanical Systems, Journal of, 1997. **6**(4): p. 363–372.
33. Yao, T.-J., X. Yang, and Y.-C. Tai, *BrF3 Dry Release Technology for Large Freestanding Parylene Microstructures and Electrostatic Actuators.* Sensors and Actuators A: Physical, 2002. **97–98**: p. 771–775.
34. Liu, J.Q., et al. *In-Situ Monitoring and Universal Modeling of Sacrificial PSG Etching Using Hydrofluoric Acid.* In Proceedings, IEEE Micro Electro Mechanical Systems Workshop (MEMS 93). Fort Lauderdale, FL: IEEE. 1993.
35. Walsh, K., J. Norville, and Y.C. Tai. *Dissolution of Photoresist Sacrificial Layers in Parylene Microchannels.* In Proceedings, IEEE International Conference on Micro Electro Mechanical Systems. Interlaken, Switzerland: IEEE. 2001.
36. Zou, J., et al., *Effect of Etch Holes on the Optical Properties of Surface Micromachined Mirrors.* IEEE/ASME Journal of Microelectromechanical Systems (JMEMS), 1999. **8**(4): p. 506–513.
37. Suh, H.-J., et al., *Dendritic Material as a Dry-Release Sacrificial Layer.* Microelectromechanical Systems, Journal of, 2000. **9**(2): p. 198–205.
38. Selby, J.S., and M.A. Shannon. *Anodic Sacrificial Layer Etch (ASLE) for Large Area Blind Cavity Release of Metallic Structures.* In Technical Digest, Solid-State Sensor and Actuator Workshop. Hilton Head Island, SC: 1998.
39. Kim, G.M., et al., *Surface Modification with Self-Assembled Monolayers for Nanoscale Replication of Photoplastic MEMS.* Microelectromechanical Systems, Journal of, 2002. **11**(3): p. 175–181.
40. Anderson, R.C., R.S. Muller, and C.W. Tobias, *Porous Polycrystalline Silicon: A New Material for MEMS.* Microelectromechanical Systems, Journal of, 1994. **3**(1): p. 10–18.
41. Dougherty, G.M., T.D. Sands, and A.P. Pisano, *Microfabrication Using One-Step LPCVD Porous Polysilicon Films.* Microelectromechanical Systems, Journal of, 2003. **12**(4): p. 418–424.
42. Mastrangelo, C.H., and C.H. Hsu, *Mechanical Stability and Adhesion of Microstructures Under Capillary Forces. I. Basic Theory.* Microelectromechanical Systems, Journal of, 1993. **2**(1): p. 33–43.

43. Srinivasan, U., et al., *Alkyltrichlorosilane-Based Self-Assembled Monolayer Films for Stiction Reduction in Silicon Micromachines*. Microelectromechanical Systems, Journal of, 1998. **7**(2): p. 252–260.
44. Abe, T., W.C. Messner, and M.L. Reed, *Effects of Elevated Temperature Treatments in Microstructure Release Procedures*. Microelectromechanical Systems, Journal of, 1995. **4**(2): p. 66–75.
45. Yee, Y., et al., *Polysilicon Surface-Modification Techniques to Reduce Sticking of Microstructures*. Sensors and Actuators A: Physical, 1996. **52**(1–3): p. 145–150.
46. Rogers, J.W., and L.M. Phinney, *Process Yields for Laser Repair of Aged, Stiction-Failed, MEMS Devices*. Microelectromechanical Systems, Journal of, 2001. **10**(2): p. 280–285.
47. Gogoi, B.P., and C.H. Mastrangelo, *Adhesion Release and Yield Enhancement of Microstructures Using Pulsed Lorentz Forces*. Microelectromechanical Systems, Journal of, 1995. **4**(4): p. 185–192.
48. Toshiyoshi, H., et al., *A Surface Micromachined Optical Scanning Array Using Photoresist Lenses Fabricated by a Thermal Reflow Process*. Journal of Lightwave Technology, 2003. **21**(7): p. 1700–1708.
49. Kim, B.H., et al., *A New Organic Modifier for Anti-Stiction*. Microelectromechanical Systems, Journal of, 2001. **10**(1): p. 33–40.
50. Man, P.F., B.P. Gogoi, and C.H. Mastrangelo, *Elimination of Post-Release Adhesion in Microstructures Using Conformal Fluorocarbon Coatings*. Microelectromechanical Systems, Journal of, 1997. **6**(1): p. 25–34.
51. Srinivasan, U., D. Liepmann, and R.T. Howe, *Microstructure to Substrate Self-Assembly Using Capillary Forces*. Microelectromechanical Systems, Journal of, 2001. **10**(1): p. 17–24.
52. Whitesides, G.M., and B. Grzybowski, *Self-Assembly at All Scales*. Science, 2002. **295**(5564): p. 2418–2421.
53. Pister, K.S.J., et al., *Microfabricated Hinges*. Sensors and Actuators A: Physical, 1992. **33**(3): p. 249–256.
54. Ross, M., and K.S.J. Pister, *Micro-Windmill for Optical Scanning and Flow Measurement*. Sensors and Actuators A: Physical, 1995. **47**(1–3): p. 576–579.
55. Dechev, N., W.L. Cleghorn, and J.K. Mills, *Microassembly of 3-D Microstructures Using a Compliant, Passive Microgripper*. Microelectromechanical Systems, Journal of, 2004. **13**(2): p. 176–189.
56. Green, P.W., R.R.A. Syms, and E.M. Yeatman, *Demonstration of Three-Dimensional Microstructure Self-Assembly*. Microelectromechanical Systems, Journal of, 1995. **4**(4): p. 170–176.
57. Syms, R.R.A., *Surface Tension Powered Self-Assembly of 3-D Micro-Optomechanical structures*. Microelectromechanical Systems, Journal of, 1999. **8**(4): p. 448–455.
58. Syms, R.R.A., et al., *Surface Tension-Powered Self-Assembly of Microstructures—the State-of-the-Art*. Microelectromechanical Systems, Journal of, 2003. **12**(4): p. 387–417.
59. Yi, Y., and C. Liu, *Magnetic Actuation of Hinged Microstructures*. IEEE/ASME Journal of Microelectromechanical Systems (JMEMS), 1999. **8**(1): p. 10–17.
60. Kaajakari, V., and A. Lal, *Thermokinetic Actuation for Batch Assembly of Microscale Hinged Structures*. Microelectromechanical Systems, Journal of, 2003. **12**(4): p. 425–432.
61. Reid, J.R., V.M. Bright, and J.T. Butler, *Automated Assembly of Flip-Up Micromirrors*. Sensors and Actuators A: Physical, 1998. **66**(1–3): p. 292–298.
62. Zou, J., et al., *Plastic Deformation Magnetic Assembly (PDMA) of Out-of-Plane Microstructures: Technology and Application*. IEEE/ASME Journal of Microelectromechanical Systems (JMEMS), 2001. **10**(2): p. 302–309.
63. Chen, J., et al., *Two-Dimensional Micromachined Flow Sensor Array for Fluid Mechanics Studies*. ASCE Journal of Aerospace Engineering, 2003. **16**(2): p. 85–97.
64. Sharpe, W.N., Jr., et al., *Effect of Specimen Size on Young's Modulus and Fracture Strength of Polysilicon*. Microelectromechanical Systems, Journal of, 2001. **10**(3): p. 317–326.

CHAPTER 12

Polymer MEMS

12.0 PREVIEW

Polymer materials are increasingly being used in MEMS for realizing structures, sensors, and actuators. Terminology related to polymers is introduced in Section 12.1. In Section 12.2, we will review seven common types of polymers in the MEMS field. Several sensor applications using polymer materials are discussed in Section 12.3 to illustrate unique techniques and challenges.

12.1 INTRODUCTION

Polymers are large, usually chainlike molecules that are built from small molecules. Long-chain polymers are composed of structural entities called *mer* units, which are successively repeated along the chain. A bulk polymer is made of many polymer chains. The physical characteristics of a polymer material depend not only on its molecular weight and the make-up of polymer chains, but also on the ways the chains are arranged.

Polymers can be classified into three major classes: fibers, plastics, and elastomers (rubbers). The major discerning characteristics of these three groups are summarized in Table 12.1. The largest number of different polymeric materials comes under the plastics classification. Polyethylene, polypropylene, polyvinyl chloride (PVC), polystyrene, fluorocarbons, epoxies, phenolics, and polyesters are all classified as plastics. Many plastic materials are manufactured by different vendors and carry different trade (common) names. For example, acrylics (polymethyl methacrylate, PMMA) are known as Acrylite, Diakon, Lucite, and Plexiglas in trade. Vendors may incorporate additive substances into polymers to adjust their physical, chemical, electrical, and thermal characteristics and to change their appearances.

According to their origin, polymers can be categorized into two groups: naturally occurring polymers and synthetic polymers. Naturally occurring polymers—those derived from plants and animals—include wood, rubber, cotton, wool, leather, and silk. Synthetic polymers are derived from petroleum products.

Polymers can be classified by their response to temperature. **Thermal plastic polymers** (thermalplasts) can be remelted and reshaped repeatedly whereas **thermal setting polymers** (thermalsets) take on a permanent shape after being melted and processed once.

TABLE 12.1 Properties of Polymers.

	Elastomers	Plastics	Fibers
Upper limit of extensibility (%)	100–1000	20–100	<10
Character of stress deformation	Completely and instantaneously elastic	Partly reversible elasticity; delayed elasticity; some permanently set	Some reversible elasticity; some delayed elasticity; some permanently set
Crystallization tendency	Amorphous (in unstressed state)	Moderate to high	Very high
Initial Young's modulus (MPa)	10^5–10^6	10^7–10^8	10^9–10^{10}

The melting of a polymer crystal corresponds to the transformation of a solid material. It can range from an ordered structure of aligned molecular chains, to a viscous liquid in which the structure is highly random. This phenomenon occurs upon heating it at the melting temperature, T_m.

The temperature at which a polymer experiences a transition from a rubbery to a rigid state is termed the glass transition temperature T_g. Glass transition occurs in amorphous and semicrystalline polymers, and it is due to a reduction in the motion of large segments of molecular chains with decreasing temperatures.

The mechanical properties of polymers differ from those of metals and semiconductors in several major aspects. An excellent review of the mechanical properties of polymers can be found in Reference [1]. Some notable facts are summarized below:

1. Polymer materials cover a wide range of materials, each with a different Young's modulus. The modulus of elasticity may be as low as several MPa for highly elastic polymeric materials, but it may run as high as 4 GPa for some of the very stiff polymers.

2. Maximum tensile strengths for polymers are on the order of 100 MPa, which is much lower than that of metal and semiconductor materials.

3. Many polymers exhibit viscoelastic behavior. When a force is applied to it, an instantaneous elastic deformation may occur, followed by viscous and time-dependent strain changes. As a result, many polymeric materials are susceptible to time-dependent deformation under a constantly maintained stress. Such deformation is called **viscoelastic creep**.

4. The mechanical properties are influenced by temperature, molecular weight, additives (many proprietary), degree of crystallinity, and heat treatment history. Mechanical properties of certain polymers can change dramatically over a narrow temperature range. For example, PMMA (acrylics) are totally brittle at 4°C but extremely ductile at 60°C. The stress–strain relation and the viscoelastic behaviors are both influenced strongly by the temperature.

Many organic polymers are dielectric insulators. However, certain polymers exhibit interesting conducting behaviors. In recent years, conducting polymer materials are being actively pursued for making transistors [2], organic thin-film displays [3], and memory [4]. Such conductive polymers include polypyrrole, polyaniline, and polyphenylene sulfide.

The mobility of charge carriers in polymers is still orders of magnitude lower than that of silicon and compound semiconductor materials. At this stage, polymer electronic devices are not able to compete with semiconductor electronics in terms of performance.

Polymers can be processed using a large number of techniques including injection molding, extrusion, thermoforming, blow molding, machining, casting, compression molding, rotational molding, powder metallurgy, sintering, dispersion coating, fluidized-bed coating, electrostatic coating, calendaring, hot forming, cold forming, vacuum forming, and vapor deposition. Many techniques can be combined with microfabrication.

12.2 POLYMERS IN MEMS

The micromachining technology for MEMS was derived from integrated circuit fabrication. Naturally, silicon has been the predominant material choice. In recent years, polymers have emerged as an important new class of materials for use in MEMS applications. There are a number of unique merits associated with polymer materials. First, the cost of the material is much lower than that of single-crystal silicon. Second, many polymer materials allow unique low-cost, batch-style fabrication and packaging techniques such as thermal micromolding, thermal embossing and injection molding. Instead of processing one wafer at a time, polymer substrates can potentially be processed in a high-throughput, roll-to-roll fashion. Third, certain polymers offer unique electrical, physical and chemical properties that are not available in silicon and silicon-derived materials. Examples of such properties include mechanical shock tolerance [5], biocompatibility, and biodegradability [6].

There are barriers for using polymers in MEMS. The viscoelastic behavior of polymers is undesirable in certain applications. Many polymer materials have lower glass-transition and melting temperatures. The low thermal stability limits fabrication methods and application potential.

A significant number of polymer materials have been introduced to MEMS applications in recent years. These materials find applications beyond handle wafers or adhesive layers; they are used as structural mechanical elements including cantilevers and membranes.

The following list includes polymers that have been used successfully and widely for MEMS applications. Some items in the list represent a family of polymers while others represent a specific product:

1. Polyimide;
2. SU-8;
3. Liquid crystal polymer;
4. Polydimethylsiloxane;
5. Polymethyl methacrylate (also known as acrylics, plexiglass, or PMMA);
6. Parylene (polyparaxylylene);
7. Polytetrafluoroethylene (Teflon) and Cytop.

A summary of pertinent electrical and mechanical properties of these polymers can be found in Table 12.2. We will discuss material processing and applications of these seven materials in greater detail in Sections 12.2.1 through 12.2.7.

12.2.1 Polyimide

Polyimides represent a family of polymers that exhibit outstanding mechanical, chemical, and thermal properties as a result of their cyclic chain bonding structure [7]. Bulk processed polyimide parts are used widely, from cars (struts and chassis in some cars) to microwave cookware. Polyimide is widely used in the microelectronics industry as an insulating material as well.

TABLE 12.2 Properties of Seven Polymer Materials.

	LCP[1]	Polyimide[2]	EPON SU-8[3]	PMMA [1]
Dielectric constant (60 Hz)	2.8	3.5	5.07	3–4
Dissipation factor (60 Hz)	0.004	0.002	0.007	0.02–0.04
Moisture absorption	<0.02%	2.8%	N/A	N/A
Glass transition temperature	145°C	360–410°C	194°C	45°C (isotactic); 105°C (syndiotactic)
Coefficient of thermal expansion	0–30 ppm/°C	20 ppm/°C	20–50 ppm/°C[4]	50–90 ppm/°C
Tensile strength	180 MPa	200–234 MPa	50 MPa	48.3–72.4 MPa
Tensile modulus	7–22 GPa	2.5–4 GPa	4–5 GPa[4]	2.24–3.24 GPa
Elongation at break	1–5%	10–150%	<1%[4]	2.0–5.5%
Density (g/cm^3)	1.4	1.42–1.53	1.2	0.9
Representative patterning methods	Laser, plasma etch	Photo definition, wet etch, plasma etch	Photo definition, plasma etch	Photo definition, plasma etch

	Parylene[5]	Perfluoro-polymers (Cytop[6])	PDMS[7]
Dielectric constant (60 Hz)	2.65–3.15	2.1–2.2	2.7
Dissipation factor (60 Hz)	0.02–0.0002	0.0007	.001
Moisture absorption	0.01–0.06%	<0.01%	0.1%
Glass transition temperature	160°C	−97–108°C[8]	−125°C[9]
Coefficient of thermal expansion	35–69 ppm/°C	125–216 ppm/°C [1]	30 ppm/°C
Tensile strength	45–75 MPa	20–35 MPa[8]	6.2 MPa
Tensile modulus	2.4–3.2 GPa	0.4–1.2 GPa	0.5–1 MPa
Elongation at break	10–200%	200–400%[8]	100%
Density (g/cm^3)	1.1–1.4	2.1	1.05
Patterning	Plasma etch	Plasma etch	Molding, plasma etch (slow)

[1] Vectra LCP, Celanese AG
[2] Kapton, Dupont
[3] Resolution Performance Products, LLC
[4] http://aveclafaux.freeservers.com/SU-8.html
[5] Parylene Coating Services Inc.
[6] Cytop, Asahi Glass Co. LTD
[7] Dow Corning, Inc.
[8] Callister, William D., "Materials Science and Engineering: An Introduction," 4th ed., New York: Wiley, 1997.
[9] Neilsen, Lawrence E., "Mechanical Properties of Polymers and Composites," 2d ed., New York: Marcel Dekker, Inc., 1994.

Polyimides are formed from the dehydrocyclization of polyamic precursors (Figure 12.1a) into cyclic polymers by incorporating aromatic groups R and R' (Figure 12.1b). These aromatic groups are chosen to affect the properties of the final polyimide. For example, by chemically altering the polyamic acid precursor to include R'' groups sensitive to UV light, as shown in Figure 12.1c, photo-patternable precursors can be made to crosslink where exposed to UV light [8].

The mechanical properties of polyimide films have been studied [9, 10]. Cured polyimide films exhibit intrinsic stress on the order of 4×10^6 Pa to 4×10^7 Pa, as measured using

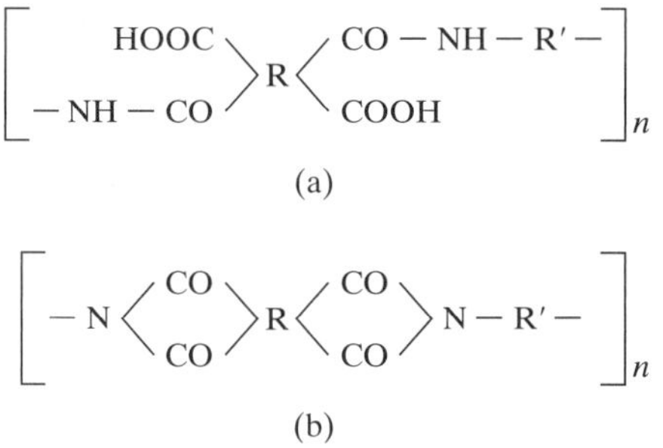

FIGURE 12.1

Polyimide chemistry. (a) Generic polyamic acid precursor, with thermally stable R and R' groups chosen for specific final properties in the (b) resultant general polyimide structure after imidization (dehydrocyclization) reaction.

suspended microfabricated polyimide strings [11]. Further, mechanical and electrical properties of polyimide may exhibit direction-dependent behavior. Many properties such as the index of refraction [12], dielectric constant [13], Young's modulus [14], thermal expansion coefficient [14], and thermal conductivity [15] vary with processing conditions.

Polyimide is commercially available as cured sheets, semi-cured sheets, or viscosity solutions for spin coating [16]. The structure of a typical commercial polyimide—HN-type Kapton—is shown in Figure 12.1d. In MEMS, polyimide is used for insulating films, substrates, mechanical elements (membranes and cantilevers), flexible joints and links [17], adhesive films, sensors [18], scanning probes [19], and stress-relief layers [5]. Polyimide materials offer many favorable characteristics in these roles including (1) chemical stability; (2) thermal stability up to around 400°C; (3) superior dielectric properties; (4) mechanical robustness and durability; and (5) low cost of materials and processing equipment.

Polyimide can be used as a structural element for sensors and actuators. Unfortunately, polyimide is neither conductive nor strain sensitive. Functional materials, such as conductors or strain gauges, need to be integrated externally. Thin-film metal strain gauges have been integrated with polyimide, exhibiting an effective gauge factor on the order of 2 to 6. An alternative is to modify the polyimide material for sensing purposes. For example, a piezoresistive composite of polyimide and carbon particles with an effective gauge factor on the order of 2 to 13 has been demonstrated [16].

12.2.2 SU-8

The SU-8 is a negative-tone, near-UV photoresist first invented by IBM in the late 1980s [20]. Its main purpose is to allow high aspect ratio features (>15) to be made in thick photosensitive polymers. The photoresist consists of EPON® Resin SU-8 (from Shell Chemical) as a main component. The EPON resin is dissolved in an organic solvent (GBL, gamma-butyrolacton), and the quantity of the solvent determines the viscosity and the range of achievable thickness. Processed layers as thick as 100 μm can be achieved; they offer tremendous new capabilities for masking, molding, and building high-aspect-ratio structures at low cost. The cost of SU-8 lithography is considerably lower than that of other techniques for realizing high-aspect-ratio microstructures, notably the LIGA process and deep reactive ion etching. SU-8 has been integrated in a number of microdevices, including microfluid devices [21], SPM probes [22], and microneedles. It can also serve as a thick sacrificial layer for surface micromachining.

12.2.3 Liquid Crystal Polymer (LCP)

The liquid crystal polymer is a thermoplast with unique structural and physical properties. LCPs are available commercially in sheets of various thickness. When flowing in the liquid crystal state during processing, the rigid segments of the molecules align next to one another in the direction of shear flow. Once this orientation is formed, their direction and structure persist, even when the LCP is cooled below its melting temperature. This characteristic differentiates LCP from most thermoplastic polymers (e.g., Kapton) whose molecule chains are randomly oriented in the solid state.

Owning to its unique structure, LCP offers a combination of electrical, thermal, mechanical and chemical properties unmatched by other engineering polymers. One of the earliest LCP films used in MEMS is a Vectra® A-950 aromatic liquid crystal polymer, produced by Hoechst Celanese Corporation [23]. The reported melting temperature of Vectra A-950 is 280°C. The specific gravity ranges from 1.37 to 1.42 kg/m^3, and the molecular weight is greater than 20,000 g/mol. The compatibility of this LCP with commonly used chemicals in micromachining was first investigated in Reference [23]. LCP is virtually unaffected by most acids, bases, and solvents for a considerably long time and over a broad temperature range. Extensive tests showed that LCP was not attacked or dissolved by at least the following chemicals common in microfabrication: (1) organic solvents including acetone and alcohol, (2) metal etchants for Al, Au, and Cr, (3) oxide etchants (49% HF and buffered HF), and (4) developers for common photoresist and SU-8 resist.

LCP films have excellent stability. They have very low moisture absorption (~0.02%) and low moisture permeability, which are better than PMMA (see Section 12.2.5) and comparable to that of glass. For other gases, including oxygen, carbon dioxide, nitrogen, argon, hydrogen, and helium, LCP also exhibits an above-average barrier performance. Further, the permeation of gases through LCP is not affected by humidity, even under elevated temperatures (e.g., 150°C). The thermal expansion coefficient of the LCP material can be controlled during the fabrication process to be both small and predictable. The LCP film also shows excellent chemical resistance.

LCP was originally used as a high-performance substrate material for a high-density printed circuit board (PCB) [23]. A number of unique processing methods have been developed for LCP, including laser drilling and via filling (for low-resistance electrical through-wafer interconnects).

LCP films are used as substrates for space and military electronics systems, both for their performance and stability. For instance, they have been explored as a high-performance carrier of radio frequency electromagnetic elements such as antennas. Results from high-frequency tests show that LCP has a fairly uniform relative dielectric constant of 3 in the range of 0.5 to 40 GHz and an extremely low loss factor of ~0.004.

It is useful to compare LCP with Kapton, another polymer sheet (belonging to the polyimide family) that has been used in MEMS in recent years. Compared with Kapton, LCP has a lower cost (50%–80% lower than Kapton) and is melt processable at lower temperatures. As a result, bonding between LCP and another substrate (e.g., glass) is easier. For example, Kapton is often bonded with an intermediate adhesive layer, but LCP films can bond to other surfaces directly by lamination.

For LCP film with uniaxial molecule orientation, its mechanical properties are anisotropic and dependent on the polymer orientation. For instance, the uniaxial LCP film can withstand less load in the transverse direction (i.e., the direction orthogonal to the orientation of its molecules) than in the longitudinal direction (i.e., the direction along the orientation of its molecules). If necessary, biaxially oriented film with equally good transverse and longitudinal direction properties can be made to correct the anisotropy behavior. A uniaxial film can be formed by bonding multiple layers of anisotropic films to the angular offsets of the crystal orientation between layers. The orientation of LCP molecules varies through the thickness of the film, while the film molecules of the two sides are oriented at opposite angles. If the angles are +45° and −45° at either side, the mechanical properties such as coefficient of thermal expansion, tensile strength, and modulus are nearly isotropic.

Commercial LCP material is supplied in sheet format. The thickness of LCP film could vary from several microns to several millimeters. Some can be provided with copper clad layers on one or both sides. The optional copper clad layer is normally ~15–20 μm thick. This copper layer is laminated in a vacuum press at a temperature around the melting point of LCP.

12.2.4 PDMS

Elastomers are materials that can sustain a large degree of deformation and recover their shape after a deforming force. **Poly (dimethylsiloxane)** (PDMS), an elastomer material belonging to the room-temperature vulcanized (RTV) silicone elastomer family, offers many advantages for general MEMS applications. It is optically transparent, electrically insulating, mechanically elastic, gas permeable, and biocompatible. The biological and medical compatibility of the material is reviewed in Reference [24]. PDMS is widely used in microfluidics. For details, see Chapter 13.

The primary processing method is molding, which is straightforward and allows fast, low-cost prototyping. A number of unique processing characteristics of PDMS are worth noting:

1. The volume of PDMS shrinks during the curing step. Compensation of dimensions at the design level should be incorporated to yield the desired dimensions.
2. Due to volume shrinking and flexibility, deposited metal thin films on cured PDMS tend to develop cracks, affecting the electrical conductivity.
3. The surface chemical properties (such as adhesion energy) can be varied by altering the mixing ratio; this alteration can be conducted through surface chemical or electrical treatment.

PDMS is commercially supplied as a viscous liquid—it can be cast or spin coated on substrates. Unfortunately, the PDMS material is not photo-definable. It therefore cannot be simply spin coated and patterned like photosensitive resists. Though UV-curable PDMS is being developed [25], the technology is not yet mature. It is possible to use plasma etching to pattern PDMS thin films. However, the etch rate is rather slow. The measured etch rate is approximately 7 nm/min at 800 W power and 100 V bias. Etching of PDMS with O_2 plasma leaves the surface and line edges rough [26].

Methods for patterning thin PDMS film on a substrate are important for MEMS applications. We discuss a process developed for producing thin-film PDMS patterns with precisely defined dimensions in Case 12.1.

Case 12.1. Precision Patterning of PDMS

The principle of a basic PDMS patterning process, modified from the screen printing method, is diagrammed in Figure 12.2.

A photoresist layer is first spin coated on top of a solid substrate (e.g., glass or silicon) and patterned by using conventional lithography processes (Figure 12.2, step 1). The thickness of the photoresist can be controlled by varying the spin rate. A viscous PDMS prepolymer solution (e.g., Dow Corning SYLGARD 184 with a 10:1 mixing ratio of base to curing agent) is poured over the wafer's front surface (step 2). A flat and smooth rubber blade is used to traverse the substrate surface while maintaining contact with the top surface of the photoresist layer (step 3). This removes excessive PDMS prepolymer, leaving PDMS only in recessed regions between elevated photoresist molds.

Due to non-ideal contact between the blade and top surfaces of photoresist, a thin ($<1\ \mu m$) and often noncontinuous residual film of PDMS may be left on top of photoresist regions (step 4). This film can be removed later by light mechanical polishing or plasma etching.

The wafer is thermally cured, allowing the PDMS to polymerize in the recessed regions (step 5). Finally, the photoresist mold is removed using acetone (step 6). The lateral dimensions of resultant PDMS patterns correspond to those of the recessed

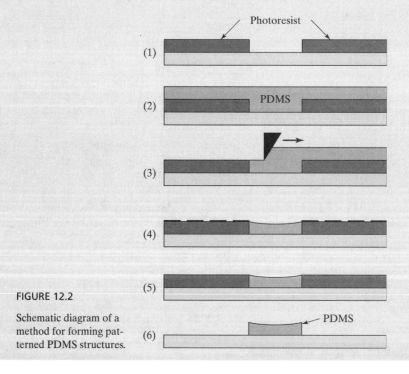

FIGURE 12.2

Schematic diagram of a method for forming patterned PDMS structures.

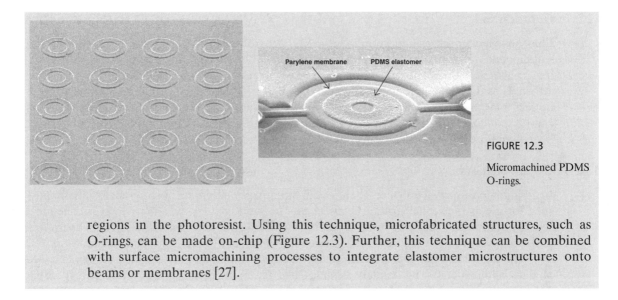

FIGURE 12.3

Micromachined PDMS O-rings.

regions in the photoresist. Using this technique, microfabricated structures, such as O-rings, can be made on-chip (Figure 12.3). Further, this technique can be combined with surface micromachining processes to integrate elastomer microstructures onto beams or membranes [27].

12.2.5 PMMA

PMMA is supplied in many different forms, including bulk, sheets, and solutions for spin coating. PMMA bulk, most commonly known by its trade name *acrylics*, has been used in making microfluidic devices. The photo-definable PMMA thin film is a widely used e-beam and X-ray lithography resist. Spin-coated PMMA has been used as a sacrificial layer as well [28]. Deep reactive ion etching processes for PMMA thin films have been demonstrated [29].

12.2.6 Parylene

Parylene is a thermalset polymer. It is the only plastic material that is deposited using a chemical vapor deposition (CVD) process. The deposition process is conducted under room temperature. A Parylene deposition system consists of a source chamber connected to a vacuum deposition chamber. A dimer (di-para-xylene) is heated inside the source chamber to approximately 150°C. It sublimates into a gaseous monomer, which then enters the vacuum chamber and coats the objects within it. Three parylene dimer variations are available from commercial vendors, including Parylene C (widely used), Parylene N (for better dielectric strength and penetration), and Parylene D (for extended temperature performance).

The Parylene film offers very useful properties for MEMS applications, including very low intrinsic stress [30], room-temperature deposition, conformal coating, chemical inertness, and etch selectivity. Parylene coating is ideal for electrical isolation, chemical isolation, preservation, and sealing.

Parylene has been used for microfluidics channels [31], valves [32], and sensors (acceleration sensors [33], pressure sensors [34], microphones [35], and shear stress sensors [34]).

The thickness of the Parylene coating is generally controlled by the amount of dimer loaded. Thickness monitors and end-of-point detectors for *in situ* Parylene thickness monitoring have been developed (for example, based on thermal transfer principles) [36].

12.2.7 Fluorocarbon

Fluoropolymers such as Teflon and Cytop [37] provide excellent chemical inertness, thermal stability, and non-flammability due to the strong C–F bond. They can be used as a surface coating, insulation, antireflection coating, or adhesion agent. Cytop is a trademarked material (by Asashi Glass Company of Japan). It exhibits as many good properties as Teflon but offers high optical transparency and good solubility in specific fluorinated solvents. Fluoropolymer films can be spin coated or deposited by PECVD method. In MEMS, Teflon and Cytop films have been used for electrical insulation, adhesive bonding, and friction reduction.

12.2.8 Other Polymers

In addition to the seven polymers mentioned, a number of emerging polymer materials are pursued for use as functional structural layers, unique sacrificial layers, adhesive layers, chemical sensors, and mechanical actuators. These include biodegradable polymers [6], wax (paraffin) [38], and polycarbonate. These three classes of polymers are briefly reviewed below.

Biodegradable polymer materials have been developed and investigated for implantable medical applications, drug delivery vehicles, and tissue engineering matrices. Biodegradable polymers such as polycaprolactone, polyglycolide, polylactide, and poly lactide-co-glycolide have been demonstrated in MEMS use. Biodegradable polymers are thermalplasts. Microstructures have been formed by micromolding for applications such as microfluid channels, reservoirs, and needles [6, 39].

Paraffin provides many interesting properties not found in other materials. For example, paraffin has a low melting temperature (40–70°) and a high volumetric expansion (14–16%). The melt temperature of paraffin can be controlled by mixing several types of paraffin with different melting temperatures. It can be selectively etched by certain organic solvents (such as acetone) very quickly at room temperature, and it offers good chemical stability against many strong acid solutions (such as HF).

The use of paraffin can lead to many interesting transduction mechanisms and microfabrication techniques. On a large scale, paraffin has been used as linear actuators for dexterous endoscopes [40]. On a small scale, paraffin-based actuators have been used for microfluid valving and pumping [38, 41] by encapsulating paraffin patches inside a volume with integrated heaters. Wax can be used as a mold for fabricating complex microstructures [42].

Paraffin can be deposited using thermal evaporation, and it can be patterned using plasma generated with an oxygen and Freon 14 gas mixture. Since the melting temperature is low (75°C for Logitech 0CON-195 or n-Hextriacotane), all steps following the paraffin deposition must use low-temperature processes or engage in active substrate cooling.

Polycarbonate is a tough, dimensionally stable, transparent thermoplast that can be used in many applications that demand good performance characteristics over a wide range of temperatures. Commercial polycarbonates are supplied in three grades: machine grade, window grade, and glass-reinforced grade. Un-notched polycarbonate has very high impact strength, excellent dielectric strength, and electrical resistivity. Polycarbonate can be processed with injection molding, extrusion, vacuum forming, and blow molding. Polycarbonate parts can be bonded easily and welded. In MEMS, polycarbonate has been used for the microfabrication of microchannels using either sacrificial etching [43] or molding. Polycarbonate sheets with ion track etched holes have found uses for filters with unique ionic filtering capabilities due to the nanometer-sized diameters and uniformity of these holes [44].

Despite the progress in recent years, a large number of polymer materials that are widely used at the macroscale are not yet exploited for MEMS applications. Many polymer materials can potentially find applications in MEMS in the future. These candidates include conductive polymers [45, 46], electroactive polymers [47] such as polypyrrole [48–50], photopatternable gelatin [51], polyurethanes, shrinkable polystyrene film [52], shape memory polymers [53], and piezoelectric polymers such as polyvinylidene fluoride (PVDF) [54, 55].

Further, there are seemingly endless ways to modify polymer materials. For example, it has been discovered that the functional, electrical, and mechanical properties of many polymers can be altered by additives such as nanoparticles [56], carbon nanotubes, and nanowires [57].

12.3 REPRESENTATIVE APPLICATIONS

Many unique material properties and fabrication techniques of polymer materials can best be understood by examining applications that involve them. We shall review four types of sensor devices, which are made using thin-film polymers or polymer bulk substrates. Selected examples of actuators (for example ones based on silicone elastomer) are discussed in Chapter 13.

12.3.1 Acceleration Sensors

Acceleration sensors can be made entirely or partially out of polymer materials using a variety of transduction principles. These generally involve depositing functional thin films on polymer substrates or microstructures.

In Case 12.2, we will review the design and fabrication process of an accelerometer that utilizes polymer springs. The polymer is used for structural purposes but not for transduction.

Case 12.2. Silicon Accelerometer with Parylene Beams

Here, we discuss a microfabricated acceleration sensor using polymer support beams [33]. The accelerometer incorporates a silicon proof mass and high-aspect-ratio Parylene beams. Foremost, the polymer beam increases the shock resistance, enabling a large deformation without failure. Because Parylene has a small Young's modulus, the spring constant is lower than it would be if it was replaced by silicon. A low spring constant translates into increased sensitivity, but it has a somewhat reduced resonant frequency.

In this design, Parylene beams are 10–40 μm wide and have aspect ratios (height over width) of 10–30. However, it is impractical to grow Parylene films with thicknesses of hundreds of micrometers. In addition, there is no high-aspect-ratio reactive ion etching process that can produce vertical etching.

An alternative process for realizing high-aspect-ratio Parylene structures is developed. It involves first creating high-aspect-ratio trenches (400 μm deep) as molds in a 500-μm-thick silicon substrate (Figure 12.4b). The wafer is oxidized by reacting with oxygen at a high temperature (Figure 12.4c). The oxidized wafer (with a conformal 2-μm-thick oxide coating) is then placed inside a Parylene deposition chamber. Parylene thin films with thicknesses of 10–20 μm fill the trenches entirely (Figure 12.4d). A global plasma etch

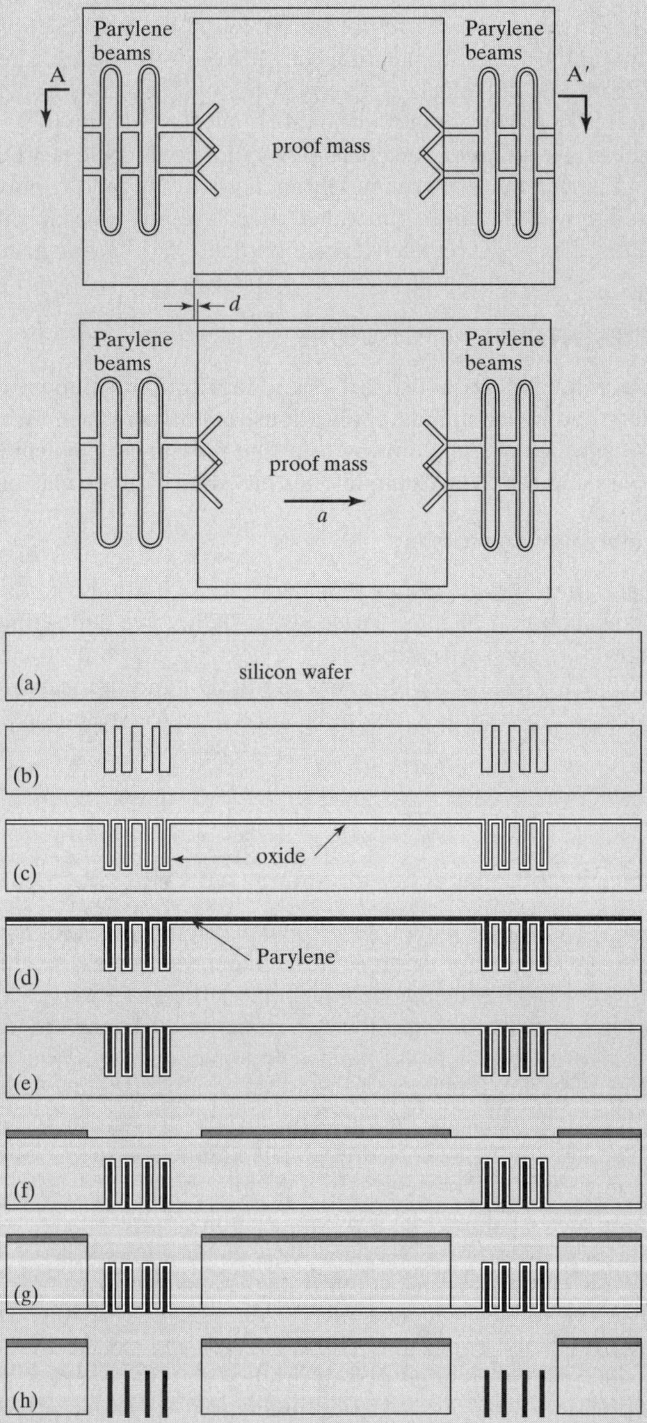

408 Chapter 12 Polymer MEMS

FIGURE 12.4
Parylene accelerometer.

is performed to remove the Parylene on the open front surface. Parylene films in the trenches are preserved because the effective thickness is much greater (Figure 12.4e).

The wafer is turned over to pattern a back-side mask layer, which is used to define the wafer with deep reactive ion etching (DRIE). The DRIE process has a very high selectivity between silicon and silicon oxide (Figure 12.4g), and it stops when it reaches the oxide layer. At the end, the oxide is removed by HF solutions to free the Parylene beams. Since the Parylene film cannot survive over-time DRIE etching, the oxide layer effectively buffers the Parylene film.

Since no active sensing layers are incorporated, the displacement of the proof mass in response to acceleration is detected using optical means. The proof mass has an area of 1.75 mm × 1.75 mm. The resonant frequency was measured to be 37 Hz. The predicted thermal mechanical noise floor is 25 nm/$\sqrt{Hz}$, while the measured noise spectrum density is 45 nm/$\sqrt{Hz}$.

12.3.2 Pressure Sensors

Pressure sensors based on the thin films of silicon, silicon nitride, and polysilicon have been discussed in previous chapters. We will discuss a surface micromachined Parylene pressure sensor in Case 12.3. The surface micromachining process and the use of metal as strain gauges completely eliminates the need to use thin-film silicon or substrates, thus reducing the cost of development and the cost of final devices.

Case 12.3. Parylene Surface Micromachined Pressure Sensor

The basic design of a surface micromachined Parylene membrane with integrated resistors is shown in Figure 12.5 [58]. The membrane, circular as shown, is elevated from the substrate surface by a distance of 0.5 to 30 μm. Strain gauge resistors for sensing membrane displacement are typically placed along the periphery of the membrane, as depicted in Figure 12.5. Metal films can serve as piezoresistors in place of doped polycrystalline silicon. However, one disadvantage lies in the fact that the resistivity of thin-film metal is much smaller compared with that of polycrystalline silicon. In order to achieve an appreciable magnitude of resistance (e.g., greater than 40 Ω), these resistors are zigzagged using alternating radial segments and tangential ones. The radial segments are primarily responsible for the displacement sensing. When a vertical force or pressure is applied on the membrane, the membrane will be deformed to induce in-plane stress in the radial direction, which is sensed by the radial segments of the strain gauge resistors.

The major design variables of a membrane device include the diameter and thickness of the membrane, the height of the underlying cavity, and the resistance of thin-film resistors. A successful design must consider the processing and performance needs simultaneously. For

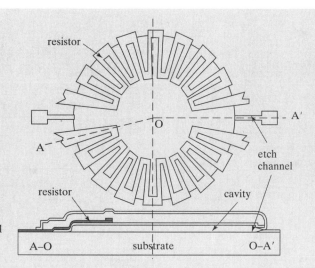

FIGURE 12.5

Top view and cross-sectional view of Parylene surface micromachined membrane with integrated metal resistors. The cross-sectional view is a composite, made along the A–O–A' line.

example, to keep the membrane from collapsing and sticking to the substrate, it is generally desirable for a membrane to be smaller, thicker, and have greater cavity heights. However, if a membrane is overly small, it may not have enough area to accommodate appreciable resistance from embedded metal resistors.

Increasing the cavity height (and membrane clearance) will generally cause difficulties with electrical continuity from the top of the membrane to the substrate level; there are also practical difficulties with building thick sacrificial layers (e.g., greater than 20 μm).

The overall process, diagrammed in Figure 12.6 and Figure 12.7, can be achieved under relatively low overall temperatures (i.e., less than 120°C). As a result, the process can be realized on a variety of substrate materials, including silicon, glass, and even polymers.

A layer of photoresist is spin coated on the front surface of a substrate and patterned photolithographically (Figure 12.6a). The spin-on photoresist is cured in a convection oven, first at 60°C for 5 minutes (to remove edge beads) and then at 110°C for 1 minute. The patterned photoresist will reflow slightly during the postdevelopment bake (110°C for 2 minutes), rounding the edges of the features to create a sloped edge. Optionally, the photoresist can be selectively thinned (to a target height of 2.5 μm) near etch/sealing holes (Figure 12.6b). The authors achieved this by additional exposure near the etch hole regions using a separate mask. This reduces the amount of Parylene needed to seal the cavity in optional step (m).

Major reasons for selecting the photoresist as the sacrificial material (as opposed to metal or silicon dioxide) include (1) the thickness of the sacrificial layer can reach the 10–20-μm range relatively easily and quickly; and (2) the edge of the photoresist sacrificial layer can be smoothed to realize gentle slopes.

A 1-μm-thick Parylene thin film is then deposited on top of the wafer surface (Figure 12.6c). The Parylene is subsequently coated with a 150-nm-thick Al thin film, which is then patterned photolithographically (Figure 12.6d). Oxygen plasma etch is used, with the thin film Al as the mask, to remove the exposed Parylene (Figure 12.6e) and to reach the underlying hard substrate.

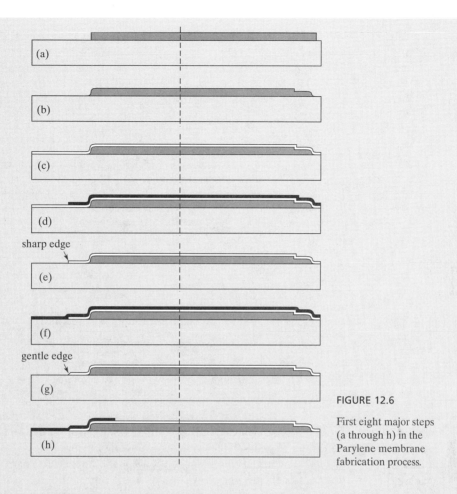

FIGURE 12.6

First eight major steps (a through h) in the Parylene membrane fabrication process.

An oxygen plasma etch step creates a rather steep transition between the substrate and the top Parylene surface. This will pose problems with the electrical continuity in the future when thin-film metal wiring traverses between the substrate and the Parylene top surface. Therefore, after removing the metal etch mask, the authors made efforts to smooth the edge of the Parylene using the process described below.

A layer of photoresist is spin coated (2500 rpm for 10 seconds) following the removal of the thin-film metal. The top profile of the photoresist after curing and reflow (5 minutes at 64°C followed by 1 minute at 110°C) is much smoother than the slope in Parylene (Figure 12.6f). A global oxygen plasma etch is performed to etch the photoresist (Figure 12.6g). Since the oxygen plasma etch rate on the photoresist and on Parylene is roughly identical (at 300 mtorr of pressure and 350 W of power), the smooth, reflown edge of the photoresist is transferred onto the Parylene after the photoresist on planar surfaces is removed.

Thin-film metal resistors embedded in the polymer membrane must be located near the surface, off the neutral axis of the membrane. After the edge profile is adjusted, a layer of 200-nm-thick Au (with a 5-nm-thick Cr thin film underneath for adhesion) is deposited

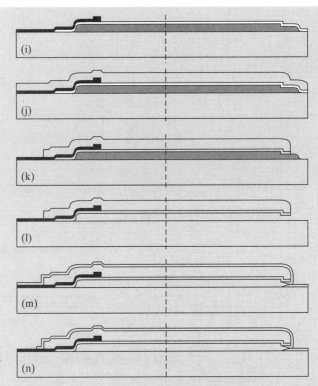

FIGURE 12.7

Final six steps (i through n) in the fabrication process of Parylene membrane.

and patterned on top of the Parylene thin film (Figure 12.6h). Gold offers better conformal coverage than the others (e.g., Ni, NiCr, and Al), allowing electrical continuity across the edges of the membrane. The underlying Parylene shields the metal resistor from direct contact and from shorting—with the bottom of the cavity if the membrane is fully displaced downward. Another layer of Parylene encapsulates the metal resistor, preventing delamination and accidental electrical shorting.

The thickness of short tangential segments is made thicker by depositing an additional layer of metal using the lift-off process (Figure 12.7i).

The entire device is then coated with another 8-μm-thick (nominal) Parylene layer (Figure 12.7j). The Parylene is patterned using a thin-film metal (300-nm-thick Al) as a mask in oxygen plasma etching. A global plasma etch is performed to pattern the newly deposited Parylene film. This exposes the end regions of the etch hole and also reopens the contact pads (Figure 12.7k). The photoresist sacrificial layer inside the cavity is removed using acetone (Figure 12.7l). This is conducted under room temperature for 3 hours (for a 400-μm-diameter membrane) or more (for larger membranes). The wafer is dried under an infrared lamp for 10 minutes. It is impractical to remove the wet chemicals from the cavity using spin-drying because the membranes can collapse and adhere to the substrates.

Optionally, the authors reported hermetically sealing the cavity by depositing another thin film of Parylene (with an approximate thickness of 2 μm). At the opening of each etch hole, the two fronts of Parylene will grow from opposite surfaces and eventually meet to

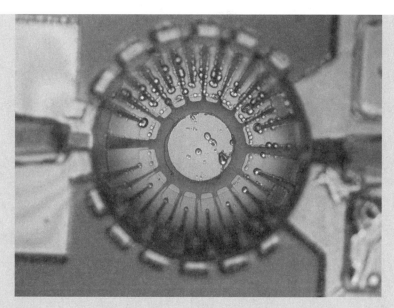

FIGURE 12.8

An optical micrograph of a membrane with a serpentine piezoresistor. The diameter of the membrane is 400 μm.

seal the cavity interior (Figure 12.7m). The pressure inside the cavity is kept at the processing pressure of the Parylene deposition (approximately 40 mtorr). The Parylene film is also deposited on the bonding pad region, unfortunately. The authors performed another sequence of masking (with metal film) and etching (using oxygen plasma) to reopen the bonding pad regions (Figure 12.7n).

An optical micrograph of a fabricated device is shown in Figure 12.8. The device has successfully measured contact pressure. Further characterization and improvement of sensitivity is needed in order to demonstrate measurement of air or liquid pressure changes.

12.3.3 Flow Sensors

Most existing micromachined sensors have been developed using single-crystal silicon substrates. An important reason for making sensors out of silicon lies in the fact that piezoresistive elements can be realized in silicon by selective doping. However, silicon devices are relatively expensive and brittle when compared with polymer and metal-based devices. A silicon beam may fracture easily in the presence of shock or contact. Flow sensors with polymer elements are reported. One example, an LCP-based flow-rate sensor, is discussed in this section (see Case 12.4).

Case 12.4. LCP Piezoresistive Flow Sensor

A flow sensor consisting of a polymer cantilever beam has been made [23]. As shown in Figure 12.9, flow imparts momentum on the cantilever and causes it to bend; this induces strain at the base of the cantilever. The strain is transduced into an electrical signal using a

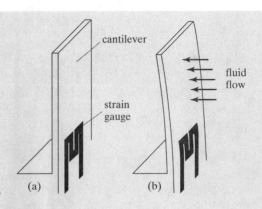

FIGURE 12.9

Schematic diagram of an LCP polymer flow sensor. (a) When there is no flow rate, the cantilever is straight. (b) Flow imparts momentum on the cantilever and causes it to bend.

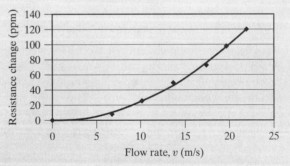

FIGURE 12.10

(a) Optical micrograph of a cantilever flow sensor. (b) Experimentally measured output characteristics as a function of flow rate.

piezoresistive sensor made of thin-film metal. While the gauge factor of doped silicon can reach 10–20, the gauge factor of thin-film metal is much lower, typically ranging from 1 to 5. However, the increased thickness and compliance of these polymer film devices has been shown to offset the reduced gauge factor of metal film strain gauges; this results in sensitivity comparable with silicon-based devices.

The flow sensor uses nickel-chrome (NiCr) strain gauges on an LCP cantilever that is 1000 μm wide and 3000 μm long. Figure 12.10a shows a micrograph of the completed device. Wind tunnel testing with flow rates from 0 to 20 m/s showed a velocity-squared relationship as expected, as seen in the quadratic trend line in Figure 12.10b.

12.3.4 Tactile Sensors

Among the various types of sensors discussed in this book—pressure, acceleration, flow, and tactile sensors—the tactile sensors have the most stringent requirements of robustness. They must be able to withstand direct contact and overloading. It is advantageous to incorporate polymers in tactile sensors to increase the level of robustness. One example of a multiple modality tactile sensor is reviewed in Case 12.5. Tactile sensors based on piezoelectric sensing using PVDF material have been made as well [59].

Case 12.5. Multimodal Polymer-Based Tactile Sensor

A multimodal flexible sensor skin has been made to mimic the functionality of biological tactile skins [60]. Biological skins are flexible and robust. They are capable of detecting multiple variables. The so-called multimodal sensory skin is capable of measuring four variables of an object: surface roughness, surface hardness, temperature, and thermal conductivity (Figure 12.11).

We selectively review the design of a hardness sensor that does not require knowledge of absolute contact force. The structure of the demonstrated hardness sensor is shown

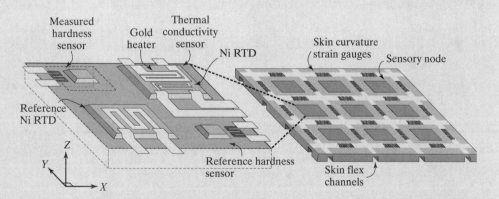

FIGURE 12.11

A sensory node incorporates four distinct sensors: reference temperature, thermal conductivity, contact force, and hardness sensors.

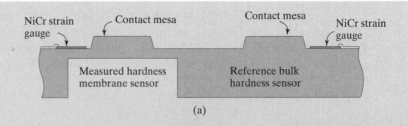
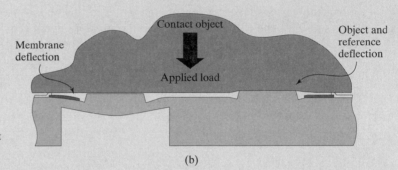

FIGURE 12.12

(a) Cross section of hardness sensor with membrane and bulk hardness sensors; (b) in contact with an object, the sensors deform with apparent pressures proportional to the contact object hardness.

in Figure 12.12a. The device consists of two membrane contact pressure sensors with very different membrane thicknesses. Each membrane sensor includes a contact mesa; on top of this contact mesa, metal strain gauges detect membrane deformation. These two sensors are close enough that, when a contact with an object is made, they are assumed to be under identical contact pressure. Under a uniform contact pressure, the thin membrane deforms more than the thick membrane (Figure 12.12b).

According to clamped–clamped-plate theory, the relation between the uniform pressure and the maximum displacement for the thin membrane is

$$q_{plate} = \frac{z_{max}Et^3}{(0.0138)b^4}, \tag{12.1}$$

where z_{max} is the peak vertical deflection in the center of the diaphragm, q_{plate} the pressure applied to the plate, b the length of the square sides, E Young's modulus, and t the plate thickness.

The reference sensor does not use a thinned diaphragm; the contact mesa and strain gauges are positioned over the full thickness of bulk polymer sheets (Figure 12.12a). The relation between a uniform pressure over the area of the contact mesa and the maximum displacement is

$$q_{bulk} = \frac{z_{max}E}{(2.24)a(1-\nu^2)}, \tag{12.2}$$

where ν is Poisson's ratio of the membrane, a the width of the contact mesa, and q_{bulk} the pressure applied to the bulk sensor contact mesa.

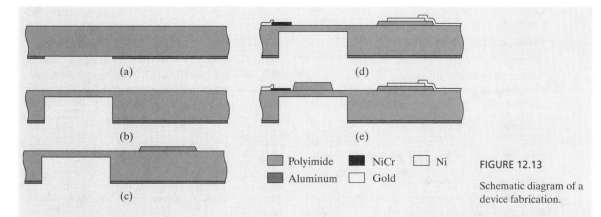

FIGURE 12.13

Schematic diagram of a device fabrication.

When the sensor skin is in contact with an object, changes in resistance are observed at both membrane sensors due to piezoresistive response (Figure 12.12). The hardness of the object in contact is correlated to the relative difference of resistive readings from these two sensors. Calibration of the hardness sensor pair is accomplished by placing a number of polymer samples in contact with the sensor skin. A range of reference samples of sorbothane and polyurethane rubber with a known hardness ranging from 10 to 80 Shore A are cut into 5 mm × 5 mm squares and pressed onto the sensor skin using a fixed mass.

The fabrication process of the multimodal tactile sensor is diagrammed in Figure 12.13. A sheet of Dupont Kapton HN200 polyimide film serves as the start substrate. An aluminum etch mask is deposited and patterned via lift-off (Figure 12.13a). The 50-μm-thick film is etched 40 μm down in an RIE plasma etcher at 350 W under 300 mtorr O_2 (Figure 12.13b) to define the thin sensor diaphragms. Next, a 2-μm layer of photo-definable polyimide (HD Microsystems HD4000) is spun on the top skin surface and patterned to define mesa platforms for temperature and thermal conductivity sensors (Figure 12.13c). Nickel thin-film resistors (500 Å Ni on 100 Å Cr) are then deposited and patterned on the contact mesas. It serves as a temperature sensitive resistor. Then, 750 Å of NiCr is deposited, and it is patterned by the lift-off method to define the strain gauges for the force, curvature, and hardness sensors. A final metal layer consisting of 1500 Å of Au on 100 Å Cr is thermally evaporated and lifted off (Figure 12.13d). The final step is to spin on, pattern, and cure the 8-μm HD4000 contact mesas for the force and hardness sensors (Figure 12.13e).

SUMMARY

This chapter broadly reviewed various polymers that have found wide use in MEMS or have demonstrated significant performance and potential. The reader should keep in mind that this is not an exhaustive list of polymers for MEMS—many existing and future materials are not included. It is my hope that this chapter serves to introduce the unique designs and fabrication aspects of polymer MEMS devices to readers, and lead to future interest and discoveries.

At the end of this chapter, the reader should be able to identify the major types of polymer materials used in MEMS technology and understand their common fabrication methods and processing conditions.

PROBLEMS

Problem 12.1: Fabrication

Form a team of three to four students and design a polymer-based accelerometer based on a principle other than optical sensing. Complete the design, fabrication process, and performance prediction. Summarize the advantages and disadvantages over MEMS accelerometers based on other materials. (*Hint:* Pay attention to mechanical properties of polymer structural layers.)

Problem 12.2: Review

For Case 12.2, estimate the force constant and the resonant frequency based on the known geometries and material properties given in Reference [33].

Problem 12.3: Design

Calculate the thermal mechanical noise floor of the sensor in Case 12.2 for the spring with the smallest force constant in the design space.

Problem 12.4: Fabrication

For the fabrication processes discussed in Case 12.2, what difference would it make to the process if, instead of oxide, silicon nitride is used to fill the DRIE trenches? Is it a better option or worse option to use oxide instead of nitride? Explain your reasoning.

Problem 12.5: Design

For Case 12.2, the sensitive axis is in the substrate plane. Develop an alternative accelerometer design, using similar materials as Case 12.2, but with the sensitive axis perpendicular to the substrate plane. Draw a detailed fabrication process. The force constant should be the same as in Case 12.2. The thickness of the Parylene should be no more than 10 μm.

Problem 12.6: Review

For Case 12.3, consider the implications for the design, material, and fabrication process if the membrane size (diameter) is reduced to 1/3 the original size. Built a list of the most important changes and concerns under each category.

Problem 12.7: Challenge

Is it feasible to make a membrane out of PDMS materials to achieve an even greater sensitivity and flexibility? Form a group of three to four students and develop the design and fabrication process of a pressure sensor with a 20-μm-thick membrane made of PDMS.

Problem 12.8: Review

For Case 12.4, derive the analytical formula of the longitudinal strain as a function of flow rate. Assume the flow rate is identical throughout the surface of the cantilever.

Problem 12.9: Challenge

For the polymer tactile sensor in Case 12.5, metal strain gauges are used, but their gauge factor is rather limited. Polysilicon and single-crystal silicon provide a greater gauge factor, but they are not compatible with the polymer substrate. Conduct the research and design of a method to integrate silicon piezoresistive elements with the polymer substrate, such that the mechanical flexibility and electronic sophistication are both achieved simultaneously.

Problem 12.10: Challenge

Form a group of three to four students and find one polymer material that has potential to be used in MEMS and that has not been covered in this text. Describe the chemical properties, mechanical properties, and other unique characteristics associated with this polymer.

Problem 12.11: Challenge
Identify one polymer material that has not been used by the MEMS community. Describe the chemical properties, mechanical properties, and other unique characteristics associated with this polymer.

REFERENCES

1. Callister, W.D., *Materials Science and Engineering, An Introduction*. 4th ed. New York: John Wiley and Sons., 1997.
2. Sundar, V.C., et al., *Elastomeric Transistor Stamps: Reversible Probing of Charge Transport in Organic Crystals*. Science, 2004. **303**(5664): p. 1644–1646.
3. Rogers, J.A., *Electronics: Toward Paperlike Displays*. Science, 2001. **291**(5508): p. 1502–1503.
4. Guizzo, E., *Organic Memory Gains Momentum*. Spectrum, IEEE, 2004. **41**(4): p. 17–18.
5. Miyajima, H., et al., *A Durable, Shock-Resistant Electromagnetic Optical Scanner with Polyimide-Based Hinges*. Microelectromechanical Systems, Journal of, 2001. **10**(3): p. 418–424.
6. Armani, D., and C. Liu, *Microfabrication Technology for Polycaprolactone, A Biodegradable Polymer*. Journal of Micromechanics and Microengineering, 2000. **10**: p. 80–84.
7. Androva, N.A., et al., *Polyimide, A New Class of Thermally Stable Polymers*. Vol. VII., Stamford, CT: Technomic, 1970.
8. Merrem, H.J., R. Klug, and H. Hartner, *New Developments in Photosensitive Polyimides*, in *Polyimides, Synthesis, Characterization, and Applications*. New York, Plenum Press: 1984. p. 919–931.
9. Bhattacharya, P.K., and K.S. Bhosale, *Relaxation of Mechanical Stress in Polyimide Films by Softbaking*. Thin Solid Films, 1996. **290–291**: p. 74–79.
10. Mapplitano, M.J., and A. Moet, *The Mechanical Behavior of Thin-Film Polyimide Films on a Silicon Substrate Under Point Loading*. Journal of Material Science, 1989. **24**(9): p. 3273–3279.
11. Kim, Y.-J., and M.G. Allen, *In Situ Measurement of Mechanical Properties of Polyimide Films Using Micromachined Resonant String Structures*. Components and Packaging Technologies, IEEE Transactions on [see also Components, Packaging and Manufacturing Technology, Part A: Packaging Technologies, IEEE Transactions on], 1999. **22**(2): p. 282–290.
12. Herminghaus, S., et al., *Large Anisotropy in Optical Properties of Thin Polyimide Films of Poly (p-phenylene biphenyltetracarboximide)*. Applied Physics Letters, 1991. **59**(9): p. 1043–1045.
13. Boese, D., et al., *Stiff Polyimides: Chain Orientation and Anisotropy of the Optical and Dielectric Properties of Thin Films*. Materials Research Society Symposium Proceedings, 1991. **227**: p. 379–386.
14. Ho, P.S., T.W. Poon, and J. Leu, *Molecular Structure and Thermal/Mechanical Properties of Polymer Thin Films*. Journal of Physics and Chemistry of Solids, 1994. **55**(10): p. 1115–1124.
15. Kurabayashi, K., et al., *Measurement of the Thermal Conductivity Anisotropy in Polyimide Films*. Microelectromechanical Systems, Journal of, 1999. **8**(2): p. 180–191.
16. Frazier, A.B., *Recent Applications of Polyimide to Micromachining Technology*. Industrial Electronics, IEEE Transactions on, 1995. **42**(5): p. 442–448.
17. Park, K.-T., and M. Esashi, *A Multilink Active Catheter with Polyimide-Based Integrated CMOS Interface Circuits*. Microelectromechanical Systems, Journal of, 1999. **8**(4): p. 349–357.
18. Dokmeci, M., and K. Najafi, *A High-Sensitivity Polyimide Capacitive Relative Humidity Sensor for Monitoring Anodically Bonded Hermetic Micropackages*. Microelectromechanical Systems, Journal of, 2001. **10**(2): p. 197–204.
19. Li, M.-H., J.J. Wu, and Y.B. Gianchandani, *Surface Micromachined Polyimide Scanning Thermocouple Probes*. Microelectromechanical Systems, Journal of, 2001. **10**(1): p. 3–9.

20. Lorenz, H., et al., *SU-8: A Low-Cost Negative Resist for MEMS*. Journal of Micromechanics and Microengineering, 1997. **7**: p. 121–124.
21. El-Ali, J., et al., *Simulation and Experimental Validation of a SU-8-Based PCR Thermocycler Chip with Integrated Heaters and Temperature Sensor*. Sensors and Actuators A: Physical, 2004. **110**(1–3): p. 3–10.
22. Zou, J., et al., *A Mould-and-Transfer Technology for Fabricating Scanning Probe Microscopy Probes*. Nanotechnology, 2004. **14**(2): p. 204–211.
23. Wang, X., J. Engel, and C. Liu, *Liquid Crystal Polymer (LCP) for MEMS: Processes and Applications*. Journal of Micromechanics and Microengineering, 2003. **13**(5): p. 628–633.
24. Belanger, M.-C., and Y. Marois, *Hemocompatibility, Biocompatibility, Inflammatory and in vivo Studies of Primary Reference Materials Low-Density Polyethylene and Polydimethylsiloxane: A Review*. Journal of Biomedical Materials Research, 2001. **58**(5): p. 467–477.
25. Ma, X., et al., *Low Temperature Bonding for Wafer Scale Packaging and Assembly of Micromachined Sensors*. Final Report 1998–99 for Micro Project 98-144, University of California at Davis.
26. Eon, D., et al., *Surface Modification of Si-Containing Polymers During Etching for Bilayer Lithography*. Microelctronic Engineering, 2002. **61–62**: p. 901–906.
27. Ryu, K., et al., *A Method for Precision Patterning of Silicone Elastomer and Its Applications*. IEEE/ASME Journal of Microelectromechanical Systems (JMEMS), 2004. **13**(4): p. 568–575.
28. Teh, W.H., et al., *Cross-Linked PMMA as a Low-Dimensional Dielectric Sacrificial Layer*. Microelectromechanical Systems, Journal of, 2003. **12**(5): p. 641–648.
29. Zhang, C., C. Yang, and D. Ding, *Deep Reactive Ion Etching of Commercial PMMA in O_2/CHF_3, and O_2/Ar-based Discharges*. Journal of Micromechanics and Microengineering, 2004. **14**(5): p. 663–666.
30. Harder, T.A., et al. *Residual Stress in Thin-Film Parylene-C*. In The Sixteenth Annual International Conference on Micro Electro Mechanical Systems. Las Vegas, NV: 2002.
31. Burns, M.A., et al., *An Integrated Nanoliter DNA Analysis Device*. Science, 1998. **282**(5388): p. 484–487.
32. Wang, X.Q., Q. Lin, and Y.-C. Tai. *A Parylene Micro Check Valve*. In The Twelfth Annual International Conference on Micro Electro Mechanical Systems. Orlando, Florida: 1999.
33. Suzuki, Y., and Y.-C. Tai. *Micromachined High-Aspect-Ratio Parylene Beam and Its Application to Low-Frequency Seismometer*. In The Sixteenth Annual International Conference on Micro Electro Mechanical Systems. Kyoto, Japan: 2003.
34. Fan, Z., et al., *Parylene Surface Micromachined Membranes for Sensor Applications*. IEEE/ASME Journal of Microelectromechanical Systems (JMEMS), 2003. In press.
35. Niu, M.-N., and E.S. Kim, *Piezoelectric Bimorph Microphone Built on Micromachined Parylene Diaphragm*. Microelectromechanical Systems, Journal of, 2003. **12**(6): p. 892–898.
36. Sutomo, W., et al., *Development of an End-Point Detector for Parylene Deposition Process*. IEEE/ASME Journal of Microelectromechanical Systems (JMEMS), 2003. **12**(1): p. 64–70.
37. Oh, K.W., et al., *A Low-Temperature Bonding Technique Using Spin-On Fluorocarbon Polymers to Assemble Microsystems*. Journal of Micromechanics and Microengineering, 2002. **12**(2): p. 187–191.
38. Carlen, E.T., and C.H. Mastrangelo, *Surface Micromachined Paraffin-Actuated Microvalve*. Microelectromechanical Systems, Journal of, 2002. **11**(5): p. 408–420.
39. Park, J.-H., et al. *Micromachined Biodegradable Microstructures*. In Micro Electro Mechanical Systems. MEMS-03 Kyoto. IEEE, The Sixteenth Annual International Conference. 2003.
40. Kabei, N., et al., *A Thermal-Expansion-Type Microactuator with Paraffin as the Expansive Material (Basic Performance of a Prototype Linear Actuator)*. JSME International Journal, Series C, 1997. **40**(4): p. 736–742.

41. Carlen, E.T., and C.H. Mastrangelo, *Electrothermally Activated Paraffin Microactuators*. Microelectromechanical Systems, Journal of, 2002. **11**(3): p. 165–174.
42. Chen, R.-H., and C.-L. Lan, *Fabrication of High-Aspect-Ratio Ceramic Microstructures by Injection Molding with the Altered Lost Mold Technique*. Microelectromechanical Systems, Journal of, 2001. **10**(1): p. 62–68.
43. Jayachandran, J.P., et al., *Air-Channel Fabrication for Microelectromechanical Systems via Sacrificial Photosensitive Polycarbonates*. Microelectromechanical Systems, Journal of, 2003. **12**(2): p. 147–159.
44. Kuo, T.-C., et al., *Gateable Nanofluidic Interconnects for Multilayered Microfluid Separation Systems*. Analytical Chemistry, 2003. **75**: p. 1861–1867.
45. Oh, K.W., C.H. Ahn, and K.P. Roenker, *Flip-Chip Packaging Using Micromachined Conductive Polymer Bumps and Alignment Pedestals for MOEMS*. Selected Topics in Quantum Electronics, IEEE Journal on, 1999. **5**(1): p. 119–126.
46. Oh, K.W., and C.H. Ahn, *A New Flip-Chip Bonding Technique Using Micromachined Conductive Polymer Bumps*. Advanced Packaging, IEEE Transactions on [see also Components, Packaging and Manufacturing Technology, Part B: Advanced Packaging, IEEE Transactions on], 1999. **22**(4): p. 586–591.
47. Bar-Cohen, Y., *Electric Flex*. Spectrum, IEEE, 2004. **41**(6): p. 28–33.
48. Smela, E., M. Kallenbach, and J. Holdenried, *Electrochemically Driven Polypyrrole Bilayers for Moving and Positioning Bulk Micromachined Silicon Plates*. IEEE/ASME Journal of Microelectromechanical Systems (JMEMS), 1999. **8**(4): p. 373–383.
49. Jager, E.W.H., E. Smela, and O. Inganas, *Microfabricating Conjugated Polymer Actuators*. Science, 2000. **290**(5496): p. 1540–1545.
50. Lu, W., et al., *Use of Ionic Liquids for pi-Conjugated Polymer Electrochemical Devices*. Science, 2002. **297**(5583): p. 983–987.
51. Yang, L.-J., et al., *Photo-Patternable Gelatin as Protection Layers in Low-Temperature Surface Micromachinings*. Sensors and Actuators A: Physical, 2003. **103**(1–2): p. 284–290.
52. Zhao, X.-M., et al., *Fabrication of Microstructures Using Shrinkable Polystyrene Films*. Sensors and Actuators A: Physical, 1998. **65**(2–3): p. 209–217.
53. Gall, K., et al., *Shape-Memory Polymers for Microelectromechanical Systems*. Microelectromechanical Systems, Journal of, 2004. **13**(3): p. 472–483.
54. Gallantree, H.B., *Review of Transducer Applications of Polyvinylidene Fluoride*. IEEE Proceedings, 1983. **130**(5): p. 219–224.
55. Manohara, M., et al., *Transfer by Direct Photo Etching of Poly(vinylidene fluoride) Using X-Rays*. Microelectromechanical Systems, Journal of, 1999. **8**(4): p. 417–422.
56. Yagyu, H., S. Hayashi, and O. Tabata, *Application of Nanoparticles Dispersed Polymer to Micropowder Blasting Mask*. Microelectromechanical Systems, Journal of, 2004. **13**(1): p. 1–6.
57. Abramson, A.R., et al., *Fabrication and Characterization of a Nanowire/Polymer-Based Nanocomposite for a Prototype Thermoelectric Device*. Microelectromechanical Systems, Journal of, 2004. **13**(3): p. 505–513.
58. Fan, Z., et al., *Parylene Surface-Micromachined Membranes for Sensor Applications*. Microelectromechanical Systems, Journal of, 2004. **13**(3): p. 484–490.
59. Kolesar, E.S., Jr., and C.S. Dyson, *Object Imaging with a Piezoelectric Robotic Tactile Sensor*. Microelectromechanical Systems, Journal of, 1995. **4**(2): p. 87–96.
60. Engel, J., J. Chen, and C. Liu, *Development of Polyimide Flexible Tactile Sensor Skin*. Journal of Micromechanics and Microengineering, 2003. **13**(3): p. 359–366.

CHAPTER 13

Microfluidics Applications

13.0 PREVIEW

Microfluidics represent a new and interdisciplinary research area. This chapter serves as an introduction to this exciting research area. Because materials for microfluidic channels, reactors, sensors, and actuators must be compatible with biochemical fluids and particles, this subject area challenges MEMS developers to incorporate new materials and develop practical, effective, and low cost solutions to sensing and actuation. In Section 13.2, we will introduce pertinent biological and chemistry concepts to device developers. Section 13.3 covers various methods for transporting fluids in a microchannel. In Section 13.4, we will review the design and fabrication technology of components under several important classes, including channels, valves, and sensors.

13.1 MOTIVATION FOR MICROFLUIDICS

Sophisticated chemical and biological analytical procedures, which are used for applications such as medical diagnosis and environmental sample screening, have traditionally been conducted in dedicated laboratories by highly trained personnel. These protocols are performed on bench tops and in test tubes and beakers. This bench-top norm has limited accessibility, long turnaround time, complex logistics (e.g., sample transportation and storage), and high costs.

Over the past few decades, integrated circuits changed the landscape of traditional electronics. An exponential increase of performance *and* a reduction of cost result when components are miniaturized, fabricated using monolithic integration methods, and connected to each other on a massive scale. Can we reduce the size of test tubes, beakers, and channels? Can we realize an integrated, low-cost system that is capable of performing a complex biological and chemical protocol without human intervention in a totally automated fashion? If the answers to these questions are "yes," then perhaps the same benefits that have been realized for the microelectronics industry can be applied to biology, chemistry, and medicine.

A microfluidics system for chemical and biological diagnosis is known as a "laboratory-on-a-chip," or a "micrototal analysis system" (μTAS). The naming of the field reveals its inspiration and motivation. The last three letters of the word "microfluidics"—"*ics*"—are identical to the last three letters of the word "microelectronics."

Much like microelectronics circuits revolutionized signal processing and communication, fluid reactions in integrated, miniaturized microfluid channels and reactors promise major changes in the practices of medical diagnosis and intervention [1], drug discovery [2], environmental monitoring [3], cell culture and manipulation of bioparticles [4, 5], gas handling and analysis (e.g., component separation [6–8] or heat transfer), heat exchange [9, 10], chemical reactors (for power and force production) [11–15], and bioterrorism defense.

Some major benefits of using microfluid platforms to replace bench-top chemistry are

1. A microfluidics system reduces dead volumes associated with a chemical assay system with large-scale chambers and connectors.
2. A microfluidics system reduces the amount of chemical assays and solutions required and thus can potentially reduce costs by saving the amount of expensive chemicals and biological samples used for a given analysis.
3. A microelectronics-style bulk fabrication will reduce the cost of sophisticated systems. Lithography and parallel fabrication reduce the difficulty of building sophisticated fluid piping and reaction networks.
4. Microfluidics can achieve a high level of multiplexity and parallel operations to increase the efficiency of chemical and biological discovery.

Microscale fluid elements also find a broader use beyond biological and chemical analysis. Many applications are being sought in areas such as optical communication [16], tactile display (e.g., refreshable Braille display [17]), IC-chip cooling [18, 19] and fluid logics [20–22]. Microfluids have also been used for performing novel microfabrication and nanofabrication [23, 24].

This chapter introduces key and basic principles, components, and applications of microfluidics; it emphasizes biological and chemical analysis applications.

13.2 ESSENTIAL BIOLOGY CONCEPTS

Microfluidic systems are used to handle and interact with biological and chemical particles and substances, including cells and polymers (e.g., DNA and proteins). This section presents a concise overview of the major characteristics of essential biological and chemistry elements that pertain to the design, fabrication, and functions of microfluidic systems.

A MEMS developer should be at least conversant about key terminology and the concepts of biology and chemistry. A few of these are reviewed below. Interested readers should refer to textbooks in the area of biology and chemistry for more details.

Cells   Cells are basic functional units of life. The function of a cell is determined by the genetic sequence it carries. A basic human cell stores genetic codes, reproduces such codes upon cell division, and manufactures protein molecules based on such codes. Cells can develop a rich variety of functional differentiation based on genetic codes.

Cells communicate with their outside environment through a highly sophisticated, compliant cell wall. The cell wall is made of a lipid bilayer lined with ion channels, which are tiny channels that allow ions (such as potassium and sodium) to pass selectively in two ways.

Bacteria and viruses are special forms of cells. Bacteria, for example, do not contain a cell nucleus. A virus, on the other hand, does not have the ability to divide and reproduce. It may only do so after infecting a host cell and taking over the reproduction mechanism.

The presence, population, and genetic variation of certain cells are indicative of medical and environmental conditions. The rapid identification of cells, bacteria, and viruses is important for medical diagnosis, environmental monitoring, and bioterrorism prevention. For example, the ability to rapidly and inexpensively determine the presence of a small number of cancerous cells in complex biological fluids would provide a powerful new weapon for the early detection of cancer and provide hope for a cure.

DNA   Life is possible only because each cell, upon division, transmits to the next generation the vital information about how it works. The substance that carries the information is a polymer called deoxyribonucleic acid (DNA), which is a large molecule with a molecular weight as high as several billion. The monomers that comprise the nucleic acids, called nucleotides, are composed of three distinct parts—a five-carbon sugar, a nitrogen-containing organic base, and a phosphoric acid molecule (H_3PO_4). Four nitrogen-containing bases are found—cytosine (C), thymine (T), adenine (A), and guanine (G). Human cells carry a total of 3 billion base pairs of nucleotide molecules. Segments of the DNA chain, called genes, regulate the production of proteins based on the specific sequence of the nucleotide arrangement of the gene. The code transmits the intended primary structure of the protein to the construction "machine" of the cell.

A single-strand DNA molecule can bind to another single strand with a complementary sequence (A to G, C to T). For example, a 10-mer DNA molecule with a sequence of AAGCCT-TAGG binds strongly with another DNA molecule that contain at least the following sequence, GGATTCCGAA. Two strands with slight mismatches can bind as well, but not as strongly as a fully complementary case. Two DNA chains with mismatches can be dissociated in so-called stringency tests, which are carried out by applying electrical potential, applying heat, or varying the concentration of salt.

While DNA molecules store the genetic information for the production of proteins, RNA molecules are responsible for transmitting this information to the ribosome, where protein synthesis actually occurs.

Synthetic DNA molecules can be made in test tubes using automated DNA sequencing machines and on a chip. As such, the role of DNA is transcending that of regulating life. DNA molecules have been explored as electrical conductors, mechanical binding agents, analytical beacons, and actuators. For example, DNA molecules can be used for recognition and nanoscale assembly [25] because of their unique hybridization scheme.

Protein   If DNA is the basic code of life, protein is the agent that carries out the intent of the code. Protein is a natural polymer. It makes up about 15% of our bodies and has a molecular weight that ranges from approximately 6000 to over 1,000,000 grams per mole. A protein molecule is made of a chain of α-amino acids. Twenty basic types of amino acids are found in life. The order of the amino acids in the protein is called the primary structure, which is conveniently indicated by using three-letter codes for the amino acids. For example, a short protein segment (called polypeptide) with the three amino acids (lysine, alanine, and leucine) is represented in short-hand notation as lys-ala-leu.

Long chain protein molecules do not maintain a straight line in nature. In fact, segments of the protein interact with each other. A second level of protein structures is the spatial folding of long protein molecules. The secondary structure is often determined by hydrogen bonding between an oxygen atom (in the carbonyl group of one amino acid) and a hydrogen atom (attached to a nitrogen atom of another amino acid). Such interactions can occur within a chain to form a spiral structure called an α-helix; the α-helix gives proteins elasticity.

A long protein molecule can assume many coiled shapes, which determine the functions of the protein. Proteins derive their functionalities from primary amino sequences as well as the way the long molecule chain folds.

The folded protein molecular structures can be broken down under certain conditions. The process is called denaturation; it occurs under heat, X-ray radiation, or nuclear radiation.

One can imagine that proteins, with 20 amino acids that can be assembled in any order or any length, have an infinite possibility of primary structures. Given the varied ways protein molecules fold, the variety of protein structures and functions become even greater.

Lock-and-Key Biological Binding Chemistry and biology are filled with examples of lock-and-key protocols—the highly selective, self-regulated assembly of two or more entities with recognition deriving from chemical bond forces and/or folded shapes of proteins. Many biological binding events are very specific and strong; they allow the chemical recognition and mechanical construction of molecular conjugates. There is no engineering equivalent of such a selective and automated selection process given its tailor-ability, accuracy of selectivity, and prevalent use. Some of the most commonly exploited biological binding protocols include

Binding between the antibody and antigens;

Binding between the biotin and streptavidin molecules; and

DNA complementary binding.

Molecular and Cellular Tags Certain cells, chemical and biological molecules, and ions, when present in a fluid environment, are too small and scattered to be detected easily. To report the location, species, binding characteristics, and environmental conditions (pH, temperature) of a biological cell or molecule, special tags (or beacons) are frequently used. Tags are designed to bind specifically to the cells or molecules of interest, and they allow the visualization, identification, selection, and capturing of such cells or molecules. Tags vary in size and operational principles. Frequently used tags include fluorescent particles and molecules, and surface-functionalized beads and particles made of magnetic, metallic, or dielectric materials.

Fluorescent tags play important roles in chemistry and biology today. They report actions and conditions at the molecule or cell level. Such tags consist of naturally occurring or engineered molecule structures that produce fluorescent signals upon excitation. The light intensity can be activated, quenched (turned off), or modulated in response to many events of interest, including the association and dissociation of chemical bonds, temperature, pH, and proximity. Such molecular probes can be purchased commercially with a large variety of choices.

We provide a few exemplary uses of biological tags and beads below. These examples should provide interested readers with a starting point to explore more deeply:

DNA and protein microarrays use fluorescent dyes to report DNA and peptide binding events between targets and probes [26];

The detection of trace amount of metal ions in water can be conducted using specialized DNA molecules and reported by fluorescent molecules [27];

Magnetic beads allow for the highly selective capturing of cells and molecules that bind to such beads [28, 29];

Functionalized gold nanoparticles allow optical [30] and electrical [31] reporting of molecule binding events to circumvent the need for fluorescent microscopes.

13.3 BASIC FLUID MECHANICS CONCEPTS

The basic function of a microfluid system is the transportation and handling of fluids. This section builds the basic fluid mechanical concepts and terms necessary for design tasks. In this section, we will distill the complex subject of fluid mechanics to the most vital concepts that are frequently used in the practice of microfluidics. Readers who wish to learn more systematically about fluid mechanics can refer to classic textbooks on fluid mechanics [32].

13.3.1 The Reynolds Number and Viscosity

The **Reynolds number** is one of several important dimensionless numbers in the fluid mechanics field. It is used to quantify the flow and thermal transfer characteristics of fluid cases involving different media, length scale, and velocity. The Reynolds number of an object in a fluid media is defined as

$$Re = \frac{\rho V L}{\mu}. \tag{13.1}$$

It is proportional to the velocity (V) and the length scale (L), and inversely proportional to the viscosity of the fluid.

Nearly all fluids of practical use are viscous in nature. The viscosity of a fluid characterizes its resistance to shear and is a measure of the fluid's adhesive/cohesive or frictional properties. There are two related viscosity terms—**dynamic viscosity** (μ) and **kinematic viscosity** (ν). These two terms are related by $\nu = \mu/\rho$.

The SI unit of the dynamic viscosity is kg/(m-s), or Pa·s. The CGS unit of the dynamic viscosity is Poise (1 Poise = 1 g/(cm-s) = 1 Dyne·s). The conversion factor is 1 kg/(m-s) = 10 poise = 1000 centipoise (cP). For reference, the dynamic viscosity of water at 20°C is approximately 1 centipoise (cP).

The SI and CGS units of the kinematic viscosity are m²/s and cm²/s, respectively. The CGS unit of the kinematic viscosity, cm²/s, is also known as Stoke (St). The conversion factor between the SI and CGS units is 1 m²/s = 10,000 St = 100 centiStoke. For reference, the density and kinematic viscosity of water at 20°C are approximately 1 g/cm³ and 1 cSt, respectively.

Example 13.1. Reynolds Number Calculation

Find the Re associated with two cases: (1) a person swimming in a swimming pool filled with molasses with a kinematic viscosity of 10,000 centiStokes (cSt), and (2) a 1.8-mm-long tadpole moving in water (with a kinematic viscosity of 1 cSt) at a velocity of 1 cm/s.

Solution. Assume the human swimmer has a length of 1.8 m and swims at a woeful velocity of 0.1 m/s in the thick liquid. The Re of the human swimmer case is

$$Re = \frac{\rho V L}{\mu} = \frac{0.1 \times 1.8}{100} = 0.0018.$$

The Re associated with the case of a tadpole in water is

$$Re = \frac{\rho V L}{\mu} = \frac{0.01 \times 1.8 \times 10^{-3}}{0.01} = 0.0018.$$

These two cases have identical *Re*! For this reason, they share a certain similarity of flow characteristics. The purpose of this exercise is to show how difficult it must be for a tadpole to swim in water at an appreciable speed even though it appears to be done elegantly and effortlessly in biology.

The Reynolds number is often used to predict the transition between laminar and turbulent flow cases. If the *Re* of a fluid flow is below a threshold value, the fluid is described as **laminar flow**—namely, the fluid flow can be described by layers that do not interfere with each other. If the *Re* of a fluid flow is greater than a threshold value, the fluid enters the **turbulent flow** regime. To illustrate this transition, simply turn the faucet knob at home or laboratory. Observe the flow of water from a faucet while turning the flow volume up (Figure 13.1). When the volume (and velocity) is small, liquid coming out of the faucet is stable and laminar. As the faucet is opened wider, the velocity of liquid flow increases. Above a certain velocity, the liquid becomes turbulent.

In microfluid systems, which encounter a generally small scale, the *Re* is typically very small. It is safe to say that most microfluid systems are dominated by laminar flow behavior.

13.3.2 Methods for Fluid Movement in Channels

Methods for moving a body of liquid on a chip in a controlled manner is a major concern when developing a microfluid system. A number of pumping methods have been demonstrated. The source of the fluid driving force can be classified into several categories:

Pressure difference. Pressure differences can be generated in a number of ways (as further outlined in Section 13.3.3). Both positive pressure (at upstream) and negative pressure (at downstream) can be used to pump a fluid. The simplest method of creating a pressure head in a laboratory is to use an elevated liquid reservoir relative to the level of the fluid outlet.

The magnetohydrodynamic (MHD) effect. This effect is the flow of electrically conducting liquid in the presence of electrical and magnetic fields [33]. In an MHD pump, an electric field and a magnetic field are both applied at the same time, transverse to the fluid channel, and perpendicular to each other. The application of an electrical field to a conducting liquid causes the fluid to move in a magnetic field.

FIGURE 13.1

Illustration of laminar and turbulent flow.

The electrohydrodynamic effect. This effect uses the interaction of an electrical field with electrical charges embedded in a dielectric fluid [34–37]. The charges or charged particles can be injected directly or by adding liquids containing a high density of ions.

Magnetorhelogical pumping. This involves moving plugs of ferrofluid using magnetic actuation. A ferrofluid is a liquid solution that contains a suspension of nanosized ferromagnetic particles [38].

Surface-tension driven flow. Surface-tension force is a relatively large force that is microscale compared with other forces such as gravity or structural restoring forces. It can be used to move liquid in a capillary or move liquid drops on a planar surface. For instance, the surface tension of a liquid and a substrate interface can be altered by applying electrical charges; this is a phenomenon called electrowetting [39, 40].

Traveling surface acoustic waves. These waves can be used to stream liquid in contact with the substrate [41].

Electroosmotic effect (EO). This effect is the motion of bulk liquid caused by the application of an electrical field parallel to a channel with a charged wall. See Section 13.3.3 for details.

13.3.3 Pressure Driven Flow

Pressure driven flow is most commonly encountered in microfluid channel flow due to simplicity and generality. High pressure can be generated on-chip using deformable membranes, whereas the membrane movement can be generated in a number of ways, including via shape-memory alloy thin films [42], piezoelectricity [43, 44], magnetostatics [45], thermopneumatics [46], thermal expansion [47], and the phase change between liquid and vapor [48]. Pressure driven flow can also be induced by vapor generation (and bubble formation) in the channel [49, 50], by osmosis exchange [51], by communicating with a pre-stored pressure source (upstream) [52], by centrifugal forces [53], or by thermal expansion of a fluid. It is also possible to use a vacuum downstream to induce pressure driven flow.

Since membrane displacement pushes fluid towards both inlets and outlets, rectification is needed to achieve a net fluid flow in one particular direction. Single-membrane pumps generally require check valves or one-way diffuser valves (called fluid diodes) [44] to ensure that the net fluid flow occurs in one preferential flow direction. Alternatively, multiple membranes can exploit spatial or temporal differences among them, such as in the case of peristaltic pumping [43].

When selecting technologies for on-chip pumps, there are several important factors of consideration:

achievable flow rate;
simplicity of fabrication;
cost of fabrication;
simplicity of control;
robustness of the membrane;
biocompatibility of membrane and channel materials;
energy consumption, which is important for portable systems.

For general cases, the volumetric flow in a microchannel is proportional to the pressure difference at its two ends. For a pipe with a circular cross section with radius r (in m) and length L (also in m), the volume flow rate Q is related to the pressure according to

$$Q = \frac{\pi r^4}{8\mu L}\Delta P. \tag{13.2}$$

For a pipe with a rectangular cross section, with a width w and height h, the relation between the volumetric flow rate and the pressure difference is described as

$$Q = \frac{wh^3}{12\mu L}\Delta P, \tag{13.3}$$

provided the ratio of w/h is relatively large. The ratio between the pressure difference and the volumetric flow rate is called the **flow resistance** of a channel.

Notice that the small cross-sectional area typically associated with most microchannels means that a significant pressure is needed to achieve a certain flow rate. Long channels may require significant pressure build-up to drive liquid inside. Pressure build-up in channels can lead to the delamination of channels and reactors.

When a fluid moves inside a channel under the pressure difference, the liquid particles next to the wall do not move with respect to the wall. The velocity of liquid molecules at the interface is said to follow the *non-slip boundary condition*. The velocity of fluid particles increases when they are further away from the wall. The velocity profile u as a function of the distance to the wall (y) is plotted in Figure 13.2. Outside a certain range of y, the velocity does not change with respect to y anymore, and it reaches a constant called the mean-stream velocity.

The non-uniform distribution of flow velocity gives rise to the term called **fluid shear stress**, as defined by

$$\tau = \mu \frac{du}{dy}. \tag{13.4}$$

The accepted formulas for estimating the boundary layer thickness (δ) are

$$\delta = \frac{5}{\sqrt{Re_x}}x \tag{13.5}$$

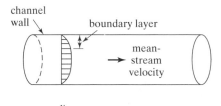

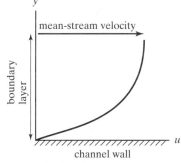

FIGURE 13.2

Velocity profile.

under the laminar flow [32] or

$$\delta = \frac{0.16}{(Re_x)^{1/7}} x \tag{13.6}$$

for turbulent flow (e.g., $Re_x > 10^6$,) with the term Re_x being the local Reynolds number $\left(Re_x = \frac{u_0 x}{\nu} \right)$, where x is the distance from the leading edge.

A generally accepted relationship for the distribution of the fluid particle velocity under low Re conditions (laminar boundary layer) is

$$u_y = f(y) = u_0 \left(\frac{2y}{\delta} - \frac{y^2}{\delta^2} \right). \tag{13.7}$$

Example 13.2. Shear Stress in Engine Oil

Suppose an SAW30 engine oil is being sheared by the relative movement of two plates separated by a distance of h. The speed of the relative movement is V. Find the shear stress in the oil when $V = 3$ m/s and $h = 2$ cm.

Solution. First, find the dynamic viscosity value in the engineering look-up table:

$$\mu = 0.29 \text{ kg/(m·s)}.$$

Assuming the velocity distribution is linearly related to the distance y, the shear stress is

$$\tau = \frac{\mu dV}{dy} = \frac{\mu V}{h} = \frac{0.29 \text{ kg/(m·s)} 3 \text{ m/s}}{0.02 \text{ m}} = 43 \text{ kg/(m·s}^2) = 43 \text{ N/m}^2.$$

13.3.4 Electrokinetic Flow

Most channel wall surfaces spontaneously develop an electric polarization when brought into contact with either weak and strong electrolyte solutions. This charge generation is caused by electrochemical reactions at the liquid/solid interfaces and, in the case of glass surfaces, the main reaction is the deprotonation of the acidic silanol groups which produce a negatively charged wall (Figure 13.3). Counter ions from the bulk liquid are attracted to the wall, and they shield these wall charges. The high capacitance charged region of ions at the interface of the liquid and wall is referred to as the **electric double layer**. The ions in the outer layer (called the Gouy–Chapman layer) is mobile and forms a net positive region of ions that spans the distance on the order of the Debye length of the solution, which is about 10 nm from the wall for symmetric univalent electrolytes at a 1-nM concentration.

When an electric field is applied parallel to the wall, ions on the liquid side will move in response to the field and drag the surrounding liquid molecules with them. The ion drag causes a net motion of bulk liquid along the channel. This phenomenon is called the **electrokinetic flow**.

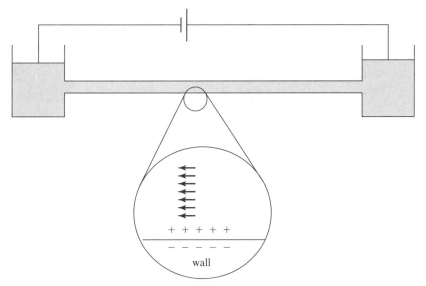

FIGURE 13.3

Electroosmosis (electrokinetic) flow.

The distribution of velocity across a channel width is different from that of pressure driven flow cases. The velocity of the fluid increases quickly from the no-slip boundary condition at the wall to a maximum value near the center of the channel. The boundary layer is very thin in this case. An approximation is generally made that the velocity distribution through the channel cross section is uniform.

Electroosmotic flow (EOF) micropumps use electrokinetic flow to transport liquids or to generate static pressure. A capillary is packed with high-density particles that form many parallel pores through which electrokinetic flows occur. The large area-to-volume ratio causes the generation of high pressure. It can be used to generate pressures in excess of 20 atm and flow rates of several μl/min under applied voltages on the order of tens to thousands of volts [54]. These pumps should ideally use deionized water as a working fluid in order to reduce the ion current, increase thermodynamic efficiency, and eliminate unwanted heating. Electrodes are generally inserted manually at opposite ends of a microfluid capillary. Alternatively, integrated planar electrodes can be used to increase the level of integration [55].

In an electroosmosis flow setup, high fields can generate electrolytic reactions and produce oxygen and hydrogen gases by dissociating H_2O. It is very important to make sure that (1) the gas generation is minimized (e.g., reducing the field strength by increasing the distance between electrodes), or (2) the formed gases are successfully evacuated before they block the channel. One of the methods is to use tailored AC waveforms [56].

13.3.5 Electrophoresis and Dielectrophoresis

The electric field acting on an individual particle is useful for bioparticle transportation, separation, and characterization. The force acting on a particle with a charge Q and polarization P is defined as

$$\vec{F} = Q\vec{E} + (\vec{P} \cdot \nabla)\vec{E}, \tag{13.8}$$

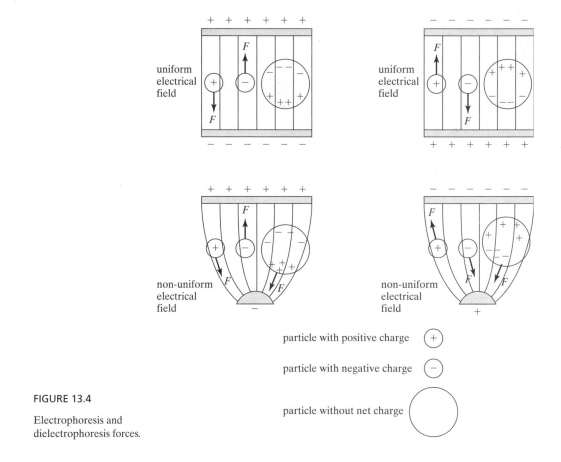

FIGURE 13.4

Electrophoresis and dielectrophoresis forces.

where the first term on the right-hand side is the **electrophoresis force** and the second term is the **dielectrophoresis force**. In a uniform electric field, the dielectrophoresis force disappears. Only particles with net charges experience the electrophoresis force. The equation is simplified to

$$\vec{F} = Q\vec{E}. \qquad (13.9)$$

If a particle does not carry net charges, the simplified dielectrophoresis force expression is

$$\vec{F} = (\vec{P} \cdot \nabla)\vec{E}. \qquad (13.10)$$

Four representative cases of charge particles (two small circles) and a neutral particle (big circle) in electric fields are diagrammed in Figure 13.4. Net forces on the particle are indicated by arrowed lines. A charged particle experiences forces when placed either in a uniform or non-uniform electrical field. The neutral particle can be polarized when placed inside an electrical field. A neutral particle, such as a cell, experiences net forces when subjected to a non-uniform electrical field. However, it experiences no net forces when inside a uniform electrical field.

The electrophoresis is widely used for separating charged biological macromolecules such as DNA, proteins, and peptides according to their sizes and charges. DNA molecules are always negatively charged, while proteins may be positively or negatively charged. Microfabricated electrophoresis devices with integrated chemical reaction stages have been fabricated [57].

A charged particle in a fluid environment will reach a steady-state velocity in a constant, uniform electrical field. Under equilibrium conditions, the electric force and the fluid friction balance:

$$F_{elec} = z_i e E = net - charge \times electric field$$
$$F_{fric} = f_i v_i. \qquad (13.11)$$

The friction coefficient is a function of the sizes of gel pores, sizes of particles, and the strength of the electric field.

This velocity is related to the electric field by

$$v_i = \frac{z_i e}{f_i} E = \mu_i E, \qquad (13.12)$$

where μ_i is referred to as the mobility of the species.

The capillary electrophoresis is a common format of electrophoresis analysis. In this method, a capillary channel is first filled with an electrophoresis gel material, which is designed with specific binding rates to molecules of interest (Figure 13.5). As a group of molecules travels through the gel matrix under the guidance of an electrical field, a number of events could occur. Some would travel unhindered. Some would be captured by the matrix permanently. Some would be captured but later released. At the end of a CE channel, molecules that have started at the same location and time arrive at different times. This method allows the analysis of constituent molecules and their relative concentration.

Dielectrophoresis takes advantage of the electrical polarization of cells in controlled non-uniform AC or DC electrical fields for the characterization and separation of living cells and organelles. DEP relies upon the attraction (or repulsion) exerted by a non-uniform electrical field upon neutral particles by virtue of their polarization. It is different from electrophoresis, which uses the effect of fields upon free or excessive charges.

The basic cause of the response to a non-uniform electrical fields by biological materials lies in the polarization effected by the applied field. The electrical polarization of cells and their components can vary. There are two main types of polarization mechanisms: bulk and interfacial (or surface connected). Each of the many possible polarization mechanisms generally has a characteristic frequency in which it cuts in. Studying the response of biological materials over a

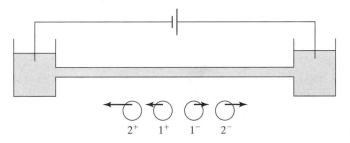

FIGURE 13.5

Capillary electrophoresis.

wide frequency range allows one to evaluate which particular mechanism operates, and how important its contribution is. Moreover, there is apparently a great sensitivity of the various polarization mechanisms to the precise physiological state of the biological materials. It is this factor that permits one to operate separations and analyze the considerable sensitivity and usefulness. Over the years, DEP has been used to separate living and dead cells, distinguish the normalcy of blood cells, and characterize the cell aging process [58].

DEP has been applied to microfluidic devices to characterize or to physically manipulate (e.g., trap) biological cells or particles [59].

13.4 DESIGN AND FABRICATION OF SELECTIVE COMPONENTS

A microfluid chip is made of many categories of components. We discuss two of the most important ones in this section: channels and valves. Other component categories include heaters, mixers, fluid reactors, and reservoirs. The design and fabrication methods for these elements must be compatible with those of channels and valves.

13.4.1 Channels

Microfluid channels are the most important components in a microfluid system, despite their relatively simple form and function compared with others (such as pumps and valves). The selection of the channel material is the starting point for any development efforts of microfluidic systems. There are several important aspects that must be taken into consideration when selecting channel materials and subsequent fabrication methods. These include

1. **Hydrophobicity of the channel wall.** Liquid moves freely in channels with hydrophilic walls via capillary action; this simplifies sample loading and priming. Glass, for example, is hydrophilic to many liquids and its properties are well known. Introducing liquid into channels with hydrophobic walls is considerably more difficult.
2. **Biocompatibility and chemical compatibility.** Ideally, the channel wall should not react with the fluid, particles, or gases within. Glass, the material for beakers and test tubes, is perhaps the most established biocompatible material and is well liked by researchers in the biological and chemistry community. However, there is a lack of micromachining methods for glass.
3. **Permeability of the channel material to air and liquid.** High permeability will cause excessive loss of fluid or, in the case of multiple channels that are placed close to each other, cross-contamination. Permeability of air or gas is often taken advantage of for air escape and the removal of trapped air bubbles.
4. **Retention of chemicals on walls.** Walls that retain chemicals may cause cross-contamination during repeated use.
5. **Optical transparency.** Optically transparent walls facilitate observation and quantitative assay analysis.
6. **Temperature of the processing.** Low-temperature processing is always desirable. High-temperature processes would narrow the choice of structural and surface coating materials.
7. **Functional complexity and development cost.** The materials for channels should be amenable to the integration of active components such as pumps and valves. The barrier to prototyping and manufacturing should be low.

13.4 Design and Fabrication of Selective Components

The materials often dictate the fabrication method and performance specifications. Material selection plays an important role in determining the channel geometries achievable. Materials and technologies for other components, such as pumps and valves, must be compatible with the channel wall material and fabrication process.

Research work in microfluidics started in two distinct research communities: the MEMS community and the analytical chemistry community. These two communities used different sets of materials.

In the early days of microfluid systems development and applications, the channels were often made of inorganic materials commonly found in MEMS studies, such as silicon, silicon dioxide, silicon nitride, polycrystalline silicon [60], or metal [61]. The fabrication processes include bulk etching (wet or dry etching), sacrificial etching, wafer-to-wafer bonding, or any combination of these steps. A few representative fabrication methods are illustrated in Figure 13.6; they make channels of various cross-sectional shapes.

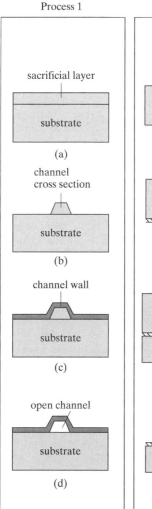
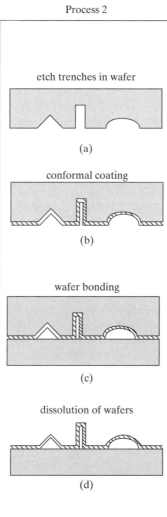
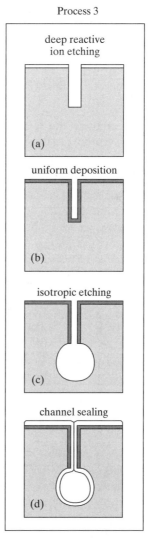

FIGURE 13.6

Representative fabrication processes of channels on silicon substrates.

In the analytical chemistry community, researchers developed channel fabrication processes based on familiar materials (glass) and simply fabrication (wafer bonding).

There are many problems associated with silicon-based microfluid devices despite their ability to generate sophisticated channel cross sections. Silicon, for example, is not an optically transparent material. Special fluid imaging and tracking methods have to be developed. The silicon microfabrication is not accessible by chemists and biologists and often proves to be very expensive and inaccessible for rapid prototyping.

Glass chips are ideal in terms of surface chemistry, optical transparency, and relative ease of construction. Successful commercial products have been launched based on glass chips; one example is the electrophoresis chip made by Agilent and Caliper Technologies. However, it remains difficult to incorporate advanced pumps, valves, and sensors into all-glass microfluid chips. Glass chips are often encapsulated permanently, making it difficult to functionalize the interior surfaces of channel walls.

Materials for channels have evolved rapidly in the past few years. Today, microfluid channels are commonly made out of the following materials which belong to two categories: organic and inorganic materials. Representative materials under each category are summarized below:

Organic polymers: Parylene, polydimethylsiloxane, acrylics, polycarbonate, biodegradable polymer, polyimide.

Inorganic materials: glass (Pyrex, specialty glasses), silicon, silicon dioxide, silicon nitride, polysilicon.

Table 13.1 compares the relative merits of several representative material systems for microfluid channels according to the criteria presented earlier.

TABLE 13.1 Comparison of Methods for Making Channels.

	Glass–glass bonding	Silicon micro-machining	PDMS bonding	Plastic bonding	Parylene surface micromachining
Hydrophobicity	Hydrophilic	Variable—treatable with coating (e.g., oxide)	Hydrophobic, may become hydrophilic, Unreliable	Variable—treatable with surface treatment	Hydrophobic
Biocompatibility	Excellent	Acceptable	Excellent	Excellent	Moderate
Wall permeability	None	None	High—for organic solvents and gas molecules	Moderate	Low
Retention of chemicals	Low	Low	High (if without special coating)	Moderate	Not known
Optical transparency	Excellent	None	Excellent	Good	Good (if on transparent substrate)
Temperature of processing	High (for fusion bonding)	High	Low	Moderate	Low
Functional complexity and cost	Moderate	High	Low	Moderate	Moderate to high

A few examples of the early work on microfluid channels for various applications are presented in the following cases. Glass microfluid channels are discussed in Case 13.1 and Case 13.2. In Case 13.3, we review a silicon microchannel integrated with a neural probe. Cases 13.4 and 13.5 deal with channels made of polymers—silicon elastomer in Case 13.4 and Parylene in Case 13.5.

Case 13.1. Gas Chromatography Channels

Chromatography involves a sample (or sample extract) being dissolved in a *mobile phase* (which may be a gas, a liquid, or a supercritical fluid). The mobile phase is then forced through an immobile, immiscible *stationary phase*. The phases are chosen such that the components of the sample have differing solubilities in each phase. A component that is quite soluble in the stationary phase will take longer to travel through it than a component which is not very soluble in the stationary phase but is very soluble in the mobile phase. As a result of these differences in mobilities, sample components will become separated from each other as they travel through the stationary phase.

The time between sample injection and an analyte peak reaching a detector at the end of the column is termed the *retention time* (t_R). Each analyte in a sample will have a different retention time. The time taken for the mobile phase itself to pass through the column is called t_M.

In 1975, a group of researchers at Stanford University unveiled an integrated gas chromatographer (GC) made from glass wafers [6] (Figure 13.7). The gas chromatographer

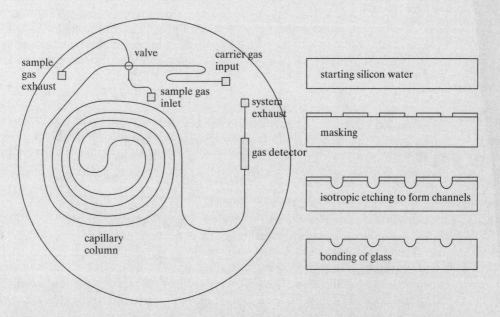

FIGURE 13.7
Schematic diagram of an integrated gas chromatography system.

device separates components within a gas mixture and analyzes the relative concentrations of gas species. Gas separation channels have semicircular cross sections, which are 200 μm across and 40 μm deep. On a 4"-diameter (100-mm) wafer, channels as long as 1.5 m were realized. The channel walls were made of glass.

The system is used for the chemical trace analysis of pollutants and toxic elements in the environment. Instead of packing a channel with solid phase materials, the wall of the channel serves as the absorption element. Under a same pressure gradient applied between the inlet and the outlet, different gas molecules exhibit different t_R. Gas species that have started at the same location and time would arrive at the exit at different times.

With the advancement of microfabrication technology, the GC chip can be further miniaturized. Long GC columns can be packed with higher efficiency using three-dimensional channels fabricated in silicon, glass, or polymer materials [8, 62, 63].

Case 13.2. Electrophoresis in Microchannels

The electrophoresis separation of multiple species in a liquid sample is a powerful technique for assaying or purification. According to our earlier discussions about electrophoresis, different species travel at different speeds under a given electrical field. In order to maximize the efficiency of electrophoresis separation, those species should begin within a close vicinity of each other, rather than being spread out over a long sample plug.

The EP separation with precision defined plug sizes can be achieved using double-T type injectors. A glass microchip for the electroseparation of biological molecules has been developed [64]. The system consists of two T-shaped junctions connected to four ports—buffer, analyte, waste, and analyte waste (Figure 13.8). The operation principle of a double-T injector is explained below and diagrammed in Figure 13.9.

First, a buffer solution is injected between the buffer and waste ports. An analyte is then injected between the analyte and analyte waste ports. A analyte plug with a precision volume is formed between the two T junctions. An EP potential is applied between the buffer and the waste ports, causing the analyte plug to move along the EP column towards the waste port. Different species within the analyte plug are detected at end of the separation column.

There are many designs and fabrication methods to implement this EP separation system. One of the simplest and earliest demonstrations is discussed below (Figure 13.10). It begins with a glass wafer—a material that is no different from conventional EP separation columns. A Cr thin film is deposited over the glass, and it is photolithographically patterned (step b). The patterns in the Cr are used to define the position and size of the channels. Glass is etched in regions not covered by Cr using a solution containing HF and NH$_4$F to a desired depth (step c). The Cr mask is removed (step d). Another glass chip is positioned on top of the glass substrate and permanently bonded. The channel is therefore entirely made of glass. Metal electrodes are inserted into ports externally.

13.4 Design and Fabrication of Selective Components

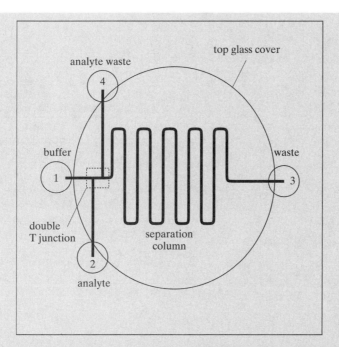

FIGURE 13.8

Schematic diagram of a glass microelectroseparation chip.

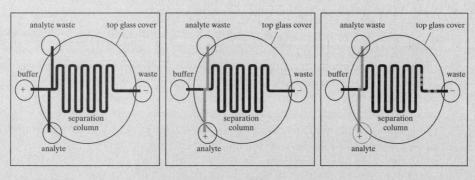

FIGURE 13.9

Operational principle of electroseparation chip.

Glass is electrochemically preferred because of its hydrophilic surfaces and familiarity. Many other incarnations of the same system can be made using a variety of other materials and construction methods. EP channels made of other materials, such as PDMS, may need to be chemically modified to provide desired functionalities over long periods of time.

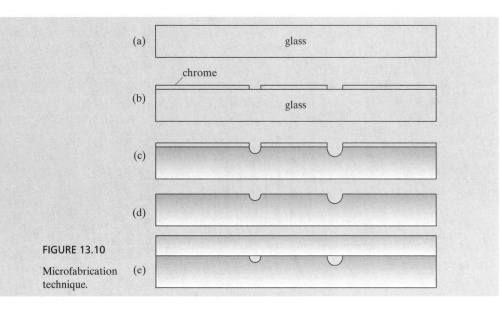

FIGURE 13.10
Microfabrication technique.

 Case 13.3. Neuron Probes with Channels

Neural physiology is one of the grandest scientific and medical quests of mankind. Understanding of neurological physiology will lead to the prevention and cure of many illnesses that severely affect the quality of life, such as Alzheimer's disease and Parkinson's disease. In order to study neurological signal processing, which is conducted in complex three-dimensional tissues with a myriad of connections, advanced engineering tools for imaging, recording, and affecting neurological behaviors are needed. One of these tools is the micromachined neurological recording probe, which can be made with a small size and a high density for minimizing unintended damages and extracting rich data.

An early pioneering work in the area of microfabricated neuron recording probes was conducted by Prof. K. D. Wise at the University of Michigan. His research group has developed many enabling engineering capabilities:

Linear arrays and two-dimensional arrays of silicon-based neuron recording probes.
Recording probes with multiple recording and accompanying stimulating sites.
Recording probes that allow chemicals to be sent to and collected from neurological tissues in conjunction with recording and electrical stimulation events.
Integrated BiCMOS circuits for locally amplifying and conditioning to preserve signal integrity.

Here, we discuss a representative work by Wise's group that incorporates microfluid channels in micromachined neural probes [65]. Such probes are used for injecting solutions. The probe must be sufficiently stiff to penetrate neural tissues. In this case, it is made of

13.4 Design and Fabrication of Selective Components   441

single-crystal silicon. We focus on the fabrication process of the silicon channel. Because the channel is long, it is impractical to use embedded sacrificial layers and later remove it. The lateral undercut of the channel would take a very long time. Also due to the length of the channel, the cross section of the channel needs to be relatively large in order to produce a sufficient flow of chemical solutions under moderate to low pressure differences (to avoid harming biological tissues). It is difficult to use the deposited sacrificial layer material to realize a large cross section because the deposition process would take a long time.

The fabrication process requires only one mask. The process starts with a <100>-oriented silicon wafer (Figure 13.11). The front surface is doped to form a 3-μm-thick region

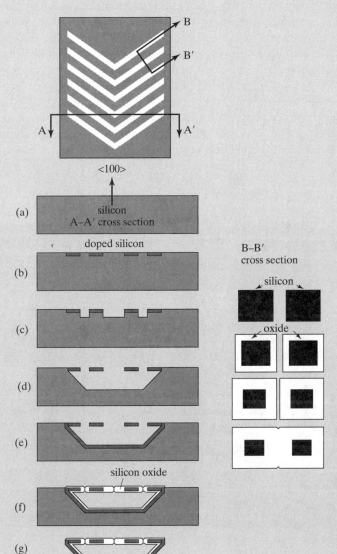

FIGURE 13.11

Fabrication process of embedded microchannels.

with high concentration (Figure 13.11b). The concentration is sufficiently high to effectively reduce the etch rate in anisotropic silicon etching solutions such as EDP. The front surface of the silicon wafer is etched using reactive ion etching, which does not discriminate against the silicon material of different doping concentrations (Figure 13.11c). The intended channel region is opened through this layer in the form of a chevron pattern, as shown in Figure 13.11d. Anisotropic silicon etching is performed to undercut materials underneath the doped regions. The etching profile underneath the Chevron shaped mask is shown in Figure 13.12. Significant undercut can be achieved to produce channels with relatively large cross-sectional areas.

A deep boron diffusion is performed to define the probe shank (Figure 13.11e). The entire inner surface of the channel reaches a concentration necessary to produce etch stop effects. The channel is sealed using thermal oxidation and LPCVD deposited dielectrics. After depositing and patterning electrodes and dielectric shields, the silicon wafer is dissolved in anisotropic etching solutions, which selectively remove bulk silicon with only background concentrations, leaving silicon shanks freestanding (Figure 13.11g).

An alternative process for realizing buried and sealed channels with a large cross section is to use porous materials such as porous silicon [66]. Porous materials with microscopic pores allow the underlying substrates to be etched. Because the continuous membrane is largely filled with small pores (up to 75% porosity), it can be hermetically sealed by a small amount of deposition [67] of thermal oxide or LPCVD materials.

FIGURE 13.12
Evolution of etching.

Case 13.4. PDMS Microfluid Channels

Channels made of polydimethylsiloxane (PDMS) are very popular because of the easy accessibility of material, rapid fabrication, and desirable performance aspects. The PDMS material can be obtained in viscous liquid precursor form from many vendors under various trade names, such as Sylgard Silicone Elastomer from Dow Corning and RTV silicone from GE Silicones. The most commonly used PDMS materials are Sylgard 184 (Dow Corning) and RTV 615 (GE Silicones).

The precursor materials consist of two parts: the base and curing agent. The two parts are mixed and then cured at room temperature, in a vacuum, or under elevated temperatures (rapid cure). Under the recommended mixing ratio, this results in a thermoset, transparent elastomeric solid. The Sylgard 184 silicone elastomer, for example, may be cured under one of the following recommended conditions: 24 hours at 23°C, 4 hours at 65°C, 1 hour at 100°C, or 15 minutes at 150°C.

The PDMS is a relatively porous material, allowing liquid and molecules to diffuse at a slow rate. Gas can diffuse through the material as well. As-cured PDMS is generally hydrophobic. The surface can be turned hydrophilic by exposing it to oxygen plasma, by treating it with chemicals (e.g., a HCl solution), or by coating it with organic polymers. In many cases, the surface will return to a hydrophobic state within 30 minutes to a few hours.

To realize precise three-dimensional features, the uncured precursor can be poured over surfaces with three-dimensional patterned features (Figure 13.13) that are made by a

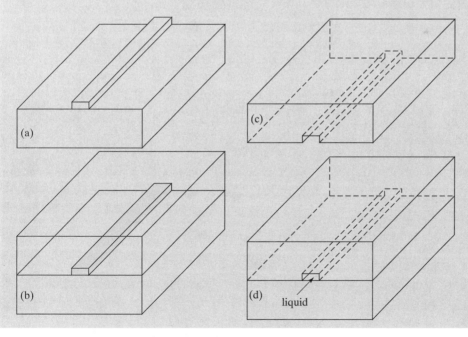

FIGURE 13.13
PDMS molding.

variety of means (including bulking etching, photoresist patterning, etc.) (step a and b). Once the elastomeric material is removed, the surface features translate into recessed or raised regions (step c). The PDMS materials can then be bonded to another piece of substrate to form an enclosed channel (step d). The matching substrate can be silicon, glass, polyimide sheet, or even another piece of PDMS. In most cases, the bonding is reversible; in other words, the two pieces can be debonded manually and closed again. If two pieces of PDMS materials are bonded, the bonding can become quite strong (permanent) if the surfaces are treated by exposure to oxygen plasma. It is also possible to integrate more than two layers and form three-dimensional microfluid circuits with complex channel geometries in three dimensions [68].

Inlet and outlet access holes can be drilled in the elastomer by punching tools. It is also possible to pour a PDMS precursor around three-dimensional structures (e.g., a bent metal wire) [69]. The metal wire can be mechanically removed or electrochemically etched to leave three-dimensional channels or inlet–outlet ports in the PDMS solid after curing [70].

The PDMS material exhibits volume shrinkage in all directions after it is removed from the mold. The dimensional change resulting from shrinking is influenced by the material, by the amount of materials poured, and by the curing method. This must be carefully calibrated for each use.

Case 13.5. Parylene Surface Micromachined Micro Channels

Surface micromachined channels have been made using photoresist as the sacrificial layer and the chemical-vapor-deposited Parylene thin film as the structural layer. The use of the Paralene-photoresist system replaces the high temperature LPCVD polysilicon/oxide system [71].

A representative fabrication process for realizing Parylene channels, monolithically connected to fluid inlet/outlet ports in the silicon substrate, is shown in Figure 13.14. The process starts with a <100>-oriented silicon wafer (step a), which is coated with a thin layer of silicon dioxide. The oxide on the back side is photolithographically patterned, and it is used as a mask for anisotropic silicon etching (step b). A layer of photoresist is spin coated and patterned (step c). The front side of the wafer is then coated with a layer of Parylene thin film (step d). A layer of polyimide is spin coated and patterned to mechanically enhance the stiffness of the channel to prevent collapsing (step e). The remaining silicon in the etched back-side holes is removed until the silicon dioxide on the front side is reached (step f). This can be accomplished through anisotropic wet etching or plasma etching. The oxide at the bottom of the cavity is then removed in a hydrofluoric acid bath (step g). The photoresist sacrificial material is removed using acetone to create open channels through the now-open inlet and outlet ports (step h).

Many components have been incorporated in such a system; they include

One time valves [72];

On-chip thermal pneumatic sources [47];

Electroosmosis pumps [56].

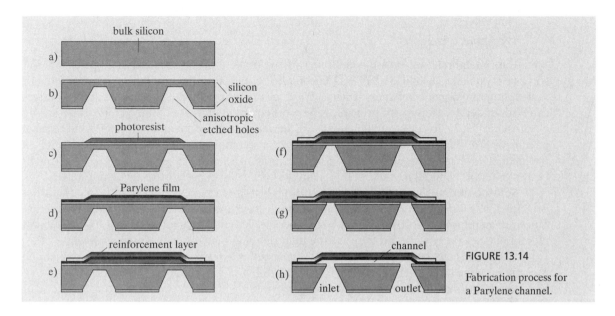

FIGURE 13.14
Fabrication process for a Parylene channel.

13.4.2 Valves

Valves are important elements in a microfluid system. They provide complex system-level functionalities for a laboratory-on-a-chip system. The following factors are generally considered when selecting or developing a micromachined valve:

The reliability of valve operation. Ideally, a valve should be leak-free during the "off" state and open during the "on" state.

The repeatability of valve operation.

The ability to withstand large pressure.

The simplicity of valve construction.

The simplicity of valve operation and control.

Biocompatibility with the fluid and biological particles.

According to the mode of operations, valves can be classified into several categories:

Cyclic valves can be operated multiple times. They can be constantly "on," meaning the valve holds its open position without the active input of power, or constantly "off," meaning the valve maintains its sealed position without active power.

One-time valves are operated only once during the life of the operation. Constantly "on" valves will seal off a channel permanently when activated. Constantly "off" valves will open once it is activated, so it can complete applications such as the collection of environmental samples.

Since the valve is critical for the performance of a microfluid system and for enabling miniaturization, many valve designs have been developed in the past. Generally, valve structures fall into the following categories:

Hard-membrane valves;
Soft-membrane valves;

Plug valves;

Threshold valves.

Hard-membrane valves use membranes of one of the following materials: single-crystal silicon, polycrystalline silicon [73], LPCVD silicon nitride, piezoelectric thin films, metal thin films, or non-elastomeric organic polymers (such as Parylene and polycarbonate). Hard-membrane valves can be operated by a variety of principles; the most common ones are based on piezoelectric [74], electrostatic [73, 75, 76], electromagnetic [1, 77], thermal bimetallic, pneumatic [17, 20], and thermal pneumatic [78, 79] actuators. Valves can be based on a combination of principles. For example, a pneumatic valve may use electrostatic force for holding closed-gap positions [17]. Hard membranes generally cannot provide a good seal in the "off" state, especially for regulating valves.

Soft-membrane valves uses valves made of elastomers such as PDMS [80]. The operation principles of elastomeric membranes are limited compared with those of hard membranes. Since the membrane is soft, it is difficult to integrate elements such as electrodes. However, soft membranes seal very well, and they are the material of choice for conventional valves.

Plug valves can be based on a variety of principles. For example, valves can be developed by exploiting the large swelling and shrinking capability of ionic hydrogels in response to chemical concentrations, pH, temperature, and electric fields [81–84], or by congregating magnetic or chemically modified particles.

Threshold valves change their "on"/"off" state depending on the pressure or flow rate. Threshold valves often leverage surface-tension principles. Burst valve is a special kind of threshold valve—its state changes from closed to open when the pressure at its head reaches a certain level.

Case 13.6. PDMS Pneumatic Valves

Soft-membrane valves using elastomeric (rubber) polymers are almost exclusively used in macroscopic valves and pumps. Their advantage is that they have a good seal against liquid or air in the "off" state. However, soft-membrane valves are more challenging from the design and fabrication point of view, as soft membranes and perhaps matching seats must be integrated into a microsystem. PDMS is a commonly used soft-membrane material because of its relatively simple processing and desirable softness. Large deformation (50–150 μm) has been reached on a membrane (1 × 1 to 2 × 2 mm^2) under the pressure input of approximately 100 mW for various working fluids, including air [79].

One representative method for making functional valves using external pneumatic control is discussed here [85, 86]. The valve involves two layers of PDMS thin film (Figure 13.15). Both the first and second layers follow the PDMS molding method discussed in the previous case. For the first layer, the thickness of the PDMS is kept as small as possible; hence the ceiling directly above a channel is very thin (Figure 13.15c). The PDMS precursor is allowed to settle and planarize before being cured.

The second layer consists of pneumatic control lines (Figure 13.15d). The first and second layers are bonded together with channels crossing each other. Oxygen plasma treatment can make the two layers bond permanently. The two-piece PDMS assembly is then bonded to

13.4 Design and Fabrication of Selective Components

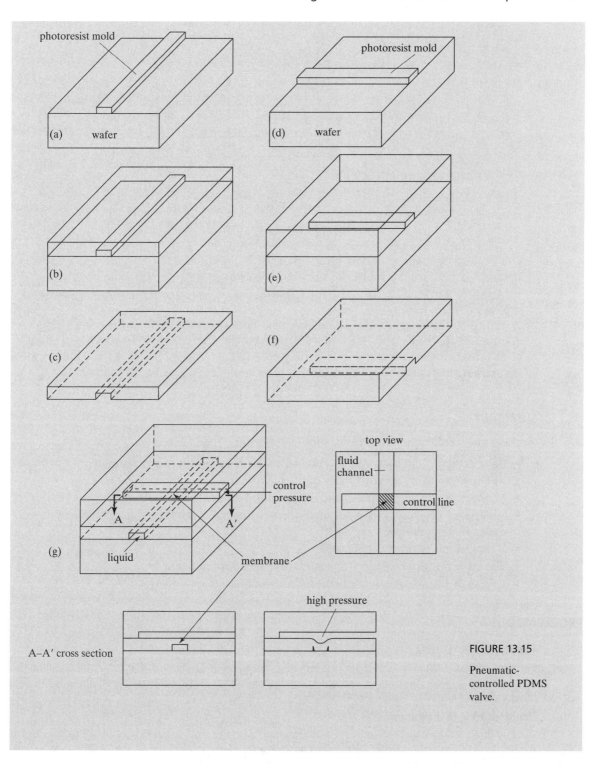

FIGURE 13.15

Pneumatic-controlled PDMS valve.

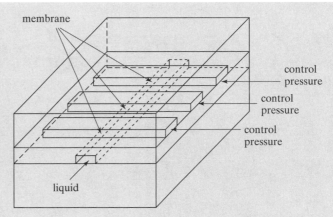

FIGURE 13.16

Fabrication process of a peristaltic pump in PDMS.

a substrate. The channel formed in the first-layer PDMS is used to transport liquid, whereas channels in the second-layer PDMS are used to convey pressure by gas or liquid.

Pressure applied in second-layer channels pushes the PDMS membrane down, sealing the channel underneath.

This method can be used to construct micropumps. One possible configuration is shown in Figure 13.16. The channel in the first layer crosses the pressure lines above it in three interaction areas, which forms three well-defined PDMS membranes. Three pressure lines working in a peristaltic fashion will continuously push liquid in two possible directions.

SUMMARY

At the end of this chapter, the reader should understand the following concepts and facts:
- Basic designs, materials, and microfabrication processes for microchannels.
- Relative simplicity and flexibility of the microfabrication of microchannels.
- Major methods for pumping fluid in microfluid channels and their principles.
- Analysis of flow resistance associated with a segment of microfluid channels under pressure-driven flow conditions.
- Basic designs of practical microvalves.
- The basic principles of electrophoresis and dielectrophoresis.
- Design of silicone elastomer microfluid channels and integrated valves.

PROBLEMS

Problem 13.1: Design

For a microfluid channel with a length of 1 mm and a cross-sectional area of 20 μm^2, find the volumetric flow and average flow speed if one end of the channel is subjected to a water column that is 5 m tall. The other end is connected to atmosphere pressure.

Problem 13.2: Fabrication

Identify three practical methods of forming the channel with the dimensions discussed in Problem 13.1 if the height of the channel is 4 μm. Part of the channel must be transparent for optical observation. Sacrificial etching is generally not practical due to the large channel length.

Problem 13.3: Design

Find the Reynolds number of the flow situation of Problem 13.1 if the width of the channel is 5 μm.

Problem 13.4: Review

Find a method to make an array of fluid channels with a length of 1–10 μm and a channel cross section of 10 nm. The cross section of the channel should be a circle or a square. Dicuss methods of patterning. Comment on practicality, efficiency, and accuracy.

Figure 13.5: Design

A segment of a microfluid channel is 10 mm long with a rectangular cross section that is 30 μm wide and 1-μm tall. What is the required pressure to achieve a volumetric flow rate of 10 nl/min?

Problem 13.6: Review

Draw the detailed fabrication process for the gas chromatography chip of Case 13.1. Justify the choice of masking layer.

Problem 13.7: Fabrication

Draw the detailed fabrication process of neuron probes with integrated fluid transport channels according to Case 13.3 [87]. Draw the process at a representative cross-section along the probe.

Problem 13.8: Design

The PDMS pneumatic valve discussed in Case 13.6 utilizes a thin elastomer membrane with a certain area defined by the crossing of the fluid and control lines. It forms reliable seals due to the contact of PDMS surfaces. Discuss at least three strategies for reducing the threshold voltage necessary to close the valve. For each strategy, discuss the effect on the fabrication process.

Problem 13.9: Fabrication

Design a complete fabrication process for making a Parylene cantilever probe with an integrated fluid delivery channel. The channel is opened at the free end of the cantilever. The probe consists of a bulk silicon micromachined handle. The handle further consists of an etched cavity that fluidically communicates with the integrated channel. The cavity serves as a fluid reservoir and inlet. Note the sidewall of the cavity and the handle can be sloped or vertical. The drawing shows a case with vertical walls. Detailed lithography, steps can be ignored in the drawing. Clearly label all layers of materials used in the process.

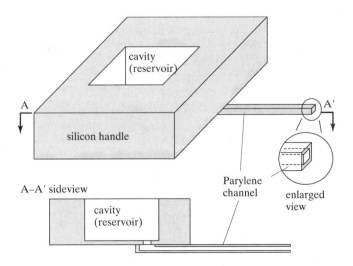

Problem 13.10: Challenge

Develop a microvalve with a footprint of no more than 1 mm^2 that can be controlled by electricity. The valve should be able to completely stop a liquid flow with a back pressure of 30 kPa. The valve must be operated with a voltage of less than 100 V. The footprint and the electric voltage should be as small as possible. The leak rate of the valve should be zero. No out-of-chip pneumatic sources should be used. The valve must be able to be repeatedly operated.

REFERENCES

1. Bae, B., et al., *Feasibility Test of an Electromagnetically Driven Valve Actuator for Glaucoma Treatment*, Microelectromechanical Systems, Journal of, 2002. **11**: p. 344–354.
2. Gwynne, P., and G. Heebner, *Drug Discovery and Biotechnology Trends—Laboratory Automation: Scientists' Little Helpers*, Science, 2004. **303**: p. 549–553.
3. Becker, T., et al., *Gas Mixture Analysis Using Silicon Microreactor Systems*, Microelectromechanical Systems, Journal of, 2000. **9**: p. 478–484.
4. Rusu, C., et al., *Direct Integration of Micromachined Pipettes in a Flow Channel for Single DNA Molecule Study by Optical Tweezers*, Microelectromechanical Systems, Journal of, 2001. **10**: p. 238–246.
5. Maharbiz, M.M., et al., *A Microfabricated Electrochemical Oxygen Generator for High-Density Cell Culture Arrays*, Microelectromechanical Systems, Journal of, 2003. **12**: p. 590–599.
6. Terry, S.C., J.H. Jerman, and J.B. Angell, *A Gas Chromatographic Air Analyzer Fabricated on a Silicon Wafer*, IEEE Transaction on Electron Devices. 1979. **ED 26**: p. 1880–1886.
7. Reston, R.R., and E.S. Kolesar, Jr., *Silicon-Micromachined Gas Chromatography System Used to Separate and Detect Ammonia and Nitrogen Dioxide. I. Design, Fabrication, and Integration of the Gas Chromatography System*, Microelectromechanical Systems, Journal of, 1994. **3**: p. 134–146.
8. Tian, W.-C., et al., *Microfabricated Preconcentrator-Focuser for a Microscale Gas Chromatograph*, Microelectromechanical Systems, Journal of, 2003. **12**: p. 264–272.
9. Harris, C., M. Despa, and K. Kelly, *Design and Fabrication of a Cross Flow Micro Heat Exchanger*, Microelectromechanical Systems, Journal of, 2000. **9**: p. 502–508.
10. Stephens, L.S., et al., *A Pin Fin Microheat Sink for Cooling Macroscale Conformal Surfaces Under the Influence of Thrust and Frictional Forces*, Microelectromechanical Systems, Journal of, 2001. **10**: p. 222–231.
11. Pattekar, A.V., and M.V. Kothare, *A Microreactor for Hydrogen Production in Micro Fuel Cell Applications*, Microelectromechanical Systems, Journal of, 2004. **13**: p. 7–18.
12. Peles, Y., et al., *Fluidic Packaging of Microengine and Microrocket Devices for High-Pressure and High-Temperature Operation*, Microelectromechanical Systems, Journal of, 2004. **13**: p. 31–40.
13. Arana, L.R., et al., *A Microfabricated Suspended-Tube Chemical Reactor for Thermally Efficient Fuel Processing*, Microelectromechanical Systems, Journal of, 2003. **12**: p. 600–612.
14. Lee, K.B., and L. Lin, *Electrolyte-Based On-Demand and Disposable Microbattery*, Microelectromechanical Systems, Journal of, 2003. **12**: p. 840–847.
15. Sammoura, F., K.B. Lee, and L. Lin, *Water-Activated Disposable and Long Shelf Life Microbatteries*, Sensors and Actuators A: Physical, 2004. **111**: p. 79–86.
16. Cattaneo, F., et al., *Digitally Tunable Microfluidic Optical Fiber Devices*, Microelectromechanical Systems, Journal of, 2003. **12**: p. 907–912.
17. Yobas, L., et al., *A Novel Integrable Microvalve for Refreshable Braille Display System*, Microelectromechanical Systems, Journal of, 2003. **12**: p. 252–263.
18. Ross, P.E., *Beat the Heat*, Spectrum, IEEE, 2004. **41**: p. 38–43.

19. Koo, J.-M., et al., *Integrated Microchannel Network for Cooling of 3D Circuit Architectures*. Presented at Proceedings of ASME International Mechanical Engineering Congress and Exposition (IMECS 03), Washington, DC: 2003.
20. Takao, H., and M. Ishida, *Microfluidic Integrated Circuits for Signal Processing using Analogous Relationship Between Pneumatic Microvalve and MOSFET*, Microelectromechanical Systems, Journal of, 2003. **12**: p. 497–505.
21. Thorsen, T., S.J. Maerkl, S.R. Quake, *Microfluidic Large-Scale Integration*, Science, 2002. **298**: p. 580–584.
22. Reyes, D.R., et al., *A Glow Discharge in Microfluidic Chips for Visible Analog Computing*, Lab on a Chip, 2002. **2**: p. 113–116.
23. Kenis, P.J.N.A., R.F. Ismagilov, and G.M. Whitesides, *Microfabrication Inside Capillaries Using Multiphase Laminar Flow Patterning*, Science, 1999. **285**: p. 83–85.
24. Goluch, E.D., et al., *Microfluidic Method for in-situ Deposition and Precision Patterning of Thin-Film Metals on Curved Surfaces*, Applied Physics Letters, 2004. **85**: p. 3629–3631.
25. Mirkin, C.A., *DNA-Based Methodology for Preparing Nanocluster Circuits, Arrays, and Diagnostic Materials*, MRS Bulletin, **25**: p. 43–54, 2000.
26. Pennisi, E., *Biotechnology: The Ultimate Gene Gizmo: Humanity on a Chip,* Science, 2003. **302**: p. 211.
27. Li, J., and Y. Lu, *A Highly Sensitive and Selective Catalytic DNA Biosensor for Lead Ions*, Journal of American Chemical Society, 2000. **122**: p. 10466–10467.
28. Nam, J.-M., C.S. Thaxton, and C.A. Mirkin, *Nanoparticle-Based Bio-Bar Codes for the Ultrasensitive Detection of Proteins*, Science, 2003. **301**: p. 1884–1886.
29. Choi, J.-W., et al., *A New Magnetic Bead-Based, Filterless Bio-Separator with Planar Electromagnet Surfaces for Integrated Bio-Detection Systems*, Sensors and Actuators B, 2000. **68**: p. 34–39.
30. Taton, T.A., C.A. Mirkin, and R.L. Letsinger, *Scanometric DNA Array Detection with Nanoparticle Probes*, Science, 2000. **289**: p. 1757–1760.
31. Park, S.-J., T.A. Taton, and C.A. Mirkin, *Array-Based Electrical Detection of DNA with Nanoparticle Probes*, Science, 2002. **295**: p. 1503–1506.
32. White, F.M., *Fluid mechanics*, 4th ed. McGraw-Hill: 1999.
33. Jang, J., and S.S. Lee, *Theoretical and Experimental Study of MHD (Magnetohydrodynamic) Micropump*, Sensors and Actuators A: Physical, 2000. **80**: p. 84–89.
34. Darabi, J., et al., *Design, Fabrication, and Testing of an Electrohydrodynamic Ion-Drag Micropump*, Microelectromechanical Systems, Journal of, 2002. **11**: p. 684–690.
35. Richter, A., et al., *A Micromachined Electrohydrodynamic (EHD) Pump*, Sensors and Actuators A: Physical, 1991. **29**: p. 159–168.
36. Fuhr, G., et al., *Microfabricated Electrohydrodynamic (EHD) Pumps for Liquids of Higher Conductivity*, Microelectromechanical Systems, Journal of, 1992. **1**: p. 141–146.
37. Yang, L.-J., J.-M. Wang, and Y.-L. Huang, *The Micro Ion Drag Pump Using Indium-Tin-Oxide (ITO) Electrodes to Resist Aging,* Sensors and Actuators A: Physical, 2004. **111**: p. 118–122.
38. Hatch, A., et al., *A Ferrofluidic Magnetic Micropump*, Microelectromechanical Systems, Journal of, 2001. **10**: p. 215–221.
39. Yun, K.-S., et al., *A Surface-Tension Driven Micropump for Low-Voltage and Low-Power Operations*, Microelectromechanical Systems, Journal of, 2002. **11**: p. 454–461.
40. Chiou, P.Y., et al., *Light Actuation of Liquid by Optoelectrowetting*, Sensors and Actuators A: Physical, 2003. **104**: p. 222–228.

41. Luginbuhl, P., et al., *Microfabricated Lamb Wave Device Based on PZT Sol-Gel Thin Film for Mechanical Transport of Solid Particles and Liquids*, Microelectromechanical Systems, Journal of, 1997. **6**: p. 337–346.
42. Benard, W.L., et al., *Thin-Film Shape-Memory Alloy Actuated Micropumps*, Microelectromechanical Systems, Journal of, 1998. **7**: p. 245–251.
43. Smits, J.G., *Piezoelectric Micropump with Three Valves Working Peristaltically*, Sensors and Actuators A: Physical, 1990. **21**: p. 203–206.
44. Olsson, A., et al., *Micromachined Flat-Walled Valveless Diffuser Pumps*, Microelectromechanical Systems, Journal of, 1997. **6**: p. 161–166.
45. Ahn, C.H., and M.G. Allen, *Fluid Micropumps Based on Rotary Magnetic Actuators*. Presented at Micro Electro Mechanical Systems, MEMS '95, Proceedings. IEEE. 1995.
46. Van de Pol, F.C.M., et al., *A Thermopneumatic Micropump Based on Microengineering Techniques*, Sensors and Actuators A: Physical, 1990. **21**: p. 198–202.
47. Handique, K., et al., *On-Chip Thermopneumatic Pressure for Discrete Drop Pumping*, Analytical Chemistry, 2001. **73**: p. 1831–1838.
48. Tsai, J.-H., and L. Lin, *Active Microfluidic Mixer and Gas Bubble Filter Driven by Thermal Bubble Micropump*, Sensors and Actuators A: Physical, 2002. **97–98**: p. 665–671.
49. Maxwell, R.B., et al., *A Microbubble-Powered Bioparticle Actuator*, Microelectromechanical Systems, Journal of, 2003. **12**: p. 630–640.
50. Tsai, J.-H., and L. Lin, *A Thermal-Bubble-Actuated Micronozzle-Diffuser Pump*, Microelectromechanical Systems, Journal of, 2002. **11**: p. 665–671.
51. Su, Y.-C., and L. Lin, *A Water-Powered Micro Drug Delivery System*, Microelectromechanical Systems, Journal of, 2004. **13**: p. 75–82.
52. Hong, C.-C., J.-W. Choi, and C.H. Ahn, *Disposable Air-Bursting Detonators as an Alternative On-Chip Power Source*. Presented at Micro Electro Mechanical Systems, 2002. The Fifteenth IEEE International Conference.
53. Madou, M.J., et al., *Design and Fabrication of CD-like Microfluidic Platforms for Diagnostics: Microfluidic Functions*, Biomedical Microdevices, 2001. **3**: p. 245–254.
54. Zeng, S., et al., *Fabrication and Characterization of Electroosmotic Micropumps*, Sensors and Actuators B: Chemical, 2001. **79**: p. 107–114.
55. Chen, C.-H., and J.G. Santiago, *A Planar Electroosmotic Micropump*, Microelectromechanical Systems, Journal of, 2002. **11**: p. 672–683.
56. Selvaganapathy, P., et al., *Bubble-Free Electrokinetic Pumping*, Microelectromechanical Systems, Journal of, 2002. **11**: p. 448–453.
57. Woolley, A.T., et al., *Functional Integration of PCR Amplification and Capillary Electrophoresis in a Microfabricated DNA Analysis Device*, Analytical Chemistry, 1996. **68**: p. 4083–4086.
58. Pohl, H.A., *Dielectrophoresis*. Cambridge University Press, 1978.
59. Mohanty, S.K., et al., *A Microsystem Using Dielectrophoresis and Electrical Impedance Spectroscopy for Cell Manipulation and Analysis*. Presented at Transducers, Solid-State Sensors, Actuators and Microsystems, 12th International Conference. 2003.
60. de Boer, M.J., et al., *Micromachining of Buried Micro Channels in Silicon*, Microelectromechanical Systems, Journal of, 2000. **9**: p. 94–103.
61. Papautsky, I., et al., *A Low-Temperature IC-Compatible Process for Fabricating Surface-Micromachined Metallic Microchannels*, Microelectromechanical Systems, Journal of, 1998. **7**: p. 267–273.
62. Lu, C.-J., et al., *Portable Gas Chromatograph with Tunable Retention and Sensor Array Detection for Determination of Complex Vapor Mixtures*, Analytical Chemistry, 2003. **75**: p. 1400–1409.

63. Hsieh, M.-D., and E.T. Zellers, *Limits of Recognition for Simple Vapor Mixtures Determined with a Microsensor Array*, Analytical Chemistry, 2004. **76**: p. 1885–1895.
64. Jacobson, S.C., et al., *Electrically Driven Deparations on a Microchip*. Presented at IEEE Solid-State Sensor and Actuator Workshop, Hilton Head Island, SC: 1994.
65. Chen, J., et al., *A Multichannel Neural Probe for Selective Chemical Delivery at the Cellular Level*, IEEE/ASME Journal of Microelectromechanical Systems (JMEMS), 1997. **44**: p. 760–769.
66. Kaltsas, G., and A.G. Nassiopoulou, *Frontside Bulk Silicon Micromachining Using Porous-Silicon Technology*, Sensors and Actuators A: Physical, 1998. **65**: p. 175–179.
67. Kaltsas, G., D.N. Pagonis, and A.G. Nassiopoulou, *Planar CMOS Compatible Process for the Fabrication of Buried Microchannels in Silicon, Using Porous-Silicon Technology*, Microelectromechanical Systems, Journal of, 2003. **12**: p. 863–872.
68. Jo, B.H., et al., *Three-Dimensional Microchannel Fabrication in Polydimethylsiloxane (PDMS) Elastomer*, IEEE/ASME Journal of Microelectromechanical Systems (JMEMS), 2000. **9**: p. 76–81.
69. Jo, B.-H., et al., *Three-Dimensional Microchannel Fabrication in Polydimethylsiloxane (PDMS) Elastomer*, Microelectromechanical Systems, Journal of, 2000. **9**: p. 76–81.
70. Chiou, C.-H., et al., *Micro devices Integrated with Microchannels and Electrospray Nozzles using PDMS Casting Techniques*, Sensors and Actuators B, 2002. **86**: p. 280–286.
71. Burns, M.A., et al., *An Integrated Nanoliter DNA Analysis Device*, Science, 1998. **282**: p. 484–487.
72. Carlen, E.T., and C.H. Mastrangelo, *Surface Micromachined Paraffin-Actuated Microvalve*, Microelectromechanical Systems, Journal of, 2002. **11**: p. 408–420.
73. Vandelli, N., et al., *Development of a MEMS Microvalve Array for Fluid Flow Control*, Microelectromechanical Systems, Journal of, 1998. **7**: p. 395–403.
74. Li, H.Q., et al., *Fabrication of a High Frequency Piezoelectric Microvalve*, Sensors and Actuators A: Physical, 2004. **111**: p. 51–56.
75. Shikida, M., et al., *Electrostatically Driven Gas Valve with High Conductance*, Microelectromechanical Systems, Journal of, 1994. **3**: p. 76–80.
76. Yobas, L., et al., *A Novel Bulk Micromachined Electrostatic Microvalve with a Curved-Compliant Structure Applicable for a Pneumatic Tactile Display*, Microelectromechanical Systems, Journal of, 2001. **10**: p. 187–196.
77. Sadler, D.J., T.M. Liakapoulos, and C.H. Ahn, *A Universal Electromagnetic Microactuator Using Magnetic Interconnection Concepts*, Microelectromechanical Systems, Journal of, 2000. **9**: p. 460–468.
78. Rich, C.A., and K.D. Wise, *A High-Flow Thermopneumatic Microvalve with Improved Efficiency and Integrated State Sensing*, Microelectromechanical Systems, Journal of, 2003. **12**: p. 201–208.
79. Yang, X., C. Grosjean, and Y.-C. Tai, *Design, Fabrication, and Testing of Micromachined Silicone Rubber Membrane Valves*, Microelectromechanical Systems, Journal of, 1999. **8**: p. 393–402.
80. Unger, M.A., et al., *Monolithic Microfabricated Valves and Pumps by Multilayer Soft Lithography*, Science, 2000. **288**: p. 113–116.
81. Lee, S., et al., *Control Mechanism of an Organic Self-Regulating Microfluidic System*, Microelectromechanical Systems, Journal of, 2003. **12**: p. 848–854.
82. Baldi, A., et al., *A Hydrogel-Actuated Environmentally Sensitive Microvalve for Active Flow Control*, Microelectromechanical Systems, Journal of, 2003. **12**: p. 613–621.
83. Richter, A., et al., *Electronically Controllable Microvalves Based on Smart Hydrogels: Magnitudes and Potential Applications*, Microelectromechanical Systems, Journal of, 2003. **12**: p. 748–753.
84. De, S.K., et al., *Equilibrium Swelling and Kinetics of pH-Responsive Hydrogels: Models, Experiments, and Simulations*, Microelectromechanical Systems, Journal of, 2002. **11**: p. 544–555.

85. Chou, H.-P., et al., *A Microfabricated Device for Sizing and Sorting of DNA Molecules*, Proc. Natl. Acad. Sci., 1999. **96**: p. 11–13.
86. Groisman, A., M. Enzelberger, and S.R. Quake, *Microfluidic Memory and Control Devices*, Science, 2003. **300**: p. 955–958.
87. Cheung, K.C., et al., *Implantable Multichannel Electrode Array Based on SOI Technology*, Microelectromechanical Systems, Journal of, 2003. **12**: p. 179–184.

CHAPTER 14

Instruments for Scanning Probe Microscopy

14.0 PREVIEW

Scanning probe microscopes are an important family of scientific instruments for surface characterization and modification. The SPM probe, which consists of a spring and a tip, is the heart of any SPM instrument. Microfabricated SPM probes are widely used in many SPM applications. In this chapter, we will first review the history and future potential of SPM technology (Section 14.1). In Section 14.2, we will discuss generic methods for making tips. Designs and fabrication methods for integrating the cantilever springs and tips of various materials are reviewed in Section 14.3. Active SPM probes with integrated sensors and actuators increase the functional reach of such instruments. A number of SPM probes with integrated position sensors and deflection actuators are covered in Section 14.4.

14.1 INTRODUCTION

14.1.1 SPM Technologies

Scanning probe microscopes (SPMs) are a family of scientific instruments used for studying the topology and surface properties of materials with ultra-high spatial resolutions. Instruments in the SPM family are collectively used to measure a diverse range of physical interaction events between a sharp tip and a sample surface. More broadly, they are a family of tools for humans to interact with materials at the atomic, nanoscopic, or molecular scales.

A generic SPM system contains the following essential components (Figure 14.1): a sharp tip, a sample, a precision robotic stage for moving the tip in an XYZ space above the sample, and electronics for data acquisition and feedback control.

The tip is a critical component. Different SPM instruments require tips of different designs and materials. Almost all tips end at a sharp apex. The radius of the curvature of a tip apex can be as small as 20 nm. The sharp tip localizes the spatial extent of the interaction with the sample and contributes to high spatial resolutions.

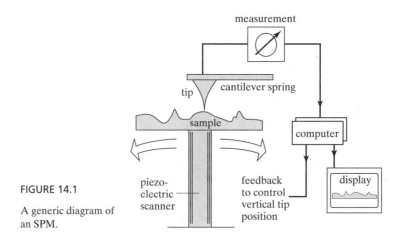

FIGURE 14.1

A generic diagram of an SPM.

The tip is supported by a spring, which is typically in the form of a cantilever beam. The cantilever serves three possible functions: (1) as a sensor to report tip displacement; (2) as an actuator to control the tip's position or temperature; and (3) as a hosting carrier of electrical leads or optical guides. Certain SPM applications call for extremely compliant springs (e.g., $k <= 0.1$ N/m) for achieving a high displacement sensitivity.

The earliest SPM probes are handmade. This leads to laborious work and poor repeatability. Micromachining technology is instrumental for the development of new SPM instruments. Miniaturization and microfabrication lead to compliant springs, wide choices of tip materials, and better repeatability. In many cases, MEMS is the only way to integrate functional tips and cantilevers.

In this chapter, we will visit several examples of microfabricated SPM probes to illustrate the interaction between design, fabrication, and performance issues. Before we touch on this topic, a broad understanding of the past, present, and future of SPM instruments is necessary. First, we will briefly review two of the earliest members of the SPM family—the scanning tunneling microscope (STM) and the atomic force microscope (AFM).

The STM is the first member of the SPM instrument family. It was invented in 1981 by Gerd Binnig and Heinrich Rohrer of IBM (Zurich). The fact that the duo were awarded the Nobel Prize in Physics only five years after the invention demonstrated its immediate and profound impact on science and technology. The STM was the first instrument used to render images of surfaces with atomic resolution. It uses a sharpened, electrically conducting tip. A bias voltage is applied between the tip and a conducting sample. When the tip is brought within about 10 Å of the sample, electrons from the sample begin to tunnel across the junction between the tip and the closest material atoms. The tunneling current is a strong function of the tip-to-sample spacing within a certain range.

In a constant-height scanning mode, the magnitude of the tunneling current is recorded and used to reconstruct the topological profile. Alternatively, the distance between the tip and the substrate may be controlled in a closed-loop fashion during the raster scan while the tunneling current is maintained as a constant. The driving signal generated in the control loop during a scan is used to derive the surface profile.

The STM can only be used to characterize conducting surfaces. To remedy this deficiency, the atomic force microscope (AFM) was invented; it allowed the atomic-resolution topology

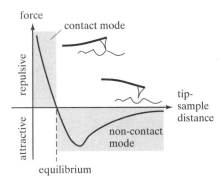

FIGURE 14.2

The sign and magnitude of atomic force is a function of interatomic spacings.

measurement of non-conductive sample surfaces. In an AFM instrument, atomic forces between a tip and the sample surface cause a cantilever to deflect. The topology is reconstructed from the force-displacement map generated during the raster scan. Several types of forces typically contribute to the deflection of an AFM cantilever. Both attractive and repulsive forces are possible (Figure 14.2).

AFM extends the capability of atomic-resolution microscopy to nonconductive families of materials, including biological materials. Today, the AFM instruments are used to characterize biological structures and monitor biological events in air or in liquid [1].

14.1.2 The Versatile SPM Family

The SPM instruments were soon extended beyond the roles of measuring tunneling current (STM) and the *van der Waals* force (AFM) alone. The AFM instrument belongs to a general scheme called the **force microscopy**. The force microscopy scheme was extended such that the tip can react to a variety of forces. SPM instruments are also used to sense a variety of physical parameters, such as temperature and light intensity. A few representative examples of new SPM instruments are discussed in the following paragraphs.

Lateral force microscopy (LFM) measures lateral deflection (twisting) of the cantilever that arises from forces parallel to the plane of the sample surface (Figure 14.3a). Lateral deflection of the cantilever usually arises from two sources: changes in surface friction and changes in slope. LFM studies are useful for imaging variations in surface friction that can arise from inhomogeneity in surface material, as well as for obtaining the edge-enhanced images of any surface.

A member of the SPM instrument family called the magnetic force microscope (MFM) can be used to image magnetic domains on the surface of a naturally occurring or engineered material. For an MFM probe, the tip is coated with a ferromagnetic thin film with a certain polarized magnetization (Figure 14.3b). Magnetic domains on surfaces (such as a magnetic disk) interact with the magnetized tip and produce repulsive or attractive forces.

SPM can be configured to measure surface charge distributions. Electrostatic force microscopy (EFM) uses tips with stored electrical charges (Figure 14.3c). The cantilever deflects when it scans over regions with static charges. EFM maps locally charged domains on the sample surface; this is similar to how MFM plots the magnetic domains of the sample surface. The magnitude of the deflection, proportional to the charge density, can be measured with the standard beam-bouncing method. EFM is used to study the spatial variations of surface charge

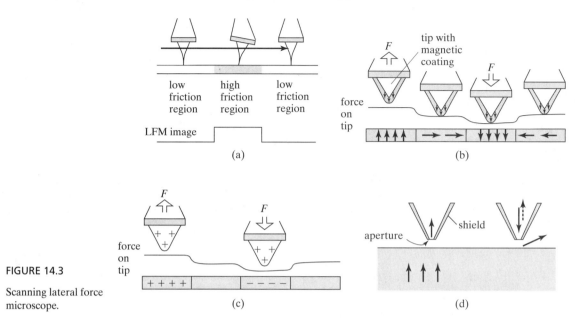

FIGURE 14.3

Scanning lateral force microscope.

density. For instance, EFM can map the electrostatic fields of an electronic circuit as the device is turned on and off. This technique is known as "voltage probing" and is a valuable tool for diagnosing behavior of circuits at the submicron scale.

Not all SPM instruments involve the measurement and interpretation of forces. The spatial resolution of many functional sensors can be drastically enhanced when the sensor element is placed on the scanning tip—provided that the sensing principle scales down well. Three examples are discussed below.

Scanning thermal microscopy (SThM) measures the thermal conductivity or temperature distribution on a surface. The heart of the scanning thermal microscope is a thermal probe with a self-heating, resistive element. The temperature of the tip is used to interpret the local temperature, thermal conductivity, or topography. Thermal probing has been achieved by thermal couples [2, 3] and Joule expansion (in scanning a Joule expansion microscope) [4].

The SPM instrument is also being used to enhance the spatial resolution of optical microscopy and spectroscopy. Conventional optical microscopy encounters limits of resolution imposed by optical diffraction (the Abbé diffraction barrier). Near-field optical imaging involves illuminating a specimen through a sub-wavelength-sized aperture while keeping the specimen within the near-field regime of the source. The Near-field Scanning Optical Microscope (NSOM) technique (Figure 14.3d) works by exciting the sample with light passing through a submicron aperture formed at the end of a single-mode drawn optical fiber. Typically, the aperture is a few tens of nanometers in diameter. Micromachined NSOM probes with integrated optical waveguides and precise apertures have been developed in the past (e.g., [5]).

Imaging of a surface magnetic field can be performed in a non-force-related manner as well. The Hall measurement is commonly used in the semiconductor industry for characterizing the doping concentration of samples. A scanning Hall-effect microscope is an alternative to the MFM for measuring surface magnetic properties.

Other impressive performances and functions achieved with SPMs include the following:

- The scanning acoustic microscope can be used to detect subsurface features for characterizing biological and solid-state materials [6].
- SPM can be used for the investigation of localized energy dissipation within a sample below its surface, for example, in applications such as mapping dopant concentrations [7].
- Single-molecule atomic force microscopy has been demonstrated for studying pathways of protein folding by resolving small forces on the order of 100 pN [8].
- Scanning force microscopes have been used to map the charge distribution in atoms [9] and even to detect single electron spin [10] events.
- Due to the flexibility of the probes and the high-sensitivity bending detection, SPM probes have been used as a tool for biological and chemical sensing in recent years [11]. Probe bending may occur by chemical deposition on the surfaces of the cantilever [12, 13] or by the interaction of chemically functionalized tips with a chemically functionalized surface of interest [14]. Multiplexed biological sensing using cantilevers integrated with microfluid channels have been demonstrated [15].

14.1.3 Extension of SPM Technologies

The role of SPM instruments has extended beyond the measurement of surface physical properties. Their applications today include nanopatterning and data storage. The SPM platform is used to satisfy the needs of addressing and modifying surfaces with molecular or atomic resolution—electrically, chemically, or mechanically. The speed of nanolithography can be improved by using parallel, sometimes independently addressable, SPM probes or probe arrays [16, 17]. Atomic resolution lithography can be based on a variety of principles, including tip-induced anodization [18], the local oxidation of single-crystal or amorphous silicon [19], organic decomposition [20], field-assisted evaporation [21], field-assisted desorption of atoms [22, 23], and local magnetization [24]. The techniques are used beyond scientific curiosity, in applications such as fabricating semiconductors and metal-oxide devices with a submicron channel length [25, 26].

One representative method for nanoscale lithography is the local surface oxidation of silicon or metal thin films by conductive SPM tips [18]. A silicon substrate can be locally oxidized by a biased tip in contact with it. The oxidant for the chemical reaction is provided by OH^- ions in the water interface that is formed between the tip and the sample (Figure 14.4). The lateral resolution of the AFM oxidation process depends strongly on the environmental humidity.

The Dip Pen Nanolithography (DPN) is a powerful technique for directly depositing biochemical molecules (including DNA and proteins) onto surfaces with small linewidths [27]. The DPN technique uses an AFM tip coated with the molecules of interest. When the tip and the

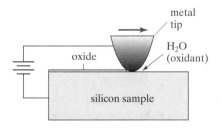

FIGURE 14.4

Local oxidation of silicon.

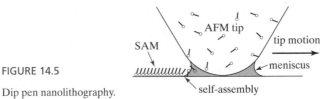

FIGURE 14.5

Dip pen nanolithography.

surface make a contact, a liquid meniscus is formed at the junction. Molecules on the tip diffuse through the meniscus interface to the writing surface, forming fine features made of the molecule (Figure 14.5). Since the SPM tip is usually very sharp, linewidths as small as 50 nm have been demonstrated.

With the proliferation of digital electronics and the Internet comes the ever-increasing demand for memory storage. SPM can be used for high-density data storage. In these applications, an SPM plays two roles—modifying surfaces (writing) and then measuring the results of the modification (reading).

In current optical disk drives using 780-nm diode lasers, data storage density is limited by the diffraction of light to ~80 Mbits/cm^2. The diffraction limited resolution can be improved by near-field scanning optical microscopy (NSOM) where a sub-wavelength sized optical source is raster scanned in close proximity to a sample to produce an image [5, 28]. Magnetooptic domains as small as 60 nm have been optically written and imaged in a 14-nm-thick Co/Pt multilayer magnetooptic film, presenting the possibility of 7 Gbits/cm^2 storage densities with this technique.

High-density thermoelectric data storage has been studied at IBM and Stanford University [29–31] and followed elsewhere [32]. It uses an array of SPM probes, each with dedicated heater, temperature sensor, and control/sense electronics.

14.2 GENERAL FABRICATION METHODS FOR TIPS

In this chapter, we will review common microfabrication methods for making SPM probes. We will focus on force microscopy probes.

When selecting materials and processes for making integrated tips, the most relevant consideration factors are

1. Apex sharpness;
2. Tip aspect ratio;
3. Conductivity;
4. Wear resistance.

The sharpness of the tip determines the resolution of the imaging. The sharpness requirement in turn determines the materials and the fabrication processes for the tip and cantilevers.

The aspect ratio is often confused with the tip sharpness. However, it is actually a separate aspect of concern. Tips with a high aspect ratio (length vs. tip diameter) can be used to resolve features in deep trenches.

The conductivity of a tip is important for certain SPM and nanolithography tasks, such as scanning tunneling microscopy and high-voltage nanolithography. AFM probes do not have to be conductive. The conductivity requirement influences the choice of materials, design, and processing techniques.

Wear resistance is important for contact mode lithography applications or industrial applications for scanning large areas. Wear-resistant tips may be made of metals, silicon, or diamonds.

Silicon anisotropic wet etching is perhaps the most widely used method for realizing integrated tips. The silicon surface is first coated with a layer of thin film such as silicon oxide (by thermal oxidation) or silicon nitride (by chemical vapor deposition). The thin-film layer is patterned photolithographically to form mask patches. Wet anisotropic etching and undercut create mesas and then sharp tips. Pyramidal-shaped tips formed under mask patches are shown in Figure 14.6.

The technique is simple and the materials are easily accessible. However, a serious drawback of this process lies in the fact that the sharpness of the tips is difficult to control, as the tip profile is unstable and transitional (UTP). Once an atomically sharp tip is formed, the mask piece will fall off and the tip is rapidly attacked from the top. The control of the tip sharpness is further complicated by spatial non-uniformity and the temporal variability of etch rates.

A common strategy for improving the quality and fabrication efficiency of anisotropically etched tips is to stop the etching process before atomically sharp tips are formed, and sharpen the tips using additional process steps. An SEM micrograph of tips with flat tops is shown in Figure 14.6a. These tips can be furthered sharpened by converting a surface silicon into silicon dioxide in an oxidation furnace and then removing the oxide [33, 34]. The principle for this sharpening method is discussed below.

The oxide growth rate is affected by the surface curvature. Oxide at the apex experiences greater stress than that on the slope. The growth rate of oxide at the apex of pyramid tips is therefore slower than that along the slopes. Tips after an oxide sharpening cycle become much sharper (Figure 14.7). The sharpness and uniformity improves by repeating the cycle a few times. Routine manufacturing can yield a curvature radius of <15 nm and cone angles of about $30°$ up to 1 μm from the apex. Ultimately, this method can result in silicon tips with a curvature radius lower than 1 nm.

An alternative method of forming high-aspect-ratio, protruding silicon tips is to use plasma etching instead of anisotropic wet etching. Plasma etching can produce anisotropic or isotropic profiles, depending on process parameters such as the gas mixture ratio, pressure, power, electrode geometry, and others. It tends to generate tips with a greater aspect ratio than

(a)

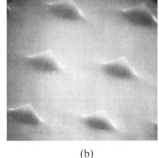

(b)

FIGURE 14.6

Array of probes before and after sharpening.

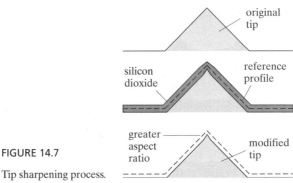

FIGURE 14.7

Tip sharpening process.

those obtained using anisotropic silicon wet etching. Although the etch process can produce spectacular tip shapes, it is not uniform over a wafer surface and it is difficult to reproduce results. In practice, etching should be stopped with a small flat top still remaining at the apex to allow those tips to be furthered sharpened by oxidation.

Tips can be formed by molding against an anisotropically etched SLSP-type cavity with four {111} slopes. We will discuss a few examples in the next section. The oxidation sharpening technique also works for inverted pyramidal cavities. The benefit of the oxidation sharpening technique can thus be translated to molded tips made of silicon nitride and metals [35].

14.3 CANTILEVERS WITH INTEGRATED TIPS

14.3.1 General Design Considerations

An SPM probe consists of a cantilever and a tip. These two parts must be considered as a whole in terms of materials selection, design, and fabrication method. The most common factors of consideration pertaining to the cantilever include

1. the spring constant of the cantilever;
2. the resonant frequency of the cantilever;
3. the intrinsic bending of the cantilever;
4. the surface roughness of cantilevers.

Many applications demand force constants within a certain range. For example, high-sensitivity AFM measurements require very soft cantilevers (with a force constant below 0.1 N/m) in order to increase the force sensitivity and to avoid scratch damage to surfaces in contact modes. Some applications, on the other hand, require stiff cantilevers for probing, imprinting, and scratching. Formulas for calculating the force constant of a cantilever have been discussed in Chapter 2. Of all the dimensional factors—length, width, and thickness—the thickness is the most influential element of design, since the force constant of a cantilever varies as t^{-2}.

The resonant frequency of a cantilever determines the maximum bandwidth for writing and displacement sensing, as well as the spatial resolution of surface imaging. SPM sensors with a greater bandwidth can be scanned over a surface at a higher speed.

Intrinsic bending of the cantilever is important. In many applications, the bending of cantilevers may upset system calibration. Worse still, intrinsic bending may affect performance. For

14.3 Cantilevers with Integrated Tips

example, if an LFM probe is bent, transverse friction force may cause the beam to warp rather than simply twist.

The surface roughness and optical reflectivity of the cantilevers are important, too. Many SPM instruments use optical levers for displacement measurement. If the beam material does not provide enough reflectivity, or if the surface is too rough, the cantilever won't operate in the SPM instrument. Additional metal coating may be needed to enhance optical reflection. However, the additional films may introduce unwanted intrinsic bending.

14.3.2 General Fabrication Strategies

General fabrication strategies for SPM cantilevers with integrated tips (but without integrated sensors or actuators) are discussed in this section.

Cantilevers made of single-crystal silicon and silicon nitride are the most widely encountered. We will compare seven schemes for realizing cantilevers with integrated tips. Schemes 1 through 3 deal with silicon nitride cantilevers, whereas schemes 4 through 7 deal with silicon cantilever. For silicon cantilevers, schemes 4 through 5 use plain silicon wafers, while schemes 6 through 7 use composite silicon wafers with buried layers (either heavily doped silicon etch stop layers or silicon oxide layers).

Two types of cantilevers are possible; one has the tip pointing away (outward) from the front surface of the substrate, and another has the tip pointing towards (inward) the substrate. Inward pointing tips are difficult to use, because the substrate may get in the way and the tip may contact the substrate first. However, the fabrication process of SPM probes with inward pointing silicon tips is more simple (scheme 2 of Figure 14.8).

Schemes 1 through 3 are diagrammed in Figure 14.8 and discussed in the following.

Scheme 1:

A plain silicon wafer is used (Figure 14.8, step 1.1) with a front surface being {100}. A mask layer is deposited and patterned, and it is used to etch protruded pyramidal tips (1.2). The tips

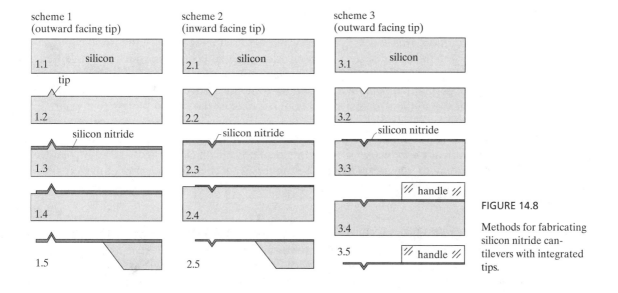

FIGURE 14.8

Methods for fabricating silicon nitride cantilevers with integrated tips.

are subsequently sharpened by oxidation. After removing the mask layer and oxide, the entire wafer is coated with silicon nitride to the desired thickness (1.3). (Note the process on the back of the wafer is not shown, for brevity.) The front surface of the silicon is coated with a photoresist layer and it is lithographically patterned. The patterned photoresist serves as a mask to etch the underlying silicon nitride (1.4). The photoresist spin coating is not straightforward, however. The photoresist thickness at the tip apex is smaller than on planar surfaces. If the thickness is insufficient, the protection layer at the tip apex may be completely removed. In such events, there is a risk that the tip may be attacked during anisotropic etching.

The nitride layer is patterned to form cantilevers with integrated silicon nitride tips. The passivation layers on the back side of the wafer are patterned, and they are used for anisotropic etching to form the chip handles (1.5). Some disadvantages associated with this process are (1) the difficulty associated with spinning the wafer and protecting the tip between steps 1.3 and 1.4; and (2) that the blanket deposition of silicon nitride reduces the sharpness of the tip and increases its radius of curvature by an amount equal to its thickness.

Scheme 2:
A {100} silicon wafer is used. A mask layer is deposited, patterned, and used for etching a pyramid (2.1). A silicon nitride layer is deposited (2.3) and patterned (2.4). The masking layer on the back is patterned. Anisotropic etching is performed to form handles (2.5). The most important improvement of scheme 2 over scheme 1 is the fact that the tip sharpness is not compromised by the silicon nitride thickness. Another important advantage of etching an inverted pyramid is that the shape of the inverted cavity is self-limiting and the process does not rely on precise time control. However, the tip is pointing inward; therefore, it provides limited application potential.

Scheme 3:
The tip molding process allows a sharp tip and uniform sharpness to be realized in a robust fashion. Can one realize an outward tip using the molding process? One method is presented here. Again, a {100} wafer is used. An inverted pyramid forms on its front surface by anisotropic etching (3.2). A silicon nitride layer is deposited and patterned (3.3) to form the cantilevers. A handle wafer is bonded with the silicon wafer (3.4) instead of being formed out of silicon substrates. Subsequently, the silicon wafer is removed by dissolving in an isotropic silicon etching solution. This process produces an outward facing tip with the tip sharpness not compromised by the thickness of the silicon nitride. In fact, this process, which was invented by Prof. Quate's group at Stanford University, is used most widely by the SPM industry to make commercial probes [36, 37]. Major disadvantages of this process are that (1) the bonding step adds complexity to the process; and (2) the removal of the silicon bulk is costly and time consuming.

SPM probes made of silicon eliminate the intrinsic stress associated with silicon nitride and can be intrinsically conductive. Cantilevers made of silicon can be realized using plain silicon wafers according to schemes 4 and 5. The processes for making such probes are noticeably longer than those for silicon nitride ones.

Scheme 4:
Starting with a {100} silicon wafer (Figure 14.9, step 4.1), a pyramidal tip is first formed and sharpened (4.2). The wafer is coated with a conformal passivation layer, such as silicon oxide or silicon nitride (4.3). If the passivation layer is silicon oxide, it can be later removed using HF solutions. If the passivation layer is silicon nitride, it can be later removed by hot phosphoric acid

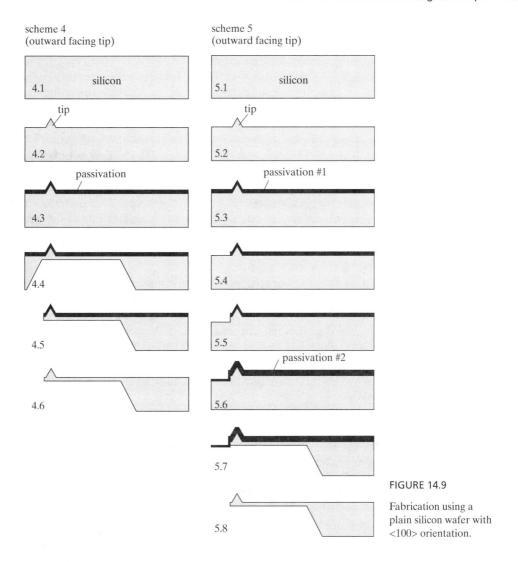

FIGURE 14.9

Fabrication using a plain silicon wafer with <100> orientation.

(H_3PO_4). A passivation layer on the back side is patterned and used as a mask for anisotropic back-side etching (4.4). The back-side etch is time controlled and targeted at a desired thickness (4.5). The layer on the front is then removed to finish the device (4.6). However, the thickness control is practically difficult.

Scheme 5:

An alternative process, scheme 5, is developed. The first few steps—5.1 through 5.3—are identical with fabrication scheme 4. A front-side etch by plasma defined the thickness of the silicon beam (steps 5.4 and 5.5). The front surface of the wafer is passivated and protected against back-side anisotropic etching. When the etching reaches the front surface, part of the wafer covered only by the passivation layer will become transparent and let light through for end-point detection. The logistics burden of carefully monitoring the time of back-side etch is relieved. However, the wafer-scale etching non-uniformity issue remains unsolved.

Using scheme 4, the realistic accuracy for controlling the thickness of the cantilever over a 4"-diameter wafer is on the order of 5–10 μm. A feasible minimum thickness of the beam should be approximately 20 μm. Using scheme 5, the accuracy is somewhat improved over scheme 4. The minimal thickness is improved to 5–10 μm on a wafer scale.

Thinner cantilevers result in a more desirable spring constant. In order to realize silicon probes with a smaller cantilever thickness and more process robustness, special silicon wafers are used. Two representative processes are described in schemes 6 and 7 (Figure 14.10).

Scheme 6 uses a silicon wafer with a buried layer of heavily doped silicon. The doping concentration is made sufficiently high for etch stop purposes. The buried layer is capped by a layer of lightly doped single-crystal silicon wafers [38]. The thickness of the lightly doped layer on top and the heavily doped layer in the middle are specified for each project. The wafer with composite layers can be formed by bonding and etch back. The process for preparing the composite wafer involves bonding two wafers; the one on the bottom has a heavily doped top layer. The top wafer is etched back or polished to leave the desired thickness of undoped silicon on the front.

A pyramidal tip is first formed through the thickness of the undoped epitaxy silicon layer (step 6.2). If the tip is formed by anisotropic wet etching, the etch will stop when the buried layer is reached. The wafer is then coated with a passivation layer (passivation #1) on the front side (6.3). The passivation layer is patterned and etched (6.4), and it is used as a mask for etching the heavily doped layer underneath by plasma etching (6.5). Another passivation layer (passivation #2) is deposited to coat the exposed silicon (6.6). A back-side wet anisotropic etching is conducted. Once the etch reaches the buried heavily doped silicon layer, the etching is automatically stopped (6.7). The passivation layers are then removed (6.8).

Using this method, the accuracy for controlling the thickness of cantilevers can be as high as 0.5 μm.

Alternatively, the process can begin with a silicon-on-insulator (SOI) wafer according to scheme 7. A silicon-on-insulator wafer consists of a buried insulator (typically oxide) sandwiched between two layers of single-crystal silicon. The wafer is formed by bonding two wafers (one plain silicon wafer and one with oxide on the front) and etching back the thickness of the plain silicon wafer (7.1). The thickness of the top single-crystal silicon can be custom specified.

The tip and the cantilever can be made in the top silicon layer. The first few steps involve etching and sharpening the pyramidal tip (7.2), covering the front side with a passivation layer (7.3), patterning the passivation layer (7.4), and etching the silicon using plasma etch. The plasma etch stops at the oxide layer readily because the etch rate selectivity by the plasma on silicon and silicon oxide is very high (7.5) [39]. The entire wafer front surface is passivated again (7.7), and back-side etching is performed using a wet anisotropic etch or deep reactive ion etch (DRIE); both offer excellent selectivity between silicon and oxide. The device fabrication ends by removing the passivation layers along with the oxide below the cantilevers (7.8).

14.3.3 Alternative Techniques

Methods discussed in Section 2.2.2 invariably call for silicon substrates and involve relatively complex processes. However, since most SPM applications do not involve semiconductor properties, there is no need to build SPM probes using silicon materials. Indeed, for certain applications, alternative materials such as metals or polymers may be sufficient or desired for performance reasons and/or cost. An efficient fabrication method for realizing SPM probes is discussed in Case 14.1. The method can be used to realize metal or polymer-based tips [40].

14.3 Cantilevers with Integrated Tips 467

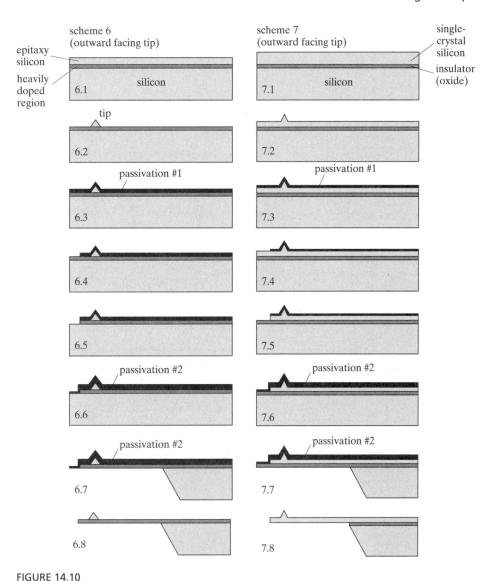

FIGURE 14.10

Silicon cantilevers with complex wafers.

Case 14.1. Mold-and-Transfer Process for SPM Fabrication

A mold-and-transfer fabrication process for realizing singular or arrayed SPM probes has been developed (Figure 14.11). A single-crystal silicon substrate with inverted pyramidal pits formed by anisotropic etching is used as the "mold" to make the SPM probe tips. The

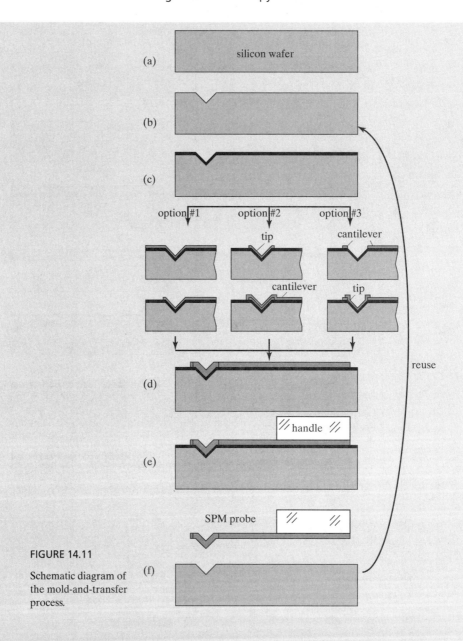

FIGURE 14.11

Schematic diagram of the mold-and-transfer process.

three-dimensional profile of the inverted pits (bounded by four {111} silicon crystal surfaces) defines the shape of the SPM tip (Figure 14.11a and b) [41]. A sacrificial layer is deposited onto the etched silicon substrate (Figure 14.11c). Materials for the tip and cantilever are then deposited and patterned. The tip layer may fill the entire pit volume or may consist of a thin shell. Note that the tip and the cantilever do not have to be made of the same materials (Figure 14.11d).

There are three process flow options between steps (c) and (d). If the tip and the cantilever are made of the same material, option #1 will be taken. The material for the tip and the cantilever may be deposited (e.g., metal, silicon nitride), electroplated (e.g., metal), or spin coated (e.g., polyimide).

When the tip and cantilever are made of different materials, there would be two optional approaches (see options 2 and 3 in Figure 14.11). According to option 2, the tip material may be deposited and patterned first, followed by the deposition and patterning of the cantilever material. According to option 3, one may first deposit a thin-film material, pattern it to provide the cantilever and an opening overlapping the inverted pits, and then deposit and pattern the tip material. The choice of process sequences must satisfy the fact that any fabrication step should not adversely affect (thermally or chemically) the material layers already present on the substrate.

Next, a handle chip is attached onto the probe cantilever to allow for easy handling of delicate SPM probes (Figure 14.11e). Finally, the sacrificial layer is completely etched to detach the SPM probe from the silicon "mold" (Figure 14.11f). Both the sacrificial layer material and etchant are carefully chosen so that the SPM probe and tip are not affected during the sacrificial layer etching. Preferably, this etching process should not attack the silicon master mold, so that it can be reused.

Compared with other SPM probe fabrication processes diagrammed in schemes 1 through 7, the mold-and-transfer process has several advantages in terms of material and performance.

1. The profile of the probe tip is defined by the inverted pyramidal pits, which are of SLSP type and bound by stable {111} surfaces. The profile is insensitive to over-time etching. High process yield for large arrays of SPM probes can be achieved (Figure 14.12).
2. This process allows the use of a rich choice of materials such as silicon nitride, metals (e.g., by evaporation, sputtering, electroplating, or electroless plating), and polymers (e.g., polyimide, polydimethylsiloxane (PDMS), photoresist, and SU-8 resists). Also, the probe tip and cantilever may be made of different materials. Some known or predicted tip and cantilever material combinations for SPM applications are listed in Table 14.1.

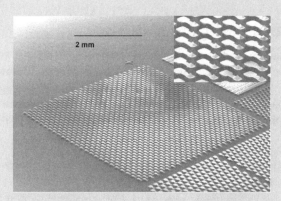

FIGURE 14.12

An array of SPM probes made using the mold-and-transfer technique. The inset figure is a magnified view of a segment of the array.

3. The silicon mold can be reused, which reduces process time and cost. This is one advantage over the scheme 3 discussed earlier.

The etched silicon pits ideally have atomically sharp ends. However, the radius of the curvature of the tips suffers from the blanket deposition of the sacrificial layer. The larger the thickness of the sacrificial layer, the greater is the radius of curvature. The thickness of the sacrificial layer is generally 100–200 nm.

This degree of sharpness is not as good as certain commercial probes; however, these moderately sharp tips prove satisfactory for many applications involving SPM probes. If sharper tips were desired, a number of methods could be used, such as minimizing the thickness of the sacrificial layer, sharpening the oxidation of the inverted pyramidal pit [35], attaching the carbon nanotubes [42], or growing "super tips" *in situ* at the tip apex.

TABLE 14.1 List of Possible Tip and Cantilever Materials.

Tip material	Cantilever material	Comments
Metal (e.g., thin-film metal including Au, Al, Cu; electroplated metal such as Ni)	Metal (e.g., thin-film metal and electroplated metal) Polymer (e.g., polyimide, SU-8) LPCVD silicon nitride or polysilicon	Use another metal as a sacrificial layer Deposit LPCVD material first
Polymer (e.g., SU-8, polyimide, photoresist)	Metal (Au, Cu, Al, Ni, etc.) Polymer (SU-8, polyimide) LPCVD silicon nitride or polysilicon	Deposit LPCVD material first
Elastomer (e.g., polydimethylsiloxane)	Polymer (polyimide, etc.) Metal LPCVD silicon nitride or polysilicon	Deposit LPCVD material first
Silicon nitride tip	Metal Polymer LPCVD silicon nitride or polysilicon	

14.4 SPM PROBES WITH SENSORS AND ACTUATORS

Presently, most SPM applications involve only a single, passive SPM probe consisting of a cantilever and a tip. However, many SPM applications demand probes with integrated sensors and actuators. Many future applications, such as data storage and nanolithography, call for arrayed SPM probes (one-dimensional or two-dimensional) to achieve high efficiency and throughput [29, 43–45]. In certain applications, probes are required to be integrated with heaters, actuators, and sensors [46, 47].

Although SPM probes have a very high spatial resolution, their field of view is generally very limited. Hence, the SPM measurement technique cannot be extended to imaging large surface areas (for example, a semiconductor mask or wafer). An array of SPM probes can increase the imaging throughput. Simultaneously sensing the probe displacement for the entire array becomes necessary.

14.4.1 SPM Probes with Sensors

Traditional SPM instruments use a laser beam bouncing off the cantilever to measure the displacement. This method is simple and low cost. However, it is impractical to use multiple laser beams to interrogate the displacement of many at high speeds. The solution for arrayed SPM sensing is for sensors to be integrated with individual probes.

One aspect of the challenge is certainly the design and materials of these sensors to achieve sufficient force sensing resolution. The beam thickness determines the sensitivity. The process complexity generally increases with decreasing beam thickness. We will discuss three probes with piezoresistive sensors. The targeted cantilever thickness values are 4.5 μm (Case 14.2), 1 μm (Case 14.3), and 0.1 μm (Case 14.4), respectively. SOI wafers are used in all three cases.

Case 14.2. Silicon Cantilevers for SPM Probe

The first example is a tipless AFM probe with a piezoresistive sensor embedded through the entire length of the cantilever [48]. The cantilever consists of two layers—an intrinsic silicon layer and a doped silicon layer (p-type with a sheet resistivity of 220 Ω), which serves as the piezoresistive strain gauge. The cantilever is 4.5 μm thick and the average depth of the doped region is 0.5 μm. The length and width of the cantilevers vary from 400 to 75 μm and from 50 to 10 μm, respectively. This corresponds to the range of spring constants being 5 to 100 N/m. The measured resonant frequency varies from 40 to 800 kHz with an air quality factor of approximately 200 to 800. The resistance of the different cantilevers ranges between 2.5 to 70 kΩ.

Resistors could be defined to occupy the region with the highest stress; however, in this paper, the resistor covers the entire cantilever. This simplifies the process by eliminating one layer of mask for patterning the doping-barrier film.

The device fabrication process starts with a silicon-on-insulator wafer manufactured using wafer-bonding techniques. The top layer is an n-type silicon with a thickness of 6 ± 1 μm and a resistivity of 15 Ωcm. The intermediate oxide layer, which is 1 μm thick, is used as an etch stop during the process of etching the bulk silicon substrate in the final release step (Figure 14.13a). First, the top silicon layer is etched down to a thickness of 2 ± 1 μm. The wafer is oxidized to provide a doping shield layer for the back side (Figure 14.13b). The researchers used wet etching to remove the oxide on the front side of the wafer. Boron is implanted on the front side at 80 keV with a dose of 10^{15} ions/cm^2. The thickness of the doped layer is controlled accurately (Figure 14.13c). The front side of the wafer is patterned using photoresist to define cantilevers. The silicon on the front side is etched using plasma unitl the oxide layer is reached (Figure 14.13d). The back side is patterned as well. The oxide on the back side is patterned photolithographically, and the oxide is etched using HF.

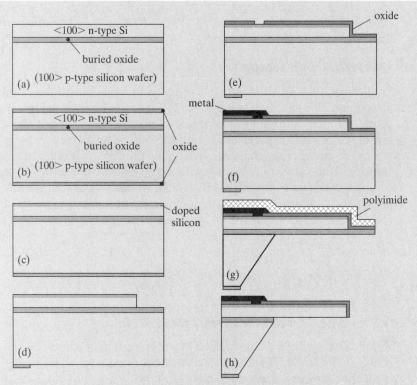

FIGURE 14.13

Fabrication process.

A thin oxide is grown, and it serves as a passivation and insulation layer. The oxide is patterned at the end of the cantilevers to provide electrical contacts (Figure 14.13e). Metal (aluminum) is sputtered, patterned, and etched to define wire leads (Figure 14.13f). Thermal annealing is performed to encourage the aluminum to form ohmic contact with the doped layer. A polyimide layer is spin coated on the front side of the wafer to serve as protection during the silicon anisotropic etch in EDP (Figure 14.13g). The anisotropic etching produces cavities that reach all the way to the buried oxide. At the end of the process, the polyimide is removed (using acetone or oxygen plasma) and the original buried oxide is removed using HF to completely release the AFM probes (Figure 14.13h). Finally, HF solutions remove the remaining oxide to yield the finished device.

The minimum detectable deflection of the cantilever depends not only on its sensitivity, but also on the noise in the measurement system. The fundamental noise limit in the measurement of resistance is given by the Johnson noise in the piezoresistor, which is

$$V_n = \sqrt{4k_B T R \Delta f}, \tag{14.1}$$

where V_n is the thermal electrical noise, k_B the Boltzmann constant, T the temperature, and Δf the bandwidth of measurement. In addition, there are other sources of noise. The 1/f noise, for example, can arise from many sources, such as carrier trapping at surface defects and contacts.

The output signal increases with the voltage applied to the Wheatstone bridge. For the purpose of increasing the sensitivity, the voltage bias should be as high as possible. However, the optimum bias voltage should not encourage the growth of 1/f noise. A bias voltage of 8 V is used, maximizing the signal-to-noise ratio and reducing the power consumption on the cantilever to a few mW. The minimum detectable deflection in the bandwidth from 10 Hz to 1 kHz varies from 0.7 to 0.1 Å$_{rms}$.

Case 14.3. Silicon Cantilevers for SPM Probe

Later, an SPM probe with a piezoresistive sensor was made by the same group as in Case 14.2 on a cantilever of only 1 μm thick. This time, a tip was integrated with the cantilever [30]. The small thickness of cantilevers and the incorporation of a tip bring new challenges compared with the previous case, including (1) precise control of the doping profile, and (2) the handling of chips after the suspended cantilever is made.

The starting material is a silicon-on-insulator wafer with a 5-μm top silicon layer (Figure 14.14a). An oxide layer is deposited and patterned to serve as the mask for forming tips. A subsequent plasma etch undercuts the mask and forms blunt tips (Figure 14.14b), which are then sharpened by oxidation (Figure 14.14c). The thickness of the remaining top silicon layer, consumed through the tip etching and the oxidation steps, is carefully designed to be approximately 1 μm. The cantilever itself is patterned by plasma etch (Figure 14.14d). A 100-nm-thick oxide is grown on the top surface to serve as a doping mask (Figure 14.14e). The oxide is selectively etched to allow dopants to reach the desired piezoresistor locations (Figure 14.14f). The authors performed the ion implantation of boron at $5 \times 10^{14}/cm^2$ dose to produce a shallow doped layer, which is activated by a 10-second rapid thermal anneal at 1000°C and a 40-minute low-temperature furnace anneal at 800°C (Figure 14.14g). Ohmic electrical contacts are made with an aluminum metallization (Figure 14.14h). A back-side etch is performed, with the front side of the wafer being covered (Figure 14.14i). The rest of the process is similar to Case 14.2.

Unlike the process in Case 14.2, the step to deposit oxide passivation is performed before the ion implantation rather than after. This is important because the oxidation is a high-temperature step that encourages the dopants to out-diffuse. With the thickness of the beam being very small, the margin for error is thinner. Therefore, the oxidation is not performed after the ion implantation as in Case 14.2. The ion-implanted dopant atoms are activated using the rapid thermal annealing method, which limits the overall expansion of the doped region.

Case 14.4. Ultra-Thin Silicon Cantilevers for SPM Probe

Another cantilever with a thickness under 1000 Å has been made [49] to satisfy the extremely high sensitivity requirement (with a target minimal detectable force of 8.6 fN/$\sqrt{Hz}$ in air). The difficulty in reducing the cantilever thickness stems from the need to precisely

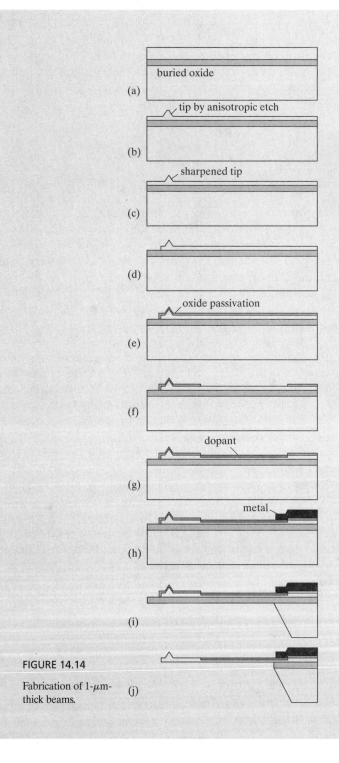

FIGURE 14.14

Fabrication of 1-μm-thick beams.

control the doping profile. Like the previous case, if the dopant is uniformly spread through the thickness, the sensitivity will be zero because of symmetry across the neutral axis. The doping must not reach more than 1/2 of the beam thickness. The methods discussed for making the cantilever and asymmetric doping do not offer enough accuracy for this small thickness anymore. Instead of doping the cantilever with ion implantation to form piezoresistors, the piezoresistors are formed by *in situ* deposition of boron-doped silicon using a low-pressure chemical vapor deposition process. Let us look at how this doping step is incorporated into the overall process flow.

The process begins with an SOI wafer with a 10-Ωcm background doping (Figure 14.15). The thickness of the silicon on the front side is 200 nm, which is approaching the limit of the SOI wafer fabrication technology (a). The silicon on the front side is selectively thinned. A few options are possible, including mechanical polishing and wet silicon etching. However, these methods do not offer the high precision required in this application. Instead, a thermal oxidation process is conducted. The oxide is formed by oxygen reacting with silicon, thus consuming the silicon from the front side. The oxidation process is rather slow and uniform, allowing the silicon thickness to be precisely controlled. The oxide is later removed, leaving

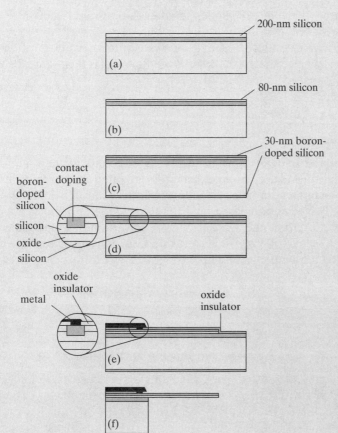

FIGURE 14.15

Fabrication process for an AFM probe with integrated position sensor.

the thickness of silicon on the front side as 80 nm (b). HCl solutions were used to clean the wafer. Although the etch rate of HCl on silicon is finite, it consumes 10 nm silicon off the front side.

A layer of boron-doped silicon is grown over the entire wafer with a thickness of 30 nm and a doping concentration of 4×10^{19} cm^{-3} (c). The wafer is coated with photoresist, which is used to define the shape of the cantilevers. The doped silicon forms the piezoresistor. However, direct contact with the metal and the piezoresistor would yield a high resistivity and rectifying contact. To produce a low resistance contact with non-rectifying characteristics, a boron implant is performed locally at 30 keV energy and 1×10^{15} cm^{-2} dose. The area of the implant is limited to a small contact pad region, with other areas covered with a cured photoresist to stop ions during implantation (d). A thermal annealing process is performed at 700°C for 3 hours in an oxygen environment. This achieves both the activation of dopants and the formation of an oxide insulator that protects the front side of the wafer.

A contact window is opened in the oxide; this is followed by the deposition and patterning of aluminum as metal leads (e). The contact is annealed in a forming gas at 400°C for 1 hour. The back-side oxide and doped silicon is patterned using a release mask. A deep reactive ion etching step removes the silicon substrate anisotropically until the buried oxide layer is reached (f). Anisotropic wet etching may not be feasible here because the passivation layers are relatively thin under low-temperature and short-duration oxide growth. Both the buried and deposited oxide layers are removed using buffered HF solutions, and the cantilever and contact regions are covered with a temporarily bonded silicon wafer with photoresist as a bonding agent. The bonding wafer is then removed in acetone, which completes the process.

14.4.2 SPM Probes with Actuators

SPM probes with integrated actuators have been developed to provide new application capabilities, such as active control and data storage. Actuators are often used to provide a transverse tip displacement. Such actuators can be based on piezoelectric, capacitive, or thermal bimetallic actuation principles. Two examples based on different actuation principles are discussed in the following paragraphs. The first one, Case 14.5, is an SPM probe with piezoelectric sensors and actuators located on the cantilever. The second one, Case 14.6, involves a thermal bimetallic actuator for displacing the cantilever and tip.

Case 14.5. SPM Probe with Piezoelectric Sensing and Actuation

A team at Stanford University led by Prof. Calvin Quate developed an atomic force microscope with integrated piezoresistive sensors and piezoelectric actuators [47]. The fabrication of the device starts with a <100>-oriented SOI wafer with a 100-μm-thick intrinsic silicon layer and a 1-μm-thick silicon dioxide layer between the top silicon and the substrate

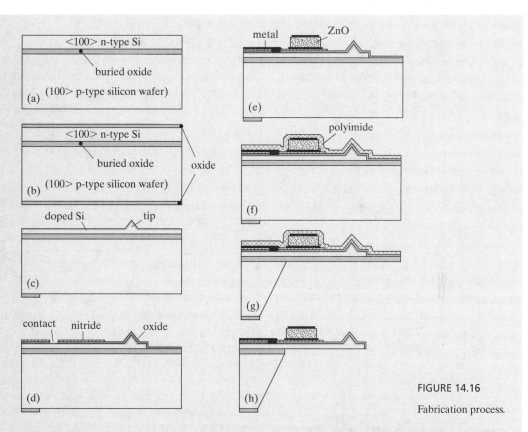

FIGURE 14.16

Fabrication process.

(Figure 14.16a). The early stage of this process overlaps with Case 14.2. A thermally grown silicon dioxide covers the front and the back (b). A double-sided lithography step is used to define oxide on the front and the back with proper registration. The patterned oxide on the front forms masks for anisotropic dry etching, whereas the patterned oxide on the back of the wafer forms masks for etching cavities in bulk silicon. Tips are made in the top-layer silicon using plasma etching. Once the tips are formed, they are sharpened by a wet oxidation at 950°C for two hours. An ion implantation at 80 keV and with a dose of 5×10^{14} ions/cm^2 is used to produce a doped layer of the piezoresistor (c).

In reality, the doping concentration will exhibit a distribution in the direction of the thickness. The effective gauge factor can be obtained by approximating the resistor in a layered fashion, with each layer having different doping concentrations. The exact solution can be facilitated by semiconductor process simulation tools (called Technology CAD or TCAD tools). For an order of magnitude estimate, it is an acceptable practice to divide the surface dose by the effective thickness to yield the volume density and find the piezoresistive coefficient based on the doping level [50].

Cantilevers are then photolithographically defined and etched. A 100-nm-thick oxide is grown on top of the cantilever. It serves many purposes, including passivation, electrical isolation, activation of dopants, and repairing surface damages caused by high-energy ion bombardment.

A 0.2-μm-thick LPCVD nitride is deposited. Its functions are discussed later. Contact holes are opened at the base of the cantilever so that the first metallization (consisting of 10 nm of Ti and 500 nm of gold) can make electrical contact with the silicon (d). This metal layer serves as a metal lead for the piezoresistor; the bottom electrode serves as the ZnO piezoelectric actuator. ZnO is deposited on the gold surface by DC magnetron sputtering to a thickness of 3.5 μm. Cr and Au layers are deposited on top of the ZnO layer again. Because ZnO reacts to photolithography chemicals, a lift-off process is used instead of development. The metal serves as the top electrode and also as a mask for etching the ZnO, which is performed using a solution consisting of 15 g $NaNO_3$, 5 ml HNO_3, and 600 ml H_2O (e).

The entire front side of the wafer is passivated with a layer of polyimide to protect against anisotropic etchants (f). Polyimide is used as the passivation because of its low temperature of deposition and the thickness of layer that can be reached. LPCVD thin films, such as silicon nitride and oxide, will not be good candidates for the passivation as their deposition temperatures approach the Curie temperature of ZnO. The bulk silicon is etched from the back side to form cavities until the buried oxide is reached (g). The oxide is removed using HF (h). Since HF solutions attack ZnO rather quickly, the LPCVD nitride layer underlying the ZnO plays the important role of shielding the ZnO during this etching step.

The cantilevers are 420 μm long with the ZnO actuator occupying 180 μm of the total length. Each leg of the cantilever is 37 μm wide and the full width of the cantilever is 85 μm. The total deflection of the cantilever is 1 μm for an applied electric field of 10^7 V/m.

The probes with integrated sensors and actuators are used for imaging surfaces. A feedback loop monitors the cantilever displacement with the piezoresistor and determines the voltage that the ZnO actuator needs to be biased with to maintain constant spacing between the tip and the sample substrate. However, the sensor and actuator are not completely decoupled. The portion of the piezoresistor underneath the ZnO piece generates a signal unrelated to the force applied to the tip. The probe needs to be electronically calibrated in order to compensate for the coupling. In a later work, the region directly beneath the ZnO actuator was heavily doped to reduce the piezoresistive sensitivity associated with that region by at least 80% [51].

Later, ZnO was used as both an actuator and a sensor, with the advantage of simplified design and fabrication by eliminating the piezoresistor element [52].

Case 14.6. SPM Nanolithography Probe with Thermal Actuation

Conventional DPN patterning is performed using a *single* AFM probe. With typical write speed of 0.1–5 μm/sec, the throughput is limited by the serial nature of the process. To increase the throughput of DPN writing, arrayed parallel probes are desired. DPN probe arrays fall into two categories: passive probes or active ones. In a passive probe array, all tips in the array move in unison and draw the same pattern. In active arrays, each probe is equipped with an actuator that allows it to be lifted away from the writing surface independently of the other probes. This allows the writing process for each probe to be turned on and off at will. The design and fabrication of typical passive probes are reviewed elsewhere [53].

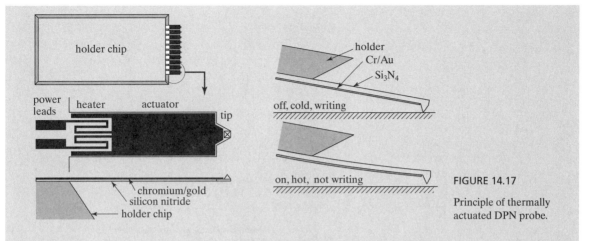

FIGURE 14.17
Principle of thermally actuated DPN probe.

There are several candidate methods for actuation: electrostatics, piezoelectricity, and thermal bending. Thermal bimetallic bending offers low-cost materials, simplicity of fabrication, and potentially large displacement. The piezoelectric actuation scheme requires complex processes, material processing expertise, and dedicated equipment. Electrostatic actuation generally exhibits a small displacement.

The schematic diagram of thermally actuated SPM probes is shown in Figure 14.17. Each probe consists of two layers: a metal layer and a silicon nitride one. Materials of these two layers have different thermal expansion coefficients. The metal layer is patterned to have thermal resistors and expansion patches. The serpentine gold wire at the base acts as the ohmic heater while the remaining gold acts with the silicon nitride beam as a bimorph thermal actuator. A tip contacts the writing surface when no heat is applied, and it pulls away from the writing surface upon provision of a heating current.

The design of an active DPN probe must satisfy several conflicting design criteria. For example, each probes must (1) generate enough force to overcome surface adhesion for lifting, (2) develop sufficient deflection to overcome surface topology; (3) not scratch the surface when pressed down to overcome array-to-surface misalignment, and (4) minimize the post-release curvature resulting from intrinsic thin-film stresses to simplify the instrument configuration. Successful geometric designs satisfying all major criteria were found through numerical analysis procedures [54].

The fabrication process for making a thermally actuated DPN (TA-DPN) probe is shown in Figure 14.18. The process begins with an oxidized <100>-oriented silicon wafer (Figure 14.18a). A protrusion tip is first made by anisotropic etching (Figure 14.18b). The apex of the pyramids are sharpened by two repetitions of thermal oxidation and oxide removal [55] (Figure 14.18c). A silicon nitride layer is then deposited by low-pressure chemical vapor deposition (LPCVD). It is patterned with reactive ion etching to form the cantilever beams (Figure 14.18d). Metal layers, including chromium and gold, are then deposited and patterned to form the metal heater leads and actuator (Figure 14.18e).

The Cr layer enhances the adhesion between the gold and the silicon substrate. Though the thickness of Cr is generally small, it nonetheless is responsible for introducing intrinsic stress and undesired bending.

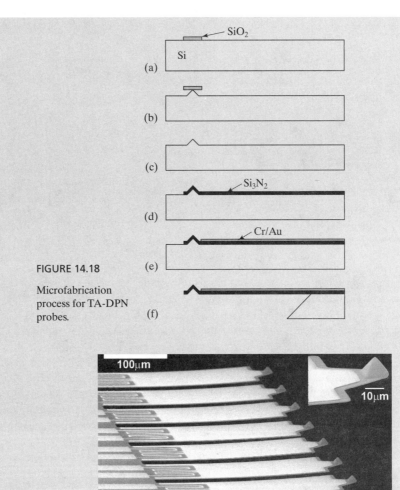

FIGURE 14.18

Microfabrication process for TA-DPN probes.

FIGURE 14.19

SEM micrograph of an active DPN array.

Finally, the beam is released from the substrate by anisotropic etching starting from the front side of the wafer (Figure 14.18f) using the silicon nitride layer as a mask.

An SEM micrograph of the resulting array design is shown in Figure 14.19. Each probe array consists of 10 individual probes on a single silicon chip. Tip-to-tip spacing is 100 μm, resulting in a 20-μm gap between individual probes. Each probe is 300 μm long (295 from the probe base to tip apex), 80 μm wide, and made of 9650-Å-thick silicon nitride and 3650-Å-thick gold with a 250-Å-thick chromium adhesion layer between them. The spring constant is 0.30 N/m by analytical calculation. The probe tips are approximately 5 μm tall.

Performance characteristics and demonstration of DPN writing has been conducted and reported [56].

SUMMARY

This chapter uses SPM as an example to illustrate design, material selection, and fabrication strategies. MEMS probes are uniquely qualified for SPM applications. Therefore, we discussed the basic principles of SPM technologies as well.

At the end of this chapter, the reader should understand the following concepts and facts:

- Basic operation principles of the scanning tunneling microscope and the atomic force microscope.
- Basic device architecture of an SPM probe and design considerations—force constant and resonant frequency.
- Eight general fabrication routes for realizing SPM probes with various materials.
- Methods for making SPM probes with silicon cantilevers under varying cantilever thicknesses.
- Motivation for including sensors and actuators on SPM probes.

PROBLEMS

Problem 14.1: Fabrication
Find a way to modify scheme 2 to produce an outward facing tip using the same steps. The modification would produce a convenient tip orientation with sharpness determined by the etch pits.

Problem 14.2: Design
Find the dimensions of a cantilever with a force constant of 0.5 N/m and a resonant frequency of 100 KHz out of single-crystal silicon, polyimide, and gold thin film ($=0.5\ \mu$m). Assume the Young's modulus of silicon is 160 GPa.

Problem 14.3: Fabrication
Develop a process for realizing an SPM probe with a gold tip and a polysilicon cantilever. Draw the cross section of the process in detail. Describe the process of each step, and clearly identify all layers involved. The shape of the tip should be similar to a pyramid with height of 5 μm or more maximize the tip sharpness.

Problem 14.4: Fabrication
Develop a fabrication process for realizing an SPM probe with a thin-film diamond tip and a polyimide cantilever. Draw the cross section of the process in detail. Investigate a method to deposit diamond on silicon wafers.

Problem 14.5: Fabrication
Develop a fabrication process for realizing an array of SPM probes with two types of SPM probes. One probe consists of a silicon nitride tip and a cantilever. A second probe consists of a conductive platinum tip and a silicon nitride cantilever. The two cantilevers have the same length. The thicknesses of the two types of probes are identical.

Problem 14.6: Fabrication
Review scheme 7 according to Figure 14.10 and determine the etchant and materials of each layer. Refer to Reference [39] and summarize the etching selectivity by the etchant on various exposed materials in each step (7.2 through 7.8).

Problem 14.7: Fabrication

According to Case 14.1, develop a process using the mold-and-transfer strategies for realizing a SPM probe with a silicon nitride tip. The cantilever should consist of a bimetallic actuator with gold and silicon nitride similar to the principle discussed in Case 14.6. Namely, the gold layer may consist of resistive heaters. The process should allow gold resistors to be connected with voltage or current supplies.

Problem 14.8: Design

For a comparison of the diffusion between Case 14.2 and Case 14.3, find the ratio of diffusivity of boron in single-crystal silicon at 1000°C and 800°C. Assuming the ion implantation results in identical surface concentration, find the ratio of the dopant concentration at a depth of 0.5 μm and time of 10 min. between these two cases.

Problem 14.9: Design

For Case 14.2, derive the equation for calculating the spring constant (Equation 1 in Ref. [48]).

Problem 14.10: Design

Find the correct expression for estimating the resonant frequency of the structure in Case 14.2. Based on known geometry [48], determine whether the experimentally measured resonant frequency agrees with the analysis. Discuss sources of any discrepancies.

Problem 14.11: Design

For Case 14.2, discuss the cross-sensitivity of the cantilever to temperature (%/°C) and to acceleration (%/g), and compare it with the force sensitivity.

Problem 14.12: Fabrication

In Case 14.4, an SPM probe with a 100-nm beam thickness is realized. However, SOI wafers are rather expensive. Discuss an alternative process to realize a 100-nm beam made of a different material, such as silicon nitride. Find the suitable displacement transduction material based on the piezoresistivity principle. Draw a fabrication process in detail. The overall thickness of the cantilever, together with displacement transduction material, should not exceed 150 nm. Discuss major performance aspects, including sensitivity, compared with the silicon beam shown in Case 14.4. Discuss the trade-offs of wafer cost and performance in this case.

Problem 14.13: Review

Based on Reference [47] in Case 14.5, find an expression of the tip displacement as a function of applied voltage. Compare the analysis result with experimental data in the paper. State your assumption.

Problem 14.14: Design

Derive an analytical expression for the thermal bimetallic active DPN probe discussed in Case 14.6, using dimensional information outlined in Reference [56].

Problem 14.15: Challenge

Develop a design and companion fabrication process of an SPM probe with thermal bimetallic actuation and an integrated displacement-sensing element. The dimensions of the probe should be identical to those in Reference [56]. Discuss coupling issues between the sensing and actuation functions.

REFERENCES

1. Horber, J.K.H., and M.J. Miles, *Scanning Probe Evolution in Biology*. Science, 2003. **302**(5647): p. 1002–1005.
2. Li, M.-H., J.J. Wu, and Y.B. Gianchandani, *Surface Micromachined Polyimide Scanning Thermocouple Probes*. Microelectromechanical Systems, Journal of, 2001. **10**(1): p. 3–9.

3. Luo, K., et al., *Sensor Nanofabrication, Performance, and Conduction Mechanisms in Scanning Thermal Microscopy.* Journal of Vacuum Science & Technology B: Microelectronics and Nanometer Structures, 1997. **15**(2): p. 349–360.
4. Varesi, J., and A. Majumdar, *Scanning Joule Expansion Microscopy at Nanometer Scales.* Applied Physics Letters, 1998. **72**(1): p. 37–39.
5. Novotny, L., D.W. Pohl, and B. Hecht, *Scanning Near-Field Optical Probe with Ultra-Small Spot Size.* Optics Letters, 1995. **20**: p. 970–972.
6. Wickramasinghe, H.K., *Contrast and Imaging Performance in the Scanning Acoustic Microscope.* Journal of Applied Physics, 1979. **50**(2): p. 664–672.
7. Stowe, T.D., et al., *Silicon Dopant Imaging by Dissipation Force Microscopy.* Applied Physics Letters, 1999. **75**(18): p. 2785–2787.
8. Fernandez, J.M., and H. Li, *Force-Clamp Spectroscopy Monitors the Folding Trajectory of a Single Protein.* Science, 2004. **303**(5664): p. 1674–1678.
9. Hembacher, S., F.J. Giessibl, and J. Mannhart, *Force Microscopy with Light-Atom Probes.* Science, 2004. **305**(5682): p. 380–383.
10. Rugar, D., et al., *Single Spin Detection by Magnetic Resonance Force Microscopy.* Nature, 2004. **430**: p. 329–332.
11. Su, M., S. Li, and V.P. Dravid, *Microcantilever Resonance-Based DNA Detection with Nanoparticle Probes.* Applied Physics Letters, 2003. **82**(20): p. 3562–3564.
12. Berger, R., et al., *Surface Stress in the Self-Assembly of Alkanethiols on Gold.* Science, 1997. **276**(5321): p. 2021–2024.
13. Wu, G., et al., *Bioassay of Prostate-Specific Antigen (PSA) Using Microcantilevers.* Nature Biotechnology, 2001. **19**(9): p. 856–860.
14. Raab, A., et al., *Antibody Recognition Imaging by Force Microscopy.* Nature Biotechnology, 1999. **17**(9): p. 901–905.
15. Yue, M., et al., *A 2-D Microcantilever Array for Multiplexed Biomolecular Analysis.* Microelectromechanical Systems, Journal of, 2004. **13**(2): p. 290–299.
16. Wilder, K., et al., *Cantilever Arrays for Lithography.* Naval Research Reviews, 1997. **VLIX**(1): p. 35–47.
17. Kramer, S., R.R. Fuierer, and C.B. Gorman, *Scanning Probe Lithography Using Self-Assembled Monolayers.* Chemical Reviews, 2003. **103**(11): p. 4367–4418.
18. Wang, D., et al., *Nanofabrication of Thin Chromium Film Deposited on Si(100) Surfaces by Tip Induced Anodization in Atomic Force Microscopy.* Applied Physics Letters, 1995. **67**(9): p. 1295–1297.
19. Minne, S.C., et al., *Atomic Force Microscope Lithography Using Amorphous Silicon as a Resist and Advanced in Parallel Operation.* Journal of Vacuum Science & Technology B: Microelectronics and Nanometer Structures, 1995. **13**(3): p. 1380–1385.
20. Park, S.W., et al., *Nanometer Scale Lithography at High Scanning Speeds with the Atomic Force Microscope Using Spin on Glass.* Applied Physics Letters, 1995. **67**(16): p. 2415–2417.
21. Koyanagi, H., et al., *Field Evaporation of Gold Atoms onto a Silicon Dioxide Film by Using an Atomic Force Microscope.* Applied Physics Letters, 1995. **67**(18): p. 2609–2611.
22. Avouris, P., et al., *STM-Induced H Atom Desorption from Si (100): Isotope Effects and Site Selectivity.* Chemical Physics Letters, 1996. **257**: p. 148–154.
23. Lyding, J.W., *UHV STM Nanofabrication: Progress, Technology Spin-Offs, and Challenges.* Proceedings of the IEEE, 1997. **85**(4): p. 589–600.
24. Ahn, C.H., K.M. Rabe, and J.-M. Triscone, *Ferroelectricity at the Nanoscale: Local Polarization in Oxide Thin Films and Heterostructures.* Science, 2004. **303**(5657): p. 488–491.
25. Minne, S.C., et al., *Fabrication of 0.1 µm Metal Oxide Semiconductor Field-Effect Transistors with the Atomic Force Microscope.* Applied Physics Letters, 1995. **66**(6): p. 703–705.

26. Snow, E.S., P.M. Campbell, and F.K. Perkins, *Nanofabrication with Proximal Probes*. Proceedings of the IEEE, 1997. **85**(4): p. 601–611.
27. Piner, R.D., et al., *"Dip-Pen" Nanolithography*. Science, 1999. **283**(5402): p. 661–663.
28. Terris, B.D., et al., *Near-Field Optical Data Storage Using a Solid Immersion Lens*. Applied Physics Letters, 1994. **65**(4): p. 388–390.
29. Vettiger, P., et al., *Ultrahigh Density, High-Data-Rate NEMS-Based AFM Storage System*. Microelectronic Engineering, 1999. **46**(1–4): p. 101–104.
30. Chui, B.W., et al., *Low-Stiffness Silicon Cantilevers for Thermal Writing and Piezoresistive Readback with the Atomic Force Microscope*. Applied Physics Letters, 1996. **69**(18): p. 2767–2769.
31. King, W.P., et al., *Design of Atomic Force Microscope Cantilevers for Combined Thermomechanical Writing and Thermal Reading in Array Operation*. Microelectromechanical Systems, Journal of, 2002. **11**(6): p. 765–774.
32. Lee, D.-W., et al., *Microprobe Array with Electrical Interconnection for Thermal Imaging and Data Storage*. Microelectromechanical Systems, Journal of, 2002. **11**(3): p. 215–221.
33. Marcus, R.B., et al., *Formation of Silicon Tips with <1 nm Radius*. Applied Physics Letters, 1990. **56**(3): p. 236–238.
34. Folch, A., M.S. Wrighton, and M.A. Schmidt, *Microfabrication of Oxidation-Sharpened Silicon Tips on Silicon Nitride Cantilevers for Atomic Force Microscopy*. Microelectromechanical Systems, Journal of, 1997. **6**(4): p. 303–306.
35. Akamine, S., and C.F. Quate, *Low Temperature Thermal Oxidation Sharpening of Microcast Tips*. Journal of Vacuum Science & Technology B, 1992. **10**(5): p. 2307–2310.
36. Albrecht, T.R., et al., *Microfabrication of Cantilever Styli for the Atomic Force Microscope*. Journal of Vacuum Science & Technology B, 1990. **8**(4): p. 3386–3396.
37. Albrecht, T.R., et al., *Method of Forming Microfabricated Cantilever Stylus with Integrated Pyramidal Tip*, in U.S. Patents and Trademarks Office. 1990, Board of Trustees of the Leland Stanford Junior University: USA.
38. Liu, C., and R. Gamble, *Mass Producible Monolithic Silicon Probes for Scanning Probe Microscopes*. Sensors and Actuators A: Physical, 1998. **71**(3): p. 233–237.
39. Williams, K.R., K. Gupta, and M. Wasilik, *Etch Rates for Micromachining Processing—Part II*. Microelectromechanical Systems, Journal of, 2003. **12**(6): p. 761–778.
40. Wang, X., et al., *Scanning Probe Contact Printing*. Langmuir, 2003. **19**(21): p. 8951–8955.
41. Petersen, K.E., *Silicon as a Mechanical Material*. Proceedings of the IEEE, 1982. **70**(5): p. 420–457.
42. Cooper, E.B., et al., *Terabit-per-Square-Inch Data Storage with the Atomic Force Microscope*. Applied Physics Letters, 1999. **75**: p. 3566–3568.
43. Lutwyche, M., et al., *5 × 5 2D AFM Cantilever Arrays a First Step Towards a Terabit Storage Device*. Sensors and Actuators A: Physical, 1999. **73**(1–2): p. 89–94.
44. Zhang, M., et al., *A MEMS Nanoplotter with High-Density Parallel Dip-Pen Nanolithography Probe Arrays*. Journal of Nanotechnology, 2002. **13**: p. 212–217.
45. Chow, E.M., et al., *Characterization of a Two-Dimensional Cantilever Array with Through-Wafer Interconnects*. Applied Physics Letters, 2002. **80**(4): p. 664–666.
46. Bullen, D., et al. *Micromachined Arrayed Dip Pen Nanolithography (DPN) Probes for Sub-100 nm Direct Chemistry Patterning*. In 16th International Conference on Micro Electro Mechanical Systems (MEMS). 2003. Kyoto, Japan.
47. Minne, S.C., S.R. Manalis, and C.F. Quate, *Parallel Atomic Force Microscopy Using Cantilevers with Integrated Piezoresistive Sensors and Integrated Piezoelectric Actuators*. Applied Physics Letters, 1995. **67**(26): p. 3918–3920.

48. Tortonese, M., R.C. Barrett, and C.F. Quate, *Atomic Resolution with an Atomic Force Microscope Using Piezoresistive Detection.* Applied Physics Letters, 1992. **62**(8): p. 834–836.
49. Harley, J.A., and T.W. Kenny, *High-Sensitivity Piezoresistive Cantilevers Under 1000-Angstrom Thick.* Applied Physics Letters, 1999. **75**(2): p. 289–291.
50. Kanda, Y., *Piezoresistance Effect of Silicon.* Sensors and Actuators A: Physical, 1991. **28**(2): p. 83–91.
51. Minne, S.C., et al., *Independent Parallel Lithography Using the Atomic Force Microscope.* Journal of Vacuum Science and Technology, 1996. **B 14**(4): p. 2456–2459.
52. Minne, S.C., et al., *Contact Imaging in the Atomic Force Microscope Using a Higher Order Flexural Mode Combined with a New Sensor.* Applied Physics Letters, 1996. **68**(10): p. 1427–1429.
53. Zhang, M., et al., *A MEMS Nanoplotter with High-Density Parallel Dip-Pen Nanolithography Probe Arrays.* Nanotechnology, 2002. **13**(2): p. 212–217.
54. Rozhok, S., R.D. Piner, and C.A. Mirkin, *Dip-Pen Nanolithography: What Controls Ink Transport?* Journal of Physical Chemistry B, 2003. **107**(3): p. 751–757.
55. Liu, C., and R. Gamble, *Mass-Producible Monolithic Silicon Probes for Scanning Probe Microscopes.* Sensors and Actuators: A Physical, 1998. **71**(3): p. 233–237.
56. Bullen, D., et al., *Parallel Dip-Pen Nanolithography with Arrays of Individually Addressable Cantilevers.* Applied Physics Letters, 2004. **84**(5): p. 789–791.

CHAPTER 15

Optical MEMS

15.0 PREVIEW

Optics is one of the earliest and most active areas to which MEMS technology has been applied [1]. Representative applications include digital light projection (DLP) [2], full-color digital displays [3], tunable reflectors for optical networking [4], integrated systems for biooptical detection [5], tunable optical sources and sensors [6–8], adaptive optics [9], fiber optics switches [10], filtering and wavelength tuning [11–13], free space communication [14], and retina raster scan display [15, 16].

There are several inherent advantages of using MEMS for optics applications. First, silicon microfabricated devices are adequately robust for optical applications because they interact only with photons, which have very little mass and exert little force on microstructures. Secondly, the packaging of optical MEMS is relatively straightforward. Optical MEMS components can be sealed in optically transparent housings that shield them from environmental elements such as particles, flow, and direct contact.

There are certainly challenges to applying MEMS technology in optics. For example, it is generally difficult to achieve a smooth mirror finish on micromachined parts, making many microoptical mirrors less than perfect. Curved refractive lenses, widely used in macroscopic optics, are generally difficult to realize using microfabrication methods. Optical MEMS devices also face competition from solid-state and optoelectronics counterparts, which have no moving mechanical elements. Although micromechanical components have relatively high mechanical resonant frequency, their response speed is still many orders of magnitude lower than optical and electrical modulation methods.

The optical MEMS field underwent a large-scale commercialization effort in the late 1990s. In the span of two years, many optical MEMS companies were created with venture capital backing to address the forecasted market need of channel switching for broadband fiber communications. The market need was not fully materialized at that time. This is one of the major factors that contributed to the later recession of commercialization efforts in this area. Nonetheless, many innovative technological solutions were created during this period and are reviewed in this chapter. The optical MEMS field played an important role in advancing the state-of-the-art technology of MEMS actuators.

Optical MEMS systems present challenges in areas of micromechanical design, microfabrication, and system integration. In Section 15.1, we will review designs of passive MEMS optical components to exemplify technology barriers and innovative microfabrication methods. Active electromechanical optical devices are very interesting from the perspective of actuator design and fabrication. In Section 15.2, we will review designs of integrated actuators for active optical MEMS devices.

15.1 PASSIVE MEMS OPTICAL COMPONENTS

One of the earliest applications of micromechanical structures for optics applications was anisotropically etched grooves for aligning optical fibers. In order to effectively couple light into the inner core of an optical fiber, the end of the fiber must be placed with a high degree of accuracy, both laterally and angularly, with respect to another fiber, an optical source, or a detector. Both the fiber cross section and the optical components are small, making it very laborious to perform free-space manual alignment.

Anisotropically etched grooves provide a miniature and precise alignment mechanism suited for the size of fibers. The schematics of using microstructures for the precision placement of fibers on a carrier (e.g., a silicon wafer) is shown in Figure 15.1 [17]. V-shaped SLSP-type grooves are etched in silicon substrates using anisotropic etching. Slanted walls of groove are formed in <111> surfaces. By placing an end segment of a fiber along the length of the groove and clamping on the fiber vertically, the position of a fiber is accurately fixed relative to the surface features on the wafer. This technique is especially useful for positioning multiple fibers. Incidentally, it is one of the earliest uses of anisotropic silicon etching. Auxiliary clamping beams have been integrated with the grooves to simplify the clamping procedure [18]. Later, the setup diagrammed in Figure 15.1 found other applications. For example, an array of short fiber segments in a linear array of grooves has been investigated for use as roller bearings for linear displacement stages [19].

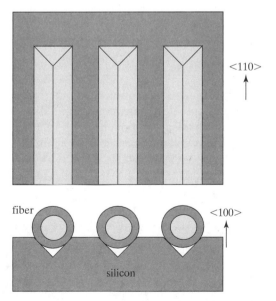

FIGURE 15.1

Passive optical alignment.

Many *passive* optical MEMS components have been developed for optical display and communication applications. Representative components and their applications include

Reflectors: display and beam steering;
Diffraction gratings: spectrometer and interferometers;
Lens elements: refractive lens, binary lens, and prisms;
Optical transmission lines: optical fiber and on-chip optical waveguides.

In this section, we discuss two representative *passive* optical devices—lenses and mirrors—to illustrate the unique design and processing issues pertaining to micromachined optical MEMS devices.

15.1.1 Lenses

The lens is the most recognizable component of an optical system. By necessity, a microoptical system must be able to incorporate lens elements with satisfactory performance. The most commonly encountered lenses in conventional optical systems are refractive lenses with two-dimensional or one-dimensional curved surfaces. Microfabrication and integration of curved *and* smooth surfaces using materials with desired optical properties (e.g., glass) face many obstacles.

A number of representative fabrication strategies for realizing the integrated refractive lens in optical microsystems are illustrated in Figure 15.2.

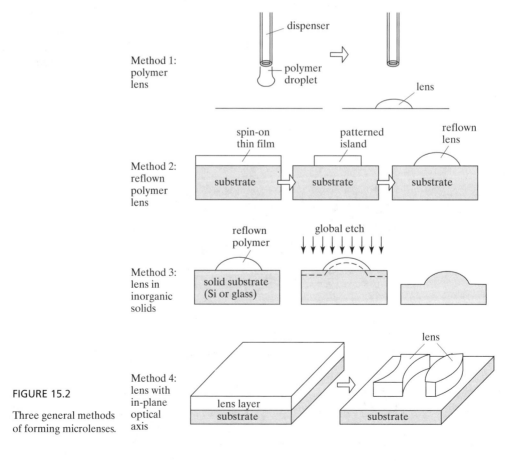

FIGURE 15.2

Three general methods of forming microlenses.

According to method 1 of Figure 15.2, curved spherical surfaces can be realized by dropping a precise amount of liquid (e.g., liquid form photoresist, polyimide, Teflon, or other polymer materials) onto a wafer surface. The droplets spread on the surface and then dry to form curved surfaces.

On the other hand, curved surfaces may be formed by the thermal reflow of polymers. According to method 2 of Figure 15.2, polymer thin films are first deposited and patterned photolithographically to form islands with a precision footprint and height. When the substrate is heated, the polymer island will reflow. The profile is controlled by the surface tension, the polymer viscosity, the thermal treatment temperature and duration, and the volume and initial footprint of the polymer islands. In Case 15.1, let us review a representative, quantitative research study on the light-focusing power of such lenses.

Case 15.1. Microfabricated Refractive Lenses

A study on the geometric profile and the optical characteristics of reflow lenses using spin-on photoresist has been conducted [20]. Before reflow, the shape of the patterned photoresist island is a column with height t and diameter D (Figure 15.3). The volume of the column is

$$V_{initial} = \frac{\pi}{4} D^2 t. \tag{15.1}$$

Assume the resist pattern reflows without changing the footprint diameter, which is a spherical shape with a radius of curvature R and a lens sag of s. The volume of the reflown lens is

$$V_{reflown} = \frac{\pi s}{24}(3D^2 + 4s^2). \tag{15.2}$$

By equating Equations (15.1) and (15.2) (ignoring the potential reduction of volume due to the loss of solvent), one can easily find the value of s:

$$s = A - \frac{1}{4}\frac{D^2}{A}, \tag{15.3}$$

where

$$A = \left(\frac{3}{4}tD^2 + \frac{1}{8}\sqrt{D^6 + 36t^2 D^4}\right)^{1/3}. \tag{15.4}$$

The focus length of such a planoconvex lens is $R/(n-1)$, where R is the radius of curvature and n the refractive index. The expression for R is

$$R = \frac{s}{2} + \frac{D^2}{8s}. \tag{15.5}$$

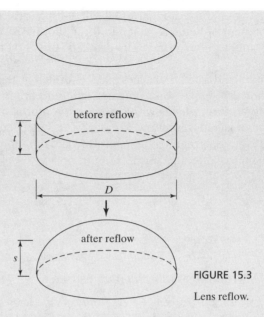

FIGURE 15.3

Lens reflow.

The formula for the focus length of the lens is therefore

$$f = \frac{1}{n-1}\left(\frac{s}{2} + \frac{D^2}{8s}\right). \tag{15.6}$$

Various techniques can be used to introduce controlled lens angles and spreading [21]. Demonstrated techniques include retention walls and introducing vapors during the reflow.

If the lens materials must be silicon and glass, alternative methods are needed. Lenses can be etched into silicon or glass by using global etching. Conventional lithography uses a flat thin-film pattern as the mask. Alternatively, the mask may exhibit a three-dimensional curved profile. This can be accomplished by using a polymer lens as the mask [22] (method 1 of Figure 15.2). Careful matching of the etch rate by plasma etching on the polymer and on the substrate material allows the faithful transfer of the 3D curved profile in the mask layer into an inorganic solid.

In certain applications, the optical path may be within the plane of the substrate. In these cases, two-dimensional lens profiles can be achieved by photolithography (method 4 of Figure 15.2). The curved profile of such lenses can be controlled to a large extent by the planar pattern. Even negative lenses can be made easily.

Curved lenses can also be made by molding. Molded lenses can be made of polymer materials or of solid materials such as silicon nitride. The lens mold can be made in a number of ways, for example, by using isotropically etched cavities in silicon [21].

Elastomeric polymers have been used for lens applications [23] (in addition to other applications, such as reconfigurable diffraction gratings [24]). Microfabricated refractive lenses with curved profiles can be made by employing a sealed compartment filled with a working liquid solution. The compartment has at least one surface made of a transparent, soft membrane. The membrane can be deflected by increasing the pressure of the liquid inside the lens or injecting more fluids.

In addition to the curved refractive lenses, lenses with planar profiles are also popular because of their reduced mass and limited thickness. Planarity of lens elements also suits microfabrication well. We will briefly discuss how a curved refractive lens can be approximated by a planar lens.

Since the refractive lens derives its light-bending power from the lens–media interface, a three-dimensional curved lens can be approximated by sections, each preserving the local graded profiles (Figure 15.4b). This type of lens can be fabricated by molding. Besides glass Fresnel lenses found in old lighthouses, plastic molded planar lenses are used as magnifiers and in overhead projectors. Because of the diffraction at the interface of curved sections, the imaging quality is not as good as with a continuous curved lens.

The segmented profile in Figure 15.4b still involves curved and smooth surfaces, which are difficult to make by microfabrication. Step-wise approximation of the curved surfaces is often used (Figure 15.4c) to be compatible with layer-by-layer microfabrication. Curved slopes can be realized using multiple photolithography steps, or by using gray-scale masks [25, 26]. The greater the number of steps, the better the optical efficiency. On the other hand, more lithography steps are involved. With each degree of approximation, the optical performance (such as transmission efficiency) decreases, whereas the ease of fabrication increases. One-level approximations using transparent materials (Figure 15.4d) or opaque materials (Figure 15.4e) are ultimately the simplest forms of Fresnel lenses. A planar lens made in polycrystalline silicon is shown in Figure 15.5.

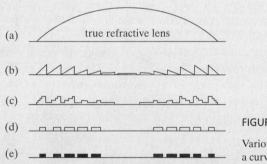

FIGURE 15.4

Various ways to approximate a curved refractive lens.

FIGURE 15.5

A microfabricated binary lens.

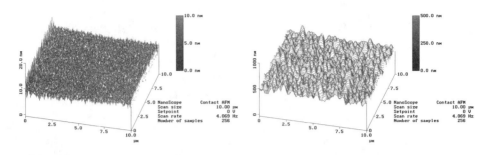

FIGURE 15.6

Surface profile of various reflector surfaces (area of scan = $10 \times 10 \ \mu m^2$).

15.1.2 Mirrors

The optical mirror is another important class of optical components. Smooth and flat reflective surfaces in MEMS are generally achieved through one of the following routes: (1) using polished front surfaces of single-crystal semiconductor wafers, with a room-mean-squared roughness of 1.2 nm; (2) using atomically smooth {111} [27, 28] silicon surfaces created by wet anisotropic etching [29]; (3) reducing mirror-plate curvatures [30, 31], or (4) using multilayer reflective coatings [31].

Many microoptical devices are made using surface micromachining technology. Polycrystalline silicon often serves as the material for the mirror plate. Polycrystalline silicon, however, exhibits finite surface roughness (its room-mean-square roughness is on the order of 7–40 nm, Figure 15.6). Although it is capable of satisfying some applications, there is room for improvement.

The processing requirement of reflective mirrors impacts their performance (reflectivity). It is time consuming to undercut and release large-area mirror plates. Etch holes are often used (Figure 15.8). A mirror surface with arrayed etch holes behaves differently than a continuous surface of the same area and material. Etch holes cause diffractions and the dissipation of energy. One past research study on the effect of etch holes on the performance of optical mirror is discussed in Case 15.2.

Case 15.2. Effect of Etch Holes on Optical Mirror Performance

The influence of etch holes on optical characteristics of reflective plates has been studied [32]. The diffraction by two-dimensional arrayed etch holes can be estimated by using the Fraunhofer (far field) diffraction theory under a collimated incident light beam. Figure 15.7a illustrates the geometric parameters in the diffraction grating. As shown in Figure 15.7a, the terms d_1, d_2 and l_1, l_2 are the spacings and the sizes of the etch holes in the x and y directions, respectively.

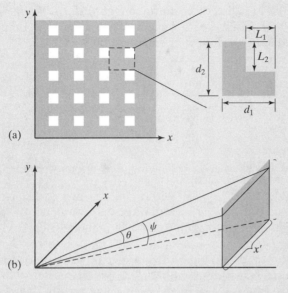

FIGURE 15.7

Schematic diagrams illustrating (a) geometric parameters of the etching holes, and (b) geometric parameters of the diffraction pattern.

FIGURE 15.8

A vertical micromirror with etch holes.

The coordinate system for analyzing the diffraction pattern of the normal incidence case is shown in Figure 15.7b. The diffraction pattern can be determined by

$$I(\theta, \psi) = I_0 \left[\frac{\sin\left(\frac{1}{2} N_1 k d_1 \sin\theta\right)}{\sin\left(\frac{1}{2} k d_1 \sin\theta\right)} \frac{\sin\left(\frac{1}{2} N_2 k d_2 \sin\psi\right)}{\sin\left(\frac{1}{2} k d_2 \sin\psi\right)} \right]^2$$

$$\sin c^2\left(\frac{1}{2} k l_1 \sin\theta\right) \sin c^2\left(\frac{1}{2} k l_2 \sin\psi\right), \quad k = \frac{2\pi}{\lambda}, \quad (15.7)$$

in which I_0 is the reflected intensity in the absence of the effect of etch holes. The terms N_1 and N_2 are the number of etch holes in the x and y directions, respectively. θ and ψ are the diffraction angles in the x and y directions, shown in Figure 15.7b. When z is much greater compared with x' and y', the position of the main maxima of the power intensity of diffraction can be approximated using the following equations (at $\theta_i = 0°$):

$$\sin \theta = \frac{m\lambda}{d_1} \approx \frac{x'}{z} \tag{15.8}$$

$$\sin \psi = \frac{n\lambda}{d_2} \approx \frac{y'}{z} \tag{15.9}$$

with m and n being the order of diffraction.

The diffraction influences the optical performance in two ways. First, the diffraction reduces the power in the zeroth order reflected beam and consequently the reflectivity of the mirror. Secondly, high-order diffraction beams can potentially cause cross talk in free-space optical systems.

To study the effects of the etch hole size and spacing on the reflectivity, two groups of polysilicon mirror surfaces with etch holes are used. One group (Group 1) contains mirror surfaces with etch holes of the same spacing (30 μm), but various sizes (6 × 6, 8 × 8, 10 × 10, 14 × 14, 15 × 15, 16 × 16, 18 × 18, 21 × 21, and 23 × 23 μm). Another group (Group 2) contains etch holes of identical sizes (diameter = 5 μm), but with different spacing (10 μm, 15 μm, and 20 μm).

The size of the etch holes will control the intensity distribution of the diffraction pattern, affecting both the power reflected from the mirror and the angular spread of the diffracted beams. Figure 15.9 presents the normalized intensity of the zeroth-order reflected beam obtained with Group 1 surfaces. The normalization was performed by dividing the intensity of the zero-order beam reflected from a polysilicon surface with etch holes by the intensity reflected from an area of the same wafer that does not have etch holes. This results from the fact that more light is transmitted through the etch holes, and more light is diffracted by etch holes as they get larger. The spacing of etch holes will also affect the reflectivity of micromachined mirrors. Figure 15.10 shows the measurement result of the reflectivity of the mirror surfaces in Group 2. As the spacing between etch holes increases, the reflectivity of the mirror surface increases.

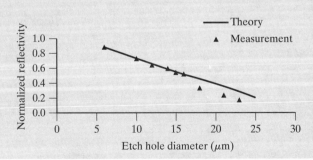

FIGURE 15.9

The reflectivity of polysilicon micromirrors with etch holes of different sizes at 632.8 nm (the spacing of etch holes is all 30 μm).

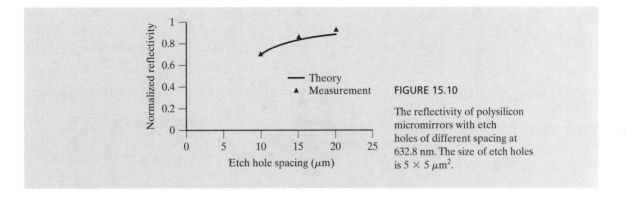

FIGURE 15.10

The reflectivity of polysilicon micromirrors with etch holes of different spacing at 632.8 nm. The size of etch holes is $5 \times 5 \ \mu m^2$.

15.2 ACTUATORS FOR ACTIVE OPTICAL MEMS

Active optical MEMS components, including reflectors, shutters, and gratings, are widely used. These elements require on-chip micromachined actuators. There are two categories of motion desired for optical MEMS applications: rotational movement and translational movement. The axis of the rotation or translation may be normal to the substrate plane or lie within the plane. Further, the range of displacement could be either large or small. A translational actuator may move by small distances (e.g., less than the optical wavelength) or large ones (e.g., a few micrometers to tens of micrometers). Likewise, a rotational actuator may move by small angles (less than 1°) or large ones (tens of degrees). The range of displacement leads to completely different designs and fabrication processes.

Out of all possible types of actuations, the ones for small motion are the most straightforward to design. For example, small in-plane displacement of micromechanical structures carrying optical elements can be generated using comb-drive actuators.

Representative designs of actuators and types of displacement achievable are summarized in Table 15.1.

In Section 15.2.1, we will review a few representative mechanisms for generating small out-of-plane translational motion.

Small angular displacement can be created by using mechanisms for the small translational movement, and by coupling translation actuators to rotational platforms. In addition, small angular displacement can be generated with magnetic actuation.

TABLE 15.1 Actuator Principles and Designs for Small Translational Movements.

Actuation principle	Specific designs	Common displacement mode
Electrostatic actuation	Transverse comb drive	Small, in-plane translation
	Longitudinal comb drive	Small, in-plane translation
	Parallel plate capacitive drive	Small, out-of-plane translation
Thermal actuation	Thermal bimetallic actuation	Small in-plane and out-of-plane translations
	Thermal expansion actuation with single material	Small, in-plane translation
Piezoelectric actuation	Piezoelectric beam actuators	Small, in-plane translation

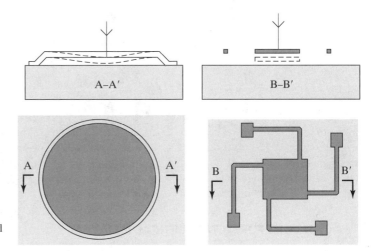

FIGURE 15.11

Two types of translational mirrors.

Actuators for large out-of-plane rotation and large in-plane translation are the most challenging, especially if the displacement is to be accomplished under steady-state conditions instead of under resonance. Resonant microactuators can achieve a large displacement by taking advantage of motion amplification at resonance. We will review design strategies and cases under a few selected actuator categories—namely, actuators for large in-plane translation (15.2.2), and large out-of-plane rotation (15.2.3).

15.2.1 Actuators for Small Out-of-Plane Translation

The translational, out-of-plane movement of mirrors can be realized using many actuation principles, including electrostatic and thermal actuation. Piston motion membranes or rigid plates are most often countered (Figure 15.11). Two piston-motion microreflectors are discussed next. One is used for modifying the phase of the reflected wave front (Case 15.3), whereas another is made using compound semiconductor thin films and is integrated on a compound semiconductor substrate (Case 15.4).

Case 15.3. Membrane-Type Translational Mirror

Adaptive optics refer to the control of optical wave-front phase in a real-time, closed-loop fashion. A typical adaptive optical system consists of a deformable mirror, a wave-front sensor, and a real-time controller used to modulate the spatial phase of the optical wave front. A research group at Boston University developed membrane-type micromirrors for this purpose [33]. The micromirrors employ a flexible silicon membrane with periodic mechanical attachments to a continuous membrane serving as the mirror (Figure 15.12). A 10×10 array of electrostatic actuators supports a continuous membrane serving as mirrors. The design using a continuous reflective surface increases the fill factor and enhanced reflectivity.

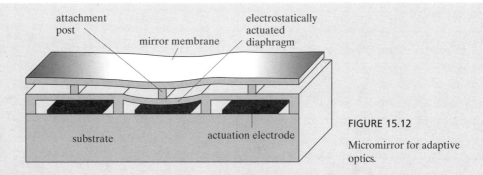

FIGURE 15.12

Micromirror for adaptive optics.

Each actuator consists of a 300 μm × 300 μm polycrystalline silicon membrane. The stroke distance is 1.9 μm (at 241 V) with a repeatability/resolution of 10 nm. The actuator can be operated between DC to 7 kHz in air.

Case 15.4. Micromirrors Integrated with Solid-State Light Sources

The resonant-cavity light emitting diode is a potential light source in applications such as wavelength division multiplexing, optical fiber communications, and free-space optical interconnects. Broad and continuous wavelength tunability is an attractive feature in many of these applications. However, because the resonant wavelength is determined by the refractive indexes and fixed layer thickness, the tuning of wavelengths beyond a few nm has never been achieved. A micromechanical solution is therefore implemented [13].

The device utilizes a deformable membrane mirror suspended by an air gap above a diode active region and bottom mirror. Applied membrane-substrate bias produces an electrostatic force which reduces the air-gap thickness and the resonant wavelength.

The device consists of a back mirror made of 22.5 period distributed Bragg reflectors (DBR) with a center wavelength near 960 nm (λ_0). The cavity consists of three 5-nm $In_{0.2}Ga_{0.8}As$/5-nm GaAs quantum wells paced within a pin diode. The moving membrane consists of a semitransparent gold reflector/electrode on top of an SiN_xH_y phase-matching layer and a GaAs cap, which is suspended above the semiconductor cavity by an air gap that can be controlled electrostatically. The top mirror provides a total reflectance of approximately 95% when the gap thickness is modulated around $3\lambda_0/4$.

The fabrication involves non-conventional materials and processes compared with established silicon micromachining practices. First, a back mirror is achieved by depositing multiple epitaxial layers using molecular beam epitaxy on n$^+$ (001) GaAs substrates. The following layers are grown on top of the back mirror: the p-i-n diode region, the 10-nm GaAs p$^+$ contact layer, and the $Al_{0.85}Ga_{0.15}As$ sacrificial layer (0.87 μm thick). This AlGaAs sacrificial layer is capped by GaAs (10 nm thick) and then SiN_xH_y (0.228 μm thick) deposited using the PECVD method. The intrinsic stress of the PECVD layer is calibrated

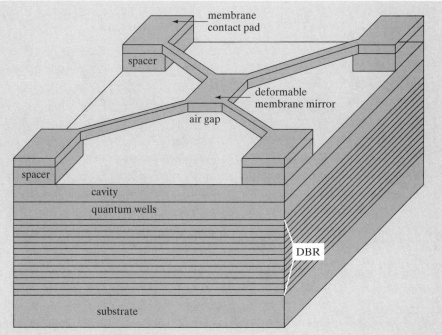

FIGURE 15.13
Schematic diagram.

(tensile, 350 MPa). An ohmic metal contact is evaporated on the back side of the contact (AuGe-Ni-Au). A Ti-Au contact pad on the membrane and gold membrane reflector were evaporated on the top mirror and patterned via lift-off. Proton implantation was then used to define the active regions of each device. The membrane and suspension cantilevers were defined in the SiN_xH_y layer. The sacrificial layer is removed by a mixture of HCl and water. During the sacrificial removal, photoresist is used around the membrane supports to form a supporting rail to prevent stiction. The photoresist is later removed using oxygen plasma.

To characterize this device, a bias current (10 mA) was injected into the active region. Light was collected over a 10° solid angle into a spectrometer. The wavelength can be tuned approximately 39 nm for a 0 to 18 V bias. The minimum spectral linewidth is approximately 1.9 nm at 957 nm, corresponding to an air-gap thickness of $3\lambda_0/4$.

15.2.2 Actuators for Large In-Plane Translation Motion

Optical elements (such as mirrors, shutters, and lenses) with their optical axes parallel to the substrate plane are sometimes mounted on in-plane translational stages. If the stage movement is small (for example, less than a few micrometers), it can be achieved using an electrostatic comb drive or thermal bimetallic actuation.

A few examples of electrostatic actuators are reviewed in Section 4.5.2. If the stage movement is large, it can be achieved using actuators with long working distances. A number of strategies for achieving long-range in-plane motions exist, including inchworm actuators with friction pads driven electrostatically [34, 35] or thermally [36], impact drives (vibromotors) driven individually [37–39] or globally (e.g., by acoustic vibration with resonance selection) [40],

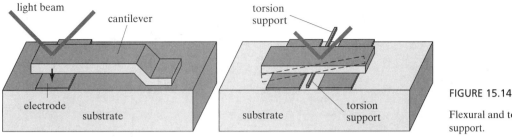

FIGURE 15.14

Flexural and torsional support.

scratch drive actuators [41, 42], and shuffle motors [43]. In the case of impact drive, the long term reliability has been studied under various impact strengths [44] and operations up to 72 continuous hours have been realized.

15.2.3 Actuators for Out-of-Plane Rotation

Rotational scanning micromirrors are used to reflect light and dynamically alter light paths. Such mirrors find many applications in industrial, civilian, and military applications. One of the most well-known devices involving scanning micromirrors is the digital light processor (DLP), which utilizes large two-dimensional arrays of digital micromirrors, each capable of rotating +/−7.5° around a common axis. However, the DLP micromirrors can only achieve bistable positioning and the rotating angle is rather limited. Many MEMS micromirrors with continuous positioning capabilities and much greater rotating angles have been developed and used for a variety of applications, ranging from optical fiber communication to retina projection systems.

To achieve greater bending angles of optical elements with surface areas greater than the DLP mirrors is more challenging, especially under practical voltage, current, or power consumption. Further, bending under static loading is more challenging than resonant bending, because resonant bending magnitude can be amplified by a factor equal to the quality factor.

Two of the most commonly used rotation mechanisms for small-angle rotations are based on flexural cantilevers or torsional bars actuated with electrostatic forces [45–47], as shown in Figure 15.14. Surface micromachining limits the gap size and the ultimate angular displacement that is achievable. It is possible to remove part of the substrate in order to provide more clearance for the angular movement of the mirror. However, the separation between the mirror and the counter electrodes increases along with driving voltages.

Comb drives generally are prepared in two configurations—lateral and transverse comb drives. Both are useful for producing in-plane motions. Can one use the comb drive to produce out-of-plane and rotational movement? One excellent example in this area is discussed in Case 15.5. For electrostatic actuators, reducing the driving voltage is important. The example in Case 15.6 achieves a large bending angle while requiring only a modest or a low-voltage bias.

Case 15.5.   Large-Angle Mirrors

One approach is to offset the planes of two sets of comb fingers. A rotational mirror based on this concept is illustrated in Figure 15.15 [7]. The fixed set of comb fingers and the movable fingers are shifted out of plane of each other. An electrical field between these two

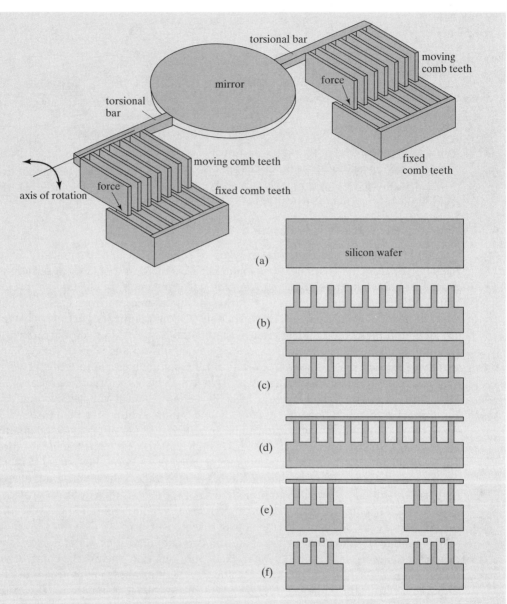

FIGURE 15.15

Micromirrors.

sets of combs will produce an out-of-plane force and a torque around the torsional axis. An angular displacement can be produced, with the magnitude depending on the geometries of the comb fingers and the frequency of operation.

The fabrication process of the device is illustrated in Figure 15.15. It starts with a silicon wafer. The front side of the wafer is patterned and etched using deep reactive ion etching to produce the fixed comb teeth (step b). A second wafer is bonded to the front side of the first wafer (step c), and then polished to a certain thickness, which corresponds to the thickness of the movable fingers (step d). The wafer assembly is turned upside down, and patterned from the back side. Deep reactive ion etching is used to remove part of the bulk silicon to allow for the clearance for mirror movement (step e). The bonded wafer is then patterned from the top to define comb fingers (step f).

There were two major reasons for selecting such a complex process flow, which involve bulk silicon etching and bonding. (1) The mirror is fabricated in single-crystal silicon material, which has very low intrinsic stress and warping. (2) The polishing step that reduces the thickness of the wafer also provides smooth finishes of the wafer.

The fabrication process discussed previously involved bulk etching and wafer bonding. The process to produce spatially offset comb fingers with accurate multilayer registration requires specialized equipment and has limited accuracy. Alternative processes combining bulk and surface micromachining to realize a vertical comb array have been demonstrated [48]. One process, called trench-refilled-with-polysilicon (TRiPs), combines deep reactive ion etching with sacrificial processes [49]. Models for designing vertical comb-drive devices have been developed as well [50].

The fabrication of the two-layer comb-drive fingers is time consuming. Another solution is to fabricate an angular vertical comb drive (AVC) in the same plane and strategically bend one set of comb fingers out of plane to achieve an angular offset [51, 52]. The schematic diagram of this method is shown in Figure 15.16, with the out-of-plane bending achieved using intrinsic stress in material layers.

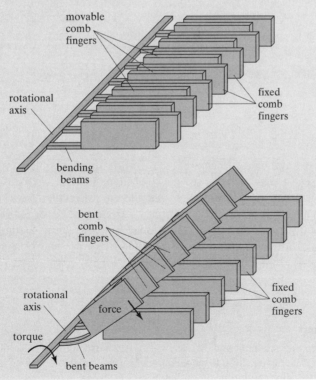

FIGURE 15.16

Comb drive with bent fingers.

Case 15.6. Low-Voltage, Large Angular Rotation

Is it possible to achieve a large, steady-state angular displacement using structures produced by surface micromachining only? Further, is it possible to achieve this type of actuation with low driving voltages (e.g., tens of volts)? The answers to these questions are "yes." Let us review one example in the following.

Certain elements such as membranes and fixed–free cantilevers can be pulled down towards the substrate by electrostatic forces. Though the vertical displacement is small, big angular displacement could result in certain regions of the structure, such as the perimeter of a clamped membrane or near the anchor of a cantilever (Figure 15.17). If the mirror is connected to these regions, a large angular steady-state mirror displacement can be achieved. For certain structures, the voltage applied needs to exceed what is required to cause a point contact or pull in. The contact regions between the suspended membrane or cantilever with the bottom electrode increases with applied voltage. This phenomenon is called the *zipping motion*. A familiar electrostatic actuator, the scratch drive actuator (SDA), utilizes the zipping motion. (See Section 4.3.5) [53].

Alternatively, large angular displacement can be realized using a multilever linkage angular motion amplifier [54, 55]. Tip displacement on the order of 14 mm can be achieved with a bias voltage of 35 V.

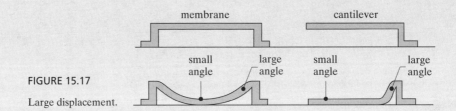

FIGURE 15.17

Large displacement.

SUMMARY

Optical MEMS is an expanding research area. It encompasses many different types of elements. This chapter did not seek to exhaustively review all of them. Instead, we focused on actuators for positioning optical elements, as optical devices present unique challenges to the state of the art of microactuation. In addition, we discussed representative methods for realizing optical mirrors and lenses, both critically important for any functional optical system. Readers who are interested in exploring further are encouraged to read related literature and books referenced in this chapter.

PROBLEMS

Problem 15.1: Review

Form a group of three to four students, and discuss two alternative technologies for the retinal display discussed in Reference [16]. Compare these three technologies (one being MEMS mirrors) in terms of cost, performance, system simplicity, resolution, and reliability.

Problem 15.2: Review

A Fabry–Perot interferometer can be used as a frequency selection device. MEMS technology can be used to realize tuning Fabry–Perot filters. Review the literature and prepare a list of major factors that affect the performance of a Fabry–Perot filter, including sharpness of filtering, range of selection, etc. Discuss how these performance concerns may dictate the fabrication process of a MEMS Fabry–Perot interferometer.

Problem 15.3: Design

Given a torsional mirror with electrostatic actuation, find the expression of the angular displacement as a function of the applied voltage. Assume the mirror plate is rigid and does not deform.

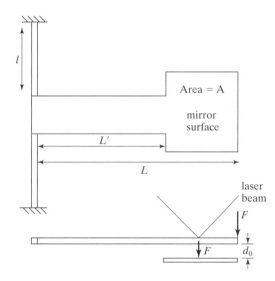

FIGURE 15.18
An electrostatically actuated micromirror.

Problem 15.4: Design

Form a team of three to four students and develop a conceptual design for an optical micromirror with a small in-plane displacement (up to 2 μm) created by using a parallel-plate capacitor. Develop the major processing steps of a companion microfabrication process.

Problem 15.5: Review

For Case 15.3, develop an analytical formula for the displacement of a single membrane under applied voltage.

Problem 15.6: Review

For the micromirror discussed in Case 15.4, it is known that the length of the cantilever is approximately 100 μm. The thickness of the suspension beam can be obtained from the description of the process—it consists of 0.228 mm of silicon nitride (SiN_xH_y) and a layer of gold conductor with unknown thickness. Let's assume the gold thickness is 20 nm and the center plate has an area of 40 × 40 μm². The sacrificial layer determines the gap size. The displacement is known to be 0.3 μm when the bias voltage is approximately 17.5 V. Find the exact width of the cantilever to match the measured performance. State all assumptions.

Problem 15.7: Design

Any micromechanical device is susceptible to vibration. For Case 15.4, find the wavelength cross-sensitivity to vertical acceleration (1 g). Does it constitute a major design concern? Find the plate position cross-sensitivity to the in-plane acceleration (1 g) along one of the edges of the plate.

Problem 15.8: Review

Draw detailed fabrication diagram for the device discussed in Case 15.4. Don't include details of lithography. Draw the cross-section across the center of the mirror. Include the length of the cantilever in the process diagram.

Problem 15.9: Review

Review the design of a large-displacement out-of-plane mirror by a thermal bimetallic actuator with pre-stressed bending [56]. Derive the analytical formula for the vertical displacement and calculate the order of magnitude based on known parameters in the paper.

Problem 15.10: Review

Review the impinging drive technology and compare it with the SDA in terms of speed, output force, and long-term reliability when they are made of the same material (e.g., polysilicon).

Problem 15.11: Design

Form a team of three to four students, and develop an actuator design that can achieve a large angular rotation in-plane (with a rotational axis perpendicular to the substrate plane). Investigate at least three different mechanisms and designs. Compare their relative advantages and disadvantages.

Problem 15.12: Design

One representative work on actuators with multiple degrees-of-freedom is discussed below, where a binary lens is attached to a stage capable of both translation and rotational (in one axis) movement [58]. The schematic diagram of the lens is shown in Figure 15.19. The entire device is made with surface micromachining technology and polycrystalline silicon as the structural material. The lens assembly consists of five polysilicon pieces that are connected with rotational hinges. The outermost pieces incorporate scratch drive actuators, allowing them to move in two directions. With multiple actuators, three modes of movement are possible. (1) If the two actuators move in opposite directions and by the same amount, the lens piece will change its elevation.

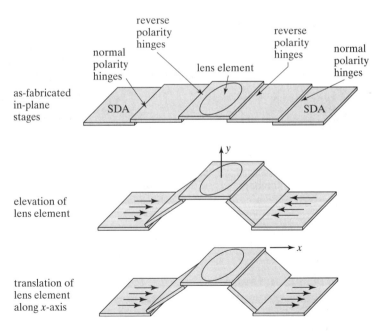

FIGURE 15.19

Principle of out-of-plane lens.

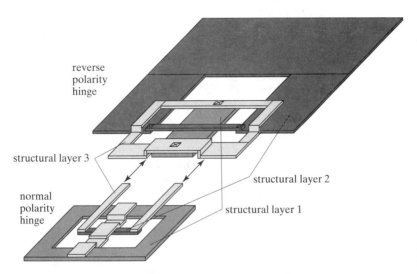

FIGURE 15.20

Detailed diagram of positive- and reverse-polarity hinges.

(2) If the two actuators move in opposite directions but by a different amount, the lens piece will rotate. (3) If the two actuators move in the same direction and by the same amount, the lens will translate.

The hinges connected to the SDA actuators are of normal polarity, whereas the hinges connected to the lens are of reverse polarity. A detailed drawing of the normal and reverse polarity hinges are shown in Figure 15.20, using a surface micromachining process that involves three structural layers.

Design a mask for the integrated lens with multiple degrees-of-freedom; focus on the actuators and parts linkage. The design must show the details of the SDA actuators and two types of hinges. Can this design be made using the MUMPS process discussed in Chapter 10? If so, design the mask according to the MUMPS design template and under the MUMPS design rules.

Problem 15.13: Challenge

Form a team and design a micromachined out-of-plane actuation with a displacement range under the DC bias capable of reaching 100 μm or greater. The displacement must be repeatable. The displacement must be able to move a smooth optical mirror in a translational path. The design objectives are (1) to minimize driving voltage, (2) to minimize driving power, and (3) to use the simplest designs possible.

REFERENCES

1. Motamedi, M.E., *Micro-Opto-Electro-Mechanical Systems*, Optical Engineering, 1994. **33**: p. 3505–3517.
2. Van Kessel, P.F., et al., *A MEMS-Based Projection Display*, Proceedings of the IEEE, 1998. **86**: p. 1687–1704.
3. Bloom, D.M., *Grating Light Valves for High Resolution Displays*. Presented at Electron Devices Meeting, 1994. Technical Digest, International, 1994.
4. Grade, J.D., H. Jerman, and T.W. Kenny, *Design of Large Deflection Electrostatic Actuators*, Microelectromechanical Systems, Journal of, 2003. **12**: p. 335–343.
5. Yue, M., et al., *A 2-D Microcantilever Array for Multiplexed Biomolecular Analysis*, Microelectromechanical Systems, Journal of, 2004. **13**: p. 290–299.
6. Syms, R.R.A., and A. Lohmann, *MOEMS Tuning Element for a Littrow External Cavity Laser*, Microelectromechanical Systems, Journal of, 2003. **12**: p. 921–928.

7. Conant, R., *Thermal and Electrostatic Microactuators*. In Electrical Engineering and Computer Sciences. Berkeley: University of California at Berkeley, 2002.
8. Lee, A.P., et al., *Vertical-Actuated Electrostatic Comb Drive with in situ Capacitive Position Correction for Application in Phase Shifting Diffraction Interferometry*, Microelectromechanical Systems, Journal of, 2003. **12**: p. 960–971.
9. Perreault, J.A., et al., *Adaptive Optic Correction Using Microelectromechanical Deformable Mirrors*, Optical Engineering, 2002. **41**: p. 561–566.
10. Chan, R.T., H. Nguyen, and M.C. Wu, *A High-Speed Low Voltage Stress-Induced Micromachined 2 × 2 Optical Switch*, IEEE Photonics Technology Letters, 1999. **11**: p. 1396–1398.
11. Vail, E.C., et al., *GaAs Micromachined Widely Tunable Fabry–Perot Filters*, Electronics Letters, 1995. **31**: p. 228–229.
12. Mateus, C.F.R., et al., *Widely Tunable Torsional Optical Filter*, IEEE Photonics Technology Letters, 2002. **14**: p. 819–821.
13. Larson, M.C., and J. S. Harris, *Broadly-Tunable Resonant-Cavity Light-Emitting Diode*, IEEE Photonics Technology Letters, 1995. **7**: p. 1267–1269.
14. Zhou, L., J.M. Kahn, and K.S.J. Pister, *Corner-Cube Retroreflectors Based on Structure-Assisted Assembly for Free-Space Optical Communication*, Microelectromechanical Systems, Journal of, 2003. **12**: p. 233–242.
15. Conant, R.A., et al., *A Raster-Scanning Full-Motion Video Display Using Polysilicon Micromachined Mirrors*, Sensors and Actuators A: Physical, 2000. **83**: p. 291–296.
16. Lewis, J.R., *In the Eye of the Beholder*, Spectrum, IEEE, 2004. **41**: p. 24–28.
17. Armiento, C.A., et al., *Passive Coupling of InGaAsP/InP Laser Array and Singlemode Fibres Using Silicon Waferboard*, Electronics Letters, 1991. **27**: p. 1109–1111.
18. Strandman, C., and Y. Backlund, *Bulk Silicon Holding Structures for Mounting of Optical Fibers in v-Grooves*, Microelectromechanical Systems, Journal of, 1997. **6**: p. 35–40.
19. Feinerman, A.D., and S.R. Thodati, *Millimeter-Scale Actuator with Fiber-Optic Roller Bearings*, Microelectromechanical Systems, Journal of, 1995. **4**: p. 28–33.
20. Toshiyoshi, H., et al., *A Surface Micromachined Optical Scanning Array Using Photoresist Lenses Fabricated by a Thermal Reflow Process*, Journal of Lightwave Technology, 2003. **21**: p. 1700–1708.
21. Fletcher, D.A., et al., *Microfabricated Silicon Solid Immersion Lens*, Microelectromechanical Systems, Journal of, 2001. **10**: p. 450–459.
22. Savander, P., *Microlens Arrays Etched into Glass and Silicon*, Optics and Lasers in Engineering, 1994. **20**: p. 97–107.
23. Grzybowski, B., et al., *Elastomeric Optical Elements with Deformable Surface Topographies: Applications to Force Measurements, Tunable Light Transmission and Light Focusing*, Sensors and Actuators A: Physical, 2000. **86**: p. 81–85.
24. Schueller, O.J.A., et al., *Reconfigurable Diffraction Gratings Based on Elastomeric Microfluidic Devices*, Sensors and Actuators A: Physical, 1999. **78**: p. 149–159.
25. Morgan, B., et al., *Development of a Deep Silicon Phase Fresnel Lens Using Gray-Scale Lithography and Deep Reactive Ion Etching*, Microelectromechanical Systems, Journal of, 2004. **13**: p. 113–120.
26. Babin, S., M. Weber, and H.W.P. Koops, *Fabrication of a Refractive Microlens Integrated onto the Monomode Fiber*. Presented at The 40th International Conference on Electron, Ion, and Photon Beam Technology and Nanofabrication, Atlanta, Georgia (USA), 1996.
27. Strandman, C., et al., *Fabrication of 45-deg Mirrors Together with Well-Defined v-Grooves Using Wet Anisotropic Etching of Silicon*, Microelectromechanical Systems, Journal of, 1995. **4**: p. 213–219.
28. John, P.M.S., et al., *Diffraction-Based Cell Detection Using a Microcontact Printed Antibody Grating*, Analytical Chemistry, 1998. **70**: p. 1008–1111.

29. Tan, S.-S., et al., *Mechanisms of Etch Hillock Formation*, Microelectromechanical Systems, Journal of, 1996. **5**: p. 66–72.
30. Min, Y.-H., and Y.-K. Kim, *Modeling, Design, Fabrication and Measurement of a Single Layer Polysilicon Micromirror with Initial Curvature Compensation*, Sensors and Actuators A: Physical, 1999. **78**: p. 8–17.
31. Cao, K., W. Liu, and J.J. Talghader, *Curvature Compensation in Micromirrors with High-Reflectivity Optical Coatings*, Microelectromechanical Systems, Journal of, 2001. **10**: p. 409–417.
32. Zou, J., et al., *Effect of Etch Holes on the Optical Properties of Surface Micromachined Mirrors*, IEEE/ASME Journal of Microelectromechanical Systems (JMEMS), 1999. **8**: p. 506–513.
33. Perreault, J.A., et al., *Adaptive Optic Correction Using Microelectromechanical Deformable Mirrors*, Optical Engineering, 2002. **41**: p. 561–566.
34. deBoer, M.P., et al., *High-Performance Surface-Micromachined Inchworm Actuator*, Microelectromechanical Systems, Journal of, 2004. **13**: p. 63–74.
35. Yeh, R., S. Hollar, and K.S.J. Pister, *Single Mask, Large Force, and Large Displacement Electrostatic Linear Inchworm Motors*, Microelectromechanical Systems, Journal of, 2002. **11**: p. 330–336.
36. Maloney, J.M., D.S. Schrelber, and D.L. DeVoe, *Large-Force, Electrothermal Linear Micromotors*, Journal of Micromechanics and Microengineering, 2004. **14**: p. 226–234.
37. Daneman, M.J., et al., *Linear Microvibromotor for Positioning Optical Components*, Microelectromechanical Systems, Journal of, 1996. **5**: p. 159–165.
38. Mita, M., et al., *A Micromachined Impact Microactuator Driven by Electrostatic Force*, Microelectromechanical Systems, Journal of, 2003. **12**: p. 37–41.
39. Ohmichi, O., Y. Yamagata, and T. Higuchi, *Micro Impact Drive Mechanisms Using Optically Excited Thermal Expansion*, Microelectromechanical Systems, Journal of, 1997. **6**: p. 200–207.
40. Saitou, K., D.-A. Wang, and S.J. Wou, *Externally Resonated Linear Microvibromotor for Microassembly*, Microelectromechanical Systems, Journal of, 2000. **9**: p. 336–346.
41. Akiyama, T., and K. Shono, *Controlled Stepwise Motion in Polysilicon Microstructures*, Microelectromechanical Systems, Journal of, 1993. **2**: p. 106–110.
42. Akiyama, T., D. Collard, and H. Fujita, *Scratch Drive Actuator with Mechanical Links for Self-Assembly of Three-Dimensional MEMS*, Microelectromechanical Systems, Journal of, 1997. **6**: p. 10–17.
43. Lammerink, T., et al., *Modeling, Design, and Testing of the Electrostatic Shuffle Motor*, Sensors and Actuators A, 1998. **70**: p. 171–178.
44. Lee, A.P., and A.P. Pisano, *Repetitive Impact Testing of Micromechanical Structures*, Sensors and Actuators A: Physical, 1993. **39**: p. 73–82.
45. Toshiyoshi, H., and H. Fujita, *Electrostatic Micro Torsion Mirrors for an Optical Switch Matrix*, Microelectromechanical Systems, Journal of, 1996. **5**: p. 231–237.
46. Hao, Z., et al., *A Design Methodology for a Bulk-Micromachined Two-Dimensional Electrostatic Torsion Micromirror*, Microelectromechanical Systems, Journal of, 2003. **12**: p. 692–701.
47. Greywall, D.S., et al., *Crystalline Silicon Tilting Mirrors for Optical Cross-Connect Switches*, Microelectromechanical Systems, Journal of, 2003. **12**: p. 708–712.
48. Krishnamoorthy, U., D. Lee, and O. Solgaard, *Self-Aligned Vertical Electrostatic Combdrives for Micromirror Actuation*, Microelectromechanical Systems, Journal of, 2003. **12**: p. 458–464.
49. Selvakumar, A., and K. Najafi, *Vertical Comb Array Microactuators*, Microelectromechanical Systems, Journal of, 2003. **12**: p. 440–449.
50. Yeh, J.-L.A., C.-Y. Hui, and N.C. Tien, *Electrostatic Model for an Asymmetric Combdrive*, Microelectromechanical Systems, Journal of, 2000. **9**: p. 126–135.
51. Nguyen, H., et al., *A Novel MEMS Tunable Capacitor Based on Angular Vertical Comb Drive Actuators*. Presented at Solid-State Sensor, Actuator, and Microsystem Workshop, Hilton Head, SC: 2002.

52. Xie, H., Y. Pan, and G.K. Fedder, *A CMOS-MEMS Mirror with Curled-Hinge Comb Drives*, Microelectromechanical Systems, Journal of, 2003. **12**: p. 450–457.
53. Donald, B.R., et al., *Power Delivery and Locomotion of Untethered Microactuators*, Microelectromechanical Systems, Journal of, 2003. **12**: p. 947–959.
54. Lin, H.-Y., et al., *Micromachined Multi-Lever Linkage Angular Motion Amplifier*. Presented at The 11th International Conference on Solid-State Sensors and Actuators, Munich, Germany: 2001.
55. Lin, H.-Y., et al., *Electrostatically-Driven-Leverage Actuator as an Engine for Out-of-Plane Motion*. Presented at The 11th International Conference on Solid-State Sensors and Actuators, Munich, Germany: 2001.
56. Helmbrecht, M.A., et al., *Micromirrors for Adaptive-Optics Arrays*. Presented at The 11th International Conference on Solid-State Sensors and Actuators, Munich, Germany: 2001.
57. Chen, R.T., H. Nguyen, and M.C. Wu, *A High-Speed Low-Voltage Stress-Induced Micromachined 2 × 2 Optical Switch*, IEEE Photonics Technology Letters, 1999. **11**: p. 1396–1398.
58. Wu, M.C., and P.R. Patterson, *Free-Space Optical MEMS*. In MEMS Handbook, J. Korvink and O. Paul, Eds. New York: William Andrew Publishing, 2004.

CHAPTER 16

MEMS Technology Management

16.0 PREVIEW

The successful development of new MEMS devices and applications requires vision, ingenuity, experience, and relentless attention to details. For a given device development task, there are many possible designs, materials, and fabrication methods, leading to various degrees of performance, ease of manufacturing, and cost. Managing a challenging MEMS research and development project requires a broad knowledge base—which the reader can gain through reading the preceding chapters of this book—*and* a regimented approach.

In this chapter, we will review the various steps of a complete R&D development cycle and the strategies associated with each step. The purpose of this review is to help readers develop a proper appreciation for the strategies and use them to improve performance, lower cost, and avoid costly mistakes in the future.

16.1 R&D STRATEGIES

A typical flow chart for approaching and executing a MEMS project is shown in Figure 16.1. A complete R&D cycle involves at least the following steps: conceptualization, review of application potentials and technology, proof of concept, design and optimization, fabrication, and testing.

The strategies and major points of caution associated with each step are reviewed in detail in the following section.

Step 1.
A MEMS project, or any technology R&D project for that matter, is carried out in one of two modes: a **technology push** or an **application pull**. The technology push focuses on the development and the refinement of the general underlying techniques for unspecified or broadly defined applications. Application pull involves the broad selection and development of technologies to fulfill a well-defined market need.

Loosely interpreted, a technology-push project is a technology in search of a problem, whereas an application-pull project is a problem in search of a solution. Both types of activities are important and necessary. A person who aspires to become successful in any field of science and technology should have a balanced mindset.

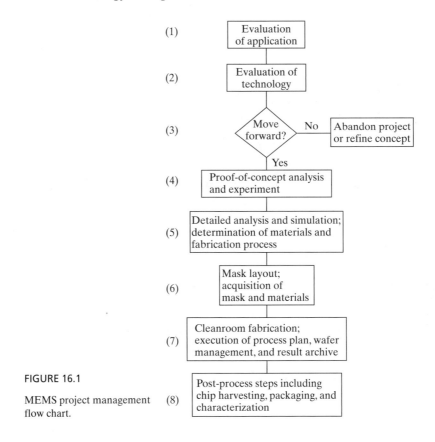

FIGURE 16.1

MEMS project management flow chart.

For readers of this book, many of whom are likely to be technology-minded, it is important to develop an application-oriented mindset. Before setting out to develop detailed technology solutions, one must polish a clear vision about what industrial, societal, military, or scientific needs the new development efforts are going to fill. Although the pursuit of scientific discovery and technology development is exciting and useful on its own, no matter what the outcome of the development effort, it is always more rewarding if the technology is uniquely capable of addressing an important and timely problem and has the potential for a large economic return or societal benefit relative to the investment in research and development.

It is very important to perform high-level due diligence about the market demand and potential market size. An educated, unbiased, and comprehensive list of pertinent performance specifications and cost objectives must be defined and refined. Important lessons can be learned from past mistakes. Don't delegate this task completely to others. Don't blindly trust expert analysis. Don't follow the crowd without going through an independent investigation and thought process. Don't skip a mundane and laborious market survey.

Performance considerations should be given not only to obvious aspects such as sensitivity, speed, and power, but also to subtle but important ones such as robustness against environmental changes, noise, drift, the size of the power supply box, and the ease of training qualified users. Many other socioeconomic factors that may affect the introduction and proliferation of a product embodying disruptive technologies should be considered as well.

Step 2.
Once the performance and cost objectives are set, it is now time to review technology options and decide whether MEMS technology can fulfill these objectives—or at least to address a niche subset competitively and economically. Here, it is important to keep in mind that micromachining technology is only a selected collection of processing and manufacturing techniques in a broad spectrum of manufacturing and assembly technologies. Correctly and broadly identifying competition and alternative solutions is critically important. Incumbent, alternative, and competitive technologies must be carefully reviewed. The technology review should consider "low technology" and "off-the-shelf" solutions as well. "Back-of-envelope" analysis should be conducted at this step to investigate scaling laws and estimate the performance of various technology solutions. The analysis should provide as much accuracy as is appropriate for this stage.

Many readers of this book will likely consider or resort to MEMS and microfabrication technology as a candidate solution to a problem. A comprehensive and balanced opinion of this field is necessary to avoid oversimplifying a project, overemphasizing the benefit, and underestimating the technological complexity. Some common misconceptions and blanket assertions about the advantages of MEMS technology are listed below:

Misconception: "Microfabrication technology leads to the miniaturization of systems."

Truth: Microfabrication enables small feature sizes and chip sizes. However, the size of a system includes that of controllers, power supplies, sensor feedback, observation apparatus, and communication interfaces. Microfabrication does not automatically lead to the miniaturization of *functional systems*. Keep in mind that although the letter *S* in MEMS stands for the word *systems*, many MEMS devices are in fact only *components* in a functional system. Much work needs to be done beyond *making the chip*.

Misconception: "Integration leads to low-cost and high-performance products."

Truth: Proponents of this argument cite the microelectronics industry as an example. Indeed, integration can potentially reduce the labor involved in assembling discrete parts. However, these benefits do not readily translate into the MEMS field, at least not at the present stage of development.

It is largely true that monolithic integration can lead to high performance, such as the high sensitivity of sensors. However, monolithic integration typically involves a lengthy and costly development effort. Flexibility of development is low. The cost of development is typically high.

It should also be noted that the integration of electronics and mechanical elements can occur at many levels, including the instrument level, board level, package level, and chip level. Chip-level monolithic integration is only one option among many.

Misconception: "Batch fabrication reduces the cost of MEMS."

Truth: There are significant setup costs for a MEMS project and a fixed cost for maintaining a MEMS fabrication facility. The batch fabrication can result in a reduced cost of chips or products only when there is a significant volume demand.

The cost for *setting up* a batch run—including designing mask making, fine-tuning the fabrication process, and calibrating—should not be underestimated. A large-area, high-resolution mask plate can cost tens of thousands of dollars to produce. The establishment of a microfabrication facility can cost hundreds of millions of dollars, due to the cleanliness requirements, automated equipment, and environmental and safety protocols. The day-to-day operation of a cleanroom facility requires a significant investment to cover consumables for environmental stabilization (e.g., cooling water and air) and processes (e.g., high-purity processing gases for thin-film deposition and etching). The cost of a MEMS device will not be automatically competitive with a device made using conventional technology. In the future, it is necessary to reduce the break-even production volume to offer customization capabilities. To achieve this goal, new designs of MEMS and new fabrication paradigms are needed.

It is also important to estimate the costs associated with *all* technological and market activities associated with the development of a *system* that may contain MEMS *chips*. Development activities beyond technology R&D may include the following: prototyping, pilot production, securing intellectual property rights, and marketing and distributing a product.

Step 3.
The decision to move forward with the project should be made carefully at this point, based on a thorough analysis and review of (1) the application and market potential, and (2) the technology competitiveness and realism. One must decide whether the project should be continued and, if so, whether a MEMS approach should be taken. If the decision is "no," congratulations! You just might have saved a lot of time and money by recognizing hidden roadblocks without actually facing them. If the decision is "yes," congratulations! You are now embarking on an exciting journey of innovation.

Step 4.
The first stage of developing a MEMS product is to *prove* that *claims and assumptions* about the unique and enabling elements of design, materials, and/or fabrication are in fact valid. Both analytical and experimental approaches may be taken. Proof-of-concept experiments, preferably using large-scale models that are made by conventional fabrication methods, should be conducted to confirm the device operation theories and key performance attributes. Though this step may seem like a time-consuming detour, it saves time in the long run by avoiding costly adjustment and abandonment in the future. This proof-of-concept analysis will also provide diagnostic clues if microfabricated devices fail to perform as planned.

Step 5.
Once the key elements of design, materials, and fabrication are validated, it is time to identify the complete set of design parameters, materials, and the full fabrication process flow.

To arrive at detailed designs, it is often necessary to perform more in-depth analyses. A scaling law analysis, analytical calculation, and numerical simulation are all appropriate. Numerical analysis may provide the most accurate analysis. Sometimes such an analysis can be carried out rather readily; in other cases, the analysis process may become a time-consuming endeavor of its own. Computer-aided analysis may also prove to have limited usefulness considering that there is much uncertainty about material properties (such as Young's modulus, intrinsic stress, etc.).

A MEMS developer should always strive to create fabrication processes that will stand the rigor of manufacturing beyond laboratory proofing of concept. Manufacturing requires repeatability, robustness, and reliability. To optimize a MEMS process in terms of robustness and yield, high etching selectivity should be realized at each step. It takes time and practice to build a working database of material reactivity with various process methods. To start, one can refer to published literature with a comprehensive characterization of multiple materials against multiple etch agents (wet or dry). Williams published etch rates on 53 materials by 35 different etchants in consistent experiments [1], which followed earlier work where a matrix of reactivity between 16 materials and 28 etchants or etching methods was shown [2]. However, it should be noted that the characteristics of material processes may vary from laboratory to laboratory, from batch to batch, from wafer to wafer, and from one part of a wafer to another.

Step 6.
Masks are drawn at this stage with full understanding of the ensuing pattern transfer process and the capabilities of pattern-generation and transfer equipment. Special attention should be paid to the achievable resolution, mask size, mask making cost, and time commitment. Don't automatically assume that the equipment that provides the highest resolution should be used, as it may prove to be too expensive or time consuming for the project or the particular phase of a project.

When drawing a mask for an exploratory device, it is a common practice to bracket around a certain set of target design values. Chip space allowing, one should incorporate more device designs beyond the exact target designs prescribed by the analysis of Step 5. The theoretical designs are often not entirely accurate due to the uncertainty of material properties, the variability of processing parameters, and the errors in analysis tools and methods. With bracketing, one can achieve a high efficiency and arrive at an accurate and optimal design.

Starting substrates should be collected at this point. Silicon wafers are commonly used. Commercial silicon wafers can be purchased with various diameters, thicknesses (often tied to diameters according to industry standards), crystal orientations, doping levels and dopant types, and smoothnesses (polished on one side or both sides). Wafers with thin-film coatings (such as oxide or nitride) are available, as well as wafers with layered structures (such as buried oxide or buried heavily doped regions).

Alternatively, wafers or non-circular substrates made of other materials are available, such as glass, quartz, sapphire, compound semiconductor materials (e.g., GaN, SiC, and GaAs), or polymers (e.g., polyimide or a liquid crystal polymer). These materials offer properties, achievable profiles, and chemical etching characteristics unavailable in silicon. Glass substrates, for example, are often used in microfluid devices. Glass can be patterned by wet etching (using HF), laser machining, blasting [3], microerosion with powder particles [4], and ultrasonic drilling.

These wafers and substrates may be modified with layered thin films (such as semiconductors or insulators). For example, sapphire wafers with single-crystal silicon on the front surface, called silicon on sapphire (SOC) wafers, are commercially available. The SOC wafer may be formed by bonding an oxidized silicon wafer with a sapphire wafer and polishing back the thickness.

Step 7.
Since cleanroom facilities are costly to operate, experiments in cleanrooms should be performed with a well-prepared plan, which should include elements for diagnosing material characteristics, validating the success of each step, and validating alternative actions. Inside a cleanroom, the researcher must be highly observant, motivated to analyze success and failures, and prepared to solve problems on the spot. It is important, especially for student readers, to keep the following practices in mind:

1. Prepare a detailed, written process plan before entering the cleanroom. Account for mundane but important steps such as cleaning, drying, labeling, and storage of the wafers. Try to minimize trial-and-error procedures in the laboratory.
2. Be a careful observer in the cleanroom. Process parameters often vary from facility to facility, from run to run, from wafer to wafer, and from one spot on a wafer to another. It is critically important to be on high alert and carefully observe a process during a run. It is a recommended practice to study the *entire* wafer after each processing step. Don't rush to proclaim a run as successful unless you have personally verified every die, using as many independent means of measurement and characterization tools as possible. For example, the height of an etched profile can be checked using an optical profiler, mechanical stylus profiler, scanning electron microscopy, and scanning probe microscopy. Within reason and upon availability, use as many

means to characterize the result of each step. Take your time—it is important to always keep in mind that some problems may run undetected until a few steps downstream or until the end. Such undetected problems may necessitate costly corrective actions in the future.

3. Invest time to carefully document your results and process conditions. A well-kept scientific notebook will take some time to prepare, but it beats unreliable and often incomplete human memory and can be an important asset in the future. Store and label used wafers and masks for material management and future retracing of steps.

For beginners and experienced researchers alike, it is important to incorporate diagnostic and validation procedures, test structures, and even test wafers. A mask should contain not only target devices, but also electrical, mechanical, and process test structures. These structures are immensely important to ensure the success of materials and to probe potential failure modes. In many cases, unique test features should be devised and inserted as an integral part of the mask. Common test features fall into three categories.

1. Electrical test structures are able to characterize conductivity, thickness, continuity over surface roughness, contact resistance between two layers, doping concentration, or current density limits.
2. Mechanical test structures are used to characterize intrinsic stress, bending, and other material properties.
3. Process test structures allow for the direct identification of process end-point conditions without resorting to complex or destructive procedures.

Oddly enough, one important part of process management, especially for the research and development stage, is to decide how many wafers to use to begin a process. Too few wafers means an accident can set the operator back to the beginning of a lengthy development cycle. Too many wafers means a lot of time may be needed to duplicate work. A careful balance based on the risk of the process, the processor's experience, the projected yield, and the final die count is required.

Process uncertainty and variability is a rule, not an exception. The rule of thumb when strategizing your research and development work is to make sure that the intended device wafers are never subjected to an unknown and uncertain process step. Even a well-characterized material or process step may behave differently when circumstances change. For example, a lithography process may have a different resolution when performed on a glass wafer instead of a reflective silicon wafer.

For this reason, companion wafers may be used throughout the wafer. These are wafers that only undergo a selective subset of process steps to test the variation of process conditions or to help the process engineer characterize exact process conditions without putting the primary wafers at risk. The companion wafer can be inserted at any point of a process and may span a single step or many steps. Because the diagnostics do not span the entire process flow, they save both time and labor.

Step 8.
Strategies and methods for harvesting dies, packaging them, and validating their functionalities seem mundane and simple. However, they are anything but trivial. Inappropriate packaging techniques may damage the process yield, reduce performance, and jeopardize reliability. Methods for die harvesting and packaging should be considered to be an integrated part of the process at a very early stage.

PROBLEMS

Problem 16.1: Review

MEMS mirrors are used for beam steering. Discuss at least one incumbent and competitive technology for NxN optical switches for steering optical signals between two strands of optical fibers. Find product specification sheets or research literature on each technology. Compare their cost, performance, and reliability compared to a MEMS optical switch.

Problem 16.2: Review

MEMS mirrors have been used for retinal scanning display. Discuss at least two technologies (other than MEMS) for direct light projection into retina for display. Compare the cost, performance, and reliability compared to a MEMS mirror based display system.

Problem 16.3: Review

Counterfeiting of medicine cause significant loss of revenue for drug companies and present dread safety concerns. Form a group of 3-4 students, and develop a list of 4 encryption technologies for authenticating the brand and origin of medicine pills. Compare their cost, performance, and reliability. Is there likely a MEMS- or nanotechnology-based method for this application?

Problem 16.4: Fabrication

Discuss strategies to make an array of pits, each 1.6 µm deep with an accuracy of +/- 0.1 µm over a 4-in-diameter wafer. The diameter of the circular shaped pit is 2.5 µm. The bottom and sidewall of each pit is covered with a thin gold film (100 nm thick). The planar surface between the cavities is covered with 100-nm-thick oxide, which must be free of any organic residue. The sidewall should have a slope with smaller than 85° angle. (A single pit is diagramed below). The pit serves as a site for bacteria attachment. The gold surface provides affinity to bacteria attachment, whereas the oxide rejects the bacteria growth. The substrate can be made of any materials.

Hint:

Some of the challenges of making this seemingly-simple device include:

(1) The need to have precise pit depth over a wafer surface;
(2) The need to have gold films on the bottom of the pit but not the top surface;
(3) The need to have uniform gold coverage on the sidewalls of the pit.

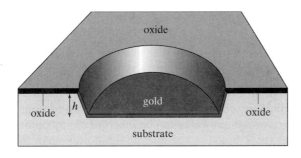

FIGURE 16.2

Diagram of a single pit.

Problem 16.5: Review

Discuss the design and principle at least one widely used test structure for measuring the contact resistance between two layers of conductors.

Problem 16.6: Review

The intrinsic stress in a thin film is very important for MEMS development. Unfortunately, the intrinsic stress is affected by process parameters. It is often important for MEMS device manufacturers to monitor the intrinsic stress level in a micro device. It is desirable to have build-in stress level indicators to quantitatively or qualitatively identify the magnitude of intrinsic stress. These test structures would allow a process engineer to gauge the stress level without breaking the wafer or performing nonreversible, intrusive procedures on other devices on the wafer.

From the literature, identify three mechanical test structures for measuring the intrinsic stress of LPCVD silicon nitride layers. Compare these methods according to the simplicity of implementation.

Problem 16.7: Design

Suppose you are involved in developing a piezoresistive accelerometer based on silicon proof mass, similar to the one illustrated in Case 1 of Chapter 10. Prepare a list of electrical, mechanical and processing test structures that you would incorporate into the mask design. (Hint: you should assume that the relation between gauge factor and doping concentration is not exactly known).

Problem 16.8: Challenge

Form a team of 3-4 students to develop a low cost gyro for the control pad of an electronics hand-held game player. The gyro will be able to receive information about rotation in three axes and allows a player to control the movement of on-screen objects or characters by gross motion of hand, instead of fine motion of fingers. Discuss the market, the opportunities for MEMS-based device, and competitions. Develop pertinent performance specifications. If the decision is made that a MEMS product can be competitive in the market place, develop necessary design, operation principle, fabrication process, and analyze performance of the MEMS gyro.

Conduct necessary studies of the market demand in your country to the point that the order of magnitude of the current market size and future growth trend is convincing. Conduct necessary studies about the breakdown of cost for various elements of a product development.

Problem 16.9: Challenge

Low tire pressure is a major contributor of low gas efficiency for automobiles and trucks and is conducive to excessive tire wear and risks of accidents. Form a team of 6-8 students to develop a low cost pressure sensor for monitoring the pressure of automobile and truck tires during driving. The sensor should be able to send information to the driver to report tire pressure in real time. Discuss the market, the opportunities of MEMS, and competitions. Develop pertinent performance specifications. If the decision is made that a MEMS product can be competitive in the market place, develop necessary design, operation principle, fabrication process, and analyze performance of the MEMS gyro.

Conduct necessary studies of the market demand in your country to the point that the order of magnitude of the current market size and future growth trend is convincing. Conduct necessary studies about the breakdown of cost for various elements of a product development.

Problem 16.10: Challenge

Hearing aids are worn by millions of people with hearing disabilities. The hearing aids should be as small as possible and as power efficient as possible, in order to integrate more functions into the ear canals. MEMS technology can be used to make microphones on small chips and can contribute to small microphone systems. Suppose you are the chief technology officer of a company focused on health care products. Conduct a market and technology survey of microphones for hearing aids and decide what key performance and cost objectives a MEMS microphone should meet.

REFERENCES

1. Williams, K.R., K. Gupta, and M. Wasilik, *Etch Rates for Micromachining Processing—Part II.* Microelectromechanical Systems, Journal of, 2003. **12**(6): p. 761–778.
2. Williams, K.R., and R.S. Muller, *Etch Rates for Micromachining Processing.* Microelectromechanical Systems, Journal of, 1996. **5**(4): p. 256–269.
3. Yagyu, H., S. Hayashi, and O. Tabata, *Application of Nanoparticles Dispersed Polymer to Micropowder Blasting Mask.* Microelectromechanical Systems, Journal of, 2004. **13**(1): p. 1–6.
4. Belloy, E., et al., *Microfabrication of High-Aspect Ratio and Complex Monolithic Structures in Glass.* Microelectromechanical Systems, Journal of, 2002. **11**(5): p. 521–527.

APPENDIX A
Material Properties

	Single-crystal silicon	PolySi LPCVD	SiN LPCVD	Si Oxide LPCVD[1]	Gold	Al	SiC	Stainless steel
E (GPa)	<100>130 <110>168 <111>187 [2,3]	120–175 [4][2]	385 [1] 254 [5]	73 [1]	78[B] [6]	70 [1]	700 [1]	200 [1] 192–200 [6][3]
Density (kg/m³)[4]	2300 [1]		3100 [1][5] 3000 [7]	2500 [1]	19,300	2700 [1]	3200 [1]	7900 [1]
Fracture strength (GPa)	0.6–7.7 [3, 8][6]	1–3 [9]	14 [1] 6.4 [5][7]	8.4 [1]	N/A	N/A	21 [1]	2.1 [1]
Yield strength (GPa)	N/A	N/A	N/A	N/A	0.25[8] [6]	0.17 [1]	N/A	N/A
Fracture toughness (MPa√m)	{100} 0.95 {110} 0.9 {111} 0.82 [6]	1 [10]	5.3[9] [6]	0.79 [6][10]			4.4–4.7[B] [6][11]	80 [6][12]
Poisson ratio	0.055–0.36 [2][13] 0.25 for <100> [6] 0.36 for <111> [6]	0.15–0.36 [11]	0.28–0.3[B] [6]	0.17[14] [6]	0.42 [6]	0.33 [6]	0.16–0.24[B] [6]	0.30 [6]
Thermal conductivity (W/mK)[15]	157 [1] 141 [6]	34 [1]	19 [1] 3.2 ± 0.5 [7][16] 4.5 [12][17] 10–33 [6]	1.4 [1]	315 [6]	236 [1] 247 [6]	350 [1] 71–490[B] [6][18]	33 [1]
Linear thermal expansion coefficient (PPM/K)	2.33 [1] 2.5 [6]	2.33 [13]	0.8 [1] 2.7–3.7[B] [6]	0.55 [1]	14.2 [6]	25 [1] 23.6 [6]	3.3 [1] 4.2–5.6[B] [6]	17.3 [1] 14.4–27 [6][19]
Thermal capacity (J/kgK)	700 [6]		700 [7]	740[B] [6]	128 [6]	900 [6]	590–1000 [6]	420–500 [6]
Seebeck coefficient (μV/K)	500–1000 [14][20]	50–150 [15, 16][21]	N/A	N/A			N/A	
Electrical resistivity at R.T. (Ωm)			N/A	N/A	2.3 × 10⁻⁸ [6]	2.6 × 10⁻⁸ [6]	N/A	5.5 × 10⁻⁷ −10 × 10⁻⁷ [6]
Piezoresistive gauge factor	On the order of 100	10–30[22]	N/A	N/A	1–4	1–4	N/A	N/A

Material	Resistivity[23] (10^{-8} Ωm)	Thermal conductivity (W/mK)	TCR (ppm/°C)	Linear thermal expansion coefficient (ppm/K)
Aluminum (Al)	2.83 [13] 2.73 [17]	237 [13]	3600 [13]	25 [13]; 23.6 [6]
Chromium (Cr)	12.9 [13] 12.7 [17]	94 [13]	3000 [13]	6.00 [13]
Copper (Cu)	1.72 [13, 17]	401 [13]	3900 [13]	16.5 [13]
Gold (Au)	2.40 [13] 2.35 [6] 2.27 [17]	318 [6]	8300 [13]	14.2 [6]
Nickel (Ni)	6.84 [13] 7.2 [17]	91 [13]	6900 [13]	13
Platinum (Pt)	10.9 [13, 17] 10.6 [6]	71 [24][6]	3927 [13]	8.8 [13]; 9.1 [6]
Silicon bulk	Doping	157 [1]; 141 [6]	Doping	2.33 [1] 2.5 [6]
Polysilicon	Doping	34 [13]	Doping	2.33 [13]
Silicon oxide	N/A	1.4 [1]	N/A	0.55 [1]; 0.4[25] [6]
Silicon nitride	N/A	19 [1] 3.2 ± 0.5 [7][26] 4.5 [12][27] 10–33 [6]	N/A	0.8 [1] 2.7–3.7[B] [6]

[1] Numbers quoted in Reference [1] are for silicon oxide fibers.
[2] Value depends strongly on sample preparation technique and growth techniques.
[3] Value depends on phase of steel: ferritic, austerritic, martensitic.
[4] The density can be calculated by knowing the atomic weight of constituent atoms and the atom packing density.
[5] Exact value may depend on the specific composition of Si_xN_y.
[6] The fracture strength is very dependent on the specimen size. The values quoted are for micrometer-sized samples. Further reduction of sample sizes will further increase the fracture strength.
[7] Value depends on temperature and specimen size.
[8] Cold worked, 60% reduction.
[9] Bulk material, sintered.
[10] Value for fused silica.
[11] Value depends on preparation method.
[12] The exact value depends on the treatment process and may vary from one brand to another.
[13] The value of Poission's ratio depends on crystal orientation.
[14] Bulk material, fused silica.
[15] Value may be influenced by sample sizes, i.e., bulk vs. thin film.
[16] Measurement made on microscale samples of low stress nitride with $Si_{1.0}N_{1.1}$.
[17] Measurement obtained on thin-film silicon nitride membrane.
[18] Value depends on preparation method.
[19] Value depends on preparation method.
[20] Actual value depends on dopant type and concentration.
[21] Actual value depends on doping concentrations and operation temperature.
[22] See Chapter 5.
[23] At room temperature, 27°C.
[24] At °C.
[25] Fused silica.
[26] Measurement made on microscale samples of low-stress nitride with $Si_{1.0}N_{1.1}$.
[27] Measurement obtained on thin-film silicon nitride membrane.

REFERENCES

1. Petersen, K.E., *Silicon as a Mechanical Material*, Proceedings of the IEEE, 1982. **70**: p. 420–457.
2. Wortman, J.J., and R.A. Evans, *Young's Modulus, Shear Modulus, and Poisson's Ratio in Silicon and Germanium*, Journal of Applied Physics, 1965. **36**: p. 153–156.
3. Yi, T., L. Li, and C.-J. Kim, *Microscale Material Testing of Single Crystalline Silicon: Process Effects on Surface Morphology and Tensile Strength*, Sensors and Actuators A: Physical, 2000. **83**: p. 172–178.
4. Sharpe, W.N., et al., *Effect of Specimen Size on Young's Modulus and Fracture Strength of Polysilicon*, Microelectromechanical Systems, Journal of, 2001. **10**: p. 317–326.
5. Sharpe, W.N., *Tensile Testing at the Micrometer Scale (Opportunities in Experimental Mechanics)*, Experimental Mechanics, 2003. **43**: p. 228–237.
6. Callister, W.D., *Materials Science and Engineering, An Introduction*, 4th ed. New York: John Wiley and Sons, 1997.
7. Mastrangelo, C.H., Y.-C. Tai, and R.S. Muller, *Thermophysical Properties of Low-Residual Stress, Silicon-Rich, LPCVD Silicon Nitride Films*, Sensors and Actuators A: Physical, 1990. **23**: p. 856–860.
8. Namazu, T., Y. Isono, and T. Tanaka, *Evaluation of Size Effect on Mechanical Properties of Single Crystal Silicon by Nanoscale Bending Test Using AFM*, Microelectromechanical Systems, Journal of, 2000. **9**: p. 450–459.
9. Bagdahn, J., W.N. Sharpe, Jr., and O. Jadaan, *Fracture Strength of Polysilicon at Stress Concentrations*, Microelectromechanical Systems, Journal of, 2003. **12**: p. 302–312.
10. Chasiotis, I., et al., *Fracture Toughness of Polycrystalline Silicon and Tetrahedral Amorphous Diamond-Like Carbon (ta-C) MEMS*. Presented at Society for Experimental Mechanics X International Congress, Costa Mesa, CA: 2004.
11. Chasiotis, I., and W.G. Knauss, *Experimentation at the Micron and Submicron Scale*, In Interfacial and Nanoscale Fracture, vol. 8, Comprehensive Structural Integrity, W. Gerberich and W. Yang, Eds. Elsevier, 2003, p. 41–87.
12. Eriksson, P., J.Y. Andersson, and G. Stemme, *Thermal Characterization of Surface-Micromachined Silicon Nitride Membranes for Thermal Infrared Detectors*, Microelectromechanical Systems, Journal of, 1997. **6**: p. 55–61.
13. Kovacs, G.T.A., *Micromachined Transducers Sourcebook*. New York: McGraw-Hill, 1998.
14. Geballe, T.H., and G.W. Hull, *Seebeck Effect in Silicon*, Physical Review, 1955. **98**: p. 940–947.
15. Von Arx, M., O. Paul, and H. Baltes, *Test Structures to Measure the Seebeck Coefficient of CMOS IC Polysilicon*, Semiconductor Manufacturing, IEEE Transactions on, 1997. **10**: p. 201–208.
16. van Herwaarden, A.W., et al., *Integrated Thermopile Sensors*, Sensors and Actuators: A Physical, 1990. **A22**: p. 621–630.
17. Lide, D.R., *Handbook of Chemistry and Physics*, CRC Press, 1994.

APPENDIX B

Frequently Used Formulas for Beams and Membranes

End constraints and loading conditions	Maximum values of stress and displacement
 Fixed–free beam under point loading at the free end.	Maximum θ occurs at the free end Max value of $\theta = \dfrac{Fl^2}{2EI}$ Maximum vertical displacement occurs at the free end Max value $d = \dfrac{Fl^3}{3EI}$
 Fixed–guided beam under point loading at the free end.	θ at the free end equals zero due to guided boundary condition Maximum vertical displacement occurs at the free end Max value $d = \dfrac{Fl^3}{12EI}$
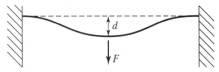 Fixed–fixed beam with point loading applied at the center.	Maximum vertical displacement occurs in the middle Max value $d = \dfrac{Wl^3}{192EI}$
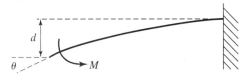 Fixed–free beam under a torque loading. The position of the torque along the length is not relevant.	Maximum θ occurs at the end of the beam Max $\theta = \dfrac{Ml}{EI}$ Maximum vertical displacement occurs at the free end Max $d = \dfrac{Ml^2}{2EI}$

The resonant frequencies of several representative resonator configurations are listed in the following table.

Case and description		Natural frequency (f_n; $n = 1, 2, \ldots$)
Uniform beam cross section, both ends fixed	Center load F, beam weight negligible	$f_1 = \dfrac{13.86}{2\pi}\sqrt{\dfrac{EIg}{Fl^3}}$
	Uniform load w per unit length including beam weight (unit of w = N/m)	$f_n = \dfrac{k_n}{2\pi}\sqrt{\dfrac{EIg}{wl^4}}$ ($k_1 = 22.4, k_2 = 61.7$)
Uniform beam cross section, one end fixed, another end free	Load F on free end, beam weight negligible	$f_1 = \dfrac{1.732}{2\pi}\sqrt{\dfrac{EIg}{Fl^3}}$
	Uniform load w per unit length including beam weight (unit of w = N/m)	$f_n = \dfrac{k_n}{2\pi}\sqrt{\dfrac{EIg}{wl^4}}$ ($k_1 = 3.52, k_2 = 22.0$)
Uniform beam cross section, one end fixed, another end guided	Uniform load w per unit length including beam weight (unit of w = N/m)	$f_b = \dfrac{k_n}{2\pi}\sqrt{\dfrac{EIg}{wl^4}}$ ($k_1 = 15.4, k_2 = 50.0$)
Circular flat plate or membrane of uniform thickness t and radius r, edge fixed	Uniform load w per unit area including own weight	$f_n = \dfrac{k_n}{2\pi}\sqrt{\dfrac{Dg}{wr^4}}$ ($K_1 = 10.2, D = Et^3/12(1 - \gamma^2)$)
Rectangular flat plate or membrane with short edge a, long edge b, and thickness t; all edges fixed	Uniform load w per unit area including own weight	$f_1 = \dfrac{k_1}{2\pi}\sqrt{\dfrac{Dg}{wa^4}}$ (D is defined above)

K_1 is tabulated for various ratios of a/b

a/b	1	0.9	0.8	0.6	0.4	0.2
K_1	36	32.7	29.9	25.9	23.6	22.6

Notation:
F: point loading force [N, Newton]
w: distributed force per unit length [N/m] or per area [N/m^2]
l: length of beam [m]
E: Young's modulus of beam material [N/m^2]
d: vertical displacement [m]
θ: angular displacement [arc angle]

To find formulas for the cases that are less frequently used, see Young, W.C., *Roark's Formulas for Stress and Strain*, 6th ed., McGraw-Hill: 1989.

Index

A

Abbé diffraction barrier, 458
Acceleration sensing, 16
Acceleration sensors, 407–409
Accelerometers, 5–6
  ADXL series accelerometer (Analog Devices), 5–6
  bulk micromachined single-crystal silicon, 227–229
  cantilever piezoelectric, 262–263
  capacitive, 119
  interferometric, 313–315
  membrane piezoelectric, 264–265
  single-crystal silicon piezoresistive, 224–227
  thermal, with no moving mass, 177–178
  torsional parallel-plate capacitive, 120–121
  tunneling, 310–311
Acceptor, 52
Acoustic sensors, 265–268
Acrylics, 405, 436
Acrylite, 397
Active tuning of the spring constant and resonant frequency, 93–94
Actuators, 14–16, 19–20, 40
  actuation methods, 19
  availability of materials, 20
  bandwidth, 20
  bidirectional magnetic beam actuator, 301–302
  bimorph actuators:
    displacement of, 163–164
    for object transport, 167–168
  cantilever piezoelectric actuator model, 249–252
  capacitive, 105
  comb drives, 143–145
  comb-drive actuator:
    with large displacement, 144–145
    for optical switching, 143–145
  design and selection criteria, 19–20
  dynamic response speed, 20
  ease of fabrication, 20
  electrostatic, 103–152
  and environmental stability, 20
  footprint, 20
  hybrid magnetic actuator with position holding, 302–303
  lateral, 169–170
  and linearity of displacement, 20
  MEMS magnetic actuator case studies, 292–303
  micromagnetic, 279
    selected principles of, 282–283
  micromagnetic actuators, selected principles of, 282–283
  optical MEMS, 495–502
    for large in-plane translation motion, 498–499
    for out-of-plane rotation, 499
    for small out-of-plane translation, 496–498
  parallel-plate, 131–133
    pull-in effect of, 110–111
  parallel-plate actuators, 131–133
  power consumption and energy efficiency, 20
  range of motion, 19
  rotational, 495
  thermal, 154
    applications, 175–200
    with a single material, 168–169
  torque and force output capacity, 19
ADXL series accelerometer (Analog Devices), 5–6
Aluminum nitride (AlN), 261
Amorphous silicon, 33
Angular vertical comb drive (AVC), 501
Anisotropic wet etching, 327–349
  chemicals for, 346–349
  design methodology summary, 345–346
  interaction of etching profiles from isolated patterns, 345
  rules of:
    complex structures, 335–345
    simple structures, 330–335
Anisotropically etched grooves, 487
Anodic bonding, 355
Antistiction methods, 383–384
Application pull, 509
Applications, 175–200
  comb drives, 138–145, 139–145
    actuators, 143–145
    inertia sensors, 139–143
  flow sensors, 178–191
  inertia sensors, 175–178
  infrared sensors, 191–194
  microfluidics, 422–454
  optical MEMS, 486
    advantages of using MEMS for, 486
  other sensors, 194–199
  parallel-plate capacitors, 116–133
    flow sensors, 127–130
    inertia sensor, 116–121
    parallel-plate actuators, 131–133
    pressure sensor, 122–127
    tactile sensors, 130–131
  piezoelectricity, 262–273
    acoustic sensors, 265–268
    flow sensors, 269–271
    inertia sensors, 262–265
    surface elastic waves, 271–273
    tactile sensors, 268–269
  piezoresistive sensors, 223–239
    flow sensors, 235–239
    inertia sensors, 224–228
    pressure sensors, 229–232
    tactile sensors, 232–234
  polymer MEMS, 407–417
    acceleration sensors, 407–409
    flow sensors, 413–415
    pressure sensors, 409–413
    tactile sensors, 415–417
Atomic force microscope (AFM), 456–457
Atomic resolution lithography, 459
Availability of materials, actuators, 20

B

Backplate, 125
Bandgap, 50
Bandwidth:
  actuators, 20
  sensors, 18
Beams:
  deflection of, 77–78
  force constants of, 80–81
  frequently used formulas for, 520–521
  magnetic beam actuation, 296–298
  neutral surface of, 76
  suspended beams and plates, 356–357
  types of, 73–75
Bidirectional magnetic beam actuator, 301–302
Bimetallic structure for infrared sensing, 192–194
Bimorph actuators, displacement of, 163–164
Bimorph artificial cilia actuator, 165–167
Binary optical lenses, 7
Binnig, Gerd, 456
Biodegradable polymer, 436
Biology concepts, 423–426
  cells, 423–424
  DNA, 424
  lock-and-key biological binding, 425
  molecular and cellular tags, 425–426
  protein, 424–425
BioMEMS, 8, 10, 11
Bonding:
  anodic, 355
  eutectic, 355
  fusion, 355
  mechanical, 355
  solder, 326

Index 523

wafer, 326
wafer-to-wafer, 354–355
Boron-doped silicon, 52
Bulk crystal silicon, manufacture of, 28–29
Bulk etching solutions and methods, properties of, 328
Bulk micromachined single-crystal silicon accelerometer, 227–229
Bulk micromachining, 34, 326–370
    definition, 326
Bulk polymers, 397
Bulk silicon etching, 326–327
Bulk silicon substrate (single-crystalline silicon), 3

C

Cady, Walter, 246
Cantilever piezoelectric accelerometer, 262–263
Cantilever piezoelectric actuator model, 249–252
Cantilevers, 73
    for data storage and retrieval, 194–195
    flexural cantilevers, stress in, 216–221
    with optical interference position sensing, 315–316
    with parallel arms, 82
Capacitive accelerometer, 119
Capacitive actuators, 105
Capacitive boundary-layer shear stress sensor, 127–130
Capacitors:
    defined, 103
    interdigitated finger capacitors, 133–139
    parallel-plate capacitors, 105–133
Carrier mobility, 54
Cells, 423–424
Channels, 309, 434–445
    comparison of methods for making, 436
    electrophoresis in microchannels, 438–440
    gas chromatography channels, 437–438
    neuron probes with, 440–442
    Parylene surface micromachined micro channels, 444–445
    PDMS microfluid channels, 443–444
Charge carrier concentration,
    calculation of, 50–54
    example of, 53–54
Chemical domain, 15
Chemical field-effect transistors (ChemFET), 49
Chemical vapor deposition (CVD), 378
Chemical vapor deposition methods (CVDs), 33–34
ChemSensing, 16

Chips, defined, 30
Chromatography, defined, 437
Circuits integration, 14
Coefficients of thermal expansion (CTE), 161–162
Cold arm, 170
Color ink-jet printing, 3
Comb drives, 139–145, 499
    applications, 138–145, 139–145
    actuators, 143–145
    inertia sensors, 139–143
    comb-drive accelerometer, 139–141
        sensitivity of, 141–142
    comb-drive actuator:
        with large displacement, 144–145
        for optical switching, 143–145
    defined, 133
Conductive polymers, 407
Conductive SPM tips, 459
Conductivity, 49–58, 94
    associated with holes, 55
    calculation of, 56–57
    carrier mobility, 54
    drift, 54
    mean free path, 54
    mean free time, 54
Convection transfer coefficients, 155
Converse effect of piezoelectricity, 245
Corner compensation, 346
Cross talk, sensors, 18
Cross-sensitivity, actuators, 20
Crystal planes and orientation, 58–61
Cured polyimide films, 400–401
Curie, Pierre and Jacques, 245
Curie point, 247
Curie temperature, 247
Cytop, 406

D

Deep reactive ion etching (DRIE), 352–353, 466
Deionized water, 29
Denaturation, 425
Dendritic polymer, 382–383
Deoxyribonucleic acid (DNA), 424
Deposition profiles, 42
Developer, 35
Diamagnetic materials, 280
Diamond thin films, 40
Diaphragm, 125
Dielectrophoresis, 431–434
Dielectrophoresis force, 432
Diffraction gratings, 7
Digital Light Processing (DLP) chip, 5–7
Digital micromirrors, 6–7
Dip Pen Nanolithography (DPN), 457–460
Direct effect of piezoelectricity, 245
Directions, summary of notation for, 60

Distributed Bragg reflectors (DBR), 496
DMD chip (Texas Instruments), 105
DNA, 424
DNA sequence identification, 17
Domain walls, 33
Donor, 52
Doping, 51
Drift, 54
    sensors, 18
Dry etching, 326
    of silicon, 349–351
Dynamic response speed, actuators, 20
Dynamic viscosity, 426
Dynamic range, sensors, 18

E

EDP (ethylene diamine pyrocatechol), 346
Elastic deformation regime, 66
Elastomeric polymers, 490–491
Elastomers, 403
Electrets, 105
Electric double layer, 430
Electrical domain, 15
Electrical properties, of metal thin films, 161
Electroactive polymers, 407
Electrohydrodynamic effect, and fluid movement in channels, 427
Electrokinetic flow, 430–431
Electromechanical coupling coefficient, 249
Electron-hole pair generation, 51
Electroosmotic effect (EO), and fluid movement in channels, 428
Electroosmotic flow (EOF) micropumps, 431
Electrophoresis, 431–434
    in microchannels, 438–440
Electrophoresis force, 432
Electroplating, 39, 287–288
    bath constitution for representative magnetic materials (table), 290
Electrostatic actuation, 19, 308, 495, 498–499
    advantages/disadvantages of, 165
Electrostatic actuator, equilibrium position of, under bias, 108–110
Electrostatic force microscopy (EFM), 457–458
Electrostatic forces, 103
Electrostatic motor, operation principle of, 104
Electrostatic sensing, 308
Electrostatic sensors and actuators, 103–152
    advantages of, 104–105
    comb-drive devices, applications of, 138–145

Electrostatic sensors and actuators (*continued*)
  disadvantages of, 105
  interdigitated finger capacitors, 133–139
  parallel-plate capacitors, 105–133
Electrowetting, 428
Elemental metals, as sacrificial layers/structural layers, 380
Energy domains, 14–17
Energy transduction, 14–15
Environmental stability, and actuators, 20
EPON® Resin SU-8 (Shell Chemical), 401
Epoxies, 397
Epson, yearly sales figures attributed to MEMS technology, 10
EPW (ethylenediamine pyrocatechol and water), 346
Equilibrium position, calculation of, 113–116
Etch holes, 382–383
Etch rate, 327
Etch rate selectivity, 327
Etch stop layer, 347
Eutectic bonding, 355
Extrinsic semiconductor material, 52

F

Fatigue, 66
Febry–Perot cavity, 313
ferromagnetic materials, 280
Feynman, Richard, 37
Fiber-based sensing, 310–311
Field-effect transistor (FET), 317–319
  displacement using the gate of, 318–319
Fixed boundary condition, 73–74
Fixed–fixed beams (bridges), 73
Fixed–free beams (cantilevers), 73
Fixed–guided beams, 73
Flexural beam bending analysis:
  beams:
    deflection of, 77–78
    types of, 73–75
    longitudinal strain under pure bending, 75–77
    under simple loading conditions, 73–83
    spring constants, finding, 78–83
Flexural cantilevers, stress in, 216–221
Flicker noise, 18
Flow resistance, 429
Flow sensors, 127–130, 178–191, 269–271, 413–415
Fluid diodes, 428
Fluid mechanics concepts, 426–434
  dielectrophoresis, 431–434
  electrokinetic flow, 430–431
  electrophoresis, 431–434
  fluid movement in channels, methods for, 427–428
  pressure driven flow, 428–430
  Reynolds number, 426–427
  viscosity, 426–427
Fluid shear stress, 429
Fluorocarbons, 397, 406
Footprint, actuators, 20
Force constant, defined, 78
Force microscopy, 457
Forced thermal convection, 155–156
Foundry process, 389–390
  design rules, 390
Free boundary conditions, 73–74
Freescale Semiconductor, yearly sales figures attributed to MEMS technology, 10
Free-space light beams, sensing with, 312
Free-space optical interconnects, between fiber bundles for dynamic routing, 7–8
Frequency modulation, 320
Fusion bonding, 355

G

Gas chromatography channels, 437–438
Gas-phase etchants, 353–354
Gas-phase etching of silicon, 353–354
Gas-phase plasma etching, 350–351
  gases and byproducts, 350
Gate oxide layer, 32
Gauge factor, 209
Gauss, 280
General thermal transfer principles, 153
Generic microfabrication process, for realizing a field-effect transistor, 30–31
Genes, 424
Glass chips, 436
Gouy–Chapman layer, 430
Grain boundaries, 33
Guided boundary conditions, 73–74
Gyroscopes, 6

H

Hall measurement, 458
Hard bake, 36
Hard magnets, 281
Heat capacity, 159
Heat sink, 158
Hewlett-Packard, 3
  yearly sales figures attributed to MEMS technology, 10
Hexamethyldisilazane (HMDS), 35
High aspect-ratio combined poly- and single-crystal silicon (HARPSS) process, 352–353
HNA, 353
Holes, 51
  conductivity associated with, 55
  etch, 382–383
  mobility of, 55
Hot arm, 170
Hot-wire anemometry (HWA), 179–182
Hybrid magnetic actuator with position holding, 302–303
Hydrophobic surface treatments, 384–385
Hyper branched polymers (HBPs), 382–383

I

IC-centric foundries, 389
In situ doping, 52
Inertia sensor, 116–121
Inertia sensors, 116–121, 175–178, 262–265
Infrared sensors, 191–194
Ink-jet printers, 3
Integrated circuits, 2
Integrated inertia sensors (Analog Devices), 5
Integration:
  circuits, 14
  microelectronics, 14
  monolithic, 14
  of mechanical and circuit elements, 39
Interdigitated fingers (IDT), 313
  defined, 133
Interferometric accelerometer, 313–315
Interferometric filters, 7
Interferometric sensing, membrane displacement sensor with, 316–317
Internal friction, 92
International conferences, 11
Intrinsic semiconductor material, 51
Intrinsic stress, 86–91
Inverse effect of piezoelectricity, 245
Ionization, 51
Ionized acceptor atoms, 52
Ionized donor atoms, 52
Isotropic etching, 327
Isotropic wet etching, 353

J

Johnson noise, 18

K

Kapton, HN-type, 401
Kelvin, Lord, 207, 245–246
Kinematic viscosity, 426
KOH (potassium hydroxide), 346–347

Index

L

"Laboratoryon-a-chip", 422
Laminar flow, 427
Large-angle mirrors, 499–501
Lateral comb drives, 499
Lateral force microscopy (LFM), 457
Lateral thermal actuators, 169–170
LCP piezoresistive flow sensor, 413–415
Lexmark, yearly sales figures attributed to MEMS technology, 10
LIGA process, 38–39
Linear expansion coefficient, 159–160
Linearity, sensors, 17
Linearity of displacement, and actuators, 20
Liquid crystal display (LCD) technology, 6–7
Liquid crystal polymer (LCP), 402–403
Lithium niobate ($LiNbO_3$), 261
Lock-and-key biological binding, 425
Long-chain polymers, 397
Longitudinal comb drive, 137
Longitudinal gauge factor, 209
Longitudinal piezoresistivity, 209
Longitudinal strain under pure bending, 75–77
Lorentz force on a current-carrying wire, 283–284
Low-pressure chemical vapor deposition (LPCVD):
  thin films by, 378–380
    alternative methods, 380
    polycrystalline silicon, 378–379
    process parameter control, 379
    silicon dioxide, 379
    silicon nitride, 379
Low-temperature silicon adhesive bonding, 355
Low-temperature silicon direct bonding, 355
LPCVD, See Low-pressure chemical vapor deposition (LPCVD)
Lucas NovaSensor, 355
Lucite, 397

M

Magnetic actuation, 19, 308
  advantage of, 287
Magnetic assisted three-dimensional assembly, 386–387, 386–388
Magnetic beam actuation, 296–298
Magnetic coil:
  design and fabrication of, 288–292
  multiple-layer planar coil, 292
  three-dimensional, 291
Magnetic domain, 15
Magnetic field density, 280
Magnetic field intensity, 279
Magnetic flux density, 280
Magnetic force microscope (MFM), 457
Magnetic induction, 280
Magnetic motor, 294–296
Magnetic polymer composite, 288
Magnetic sensing and actuation, 279–306
  essential concepts and principles, 279–287
Magnetic torque, 284
Magnetism, study of, 279
Magnetization, 279
  bidirectional magnetic beam actuator, 301–302
  hybrid magnetic actuator with position holding, 302–303
  Lorentz force on a current-carrying wire, 283–284
  magnetic beam actuation, 296–298
  magnetic motor, 294–296
  MEMS magnetic actuator case studies, 292–303
  micromagnetic actuators, selected principles of, 282–283
  micromagnetic components, fabrication of, 287–292
  multi-axis plate torsion using on-chip inductors, 300–301
  non-uniform magnetic field, 284–285
  plate torsion with Lorentz force actuation, 298–299
  shape anisotropy, 285–286
  uniform magnetic field, 284–285
  unit analysis, 284
Magnetization hysteresis curve, 280, 281–282
Magnetohydrodynamic (MHD) effect, and fluid movement in channels, 427
Magnetorhelogical pumping, and fluid movement in channels, 427
Majority carrier, 52
Mass fabrication with precision, 14
Material properties, 517–519
Material safety data sheet (MSDS), 327
Mean free path, 54
Mean free time, 54
Mean-stream velocity, 429
Mechanical bonding, 355
Mechanical depolarization, 246
Mechanical domain, 15
Mechatronics, 11
Membrane capacitive condenser microphone, 124–127
Membrane parallel-plate pressure sensor, 122–124
Membrane piezoelectric accelerometer, 264–265
Membranes:
  frequently used formulas for, 520–521
  stress in, 221–223
Membrane-type translational mirror, 496–497
MEMS technology management, 509–516
MEMS-centric foundries, 389
Mer units, 397
Metal piezoresistive flow-rate sensor, 237–239
Michelson interferometer, 313
Microchannels, electrophoresis in, 438–440
Microelectromechanical systems (MEMS), See also Microfabrication
  actuators, 14–16, 19–20
    actuation methods, 19
    design and selection criteria, 19–20
  applications, 1
  BioMEMS, 8, 10, 11
  community of researchers, growth of, 11
  conventional macroscale machining, compared to MEMS process, 30
  dedicated MEMS services, 389
  energy domains, 14–17
  history of development, 1–2
  intrinsic characteristics of, 11–20
  mass fabrication with precision, 14
  MEMS technology management, 509–516
  microelectronics integration, 14
  miniaturization, 12–14
  multi-user MEMS processes (or MUMPS), 389
  optical MEMS, 7–8, 11, 486–508
  polymer MEMS, 397–421
  radio frequency (RF) MEMS, 8, 11
  representative major branches of mems technology, 8
  sensors, 14–16
    categories of, 17
    characteristics of, 17–19
  silicon-based MEMS processes, 33–39
  suggested courses and books, 94
  transducers, 14–17
  yearly sales figures attributed to MEMS technology, 10–11
Microelectronics fabrication process, 30–33
Microelectronics integration, 14
Microfabricated neuron probes, 8
Microfabrication, 28–47, 456
  chemical and temperature compatibility, 41–43
  microelectronics fabrication process, 30–33
  nanostructure patterning techniques, 40
  new materials and fabrication process, 39–40

Microfabrication (*continued*)
  overview of, 28–30
  points of consideration for
      processing, 40–43
  silicon-based MEMS processes, 33–39
  technology, 2
Microfluidics, 8, 11
  applications, 422–454
  basic fluid mechanics concepts,
      426–434
    dielectrophoresis, 431–434
    electrokinetic flow, 430–431
    electrophoresis, 431–434
    fluid movement in channels,
        methods for, 427–428
    pressure driven flow, 428–430
    Reynolds number, 426–427
    viscosity, 426–427
  channels, 434–445
    comparison of methods for
        making, 436
    electrophoresis in microchannels,
        438–440
    gas chromatography channels,
        437–438
    neuron probes with, 440–442
    Parylene surface micromachined
        micro channels, 444–445
    PDMS microfluid channels, 443–444
  defined, 422
  essential biology concepts, 423–426
  microfluid platforms,, benefits of, 423
  motivation for, 422–423
  valves, 445–448
    PDMS pneumatic valves, 446–448
Microlenses, method of forming, 488
Micromachined optical switches, 8
Micromachined pressure sensors, 2
  process for, 34–35
Micromachining, defined, 3
Micromagnetic actuators, 279
  selected principles of, 282–283
Micromagnetic components:
  fabrication of, 287–292
    deposition of magnetic materials,
        287–288
    magnetic coil, design and fabrication
        of, 288–292
Micromechanical devices, 2
Micromirrors:
  digital, 6–7
  integrated with solid-state light
      sources, 497
  rotational scanning, 499
Micromotor fabrication, 374
Micromotor fabrication process:
  first pass, 372–373
  second pass, 374–375
  third pass, 375–376
Microoptoelectromechanical systems
    and devices (MOEMS), 7

Microrotary motors, 3
Micrototal analysis system, 422
Miller Indices, 58–60
Miniaturization, 2, 12–14, 456
Mobility, 430
Modulus of elasticity, 65
Molecular and cellular tags, 425–426
Moment, 62
Moments of inertia, 77
Monolithic integration, 14
Moore, Gordon, 1
Moore's Law, 1
Multi-axis capacitive tactile sensor,
    130–131
Multi-axis piezoresistive tactile se,
    232–234
Multi-axis plate torsion using on-chip
    inductors, 300–301
Multimodal polymer-based tactile
    sensor, 415–417
Multi-User MEMS Processes
    (or MUMPS), 389

N

Nano electromechanical systems
    (NEMS), 8, 13–14
Nano whittling, 40
Nanoimprint lithography, 40
Nanosphere lithography, 40
Nathanson, Harvey, 2
Native oxide, 326, 354
Natural (passive) thermal convection,
    155–156
Near-field scanning optical microscopy
    (NSOM), 460
Neuron probes, 440–442
Neutral axis, 76
Neutral surface, of beams, 76
Newton's Laws, 116
Non-slip boundary condition, 429
Non-uniform magnetic field, 284–285
Normal strain, 64
Normal stress, 64
N-type material, 52
Nucleotides, 424

O

Olfactory sensing, 16
1/f noise, 18
Optical communication, and
    microfluidics, 423
Optical diffraction, 458
Optical fibers, and sensors, 311
Optical interferometry, position sensing
    with, 313
Optical lever, 312
Optical MEMS, 7–8, 11, 486–508
    actuators, 495–502

    for large in-plane translation
        motion, 498–499
    for out-of-plane rotation, 499
    for small out-of-plane translation,
        496–498
  applications, 486
    advantages of using MEMS
        for, 486
  large-angle mirrors, 499–501
  large-scale commercialization effort
      in field, 486
  lenses, 488–492
    microfabricated refractive lenses,
        489–491
  low-voltage, large angular
      rotation, 502
  membrane-type translational mirror,
      496–497
  microlenses, method of forming, 488
  micromirrors integrated with solid-
      state light sources, 497
  mirrors, 492–495
    optical mirror performance, effect
        of etch holes on, 492–495
  passive MEMS optical components,
      487–494
  passive optical alignment, 487
  passive optical MEMS
      components, 488
Optical scanning mirrors, 7
Optical sensing, 311–317
Organic polymer materials, 40
Organic semiconductors, 50

P

Paraffin, 406
Parallel-plate actuators, 131–133
  pull-in effect of, 110–111
Parallel-plate capacitors, 105–133
  applications, 116–133
    flow sensors, 127–130
    inertia sensor, 116–121
    parallel-plate actuators, 131–133
    pressure sensor, 122–127
    tactile sensors, 130–131
  capacitance sensor response, 117
  capacitance value, calculating,
      107–108
  electrostatic actuator, equilibrium
      position of, under bias, 108–110
  parallel plates, capacitance of,
      105–107
  parallel-plate capacitive
      accelerometer, 118–119
  pull-in effect of, 110–116
  torsional parallel-plate capacitive
      accelerometer, 120–121
Paramagnetic materials, 280
Parylene, 39, 405, 436

Parylene deposition, process monitor for, 196–199
Parylene sacrificial layer, 380–381
Parylene surface micromachined micro channels, 444–445
Parylene surface micromachined pressure sensor, 409–413
Passive MEMS optical components, 487–494
  passive optical alignment, 487
  passive optical MEMS components, 488
Passive thermal convection, 155
PDMS microfluid channels, 443–444
PDMS pneumatic valves, 446–448
Peltier coefficient, 173
Permalloy, 288
Permanent magnet, 282
Petersen, Kurt, 2–3
Phase modulators, 7
Phenolics, 397
Phosphosilicate glass (PSG), 377, 379
Photolithographic patterning method, 29
Photolithography, and MEMS technology, 14
Photopatternable gelatin, 407
Photoresists, 32, 35–36
Photosensitive polymer (photoresist), 329
Photovoltaic electrochemical etch stop technique (PHET), 347
Piezoelectric actuation, 19, 308, 495
Piezoelectric flow-rate sensor, 270–271
Piezoelectric materials, properties of, 252–262
  PVDF, 256
  PZT, 254–256
  quartz, 253–254
  ZnO material, 256–261
Piezoelectric polymers, 407
Piezoelectric sensing and actuation, 245–278, 308
  background, 245–246
Piezoelectricity:
  applications, 262–273
    acoustic sensors, 265–268
    flow sensors, 269–271
    inertia sensors, 262–265
    surface elastic waves, 271–273
    tactile sensors, 268–269
  cantilever piezoelectric accelerometer, 262–263
  cantilever piezoelectric actuator model, 249–252
  direct effect of, 245
  inverse effect of, 245, 248
  mathematical description of piezoelectric effects, 247–249
  membrane piezoelectric accelerometer, 264–265

piezoelectric beam, bending of, 251–252
piezoelectric coefficient matrix, 247
piezoelectric crystals, 246
piezoelectric effects:
  mathematical description of, 247–249
  orientation dependence of, 247
piezoelectric flow-rate sensor, 270–271
piezoelectric materials, properties of, 252–262
poling, 246
polymer piezoelectric tactile sensor, 268–269
PZT piezoelectric acoustic sensor, 265–267
PZT piezoelectric microphone, 267–268
time dependence of the properties of piezoelectric elements, 247
Piezoresistive flow shear stress sensor, 235–237
Piezoresistive sensing, 308
Piezoresistive sensors, 207–244
  applications, 223–239
    flow sensors, 235–239
    inertia sensors, 224–228
    pressure sensors, 229–232
    tactile sensors, 232–234
  bulk micromachined single-crystal silicon accelerometer, 227–229
  metal piezoresistive flow-rate sensor, 237–239
  piezoresistive flow shear stress sensor, 235–237
  single-crystal silicon piezoresistive accelerometers, 224–227
  surface micromachined piezoresistive pressure sensor, 230–232
Piezoresistivity:
  defined, 207
  origin and expression of, 207–209
  piezoresistive sensor materials, 211–215
    metal strain gauges, 211
    polycrystalline silicon, 215
    single-crystal silicon, 211–215
  stress analysis of mechanical elements, 215–223
Pink noise, 18
Planes, summary of notation for, 60
Plasma etching, 351
Plasma-enhanced chemical-vapor deposition (PECVD), 34, 378
Plastic deformation magnetic assembly (PDMA), 387–388
Plate torsion with Lorentz force actuation, 298–299
Plexiglas, 397
PMMA, 405

Poisson's ratio, defined, 65
Poling, 246
Poly (dimethylsiloxane) (PDMS), 403–405
  precision patterning of, 404–405
Polycarbonate, 406, 436
Polycrystalline germanium, 380
Polycrystalline silicon, 33, 378–379
Polydimethylsiloxane (PDMS), 436, 443
Polyesters, 397
Polyethylene, 397
Polyimides, 39, 399–401, 436
Polymer materials, 39–40
Polymer MEMS, 397–421
  applications, 407–417
    acceleration sensors, 407–409
    flow sensors, 413–415
    pressure sensors, 409–413
    tactile sensors, 415–417
  LCP piezoresistive flow sensor, 413–415
  multimodal polymer-based tactile sensor, 415–417
  Parylene surface micromachined pressure sensor, 409–413
  silicon accelerometer with Parylene beam, 407–409
Polymer piezoelectric tactile sensor, 268–269
Polymer thin films, 380
Polymers, 380
  biodegradable, 436
  biodegradable polymer materials, 406
  bulk, 397
  conductive, 407
  dendritic, 382–383
  elastomeric, 490–491
  electroactive, 407
  fluorocarbons, 406
  liquid crystal polymer (LCP), 402–403
  long-chain, 397
  mechanical properties of, 398
  mobility of charge carriers in, 398
  naturally occurring, 397
  organic, 398, 436
  paraffin, 406
  Parylene, 405
  photopatternable gelatin, 407
  piezoelectric polymers, 407
  PMMA, 405
  poly (dimethylsiloxane) (PDMS), 403–405
  polycarbonate, 406
  polyimides, 399–401
  polypyrrole, 407
  polyurethanes, 407
  polyvinylidene fluoride, 407
  properties of, 398, 400
  shape memory, 407
  shrinkable polystyrene film, 407

Polymers (*continued*)
  SU-8, 401
  synthetic, 397
  viscoelastic behavior of polymers, 399
Polymethylmethacrylate (PMMA), 39, 105
Polypeptide, 424
Polypropylene, 397
Polypyrrole, 407
Polysilicon, 436
Polysilicon (polySi/poly), 33
Polystyrene, 397
Polyurethanes, 407
Polyvinyl chloride (PVC), 397
Polyvinylidene fluoride, 407
Polyvinylidenfluoride (PVDF), 256
Porous silicon, 383
Pressure differences, and fluid movement in channels, 427
Pressure driven flow, 428–430
Pressure sensor fabrication process, 36–37
Pressure sensors, 122–127, 409–413
Process monitors, 90
Process parameter control, 379
Process yield, 30
Projection display, 6
Protein, 424–425
Pull-in voltage, 111
Pure bending, 75
PVDF, 256
PZT, 254–256
PZT piezoelectric acoustic sensor, 265–267
PZT piezoelectric microphone, 267–268

Q

Quality factor, 91–93
Quartz, 253–254
Quate, Calvin, 464, 476

R

Radiation, 155
Radiative domain, 15
Radio frequency resonance sensing, 320–321
Radio Frequency (RF) MEMS, 8, 11
"Raindrop-on-a-tin-roof" noise, 18
Reactive ion etching (RIE), 351
Recombination, 51
Reflux system, 346
Reliability, sensors, 18
Remnance, 281
Resist, 32
Resistance value, determining, 207–208
Resonance mode pressure sensor, 320–321

Resonant frequency, 91–93
  active tuning of, 93–94
  example, 93
Resonant gate transistor (RGT), 2
Resonant microactuators, 496
Resonators, 13–14
Responsivity, sensors, 17–18
Reynolds number, 426–427
Robert Bosch GmbH, 352
Roll-off, 91
Roll-to-roll printing, 40
Rotational acceleration sensors, 6
Rotational actuator, 495
Rotational scanning micromirrors, 499
RTV silicone (GE Silicones), 443

S

Sacrificial etch, acceleration of, 381–383
Sacrificial etching process, 371–372
Sacrificial layer, 37
Sacrificial surface micromachining, 37–38
Saturation magnetization, 281
Scaling laws, 12
  of area-to-volume ratio, 13
  of a spring constant, 12–13
Scanning acoustic microscope, 459
Scanning force microscopes, 459
Scanning Hall-effect microscope, 458
Scanning probe microscopy (SPM), 455–485
  cantilevers with integrated tips, 462–470
    alternative techniques, 466–470
    general design considerations, 462–463
    general fabrication strategies, 463–466
  force microscopy, 457
  general fabrication methods for tips, 460–461
  mold-and-transfer process for SPM fabrication, 466–470
  SPM family, 457–459
    conductive SPM tips, 459
    electrostatic force microscopy (EFM), 457–458
    lateral force microscopy (LFM), 457
    magnetic force microscope (MFM), 457
    near-field scanning optical microscopy (NSOM), 460
    scanning Hall-effect microscope, 458
  SPM probes, 470–480
    with actuators, 476–480
    nanolithography probe with thermal actuation, 478–480

    with piezoelectric sensing and actuation, 476–478
    with sensors, 471–476
    silicon cantilevers for, 471–473
    ultra-thin silicon cantilevers for, 473–476
  tips, 455–456, 459
    general fabrication methods for, 460–462
Scanning thermal microscopy (SThM), 458
Scanning tunneling microscope (STM), 456
Scratch-drive actuator (SDA), 132
Screen printing, 254
Seebeck coefficient, 170–171
  of common industrial thermal couple materials., 172
  of silicon, 172
Seedlayer, 288
Selectivity, etch rate, 327
Self-assembled monolayer (SAM), 384
Self-heating, 174–175
Self-limiting stable profile (SLSP), 332
Semiconductor processing equipment, 33
Semiconductor silicon, 50
Semiconductors:
  bandgap, 50
  charge carrier concentration, calculation of, 50–54
  conductivity of, 49–58, 94
  extrinsic semiconductor material, 52
  hole, 51
  intrinsic semiconductor material, 51
  majority carrier, 52
  materials, 49–50
  n-type material, 52
  organic, 50
  p-type material, 52
Sensing and actuation methods, relative advantages and disadvantages of electrostatic sensing, comparison of, 308
Sensitivity, sensors, 17
Sensors, 14–16, 40
  acceleration, 407–409
  acoustic, 265–268
  capacitive boundary-layer shear stress, 127–130
  categories of, 17
  characteristics of, 17–19
  cross talk, 18
  development cost and time, 19
  dynamic range, 18
  electrostatic, 103–152
  flow, 127–130, 178–191, 269–271, 413–415
  inertia, 116–121, 175–178, 262–265
  infrared, 191–194

membrane parallel-plate pressure, 122–124
metal piezoresistive flow-rate, 237–239
multi-axis capacitive tactile, 130–131
multimodal polymer-based tactile, 415–417
Parylene surface micromachined pressure, 409–413
piezoresistive, 207–244
polymer piezoelectric tactile, 268–269
pressure, 122–127, 409–413
PZT piezoelectric acoustic, 265–267
resonance mode pressure, 320–321
rotational acceleration, 6
sensitivity, 17
tactile, 130–131, 268–269, 415–417
thermal transfer shear stress, 183–191
ZnO piezoelectric force, 257
Shape anisotropy, 285–286
Shape memory polymers, 407
Shear modulus of elasticity, 66
Shear stress in engine oil, 430
Sheet resistivity, 56
  example of, 57–58
Shot noise, 18
Shrinkable polystyrene film, 407
Signal-to-noise ratio (SNR), 18
Silicon, 3, 39–40, 436
  amorphous, 33
  boron-doped, 52
  forms of, 33
  Poisson's ratio of, 69
  as a sacrificial material, 380
  Seebeck coefficient of, 172
  as semiconductor material, 50
  Young's modulus of, 69
Silicon accelerometer with Parylene beam, 407–409
Silicon anisotropic etching, 326–370
Silicon carbide, 40
Silicon dioxide, 379, 436
Silicon micromachining technology, 3
Silicon nitride, 379, 436
Silicon on sapphire (SOC) wafers, 513
Silicon wafer process, 28–29
Silicon-based MEMS processes, 33–39
Single-crystal silicon, 33
Single-crystal silicon piezoresistive accelerometers, 224–227
Single-molecule atomic force microscopy, 459
Single-unit recording, 8
Soft bake, 35
Soft magnets, 281
Solder bonding, 355
Specific heat, 159
Spring constants:
  active tuning of, 93–94
  defined, 78

Squeezed film damping, 92
Stable transitional profiles, 332
Stators, 104
Step-and-repeat process, 30
Stepper, 30
Stiction methods, 383–384
Stiffness matrix, 66
Stoney's formula, 90
Strain, *See also* Stress and strain
  longitudinal strain under pure bending, 75–77
  normal, 64
Stress:
  defined, 64
  in flexural cantilevers, 216–221
  fluid shear, 429
  intrinsic, 86–91
  in membrane, 221–223
  normal, 64
Stress and strain, 61–72
  definitions of, 63–66
  general scalar relation between tensile stress and strain, 66–68
  general stress–strain relations, 70–72
  mechanical properties of silicon/related thin films, 68–69
  Newton's laws of motion, 61–63
  stiffness matrix, 72
Stringency tests, 424
Surface finish and defects, etching, 328
Surface force, 103
Surface micromachined hinged structures, 38
Surface micromachined piezoresistive pressure sensor, 230–232
Surface micromachining, 34, 37–38, 371–396
  basic process, 371–376
  foundry process, 389–390
    design rules, 390
  magnetic assisted three-dimensional assembly, 386–388
  micromotor fabrication process:
    first pass, 372–373
    second pass, 374–375
    third pass, 375–376
  plastic deformation magnetic assembly (PDMA), 387–388
  sacrificial etch, acceleration of, 381–383
  sacrificial etching process, 371–372
  stiction and antistiction methods, 383–384
  structural and sacrificial materials, 376–381
    material selection, 376–378
    thin films by low-pressure chemical vapor deposition, 378–380

three-dimensional MEMS, assembly of, 385–388
Surface profilometer, 43
Surface-tension driven flow, and fluid movement in channels, 428
Suspended beams and plates, 356–357
Suspended membranes, 357–360
SU-8, 401
Sylgard Silicone Elastomer (Dow Corning), 443

T

Tactile sensors, 130–131, 268–269, 415–417
Technology CAD, 477
Teflon, 105, 406
Temperature coefficient of resistance (TCR), 173
Temperature sensing, 153
Tesla, 280
Tetramethyl ammonium hydroxide (TMAH), 348
Texas Instruments, 352
  Digital Light Processing (DLP) chip (Texas Instruments), 5–7
  DMD chip, 10
Thermal accelerometer, with no moving mass, 177–178
Thermal actuation, 308, 495
Thermal actuators, 154
  with a single material, 168–169
Thermal bimetallic actuation, 19
Thermal bimorph actuation, advantages/disadvantages of, 165
Thermal bimorph principle, 161–165
Thermal conduction, 155–156
Thermal couples, 170–173
Thermal domain, 15
Thermal elastic energy dissipation (TED), 92
Thermal expansion:
  linear expansion coefficient, 159–160
  sensors and actuators based on, 159–170
  volumetric thermal expansion coefficient (TCE), 159
Thermal pile, 172
Thermal plastic polymers (thermalplasts), 397
Thermal properties of common materials, 160
  of metal thin films, 161
Thermal resistance, 156–157
  of a suspended bridge, 158–159
Thermal resistors, 173–175
  temperature coefficient of resistance (TCR), 173
Thermal sensing, 308

Thermal sensing and actuation, 153–206
Thermal sensors and actuators, 153
    applications, 175–200
        flow sensors, 178–191
        inertia sensors, 175–178
        infrared sensors, 191–194
        other sensors, 194–199
Thermal setting polymers (thermalsets), 397
Thermal transfer, fundamentals of, 154–159
Thermal transfer principle, accelerometer based on, 176–177
Thermal transfer shear stress sensor, 183–191
Thermal–mechanical noise, 18
Thermistor, defined, 174
Thermistor resistance, example of, 174
Thermoelectric power, 170
Thermopower, 170
Thin film silicon, 3
Thin-film piezoelectric materials, 246
Thomson, William, *See* Kelvin, Lord
Torque and force output capacity, actuators, 19
Torsional bars, deformation of, 85–86
Torsional deflections, 83–86
Torsional moment of inertia, 84
Torsional parallel-plate capacitive accelerometer, 120–121
Transducers, 14–17
Translational actuator, 495
Transverse comb drives, 499
Transverse gauge factor, 209
Transverse piezoresistivity, 209
Traveling surface acoustic waves: and fluid movement in channels, 428
Trench-refill polysilicon technology (TRiPs), 352, 501
Tunable optical mirrors, 7
Tunneling, 309
Tunneling accelerometer, 310–311
Tunneling sensing, 309–311

U

Undercut, 327
Uniform magnetic field, 284–285
Unstable transitional profiles, 332
UV-curable PDMS, 403

V

Valves, 445–448
    PDMS pneumatic valves, 446–448

Vertical translational plates, 82–83
Viscoelastic creep, 398
Viscosity, 426–427
Volumetric thermal expansion coefficient (TCE), 159

W

Wafer-to-wafer bonding, 354–355
Wagon wheel pattern, 348, 351
Wet silicon etch, 326
Wise, K. D., 440

Y

Young's modulus, 65, 512

Z

Zipping motion, 502
ZnO material, 256–261
    aluminum nitride (AlN), 261
    ZnO piezoelectric actuator, 258–261
    ZnO piezoelectric force sensor, 257
ZnO thin films, 383